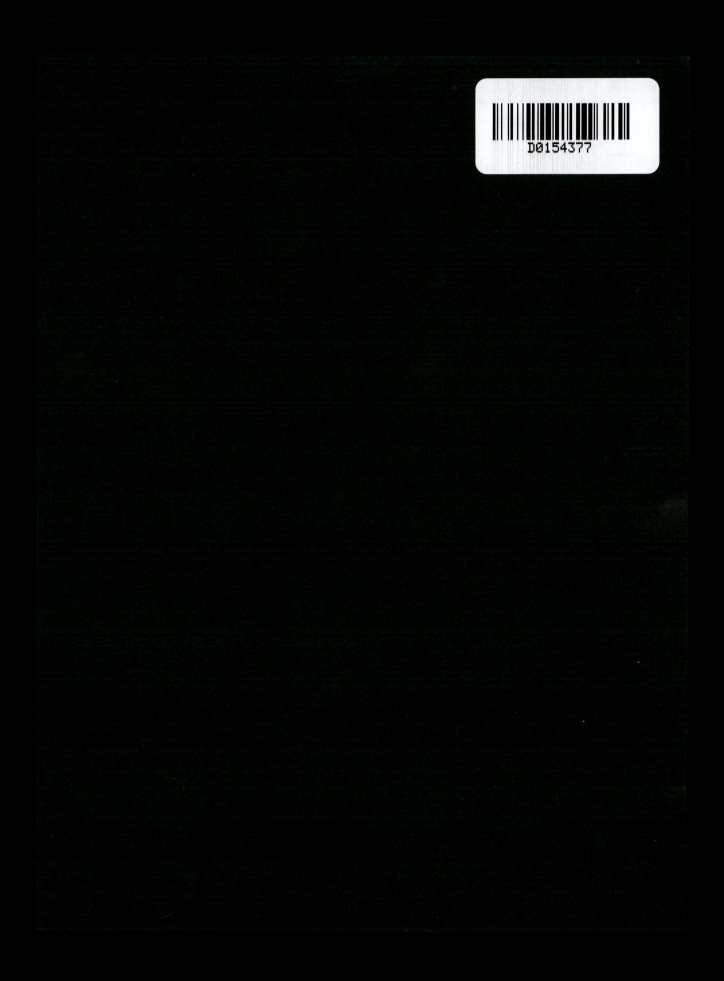

Sixth Edition

Business Research for Decision Making

Duane Davis
University of Central Florida

Sixth Edition

Business Research for Decision Making

Duane Davis
University of Central Florida

THOMSON

SOUTH-WESTERN

Australia · Brazil · Canada · Mexico · Singapore · Spain · United Kingdom · United States

THOMSON

™

SOUTH-WESTERN

Business Research for Decision Making, 6th Edition
Duane Davis

Publisher:
Curt Hinrichs

Assistant Editor:
Anne Day

Editorial Assistant:
Katherine Brayton

Marketing Manager:
Larry Qualls

Adverising Project Manager:
Nathaniel Bergson-Michelson

Permissions Editor:
Joohee Lee

Technology Project Manager:
Burke Taft

Project Manager, Editorial Production:
Kelsey McGee

Web Coordinator:
Scott Cook

Print/ Media Buyer:
Judy Inouye

Production House:
International Typesetting and Composition

Art Director:
Anne Marie Rekow

Cover Designer:
John Walker

Cover/Interior Image:
Archway in Rhodes, Greece. CORBIS

Production Service:
Matrix Productions

Printer:
RR Donnelley
Crawfordsville, IN.

Library of Congress Control Number:
2004105800

For more information about our products, contact us at:

Thomson Learning Academic Resource Center

1-800-423-0563

Thomson Higher Education
5191 Natorp Boulevard
Mason, OH 45040
USA

Asia (including India)
Thomson Learning
5 Shenton Way
#01-01 UIC Building
Singapore 068808

Australia/New Zealand
Thomson Learning Australia
102 Dodds Street
Southbank, Victoria 3006
Australia

Canada
Thomson Nelson
1120 Birchmount Road
Toronto, Ontario
M1K 5G4
Canada

Latin America
Thomson Learning
Seneca, 53
Colonia Polanco
11560 Mexico
D.F.Mexico

UK/Europe/Middle East/Africa
Thomson Learning
High Holborn House
50/51 Bedford Row
London WC1R 4LR
United Kingdom

Spain (including Portugal)
Thomson Paraninfo
Calle Magallanes, 25
28015 Madrid, Spain

To Colleen,
For her love and understanding

ABOUT THE AUTHOR

Duane Davis, teacher and researcher at the University of Central Florida, holds his Doctorate in Marketing from the University of Kentucky, his MBA from Southern Illinois University, and his BS in Marketing from Northern Illinois University. A Fulbright Scholar, he has taught at the Instituto Politecnico De Faro, Universidado Do Algarve, and the Instituto Superior De Estudos Empresariaia, Universidade Do Porto in Portugal. His professional interests include services marketing, marketing strategy, and business research.

Professor Davis has taught at a number of universities, including Pepperdine University, the University of Maryland, and Northern Illinois University. He has published numerous articles in such outlets as the *Journal of Retailing*, the *Journal of Academy of Marketing Science,* the *Journal of Travel Research,* the *Journal of Marketing Education,* the *Journal of International Business Studies,* the *Journal of Personal Selling and Sales Management.*

In addition to his academic work, Dr. Davis has worked in the private sector and has consulted for a wide variety of clients in both the private and public sectors. One of his major projects was the development of the Dick Pope Sr., Institute for Tourism Studies at the University of Central Florida.

BRIEF CONTENTS

About the Author   vi

Preface   xix

PART ONE   BUSINESS RESEARCH, KNOWLEDGE MANAGEMENT AND SCIENTIFIC INQUIRY

CHAPTER **1**   Business Research and Decision Making   1

CHAPTER **2**   Information, Research, and Knowledge Management   23

CHAPTER **3**   Scientific Inquiry   43

PART TWO   BEGINNING THE BUSINESS RESEARCH PROCESS

CHAPTER **4**   Secondary Data Collection in Business Inquiry   69

CHAPTER **5**   Problem and Proposal Development and Management   99

PART THREE   RESEARCH DESIGN

CHAPTER **6**   Fundamentals of Research Design   133

CHAPTER **7**   Foundations of Measurement   173

CHAPTER **8**   Scaling and Instrument Design   199

CHAPTER **9**   Sampling Design   227

PART FOUR   PRIMARY DATA COLLECTION (PDC)

CHAPTER **10**   PDC Using Survey Instruments   269

CHAPTER **11**   PDC Using Observation, In-Depth Interviews, and Other Qualitative Techniques   305

PART FIVE   ANALYTICAL PROCEDURES

CHAPTER **12**   Planning for Data Analysis   325

CHAPTER **13**   Basis Analytical Methods   351

CHAPTER **14**   Analysis of Variance and Regression Techniques   383

CHAPTER **15**   Advanced Multivariate Analysis   431

PART SIX   **RESEARCH REPORTING AND EVALUATION**

CHAPTER **16**   Ethical Considerations in Business Research   459

CHAPTER **17**   Research Reporting and Evaluation   483

Research Cases   503

APPENDIX **A**   A Practitioner's Guide to Secondary Business Information Sources   543

APPENDIX **B**   Selected Statistical Tables   557

Index   573

CONTENTS

About the Author   vi

Preface   xix

PART ONE   **BUSINESS RESEARCH, KNOWLEDGE MANAGEMENT, AND SCIENTIFIC INQUIRY**

CHAPTER 1   **Business Research and Decision Making**   **1**

Overview   **2**

The Nature of Decision Making   **2**

Levels of Decision Making   3

Decision-Making Processes   5

The Role of Research in Decision Making   **7**

Business Research Defined   7

Business Research and Ethics   8

Role in Decision Making   9

Research and the Global Marketplace   11

The Manager–Researcher Relationship   **12**

Plan of the Book   **14**

The Business Research Process   15

Summary   **19**

Discussion Questions   **19**

Notes   **20**

Suggested Reading   **21**

CHAPTER 2   **Information, Research, and Knowledge Management**   **23**

Overview   **24**

Understanding the Basics of KM   **24**

So What Is KM   **26**

Definition   26

Expected Benefits   27

Information Systems (IS) and KM   **31**

Technology and ISM   31

Decision Support Systems (DSS)   33

Database Management (DBM) and
    Online Analytical Processing (OLAP)   35

Business Research (BR) and KM   **36**

Summary   **39**

Discussion Questions   **39**

Notes   40
Suggested Reading   41

CHAPTER 3   **Scientific Inquiry**   **43**

Overview   44
Definitions   45
Observations, Facts   45
Concepts, Constructs, Definitions, Variables   45
Problems, Hypotheses, Laws   47
Theories, Models   49
Methods of Theory Construction   53
Model-Based Theory   54
Deductive Theory   54
Functional Theory   55
Inductive Theory   55
Relevance of Science in Business Research   55
Science Versus Nonscience   56
Scientific Method   57
State of the Art in Business Research   57
Levels of Scientific Endeavor   60
Where Do We Go from Here?   62
Summary   63
Discussion Questions   63
Notes   64
Suggested Reading   66

PART TWO   **BEGINNING THE BUSINESS RESEARCH PROCESS**

CHAPTER 4   **Secondary Data Collection in Business Inquiry**   **69**

Overview   70
Secondary Data in Business Research   70
Uses of Secondary Data   71
Development of a Search Strategy   74
Data Retrieval Methods   74
Data Locations   77
Internet Searching   79
The Internet   85
Background   86
An Example—Researching Harley-Davidson on the Web   87
Managerial Considerations   91
Summary   94
Discussion Questions   95
Notes   96
Suggested Reading   97

CHAPTER 5   **Problem and Proposal Development and Management**   **99**

Overview   **100**

Problem Identification and Formulation   **101**

Problem Identification   101

Problem Formulation   103

The Question of Value   105

Proposal Development   **107**

A Typology of Business-Related Research Proposals   108

Structure of a Proposal   110

A Sample Research Proposal   111

Managerial Considerations   **117**

Proposal Evaluation Mechanisms   118

Other Control Mechanisms   120

Summary   **123**

Discussion Questions   **123**

Notes   **124**

Suggested Reading   **125**

Appendix: Methods of Assessing the Value of Research Information   **126**

PART THREE   **RESEARCH DESIGN**

CHAPTER 6   **Fundamentals of Research Design**   **133**

Overview   **134**

The Nature of Research Design   **134**

Error Reduction Through Design   **136**

Potential Sources of Error in the Research Design Process   136

Managerial Strategies for Dealing with Error   139

Major Types of Designs   **140**

Ex Post Facto Designs   144

Experimental Designs   147

Specific Design Configurations   **149**

Validity Concerns   149

Specific Designs   151

Online Research Design Issues   **157**

Managerial Considerations   **158**

Summary   **160**

Discussion Questions   **161**

Notes   **162**

Suggested Reading   **164**

Appendix: Advanced Experimental Designs   **165**

CHAPTER 7   **Foundations of Measurement**   **173**

Overview   **174**

The Nature of Measurement   **174**

Components of Measurement   175
The Measurement Process   177
Levels of Measurement   179
Nominal   180
Ordinal   182
Interval   183
Ratio   184
Evaluation of Measurement Scales   184
Validity   185
Reliability   188
Managerial Considerations   192
Summary   194
Discussion Questions   195
Notes   196
Suggested Reading   197

CHAPTER 8   **Scaling and Instrument Design   199**

Overview   200
The Nature of Instrument Design   200
Scale Development   202
Item Phrasing   203
Response Formats   209
Frequently Used Scaling Techniques   210
Instrument Design   213
Scale Sequencing and Layout   213
Online Design Aids   214
Pretesting and Correcting   219
Managerial Considerations   220
Summary   222
Discussion Questions   223
Notes   224
Suggested Reading   225

CHAPTER 9   **Sampling Design   227**

Overview   228
The Nature of Sampling   228
Terminology   228
The Rationale for Sampling   231
The Sampling Process   232
An Introduction to the Philosophy of Sampling   235
Sample Designs   237
Sample Design Choice Considerations   237
Probability Designs   242
Nonprobability Designs   250

Practical Considerations in Sampling   251
Incidence and Response Rates   255
Internationalization of the Marketplace   256
Online Sampling Design Issues   257
Managerial Considerations   259
Summary   260
Discussion Questions   261
Notes   262
Suggested Reading   263
Appendix: A Computational Example Illustrating the Properties
     of the Central Limit Theorem   265

PART FOUR   PRIMARY DATA COLLECTION (PDC)

CHAPTER 10   PDC Using Survey Instruments   269

Overview   270
The Nature of Primary Data Collection (PDC)   270
Active PDC Using Survey Instruments   274
Offline Methods   274
Online Methods   284
A Comparison of Collection Methods   291
PDC Vendors   292
Panel Vendors   296
Managerial Considerations   298
Summary   300
Discussion Questions   301
Notes   302
Suggested Reading   304

CHAPTER 11   PDC Using Observation, In-Depth Interviews, and Other
Qualitative Techniques   305

Overview   306
Nature and Uses of Qualitative Research   306
Passive PDC Using Observation   308
Active PDC Using Qualitative Research Techniques   310
Individual In-Depth Interviews   312
Focus Groups   313
Other Qualitative Research Techniques   316
Qualitative Research Vendors   316
Managerial Considerations   318
Summary   320
Discussion Questions   321
Notes   322
Suggested Reading   323

PART FIVE   ANALYTICAL PROCEDURES

CHAPTER 12   Planning for Data Analysis   325

Overview   326
Planning Issues   326
Selecting the Appropriate Analytical Software
    or Online Research Supplier   328
Analytical Software   328
Online Research Suppliers   333
The Preanalytical Process   333
Step 1: Data Editing   335
Step 2: Variable Development   336
Step 3: Data Coding   337
Step 4: Error Check   339
Step 5: Data Structure Generation   339
Step 6: Preanalytical Computer Check   341
Step 7: Tabulation   341
Basic Analytical Framework for Business Research   344
Managerial Considerations   346
Summary   348
Discussion Questions   348
Notes   349
Suggested Reading   350

CHAPTER 13   Basic Analytical Methods   351

Overview   352
Classification of Analytical Methods by Purpose   352
Exploratory Data Analysis   360
Basic Methods of Assessing Association   360
Crosstabulation   361
Contingency Correlation   362
Spearman Rank Correlation   364
Pearson's r 365
Basic Methods of Assessing Differences   368
Chi-Square (χ^2) Test   368
Z-Test for Differences in Proportions   370
t-Test for Differences in Means   371
Summary   374
Discussion Questions   375
Notes   377
Suggested Reading   378
Appendix: Additional Basic Analytical Techniques   379

CHAPTER 14   **Analysis of Variance and Regression Techniques**   **383**

Overview   384
The Nature of Variance Decomposition   384
Linear Models   387
One-Way Analysis of Variance (ANOVA)   387
Two-Way Analysis of Variance (ANOVA)   395
Linear Regression   403
Analysis of Covariance   419
Nonparametric ANOVA   420
Summary   422
Discussion Questions   422
Notes   425
Suggested Reading   426
Appendix: The Use of Dummy and Effect Coding to Examine Group
       Differences Using Multiple Linear Regression   427

CHAPTER 15   **Advanced Multivariate Analysis**   **431**

Overview   432
Introduction to Multivariate Analysis   432
Selection of a Multivariate Technique   433
Analysis of Dependency   435
Multivariate Analysis of Variance   435
Multiple Discriminant Analysis   438
Conjoint Analysis   440
Covariance Structure Analysis   442
Analysis of Interdependency   444
Factor Analysis   444
Cluster Analysis   448
Multidimensional Scaling   450
Comment on Multivariate Techniques   454
Summary   454
Discussion Questions   455
Notes   456
Suggested Reading   457

PART SIX   **RESEARCH REPORTING AND EVALUATION**

CHAPTER 16   **Ethical Considerations in Business Research**   **459**

Overview   460
Ethical Issues in Business Research   460
Societal Rights   462
Subjects' Rights   463
Clients'/Managers' Rights   465

Researchers' Rights   467
Ethical Issues in a High-Technology Environment   468
Protection of Subjects' Rights   469
Quality of Research   469
Research Versus Direct Marketing   470
Codes of Ethics   470
Managerial Considerations   471
Summary   472
Discussion Questions   473
Notes   474
Suggested Reading   475
Appendix: Excerpts from the ICC/ESOMAR International Code
        of Marketing and Social Research Practice   476

CHAPTER 17   Research Reporting and Evaluation   483

Overview   484
Written Research Reports   484
The Outline   485
Guidelines   490
Oral Presentations   493
Research Evaluation   495
Communication Challenges in the Future   497
Managerial Considerations   498
Summary   499
Discussion Questions   500
Notes   501
Suggested Reading   501

Research Cases   503

Amtech, Inc.—Training Needs Analysis   504
Checker's Pizza—Employee Retention   511
Fastest Courier in the West—Selecting a Service   515
Glenco Bonus Program—Employee Absenteeism   519
Ryder Appraisal District—Property Tax Assessment   526
The Keels Agency—Advertising Media Selection   530
River End Golf and Entertainment Center—Market
    Assessment Study   535

APPENDIX A   A Practitioner's Guide to Secondary Business
            Information Sources   543

Overview   544
Research Guides and Bibliographies   544
Encyclopedias, Directories, Dictionaries, and Handbooks   546
Indexes   547
Indexes of Periodicals   547

Indexes of Newspapers   548

Indexes of Documents   548

Statistical Sources   548

References for Individual Companies   551

Other Sources: International Business   551

Handbooks   551

Almanacs, Dictionaries, and Encyclopedias   552

Directories   552

Yearbooks   552

**A Note on the Standard Industrial Classification (SIC) and the North
American Industry Classification System (NAICS)   553**

APPENDIX B   **Selected Statistical Tables   557**

B.1 Areas Under the Normal Curve   558

B.2 Distribution of t 560

B.3 F Distribution   562

B.4 Distribution of χ^2 571

Index   573

PREFACE

The role of business research in organizations of all types has never been more critical. Today, markets are more competitive, more complex and offer less room for error. *Business Research for Decision Making* provides the manager or future manager with the skills and knowledge necessary to acquire and distill the information needed to make good decisions.

The sixth edition of *Business Research for Decision Making* is the culmination of a time-tested approach to business research and the integration of the latest advances that are affecting this dynamic field. As with previous editions, I have strived to make this text the most readable, comprehensive, and up-to-date treatment of business research methods available today. While the fundamental processes of conducting business research have changed little, the tools and techniques available to the manager have never been better or more easily available. The role of the Internet and the way in which we conduct research is at the forefront of many advances in the field. However, it is important to note that the Internet is not business research, but rather an enabling technology that has profoundly changed how we conduct, implement, and report the information-gathering activities in the conduct of research. This text examines these changes throughout and offers practical advice on how to harness this technology.

The primary focus of the text remains on how one can perform information-gathering activities in a rapidly changing business environment. Its applied managerial orientation organized around the business research process itself gets the reader involved in both the why and how of conducting and evaluating research in real-life settings. Although the content has evolved to keep pace with current practice, the text remains highly focused and flexible enough to accommodate the vast differences in teaching styles found in the field of business research.

As before, the fundamental concepts and techniques of research are presented in a straightforward and clear manner for those who are or will be information producers and consumers. Contemporary applications are used in conjunction with many practical examples to illustrate significant points. I have found this approach successful in teaching advanced undergraduates and MBA students who will need to specify, evaluate, and integrate research into a decision-making framework. The text has also been used successfully as a core text in doctoral programs when accompanied by more advanced statistical and methodological readings. It has been gratifying to see the text's wide acceptance internationally, and I have made efforts to reflect the cultural and geographical issues present in these environments.

Research in business has its primary role in creating information for decision-making purposes. Flattening organizational structures, cross-functional cooperation, and the growth of entrepreneurial activity all require managers to become more "hands on" in dealing with information for decision-making purposes. Organizations recognize that information and research is essential in the formulation of the collective knowledge of the organization. Knowledge is the key organizational asset in the highly competitive global marketplace.

Therefore, the astute manager should have a solid background in the basic concepts, techniques, and uses of research in the business environment. This book is designed to provide the manager with these fundamental skills by

- **Positioning research as a means to more effective decision making, not as an end in itself.** I emphasize a user-oriented approach throughout the text. Numerous real research examples and exhibits reinforce this approach and serve to increase the practical value and usefulness of the book. Managerial considerations are dealt with in each chapter to ensure that this applied focus is carried throughout.

- **Discussing the major challenges of integrating research into the decision-making process.** This is accomplished by focusing on converting data into information and thoroughly examining the role and concerns of managers in conducting global research.

- **Fully integrating the advances that are quickly reshaping how research is conducted in a fast-changing interconnected world.** Here the Internet advancements have been inserted and illustrated with contemporary examples throughout the text. Discussions of the future implications of these technological innovations are undertaken where they are appropriate. Modern tools and techniques such as data mining are integrated throughout.

- **Presenting contemporary research issues,** including

  - Investigation of ethical issues in the conduct of research, including the ethical challenges being created by technological advances.

  - Examining the advances and impact of the growing field of knowledge management and how this development is affecting the conduct and integration of the business research function into the contemporary organization.

  - Exploring the advances and concepts such as data warehousing, modeling, and data mining that are becoming mainstream in the researcher's arsenal of tools.

These features and the straightforward writing style make this book invaluable reading for both present and future managers.

NEW TO THE SIXTH EDITION

The goals of the sixth edition were to thoroughly update advances in online research (particularly as it applies to the Internet), position the function and role of business research into the growing field of knowledge management, increase attention to the globalization of the research function, and make the text as managerially relevant as possible. With these goals in mind, the following general changes have been made in this edition.

- New technological advances in all aspects of the business research process are interwoven throughout the text. Computer- and Internet-aided examples of online searching, questionnaire development, data acquisition, data analysis, and data reporting are all discussed in the context of conducting research for decision-making purposes. Internet-based exercises have been added to the end of each chapter to encourage active exploration of this medium.

- A new Chapter 3 has been added to clearly focus on the advances being made in the field of knowledge management. This new chapter looks at the basics of KM then leads the reader through this emerging discipline. The chapter examines the role and contribution of business research in this often confusing and still developing field of study.

- The nuances of conducting research in the global marketplace are further strengthened in this edition. More real-life examples and discussions direct the reader to consider the impact of this trend toward globalization on the information- gathering function.

- Emerging issues such as ethical concerns involved with the direct marketing/research quagmire and the accelerated growth of data-modeling and mining applications due to the extensive development of data warehouses and data marts are now explored in Chapter 3. The old Chapter 12 has been eliminated, and the important topics in that chapter have been redistributed in other parts of the text.

- As in previous editions, more practical examples and relevant Web sites have been added to illustrate various points throughout the text—actual situations and problems faced by managers in the conduct of their business.

- Keeping with the managerial focus of the text, the references and readings sections have been thoroughly updated at the end of every chapter to help make this sixth edition a more valuable research reference.

- Seven new research cases, all with datasets that can be analyzed, have been added in the back of the text. Six of these cases, by Brian Gray of the University of Alabama and Lawrence Peters of

Texas Christian University are based on real problems and exhibit some of the complexities involved in conducting business research in organizations. The seventh case is an actual business situation in which a market assessment of a golfing facility is undertaken. These cases can be used at various points in the text to illustrate important concepts or may be assigned to students.

■ Finally, the number of chapters in the text remains the same— seventeen—as in the fifth edition. However, the chapters have been slightly rearranged, with topics added or deleted to align contemporary research practices clearly with the research process organizing mechanism. As noted previously, Chapter 3 is completely new with substantial modifications occurring in most other chapters in the text.

ORGANIZATION

The seventeen chapters of the book are organized into six parts, progressing logically through the activities that are undertaken during the business research process. Part One examines the role of business research, knowledge management, and scientific inquiry in management's decision-making process. Here, the major issues are identified and the organizing mechanism of the book is presented. Part Two covers the beginning stages of conducting business research: secondary data searching and subsequent problem and proposal development. Chapter 4 includes an important section dealing with searching for secondary data on the Internet. Part Three, on research design, covers the topics of design configuration, measurement, scaling, and sampling. Significant additions have been made in this part to reflect recent advances in Internet technology and software development in the research design area.

Part Four deals with primary data collection. Chapter 10 addresses quantitative research using survey instruments, and Chapter 11 now focuses strictly on qualitative research data collection procedures.

Part Five covers the basic analytical procedures that researchers use to prepare and analyze their data. Chapter 12 now deals with planning for data analysis. Chapters 13 through 15 deal with specific statistical modeling and analytical procedures frequently used in the analysis process.

Part Six deals with the issues or research reporting and evaluation. Chapter 16 deals with contemporary ethical considerations in business research, and Chapter 17 covers actual research reporting and evaluation. Finally, the text ends with a set of seven research cases and two appendices. One appendix provides a guide to information sources and another provides four commonly used statistical tables.

YOUR COURSE: DESIGN IT YOUR WAY

Given the variety of approaches, student backgrounds, and objectives in teaching a business research course, I have put together a number of additional resources and packaging options, all of which allow instructors to tailor the print and software components available with this text to accommodate their own teaching styles.

The basic package includes this text, the Instructor's Resource CD (including many free add-ins), and InfoTrac® College Edition.

- ■ The **Instructor's Resource CD** includes course outlines, chapter outlines, problem sets, discussion questions, and case notes to name a few resources. **PowerPoint slides** for classroom presentations are also available to adopters. **The Web site** for *Business Research for Decision Making*, which may be accessed through http://www.swlearning.com, is also available to users of the text. This includes downloadable data for the cases and problems that also appear on the Instructor's Resource CD.

- ■ A free subscription to **InfoTrac College Edition** (ICE), a comprehensive and powerful reference resource for student and consumer research, is included. The ICE combines tens of thousands of articles from major journals, encyclopedias, reference books, magazines, pamphlets, and other sources on a single Web site with fast and easy search tools for all your reference needs. ICE is a one-stop reference source for students doing homework, making vacation plans, keeping up with current events, or researching any topic from important health issues to hobbies to making large purchasing decisions.

Also, other options exist for the adopter of this text. These options provide instructor and student with unprecedented flexibility. The instructor can pick and choose problem sets and software to match his or her particular course needs. The options are:

- ■ **Case Studies.** While this new edition contains seven case studies in the back of the text for those who wish to incorporate additional case exercises into the course, William Carlson's *Cases in Managerial Data Analysis* provides dozens of additional case exercises. These may also be custom-published to your preference. Ask your Thomson Learning representative.

- ■ **Software options** exist for those who wish to incorporate a commercial statistics package into the course. If you wish to do so, contact your Thomson Learning representative for more information and pricing on these outstanding values.

- ■ **SPSS® 12 for Windows: Student Version.** Again, we are happy to offer this feature with the sixth edition. SPSS for Windows Student Version is a fully functional, statistics software based on SPSS's best-selling statistical package. It includes a comprehensive tutorial, statistics coach, and sample data sets. The student version can handle up to 50 variables and 1500 cases. It is available on CD at a great price.

- ■ EXCEL®. Because Microsoft® Excel has become an important tool in business, we integrate its use in this book. However, Excel has limited capabilities, and to use it in the course effectively, it must be supplemented with other materials. For this reason, we offer the best-selling add-in, StatPlus, accompanied by an excellent paperback, *Data Analysis with Microsoft Excel*, by Kenneth Berk and Patrick Carey. If you wish to use Excel in the course, we highly recommend this special package. This provides both student and teacher a great deal of versatility.

- ■ Other packages such as SAS Institute's JMP®-IN and Minitab®: Student Edition can also be ordered with the text.

If you'd like to customize this book for your course, call your Thomson Learning representative to inquire about a package tailored for your particular needs. *Have it your way!*

ACKNOWLEDGEMENTS

Many people have helped in the development of this book. Academic colleagues have given new perspectives on the value of information and its contribution. Their influences are significant in understanding how information should be treated in an emerging world.

The framework provided by academics must be tested in the marketplace. My experiences as a business research professional over more than 25 years have resulted in a tremendous amount of self-improvement and, more importantly, great friends.

Special thanks go to Colleen for her continued love and support and to the Fulbright Association for the opportunity to raise my consciousness in the international research arena. My international consulting experience has also resulted in great friends and priceless insight into the difficulties and joys of conducting research in very diverse environments. I would also be remiss if I failed to thank Eddie and Rye Myers for their unique contributions to the instructor's resources for this edition. For the six excellent business research cases, I thank Brian Gray of the University of Alabama and Lawrence Peters of Texas Christian University.

My sincere thanks, also, to J. Douglas Barrett of the University of Northern Alabama, Alisa Fleming of the University of Phoenix, Joe Flowers of Indiana Wesleyan University, Joann Frantino of the University of Phoenix, and Dennis Mathern of the University of Findlay, each of whom reviewed this edition, and to the long list of reviewers of previous editions whose well-considered comments were valuable and appreciated.

Duane Davis

1

Business Research and Decision Making

Overview

The Nature of Decision Making
Levels of Decision Making
Decision-Making Processes

The Role of Research in Decision Making
Business Research Defined
Business Research and Ethics
Role in Decision Making
Research and the Global Marketplace

The Manager–Researcher Relationship

Plan of the Book
The Business Research Process

Summary

Discussion Questions

Notes

Suggested Reading

OVERVIEW

The very essence of managerial action in any modern organization is decision making. Business decision making involves the formulation, evaluation, and selection of alternatives to solve managerial problems. The quality of decisions is usually closely tied to the availability of usable information at the time the decision is made. Business research is a primary means of obtaining usable information for decision making. Thus, this text is concerned with helping managers and potential managers make better decisions by providing them with a framework for the evaluation of business research.

Specifically, managers must wade through an ever-increasing amount of data, information sources, and studies that are often confusing and outright conflicting. This state of affairs, coupled with the increased complexity of the decision-making environment, suggests that management in all functional areas must gain a better understanding of the process of obtaining and using information for decision-making responsibilities. Because the size and complexity of management information systems (MIS) and their associated decision support systems (DSS) continue to increase in today's organizations, managers must monitor the input to such systems. The data for decision making must be of sufficient quality and appropriate type to help solve the problems of the organization. Only by developing an understanding of the role and nature of business research can managers effectively accomplish this task.

This introductory chapter provides the framework for examining research's role in decision making. First, the nature of decision making is examined. The levels of decision making in an organization are identified, and a generalized model of the decision-making process is presented. Business research is defined, and its role in the process is examined. With these topics outlined, the manager–researcher relationship is examined for potential conflicts. Management's role in the relationship is highlighted. The major components in the manager's and researcher's respective spheres are then diagrammed. This diagram subsequently provides the framework and the rationale for the plan of the text. The chapter ends by expanding this framework in the context of the business research process, the organizing framework for this text.

THE NATURE OF DECISION MAKING

Decisions are made by everybody, every day. What do I do today? Should I leave, or should I stay? Some decisions are explicit, and some are implicit. By inaction, you have made a decision. In other words, success largely depends on what you do and do not do. (Of course, a bit of luck in the lottery helps, but that also entails a decision.) The necessity of making decisions often seems to be the only constant of managerial action; the process, the environment, the individual, and the organization seem to change continually.

Furthermore, the diversity of decision-making activities in organizations is boundless. The same decision maker may use different decision styles at different times; different decision makers may not use the same decision style to reach a decision for a particular situation.[1]

For example, imagine two managers in an organization, each with a separate problem. One manager must make a decision whether or not to enter a foreign market with a new product; the other must decide how much raw material to order from a particular supplier. Each decision has a different degree of complexity, different informational requirements, and a different time frame. In fact, the immediate decision may be to obtain more information (which requires another decision—i.e., what information is needed to make the original decision). These two decision makers will most likely follow different routes to arrive at their respective decisions. In fact, the next time each manager needs to make a similar decision, the process and requirements of the decision-making situation may drastically differ.

However, this does not mean that one should not attempt to understand decision processes. Quite to the contrary, one must continually examine how decisions are made, in the context of a complex decision-making environment. The study of decision making can lead to better decisions. Research is one way to make better decisions, so an exploration of the general nature and structure of decision making is a proper beginning to the study of business research.

The general nature and structure of decision making are well discussed in the literature.[2] Because decisions and decision styles are remarkably diverse, there are various information needs within an organization. Furthermore, the types of information needs in a specific decision situation appear to be related to the levels of decision making in an organization.

Levels of Decision Making

One approach to decision-making structure in the organization is that which differentiates the strategic, tactical, and technical levels.[3] Table 1.1 outlines the three major decision levels and their associated attributes. Strategic decisions are largely unstructured. They are characterized by a great amount of uncertainty and are nonroutine in nature. The information needs for strategic decisions are primarily external to the organization and are future-oriented. Strategic decisions are those that affect the general direction of the organization. Strategic decisions at the corporate level include those dealing with such issues as diversification, product or market development, and divestiture.

TABLE 1.1 Levels of Decision Making and Characteristic Attributes

Levels of Decision Making	Exemplary Types of Information Needed	Relative Programmability	Planning-Control Organization

Strategic	1. External information
	  a. competitive actions
	  b. customer actions
	  c. availability of process
	  d. demographic studies
	2. Predictive information (long-term trends)
	3. Simulated what-if information
		Low	Planning—heavy reliance on external information

Tactical	1. Descriptive-historical information 2. Performance-current information 3. Predictive information (short-term) 4. Simulated what-if information	Limited programmability	Mixture of planning/control orientation
Technical	1. Descriptive-historical information 2. Performance-current information	High	Control—heavy reliance on internal information

Tactical decisions are concerned with the implementation of the broad strategic decisions just outlined. These decisions are operations-oriented, in that they deal more with issues of a considerably shorter time frame. Planning and control activities are both important at this level. The information needs are descriptive/historical in nature and are more often internal. Examples of tactical decision making are budget allocations, personnel assignments, minor resource commitments, promotional mix decisions, and other short-term internal assignments. Tactical decisions have limited programmability because they lack consistent structure in the problem situation.

The final group consists of technical decisions. Technical decisions are of a routine nature and deal with the control of specific tasks. The information needed to perform this function is primarily descriptive/historical in nature and is usually supplemented with current performance information. Little, if any, external information is needed to make decisions at this level. The decisions are highly programmable and are most amenable to mathematical modeling and standardization. Examples of this type of decision making are quality control, payroll, scheduling, transportation, and credit acceptance or rejection.

Although the distinctions among these levels are by no means clear-cut, the classification does serve to highlight important differences in information needs in the modern business organization. It is necessary to understand differences in the needs of managers in a given organization because the quality of the decisions they make is directly tied to the information they receive. More specifically, although good information of the appropriate

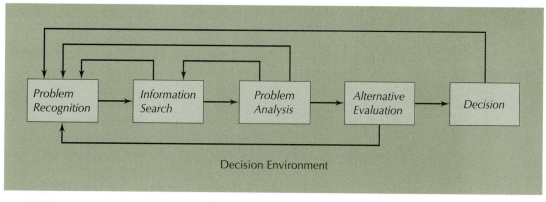

Decision Environment

FIGURE **1.1**

Major Activities in the
Decision-Making
Process

type and quality cannot ensure that the correct decision will be made, its
availability increases the probability of a better decision.

Decision-Making Processes

Given the complexity of the decision-making task, no one universal framework is capable of representing all decision paths. Instead, one must view
the process in the context of the decision itself. In a Western view of the
world (primarily North America and Europe), most decisions, and the subsequent means to obtain information, are viewed as a linear and sequential process.[4] In other words, the decision process is broken down into a
series of interrelated activities that leads to a choice among alternatives.[5]
Figure 1.1 outlines five major activities that can be identified in these decision situations. These activities are defined and interrelated as follows:

> **Problem recognition.** The process begins when management
> acknowledges that some situation exists, or will exist, that needs to
> be acted upon in the near future. Problem recognition can be as
> simple as deciding to introduce a new product or as complex as the
> need to develop a business plan for an organization. In the latter
> case, multiple subproblems will require research. Internal and exter-
> nal environments need to be scanned and assessed, opportunities
> and problems identified, and subsequent strategies developed to
> complete the business plan.[6]
>
> Problem recognition can be planned, as in recognizing the
> need for a comprehensive business plan due in six months, or it
> can result from extensive monitoring and study of an organiza-
> tion's operations and customers, as when a complete marketing
> audit reveals changing customer preferences. Either way, it is a
> decision situation that has multiple consequences for the decision
> maker.

Information search. Information search is the second major stage of the decision-making process. Although information searching may also occur before problem recognition, its purpose here is to collect information pertaining to the particular problem situation identified. One particularly advantageous research strategy that has helped researchers tremendously is the use of Internet searches of online databases and other secondary information services for problem solving. Most companies now use a combination of technologically sophisticated search and data collection strategies to help identify and solve their problems. Throughout the rest of this text, explanations and actual examples will highlight the nature and role of these technologies in business research.

Here, the goal of the searching activity is to assimilate the information into a format that is conducive to analysis of the problem at hand. Assumptions and alternative scenarios are generally posed for subsequent problem analysis. The information generated at this stage may substantially redefine the problem as it was initially formulated.

Problem analysis. Once the relevant search activities are completed and the information is assimilated, the problem is thoroughly analyzed. Here, the analysis of information generally reveals the areas of concern and the major factors that affect the problem.

Problem analysis may also affect recognition and necessitate the gathering of more information to adequately define the situation. For example, National Semiconductor, a $1.6 billion company selling semiconductor chips, integrated circuits, and similar products, was faced with losing its market share in a highly competitive marketplace.[7] Management recognized that a symptom (declining market share) was indicative of some problem, as yet unidentified. This led to an extensive internal and external information search for further problem clarification. The research revealed that 70% of all calls to National Semiconductor left the customer dissatisfied. Subsequent research asked customers what they expected and needed. The findings were simple: better, faster service.

Alternative evaluation. In this stage of the decision-making process, possible alternative courses of action are enumerated and evaluated according to the criteria set forth by the decision maker. The alternatives may be stated either implicitly or explicitly. It should be noted here that even though many managers do not consider inaction to be a decision alternative, it is indeed one. For example, National Semiconductor could have chosen to do nothing with the findings of its problem analysis. Instead, National chose to understand its customers more fully. Therefore, the company built a comprehensive database so that its customer contact people could make customer service faster and more efficient. In other words, alternative evaluation proceeded, so that ultimately a more informed decision could be made in the next stage of the process.

Decision. The final stage in the process is the selection of an alternative to address the problem identified. In the National Semiconductor example, the decision was to build a customer service database to improve its service level to customers. The decision is presumably the one the decision maker finds most appealing in light of all the information available to him or her. The decision in turn may create new problem situations that will need to be addressed. For example, the decision to build a database necessitates decisions as to what information the database should include, what equipment will be necessary to create the database, and whether there are multiple uses for such a database.

In this process framework, the stages are by no means clearly delineated or static. The various activities may be occurring simultaneously or may be so programmed that decisions are made almost instantaneously by computers. Such decisions are largely of a technical nature, relying on current performance and/or historical information such as quality control and customer billing.

This process view, although extremely useful, does not incorporate chaotic decision models and other "garbage can" models of decision making.[8] Essentially, these models extend the basic process model to recognize factors that prevent a simple progression from one activity to another. In these decision-making modes, processes are neither formal nor sequential. The process amounts to making decisions with uncertain information from a variety of information sources. This decision situation thus creates a dynamic, open system subject to dead ends, feedback loops, and outside interference. Such decision modes are usually found when managers are under extreme time pressure, in highly volatile or uncertain situations.

Keeping these processes in mind, we will focus in this text on decisions that are of a strategic and tactical nature. As noted earlier, these decisions have relatively limited programmability and are therefore more complex than merely technical decisions. The problems are nonroutine and generate a variety of informational needs. The problem situations for strategic and tactical decisions are more complex to analyze and require greater managerial input. The decisions and outcomes are probabilistic in nature and are characterized by some degree of uncertainty.

THE ROLE OF RESEARCH IN DECISION MAKING

Business Research Defined

Research is defined by authorities in many different ways. The social sciences (from which most business research methodologies originally come) tend to take a rather strict view of what is and is not research. One prominent author defines scientific research as follows:

> *Scientific research* is systematic, controlled, empirical, and critical investigation of hypothetical propositions about the presumed relations among phenomena.[9]

This definition emphasizes that scientific research is a particular type of critical investigation, studying specific relationships among phenomena.

The only problem, from a managerial viewpoint, in using this definition is that it excludes some types of investigations that may need to be executed from a decision-making perspective. For example, when Anheuser-Busch was contemplating entering the "soft drink" market with its low-alcohol lemon-lime drink, and Coca-Cola was considering expanding into pop wines, their respective managements needed a wide variety of information that was largely of a descriptive nature. The information needed included a description of the current markets, the major competitors, the factors affecting success and failure in the marketplace, and a thorough understanding of consumers' needs and wants in each product market area. Under the definition just quoted, these types of investigations are excluded from the domain of scientific research.

The applied decision-making orientation of managers in business demands that a wider range of studies and investigations be included under the designation of business research. The requirement regarding "hypothetical propositions about the presumed relations among phenomena" excludes many types of studies that are undertaken in an attempt to better understand problem situations that the decision maker must face. Therefore, the definition of business research that is utilized in this text is:

> *Business research* is systematic, controlled, empirical, and critical investigation of phenomena of interest to managerial decision makers.

Under this framework, *scientific business research* becomes a specialized type of investigation that is characterized by the testing of hypothetical relationships.

This definition reflects two important issues. First, it states that business research is a systematic and critical investigation of empirical phenomena controlled by the researcher. Thus, business research is not a random process, but rather a specific type of inquiry that requires the researcher to follow a method with specified characteristics to ensure the best results possible, given the conditions of the study. The second issue addressed in the definition is the scope of investigation. Here, the scope is defined to include any phenomena of interest to managerial decision makers. This position is taken out of necessity, to include the diverse information needs of decision makers. Any attempt to enumerate a laundry list of subjects that would fall within the scope of business research would be frivolous.

Business Research and Ethics

One topic that is of utmost importance to the study of business research is ethics. *Research ethics* is concerned with the proper conduct of the research process in business inquiry. Serious abuses of ethics have occurred in business in general and in research in particular.[10] This state of affairs is disturbing because it affects the quality of decision making for the manager and the rights of individuals and organizations in the research process.

Researchers and managers involved in business research should be concerned about ethical issues in conducting research for a number of reasons:

Unethical research is morally wrong. This is the most basic reason. Even though unethical research may not be illegal, it is not the right thing to do.

Unethical research can affect the image of the firm and management. Unethical firms and individuals can be hurt in the marketplace if they get a reputation for unethical actions. In other words, it is not in the long-term interest of the organization or the individual to conduct unethical research.

Unethical research can lead to poor-quality data and, ultimately, to poor decisions. In the interest of obtaining the highest quality data possible, one should conduct ethical research. Unethical behavior can lead to questionable data. Ultimately, individuals and organizations will refuse to participate in future research efforts of the firm.

New, potentially damaging ethical challenges are being created because of the explosive growth of technological capabilities in the research arena. The widespread use and availability of almost instantaneous access to information and two-way communication to customers (both internal and external) is creating a whole new set of ethical challenges for researchers and managers alike. Issues such as privacy, the blurring of the line between research and direct and/or relationship marketing, and use policies for technological resources are all significant issues that need to be addressed to protect the sanctity of the firm and the research function.[11]

The importance of ethics in business research cannot be overstated. Therefore, one whole chapter in this text is devoted to this issue. Chapter 16 deals with the major ethical considerations in business research and presents the rationale for the development of codes of ethics to help guide researchers.

Role in Decision Making

The role of business research in decision making is relatively simple: providing the decision maker with relevant and useful information with which to make decisions. However, as Figure 1.2 indicates, research is only one of four major means of obtaining information. The others are through authority, intuition, and experience.

Authority. This type of information is obtained from credible individuals who have some expertise in the decision being made. This is an

FIGURE 1.2

Four Major Methods of
Obtaining Information
in the Decision-Making
Process

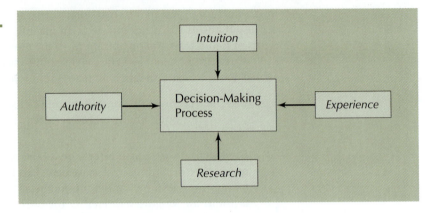

extremely important source of information in business organizations
because the searching time for information can be drastically reduced.

Intuition. This type of information is accepted by the decision
maker because it appears to be true. The information does not necessarily agree with established facts. Intuition tends to be described
as "gut feeling" and "management's sixth sense." Many managerial
decisions appear to be made from such quick flashes of insight.

Experience. The third major means of gaining information is from
experience. Experience-based information is drawn from past situations that were similar in nature to the current problem. The information is empirically based, in that past observations form the basis
for it.

Each of the four methods, including research, plays an important role in
the overall decision-making process. Some methods are used exclusively on
some decisions and only partly on others, but they all contribute to the
overall effectiveness of managerial decision making. The major difference
between the three other means of obtaining information and business
research is the purposeful and systematic nature of the latter. Research is
specifically designed to answer managerial problems within a systematic
and controlled framework, so as to ensure the best possible information for
decision making. This difference is noteworthy, because if the research
process is correctly implemented, the ultimate result should be more
dependable information on which the decision maker can act.

The results of properly conducted research can be profitable indeed. One
case in point is the experience AT&T had in copy-testing its "Reach Out" ad
campaign, which was designed to increase long-distance calling among
light users.[12] An attitude survey was used to discover that light users were
overestimating the costs of long-distance calls by as much as 50%. As a
result, AT&T developed an advertising campaign called "Cost of Visit,"
focusing on the economy of making calls in off-peak hours. This campaign

was subsequently tested for effectiveness with an experimental design. From this research, the company estimated that the new ad copy could generate additional revenue of nearly $100 million over a five-year period at no additional cost to AT&T.

On the other hand, inadequate research, or research that is not effectively integrated into the decision-making process, can have drastic consequences for the firm. New product failures—"flops"—ranging from frozen egg rolls to automobiles, are all too common.[13] For example, RCA's Videodisc lost $500 million, IBM's PC-jr ate $40 million in marketing expenditures, and Dupont's Corfam shoe entry failure cost was in the neighborhood of $90 million. Conducting research might not have prevented these failures, but it could have decreased the probability of failure.

In addition, technological advancements continue to make systematic research a viable alternative to the other means of obtaining information.[14] It is hard to believe that just over 30 years ago there was no microcomputer, or that microcomputer systems are currently over doubling their processing power every year.[15] This radical change in information technology is providing the impetus for reengineering the research function, making it easier and more cost-effective to use research in the decision-making process.

The explosive growth of the information industry, coupled with the unprecedented acceptance and dissemination of new technology, is creating both the opportunity and need to search for the best possible information before making a decision. Consider some key facts. Since 1991, businesses have spent more on computer and communications hardware than on capital investments such as factories, buildings, and other durables. Similarly, it is now estimated that the world's fund of information is doubling every 2 to 2-1/2 years, and scientific information doubles every 5 years. With this type of explosive growth, systematic data searching in a timely, cost-efficient, and comprehensive manner will be essential for organizations to remain competitive in a highly volatile and complex environment.[16]

Research and the Global Marketplace

Global business opportunities are expanding at an ever-increasing rate. It seems that advances in information technology have speeded up the integration of global markets and competition. Economic integration efforts such as the European Union (EU), North American Free Trade Agreement (NAFTA), Latin American Integration Association (LAIA), and the Association of Southeast Asian Nations (ASEAN) have created large multinational markets, open to astute business operations that are willing to take advantage of these opportunities. The further integration of these alliances into a world trading association seems inevitable in the future.

One may ask: What does this have to do with business research? Plenty. Increased research activities will be necessary to make more informed decisions on little-known foreign markets. Therefore, the application of research methods in an international setting is extremely relevant to this text.

Although the techniques of international research are the same in domestic research, the environment in which they are applied is very different. Researchers and managers must recognize the differences inherent in international research, then make the necessary adjustments in methodology and implementation to ensure the highest quality information for decision making.

The differences between domestic and international research center on four major factors:[17]

New parameters. These parameters include duties, foreign currencies, different distributive structures, international documentation, labor regulations, and structural differences in the way business is done.

New environmental factors. The business and task environments are often very different. Such factors as language, culture, religion, political system, and level of technology all affect the design and implementation of research.

Number of factors involved. We must learn and adjust to a huge number of factors. Think of the possible combinations of factors involved in a ten-country study. The research design and implementation problems can become mind-boggling.

Broader competition. When a company enters the international arena, the nature of the competition broadens, including a whole host of new firms.

Because of these differences, we must adapt and modify our research paradigms so as to obtain the best-quality information possible. For this reason, international examples and issues are integrated throughout the text to acquaint the reader with the pitfalls and nuances of conducting research in international settings.

Information from both international and domestic research is utilized in all levels and areas of the decision-making process. Research can be useful in such diverse activities as planning for the financial needs of the organization in the future or testing the effectiveness of alternative quality-control mechanisms on an assembly line. For the research to have maximum value, there must be a mutual understanding between the user (the manager) and the doer (the researcher) of the research. Therefore, the manager–researcher interface is examined next, to identify the major components in the relationship.

THE MANAGER–RESEARCHER RELATIONSHIP

The manager–researcher relationship is characterized by a number of potential conflicts.[18] These conflicts arise regardless of whether the manager and researcher are the same person or specialized roles exist (i.e., a professional manager and a professional researcher). It is important for the

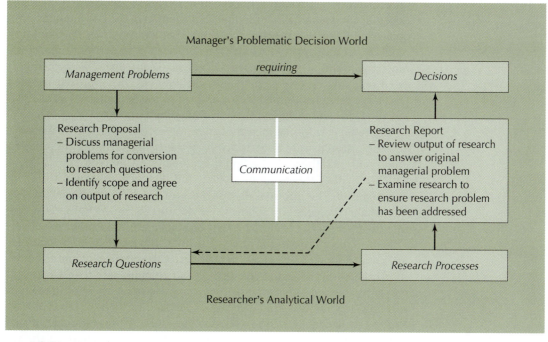

Manager's Problematic Decision World

| Management Problems | requiring → | Decisions |

Research Proposal
- Discuss managerial problems for conversion to research questions
- Identify scope and agree on output of research

Communication

Research Report
- Review output of research to answer original managerial problem
- Examine research to ensure research problem has been addressed

| Research Questions | | Research Processes |

Researcher's Analytical World

FIGURE 1.3

Manager's and Researcher's Decision Worlds

manager to fully understand these conflicts so that communication between the two parties can be maximized.[19] Figure 1.3 illustrates the different worlds in which the manager and researcher operate.

The manager's world can be characterized as a problematic decision world. Essentially, managers are faced with a number of problems that they must solve through specific decisions. Managers want quick, low-cost, timely, and accurate information to solve managerial problems. The information should be as definitive as possible to allow quick decisions with minimal uncertainty. Managers desire nontechnical presentations of findings, with few statistics and qualifiers on the conclusions in the study. Overall, managers want answers to management problems as quickly and inexpensively as humanly possible.

On the other hand, the researcher's sphere of activities can be described as an analytical world. Researchers desire readily researchable questions that can be studied with sound methodologies and adequate resources in terms of time, money, and personnel. The researcher's world uses a specialized vocabulary that is largely foreign to the practicing manager. Researchers want to present the findings of the study with qualifications and statistical evidence to support the conclusions.

These two scenarios point out some of the potential major conflicts in the manager–researcher interface. Adequate communication is essential to keep these conflicts from affecting the results, and ultimate value, of the studies undertaken. When sales of Chrysler automobiles lagged in 1989, President Robert A. Lutz wanted a more complete integration of the new product development function because the process took too long and was not producing results the company could act on.[20] The process was revamped so that cross-functional teams of engineers, researchers, marketers, and stylists communicated regularly and frequently with each other. This approach reduced product development time by 40% and led to the development of the successful LH line of sedans and the new Dodge Ram trucks.

The two main formal means of communication between the manager and the researcher are the *research proposal,* which outlines the problem under study and the plan of action to be undertaken, and the *research report,* which describes the method, results, and conclusions drawn from the investigation. The informal discussions and other communications that immediately precede and follow these documents are critical in ensuring that the research has relevance to the original managerial problem stated.

Management's role in the research relationship is difficult. The manager has a dual responsibility: to accurately specify the problem of interest, then subsequently evaluate the research output to make sure that it is of the proper type and of sufficient quality to make decisions. Therefore, the manager must have at least a rudimentary understanding of the research process and the basic concepts and methodologies used by the researcher. If this understanding is lacking, the manager is wholly dependent on the researcher for the generation of useful decision-making data.

Similarly, managers must constantly evaluate research they have not requested but that may be relevant to the situation at hand. The worth of the study cannot be evaluated unless the manager has basic research competency. This text is designed to provide the user of research with this basic competency.

PLAN OF THE BOOK

This text takes an applied user orientation (applying concepts to aid business decision making) as the primary organizing mechanism for the chapters that follow. Figure 1.4 outlines the plan of the book. As the diagram illustrates, the approach recognizes the two separate worlds in which the manager and researcher operate. The focus of the text will be on the manager's role as specifier and evaluator of research. Subsequent chapters will more specifically evaluate the primary concepts, techniques, and mechanisms of communication that managers need to understand so as to direct the research process for decision-making purposes.

The chapters are presented in the following fashion. First, Chapter 2 outlines business research's relationship with the growing fields of information systems and knowledge management in the 21st century organization. Here, research is positioned as an important component of knowledge

FIGURE **1.4**

Plan of the Book

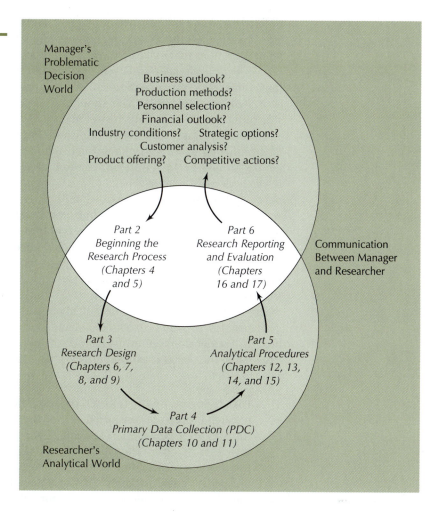

management and a necessary tool for contemporary managers in an increasingly interconnected environment. Next, Chapter 3 discusses the scientific method and the major concepts that must be understood before the manager can begin to enter the researcher's analytical world. This chapter sets the foundation and provides the building blocks on which the two parties can begin. Given this groundwork, the remainder of the text is presented in five major parts, following the sequence of the business research process.

The Business Research Process

The *business research process* is a set of operations that aids the researcher in the systematic gathering, recording, and analysis of data to help solve decision-making problems. The process provides the researcher with a somewhat orderly means to investigate problems. Like the decision processes discussed earlier, it is not a static, highly structured, linear process. Rather, the set of operations that make up the research process are general areas of concern in *most* undertakings. Given the diversity of business problems, it is impossible to identify one means to solve all the questions that need to be addressed. The research process should be viewed as a fluid guide to help structure research projects to answer management's questions.

For example, a researcher attempting to study CEO attitudes toward environmental scanning and a researcher studying why a particular product failed have entirely different research objectives and purposes. Both researchers could use the general research process framework to help them obtain information for the problem at hand. However, neither researcher would probably follow the exact same procedure.

Following the basic research process allows the researcher to create a structure and plan to solve the problem selected for study. Different researchers may follow the general steps in the research process and use totally different methods and techniques to solve the same problem. The value of the process is that it forces critical thinking about important issues in the design and implementation of any research project.

Figure 1.5 presents an expanded outline of the business research process, further developing the plan of the book as shown in Figure 1.4. Essentially, the text is organized into five parts, each composed of at least two chapters:

> **Part 2: Beginning the research process.** The preliminary steps in the research process are considered in two chapters. Chapter 4 presents the topic of secondary data searching, which is used to understand the current problem situation. Here, and throughout the text, the Internet and other online alternatives are examined as essential in the researcher's arsenal of tools. Chapter 5 then deals with the subsequent topics of problem identification and formulation, the assessment of the value of the research information, and the ultimate development of the research proposal. A *research proposal* is a formalized offer or bid for research, specifying the problem to be studied and the plan of action to solve the stated problem. The research proposal directs the researcher's efforts; through it, the manager and researcher agree on what is to be accomplished by the research undertaken.
>
> **Part 3: Research design.** The *research design* is the structure of the research project to solve a particular problem. Design is largely concerned with controlling potential sources of error in the study. The major topics covered in this section of the text are the fundamentals of design (Chapter 6), the foundations of measurement (Chapter 7), scaling and instrument design (Chapter 8), and sampling (Chapter 9).
>
> **Part 4: Data collection.** Once the structure of the study is set, the researcher moves on to actual data collection. Although secondary

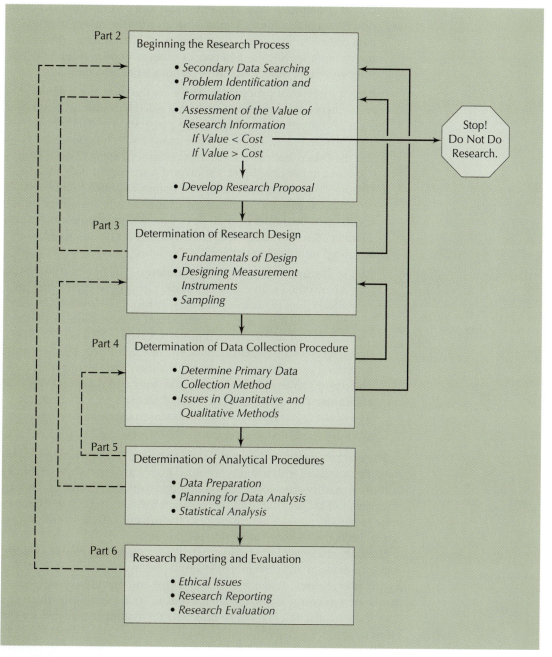

FIGURE 1.5

An Expanded Look at
the Business Research
Process

data is also utilized here, this section focuses on primary data collection (PDC) procedures. Chapter 10 deals with the nature of primary data collection in general, then focuses on PDC using survey instruments more specifically. Chapter 11 examines observation, in-depth interviews, and other qualitative techniques as PDC methods.

Part 5: Analytical procedures. These are the tools and techniques researchers use to summarize data and reach conclusions. The major areas of concern in this stage are the preparation of data for analysis and the selection of an appropriate analytical technique to summarize the data collected to solve the problem. Chapter 12 deals with the issues of basic data preparation and planning for data analysis. Chapter 13 presents basic analytical frameworks used in business research. Chapter 14 presents a discussion of basic linear models, focusing on the topics of analysis of variance and regression. Finally, Chapter 15 surveys multivariate analysis, with brief discussions of advanced multivariate statistical procedures.

Part 6: Research reporting and evaluation. The final section of the text is concerned with communication. Chapter 16 deals with critical ethical issues in business research. This chapter is placed in this section because it is particularly important to evaluation and subsequent design of future research projects. Chapter 17 deals with the research report and its evaluation. The research report is the formal communication device between the manager and the researcher. The report conveys the findings of the study and the means used to obtain those findings. The chapter and text end with an examination of the major managerial considerations in research reporting, given today's fast-changing global marketplace.

The diagram in Figure 1.5 might seem to suggest that each step is a discrete phenomenon with a highly specified progression, but this is a simplification. In reality, the process is highly interdependent; decisions made in early stages of the research process constrain the later stages. For example, if the researcher chooses a measurement instrument with specified properties, it limits the choice of analytical techniques.

Similarly, the process is recursive, in that later stages may affect or modify decisions made in earlier steps. For example, an investigator contemplating a particular research design may discover that the data collection requirements are too costly. He or she then needs to find a design in which the cost of data collection is more in line with the resources available for the project.

The interdependency and recursiveness are represented in the diagram by the solid and dashed arrows. As we progress through the rest of the research process, you should be aware of the implications of making a particular decision about the use of a technique in any of the stages. This type of thinking is necessary to ensure that the originally stated research problem is solved in the best way possible.

SUMMARY

This chapter develops the framework for studying business research. The chapter begins with the proposition that the essence of managerial action is decision making. Decision making in business organizations is examined in terms of both the level (strategic, tactical, and technical) and the process by which decisions are made.

Business research is defined as systematic, controlled, empirical, and critical investigation of phenomena of interest to managerial decision makers. Business research is the only means of obtaining information that is designed to answer managerial problems within a systematic and controlled framework.

The manager–researcher relationship is examined with attention to the conflicts that are inherent in the manager–researcher interface. The role of the manager in the research process is as both a specifier and an evaluator of research projects. Thus, managers must have a basic research competence so that they can adequately judge the worth of the specific research proposals and findings that are presented to them.

Finally, the orientation and plan of the book are presented. The text is managerial in nature, with an applied user orientation. The book is organized around the research process as viewed by the practicing manager.

DISCUSSION QUESTIONS

1. What is business research? What is the role of business research in the decision-making process?

2. What distinguishes business research from other types of research endeavors? What are the similarities in these types of research?

3. What environmental factors have contributed to the growing importance of research in the contemporary business setting?

4. Compare and contrast alternative means of obtaining information for decision-making purposes. What makes business research so different from these other information sources?

5. What are the major levels of decision making within any organization? What role does research play in each of these levels?

6. What are the important activities in the decision-making process? Discuss and evaluate the usefulness of research in each of these activities.

7. What are some of the potential conflicts in the manager–researcher relationship? What are some possible means to reduce sources of conflict?

8. What role do managers play in the business research relationship? What implications does this role have in terms of the involvement of the manager in the business research process?

9. Distinguish between a "user" and "doer" orientation in the study of business research. What focus does a user orientation have that a doer orientation does not?

10. Compare and contrast the manager's problematic decision world with the researcher's analytical world.

11. What is the business research process? Why is the process so important in the conduct of business research?

12. Multinational research is becoming increasingly important. What are the forces driving this trend? What are some issues that one must consider when designing and conducting multinational research?

13. One of the difficulties of fully comprehending the research process is that no one process can be used to solve all business research problems. Discuss this statement.

NOTES

1. Two very good articles discussing the diversity of decision making and the associated implications are Mohan R. Limaye and David A. Victor, "Cross Cultural Business Communication Research: State of the Art and Hypotheses for the 1990s," *Journal of Business Communication* (Summer 1991), *28*:3, pp. 277–299; and Susan Goodman, "Information Needs for Management Decision-Making," *Records Management Quarterly* (October 1993), *27*:4, pp. 12–23.

2. For example, see Herbert A. Simon, *The New Science of Management Decisions,* 2nd ed. (Englewood Cliffs, NJ: Prentice-Hall, 1977); R. Hogarth, *Judgement and Choice,* 2nd ed. (New York: Wiley, 1987); and Robert T. Clemen and Terence Reilly, *Making Hard Decisions* (Pacific Grove, CA: Duxbury, 2001) for different views on the decision-making process.

3. Originally outlined by R. N. Anthony, *Planning and Control Systems: A Framework for Analysis* (Boston: Harvard University Graduate School of Business Administration, 1965); and subsequently discussed by G. A. Gorry and M. S. Morton, "A Framework for Management Information Systems," *Sloan Management Review* (Fall 1971), pp. 55–70; and J. G. Burch, F. R. Strater, and G. Grudnitski, *Information Systems: Theory and Practice* (New York: Wiley, 1979), pp. 40–64.

4. See, for example, Goodman, op. cit. note 1, p. 13; and Limaye and D. A. Victor, op. cit. note 1, pp. 283–285.

5. For more thorough discussions of these processes, see Simon, op. cit. note 2, p. 39; and K. J. Radford, *Information Systems for Strategic Decisions* (Reston, VA: Reston Publishing, 1978), pp. 68–73.

6. Two very good articles on scanning are J. F. Preble, A. R. Pradeep, and A. Reichel, "The Environmental Scanning Practices of U.S. Multinationals in the Late 1980's," *Management International Review* (1988), *28*, pp. 4–14; and E. Auster and Chun Wei Choo, "Environmental Scanning by CEOs in Two Canadian Industries," *Journal of the American Society for Information Science* (May 1993), *44*:4, pp. 194–203.

7. Susan Krafft, "The Big Merge," *American Demographics* (June 1991), p. 46.

8. See John P. Kotter, "What Effective General Managers Really Do," *Harvard Business Review* (November–December 1982), *60*, pp. 156–167; Carol Sanders and Jack W. Jones, "Temporal Sequences in Information Acquisition for Decision Making," *Academy of Management Review* (1990), *15*:1, pp. 29–46; and S. A. Goodman, op. cit. note 1, pp. 13–16.

9. Fred N. Kerlinger, *Foundations of Behavior Research,* 3rd ed. (New York: Holt, Rinehart, and Winston, 1986), p. 10.

10. See, for example, Philip Maher, "Corporate Espionage: When Market Research Goes Too Far," *Business Marketing* (October 1984), pp. 51ff; Al Ossip, "Ethics—Everyday Choices in Marketing Research," *Journal of Advertising Research* (October– November 1985), pp. 10–12; and Walter Shapiro, "What's Wrong," *Time* (25 May 1987), pp. 14–17.

11. See, for example, David O. Schwartz, "Sharing Responsibility for E-Commerce and the Privacy Issue," *Direct Marketing* (June 1998), pp. 48–52; and Melissa Everett Nicefaro, "Internet Use Policies," *Online* (September/October 1998), pp. 31–33.

12. Alan Kuritsky, Emily Bassman, John D. C. Little, and Alvin J. Silk, "The Development, Testing and Execution of a New Marketing Strategy at AT&T Long Lines," *Interfaces* (December 1982), p. 22.

13. See "Flops," *Business Week* (16 August 1993), pp. 76–80, 82.

14. For example, see Tim Powell, "Information Technology Helps Reengineer Research," *Marketing News* (28 February 1994), 28:5, pp. 11–14; and B. Ives, S. L. Jarvenpaa, and R. O. Mason, "Global Business Drivers: Aligning Information Technology to Global Business Strategy," *IBM Systems Journal* (1993), 32:1, pp. 143–161.

15. Abstracted from remarks by John S. Mayo, president of AT&T Bell Laboratories, in a speech entitled "Telecommunications Technology and Services in the Year 2010," delivered at the AT&T Bell Laboratories Technology Symposium in Toronto, Canada, 13 October 1993.

16. Graham T. T. Molitor, "Trends and Forecasts for the New Millennium," *The Futurist* (August–September 1998), pp. 53–59.

17. See Michael R. Czinkota and Ilkka A. Ronkainen, *International Marketing*, 7th ed. (Mason, OH: Southwestern, 2004), pp. 188–189.

18. For other discussions of this relationship, see Paul D. Leedy, *How to Read Research and Understand It* (New York: Macmillan, 1981), pp. 1–7; and Forrest S. Carter, "Decision Structuring to Reduce Management-Research Conflicts," *MSU Business Topics* (Spring 1981), pp. 40–42.

19. See Joseph A. Bellizzi, "Recognizing 4 Researcher–Research User Differences Can Be First Step Toward Removing Barriers to Good Research," *Marketing News 14*:23, part 2 (15 May 1981), p. 6.

20. Op. cit. note 13, p. 162.

SUGGESTED READING

The Nature of Decision Making

Clemen, Robert T., and Terence Reilly, *Making Hard Decisions* (Pacific Grove, CA: Duxbury, 2001).

Provides a useful discussion of the nature of decision analysis and provides a rationale for studying the decision-making process, particularly in Chapter 1. The integration of software to aid in the process is very insightful.

Goodman, S. A., "Information Needs for Management Decision-Making," *Records Management Quarterly* (October 1993), 27:4, pp. 12–23.

This article reviews the decision-making process and explores theories relating to the use of information for management decision making. A well-written article highlighting the nature of decision making.

Simon, H. A., *The New Science of Management Decisions,* 2nd ed. (Englewood Cliffs, NJ: Prentice-Hall, 1977).

Simon is one of the founding fathers of managerial decision theory. A must for future users of research in a business setting.

The Role of Research in Decision Making

Andreason, A. R., "Backward Marketing Research," *Harvard Business Review* (May–June 1985), pp. 176–182.

This article suggests an approach in which both managers and researchers can make research more relevant to decision making in an organization. The proposed "backward" design helps create more usable research information.

Limaye, M. R., and D. A. Victor, "Cross Cultural Business Communication Research: State of the Art and Hypotheses for the 1990s," *Journal of Business Communication* (Summer 1991), 28:3, pp. 277–299.

This article examines cross-cultural communication and decision-making paradigms with implications for research. A good article to begin a discussion of how research is used and conducted in cross-cultural settings.

Richie, J. R. Brent, "Roles of Research in the Management Process," *MSU Business Topics* (Summer 1976), pp. 13–22.

Stresses that effective research strategies should correspond to the nature and level of management decisions. An excellent classification scheme is used, relating research methodologies to the management process.

The Manager–Researcher Relationship

Druckman, Daniel, "Frameworks, Techniques and Theory: Contributions of Research Consulting in Social Science," *The American Behavioral Scientist* (August 2000), 43:10, pp. 1635–1666.

An interesting discussion of the different roles played in the manager/researcher/consultant relationship. Insightful in its approach and contributions

"Flops," *Business Week* (16 August 1993), pp. 76–80, 82.

 This article is a personal favorite. Although research is not the focal point of the article, one can clearly see the linkage in the manager–researcher interface. Inadequate and ill-communicated research can and does lead to product failures.

Hartley, Robert F., *Marketing Mistakes and Successes,* 9th ed. (New York: John Wiley, 2003).

 Hartley presents a set of cases that specifically deal with marketing mistakes that were made because of miscommunication or mishandling at the manager/researcher interface. Interesting reading.

Parent, L. V., "Corporate Planning/Market Research: The Connection." *Managerial Planning* (March–April 1985), pp. 22–26.

 This article discusses the essence of the relationship between corporate planners and marketing research. The article describes and illustrates the role of each in the accomplishment of the firm's objectives.

Plan of the Book

Harris, G. W., "Living with Murphy's Law," *Research-Technology Management* (January–February 1994), 37:1, pp. 10–13.

 This article deals mainly with avoiding "Murphy's gifts" in the R&D process, and it points out the necessity to plan earlier on in the research endeavor. In other words, if anything can go wrong, it will go wrong, and the R&D process had better be flexible enough to handle the unexpected.

Sekaran, Uma, *Research Methods for Business: A Skill Building Approach,* 4th ed. (New York: John Wiley, 2002).

 This supplementary text provides a very good how-to approach in understanding the business research process. The applied nature is its best asset.

2

Information, Research, and Knowledge Management

Overview

Understanding the Basics of KM

So What is KM?
The Definition
Expected Benefits of KM

Information Systems (IS) and KM
Technology and ISM
Decision Support Systems (DSS)

Database Management (DBM) and
Online Analytical Processing
(OLAP)

Business Research (BR) and KM

Summary

Discussion Questions

Notes

Suggested Reading

OVERVIEW

As noted in Chapter 1, the management and use of data and information in organizations is becoming increasingly complex and intertwined. Similarly, the macro environment managers work in is characterized by rapid change, explosive information growth, increased data access and exchange capabilities, and increased global competition. This state of affairs makes it imperative that integrative paradigms be developed so the knowledge assets that are at the disposal of the firm not be wasted, but instead utilized to their fullest extent. Not to do this puts the organization at best at a competitive disadvantage, with its very survival in a highly competitive global marketplace at risk. Enter knowledge management (KM).

KM is a relatively new discipline although its roots are deep. It was not until the mid-1990s that the field began to coalesce and information technology capabilities had evolved sufficiently to make KM a serious possibility in proactive organizations. KM is concerned with facilitating an organization's use of all its intellectual assets to accomplish its objectives. It is an integrative paradigm that attempts to converge a set of diverse disciplines.

The primary purpose of this chapter is to briefly examine KM and to put into perspective the role of business research. To do this, the chapter begins with a discussion of the basics and terminology underlying the conversion of data to information, knowledge, and ultimately wisdom. Following this, are a definition and discussion of the benefits of KM.

One cannot fully understand the role of research in this field without examining how the components of information systems management (ISM) interact with KM. As a result, the next section explores how such topics as technology, decision support systems (DSS), database management (DBM), and online analytical processing (OLAP) fit into the picture.

Finally, the chapter concludes with a discussion turning on how business research (BR) interacts in today's KM-oriented organization. This section gives the reader a perspective on the rest of the text.

UNDERSTANDING THE BASICS OF KM

In order to begin to understand the complex field of knowledge management (KM), one must understand the basics of the transformation of data into a useable format that can facilitate decision making in an organization. A useful way to approach this is to use Figure 2.1, which depicts this transformation process as a pyramid that takes *data*, or simple facts or recorded observations, through the various stages useful from a managerial perspective to the ultimate use and value of data, which is the creation of wisdom. However, for data to be ultimately converted into the more

FIGURE **2.1**

Pyramid of Data
Transformation

Adapted from "Knowledge
Management-An Overview,"
downloaded from http://www.
knowledgeboard.com/library/
km_an_overview.pdf

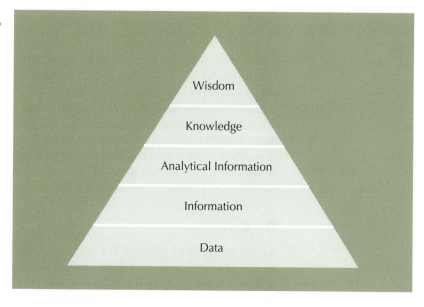

abstract and general notion of wisdom, it goes through a purifying and enhancing series of stages.[1]

Data is converted into *information* when it becomes useful for an activity such as decision making. Therefore, although the receiver may be inundated by data, it does not become information until it is meaningful. Information can subsequently be further refined and enhanced into analytic information. *Analytic information* is information that is extended and/or enhanced by the examination of relationships and patterns it forms when it is combined in unique ways, which makes the information more useful than the data/information from which it was derived. This task is often accomplished by analytical processing and the judicious use of statistics. However, it can also be accomplished through creative and critical thinking.

As information and analytical information are assimilated and applied along with previous experiences and intuition, knowledge can result. *Knowledge* is then the creative blend of the various types of information, experiences, and intuition that forms the basis for understanding and/or acting on new information or new decision-making contexts. Knowledge is at the heart of 21st century managing and organization. The more knowledgeable an individual or organization is, the more likely they will be able to perceive and respond to challenges in their environments.

At the top of the pyramid is wisdom. *Wisdom* is the accumulated ability and willingness to apply knowledge with good judgment. It is the culmination of the accumulation and transformation process of inputs from all sources and the experiences gained through the entity's lifetime activities. Great wisdom implies the prudent application of knowledge.

SO WHAT IS KM?

A fair question indeed, but one with no easy answer. There is no universally accepted definition of KM, any more than there is for the concept of knowledge itself. It is a relatively new discipline emerging from a convergence of diverse disciplines that are concerned with organizational culture, organizational learning, research, information systems, and human resources, to name but a few disciplines relevant to this new field.[2] However, the skyrocketing interest in the field has produced a wealth of resources that one can use to understand it. Table 2.1 identifies several web sites very useful for understanding and keeping up with the fast development of KM.

The Definition

For the purposes of this text, the working definition of *knowledge management* is the facilitating of how an organization creates, generates, stores, communicates, and utilizes its intellectual assets to accomplish its objectives. This definition both recognizes the diverse nature of the field and describes an objective of this management process.

Several items in this definition are noteworthy:

- KM is multidisciplinary. For example, information technology (IT), communication, psychology, sociology, anthropology, research, and human resources all have a stake in how effectively the firm uses its intellectual assets. To believe otherwise is indeed to think simplistically.

TABLE 2.1 Useful Web Sites for Understanding Knowledge Management

http://www.kmresource.com	IKM's knowledge management portal is a highly useful resource for understanding this complex field.
http://www.knowledgeboard.com	The European KM Forum aims to build up a KM community in Europe and to support and identify commonality in KM terminology, application, and implementation.
http://www.cio.com	Billed as the resource for information executives. Online since 1995. Worth a visit.
http://www.metakm.com	This is a knowledge management portal for practitioners, researchers, and solution providers. Provides an interesting array of content on the field of KM.
http://krii.com	A private web site of the Knowledge Research Institute. Interesting readings are available.
http://www.brint.com	BRINT Institute is an interesting knowledge resource and provides a wealth of information related to KM.

- No one discipline owns KM. IT, although a significant part of any KM endeavor, does not own KM. It is truly organizationally owned and facilitated by all its component parts.

- People and learning are central to the facilitation process. The definition implies that knowledge of all types (both documented, or *tacit*, and *implied*, or subjective, knowledge) is created, processed, and integrated into a firm's culture. Learning must occur for this to happen, and people are central to this process.

- The definition is "fuzzy" at best. It is broad and all-encompassing to ensure that any activity that facilitates how any organization can improve the use of its intellectual assets will be included under KM, so as to ensure optimal use of these assets.

These points highlight both the complexity and dynamic nature of the field. Given this, one might ask how it is being accepted in the organizational marketplace. In a survey by Pricewaterhouse Coopers and the World Economic Forum, 95% of CEOs saw KM as an essential ingredient for the success of their companies. Furthermore, a Conference Board survey of 200 senior executives reported that over 80% had some KM activities underway.[3]

Expected Benefits of KM

The expected benefits of a properly developed and implemented KM system are many.[4] These benefits will take root because organizations and the individuals in them will begin to create and use the intellectual assets inherently available to them more effectively. The potential benefits include, but are not limited to

- Encouraging innovation through improved information and knowledge-handling abilities.

- Increasing competitiveness because overall organizational knowledge will increase the responsiveness of the organization to the market and environmental changes.

- Improving customer service by encouraging relationship development and meeting customer expectations.

- Enhancing employee effectiveness and morale, thus potentially reducing operational costs associated with turnover and retention.

- Creating more effectively an organizational memory of valuable knowledge that would be lost when key personnel leave the organization.

These key benefits and others are being actively pursued by organizations in a variety of situations. To illustrate one organization's efforts, Exhibit 2.1

EXHIBIT **2.1**

Vision and Strategy
for Knowledge
Management and IM/IT
for Health Canada
December 1998
Table of Contents

From www.hc-sc.ca/
iacb-dgiac/km-gs/
english/vision_en/
tableofcontent.htm

I. VISION

I.1 Narrowing the Gap Between the Status Quo and our Desired Future
I.2 The Vision Statement
I.3 Working Definitions of Knowledge Management Terms
I.4 Rationale

I.1 Narrowing the Gap Between the Status Quo and our Desired Future

Health Canada needs to strategically and aggressively narrow the gap between the status quo as described in the adjacent text box, and its desired future as follows:

The Status Quo—At Health Canada we do not,

- know what our employees know
- know what information we have
- know what information we need
- have a coordinated approach to the capturing of employees' knowledge, or
- have a guiding blueprint for investments in knowledge, information, applications or technology.

- knowledge is recognized as a tangible, mission-critical resource; knowledge management is integrated into business initiatives and processes as a means of fulfilling business requirements;
- Health Canada's Management supports and invests in knowledge initiatives which continue to build knowledge and IM/IT infrastructure, tools and services to support the department's business lines;
- Health Canada employees at all levels have fully endorsed and adopted a knowledge and learning culture;
- Health Canada is a recognized leader in the development and implementation of a Canadian health infostructure which is built on common infrastructure and standards where logical and cost-effective;
- Health Canada systematically creates knowledge and influences the conduct of research where there are knowledge gaps, shares this knowledge securely and uses new and existing knowledge effectively in taking evidence-based decisions; and
- Health Canada is a respected business partner and active participant in communities of practice, nationally and internationally.

Health Infostructure is defined as:

The application of communications and information technology in the health sector to allow the Canadian public, patients and care givers, as well as health care providers, health managers, health policy makers and health researchers to communicate with each other, share information and make informed decisions about their own health, the health of others, and Canada's health system.

I.2 The Vision Statement

Health Canada's Mission:
To help the people of Canada maintain and improve their health.

EXHIBIT 2.1

Continued

Health Canada analyses, creates, shares and uses health knowledge to maintain and improve the health of the people of Canada:

- through its knowledge management processes and strategies which are tailored to advance the business lines of the department;
- as a model knowledge organization; and
- as a leader, facilitator and partner, in the development of a Canadian health infostructure, responding to national and international trends and opportunities.

I.3 Working Definitions of Knowledge Management Terms

The term 'knowledge' is used extensively in this document and often appears in association with, and can be confused with, the terms, 'information' and 'data'. To clarify the distinction, 'knowledge' is defined as follows:

Working Definitions of KM Terms:

Data are facts, observations, or measures that have been recorded but not put into any meaningful context. A single musical note is data.

Information is data that has been arranged in a systematic way to yield order and meaning. A series of notes arranged into a tune is information.

Knowledge is information in the mind, in a context which allows it to be transformed into action. A musician is able to play a tune because of his knowledge.

Knowledge is a fluid mix of framed experience, values, contextual information, and expert insight that provide a framework for evaluating and incorporating new experiences and information. It originates and is applied in the minds of knowers. T.H. Davenport and L. Prusak, Working Knowledge.

The two key distinctions to be made among knowledge, information and data are firstly, that knowledge exists in the mind, and secondly, that knowledge is a framework for evaluation. If, for example, a heart surgeon writes down instructions for performing a new transplant procedure on a piece of paper, the contents of the paper (i.e., information) will become knowledge when read by another heart surgeon who understands the context and how to apply it. It remains information when read by a non-surgeon who understands only the general concept of a heart transplant, and it becomes data when viewed by a person who does not speak the language in which it is written.

It is because knowledge is contextual that knowledge management initiatives have to be described as sets of information management/information technology (IM/IT), learning, and business initiatives. These initiatives aim to get information to the people who need it, give them the tools and freedom to analyze it and fill gaps, and give them a framework in which to apply it. The actual transformation of information into knowledge occurs at all stages in people's minds and becomes evident only in the decisions they make and the actions they take. Knowledge management, therefore, aims to create the same fluid mix of framed experience, values, contextual information, and expert insight within the organization that exists within the individual mind, thereby providing the organization with a framework for evaluating and incorporating new experience and information. The transformation to a knowledge management culture would consequently only become evident in the decisions the organization makes.

The Giga Information Group (March 1998) has advised its clients that knowledge management will never possess the kind of crisp definition afforded its individual

(Continued)

EXHIBIT **2.1**

Continued

components. Knowledge management tends to be defined by the organization applying it and by the organization itself.

Health Canada has defined knowledge management as per the adjacent text box, to provide a common understanding of the way in which knowledge management will be applied to meet today's needs and prepare people for a future, more knowledge-based health system and society. Knowledge management will be different and applied differently in each of the business lines of Health Canada. In some cases, it is the creation and dissemination of knowledge that constitutes the business line, and in other cases the business line relies on ready access to knowledge.

As program operations should influence knowledge management, so should knowledge management be a strength called upon in the design and delivery of the business of the department. Knowledge management, with IM and IT as "enablers," should add significant value to the management of the department's business, by providing techniques for comprehensive evidence generation and assessment, more economical delivery, better service and improved efficiency.

**Health Canada Operational Definition
of Knowledge Management**

A departmental strategy for ensuring that health knowledge is identified, captured, created, shared, analyzed, used and disseminated to improve and maintain the health of Canadians.

Health knowledge is defined to encompass information, skills, expertise and experience related to and supporting health and the health system, nationally and internationally.

**I.4 Rationale—The Business Case for Knowledge Management
at Health Canada**

Better health decisions will be made if health knowledge is created strategically, shared effectively and managed efficiently. Health consumers can hold providers to account, and providers can better serve consumers, when knowledge provides the evidence base for health decisions. Strategic investments in knowledge and commitment to knowledge culture, then, are instrumental in maintaining and improving the health of Canadians.

**Health Canada's Strategic Learning and Development
Policy—The Foreword**

The world is moving into a "knowledge economy" where the performance of organizations will depend more and more on the acquisition, sharing and application of knowledge. This is why Health Canada considers Learning and Development activities, aimed at enhancing knowledge and skills, to be a priority and an essential business investment that contribute to the attainment of departmental objectives and of employee career goals. It is through the development of knowledge that Health Canada can achieve enhanced service to the public and greater organizational performance.

Health Canada is prepared to provide leadership in the development of a national health infostructure, building on existing and future Canadian and international infostructures, to strengthen the ability of people to make informed choices about their own health, the health of others and Canada's health system.

presents the "Vision and Strategy for Knowledge Management and IM/IT for Health Canada," which describes what Health Canada wants to achieve, how the organization will achieve it, and the rationale for such an undertaking.[5]

INFORMATION SYSTEMS (IS) AND KM

Since this text is focused on the role of business research in making better decisions and ultimately improving the intellectual assets of the firm, it must include an examination of the contexts in which research provides value to IS and KM. Figure 2.2 is a visual representation of the "fuzzy" inter-relationships of KM, IS, and other components in building, using, and communicating knowledge within the firm.

The figure places IS at the center of the KM function of the organization, but it is not the KM function itself. Just as IS is not KM, neither is IT. Both are facilitating components essential to enabling the KM function in an organization. Traditionally, information systems management (ISM) has been portrayed as concerned with the collection, creation, and distribution of useful data.[6] However, the collection of data has generally not included the collection of primary data.[7] Instead, the collection was a business research and/or intelligence function that was somewhat removed from the IS function itself.

Overall, the convergence of disciplines is making it very difficult indeed to define clearly traditional areas of specialty. KM is requiring anyone concerned with the effective use of the intellectual assets of the firm to search for ways to accomplish this goal more effectively. This in turn requires a multidisciplinary approach that often does not fit conveniently with a traditional functional view of the world.

Technology and ISM

Technology is continuing to have a tremendous impact on developments in the field of KM. In particular, computer penetration, interconnectivity, and capabilities are drastically changing the way managers, information specialists, and researchers collect, analyze, access, use, and report information. More research information can be accessed faster and in a more understandable form than ever before. For example:

- 3Com, a company that focuses on providing networking products and solutions, was faced with the problem of integrating many technical databases composed of *data silos* (in-depth data storage devices, usually project or technically based) created by their engineers. In addition, the firm sought to record, integrate, and communicate the combined knowledge of these and future resources to maximize their value to the firm. This effort led to a six-month payback, an 80%

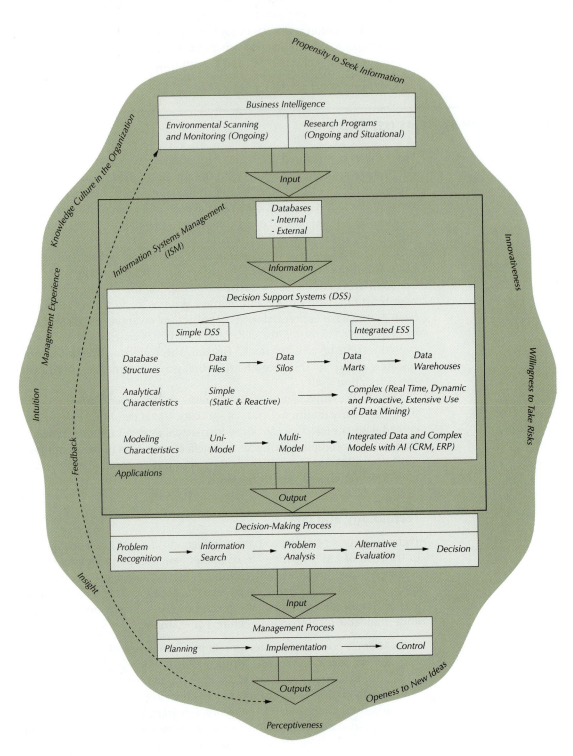

FIGURE 2.2

The Fuzzy World of Knowledge Management (KM), Information Systems (IS) and Business Research (BR)

reduced response time to questions, and over 30% of all 3Com networking customer support issues being now solved online.[8]

- Reuters is one of the world's leading information brokers, supplying online data to fund managers, dealing rooms, assurance companies, and corporations.[9] In the 19th century, Paul Julius Reuter used pigeons to deliver financial news and financial information. Today, computers allow the company to collect information about more than 180 exchanges and over-the-counter markets from more than 4000 customers who contribute information directly and from a network of approximately 1200 journalists, photographers, and camera operators. Reuters can provide information within three seconds to customers around the world. Research functions are immediate and targeted to the individual user.

- High-capacity computers have spawned significant interest in maximizing the value of large databases, an approach commonly called *data mining*. Data mining is usually considered the seeking of information in databases, which is formally known as Knowledge Discovery in Databases, or KDD.[10] The steps in the KDD process include establishing data access, selecting a subset of data for analysis, cleaning the data, seeking information within the data, and reporting the information discovered. Already such organizations as IBM, Microsoft, NCR, Disney's Buena Vista Home Video, and GTE are all making significant new advances and investments in this technology.

These examples illustrate just a few of the ways computer technology is affecting the information function in the organization, an impact that is expected to become even more significant in the future. The rest of this text will explore how technology is changing all aspects of information and knowledge management in the firm, particularly as it relates to the research function.

Decision Support Systems (DSS)

DSS are systems that are designed to support decision making; generally, they include a systematic collection of data, tools, and techniques that an organization uses to gather information for decision-making purposes.[11] They can be very simple in nature or very complex. Figure 2.2 illustrates this continuum.

In its simplest form, the DSS can be a simplistic model such as break-even analysis from a limited database. In its most sophisticated form, the DSS is part of an integrated data warehouse that uses artificial intelligence (AI) research to refine the DSS into an *expert support system (ESS)* accompanied by graphical and statistical modeling tools for analysis and reporting purposes. Essentially, AI researchers attempt to develop computer systems that produce output as good, or almost as good, as humans—intelligent decisions. An ESS is simply the integration of a DSS with AI to form an "intelligent" DSS that is designed to mimic human decision making.[12] A *data warehouse,* as will be discussed later, is a large, centralized, enterprisewide database that is a collection of all the demographic and transactional data about consumers and businesses that a company collects, stores, and reviews.[13]

Models are extremely important to decision makers because they form the basis for the development of *decision support systems (DSS).* A model can be defined as any highly formalized representation of a theoretical system, usually designated through the use of symbols. In this chapter, a model is further specified as representing a theoretical or highly applicable decision system. Models can be tested, examined, and analyzed by those who created them. In essence, modeling is an organizing process for the business researcher and decision maker. Models are thus used to aid decision makers. However, for a model to be considered useful from a decision-making perspective, it must be able to transform data into information.

Since the early 1950s, there has been a significant increase in the development of models. During the decades that followed, their use as an aid to decision making was generally hindered by complex symbolism and limitations in hardware and software capabilities. However, since the 1990s, ever accelerating capabilities in data storage, movement, and analysis have hastened the acceptance and demand of increased modeling and analytical capabilities to maximize the value of the huge amount of information being amassed by organizations. Decreasing computing and storage costs now make the building of large data warehouses for data analysis by end-users affordable.

As one can see, modeling and DSS are central to the process of converting data into information and ultimately knowledge. This is well documented in business. For example, Taco Bell Corporation relied on an integrated set of operations research models—including forecasting to predict consumer arrivals, simulation to determine the optimum labor required to provide desired customer service, and optimization to schedule and allocate crew members—to realize payroll savings of over $53 million dollars from 1993 through 1997.[14] Similarly, pharmaceutical maker Hoescht-Rousel of Warren, New Jersey, cut selling costs by about 15%, saving about $10 million, by developing a sales force allocation model. Numerous other modeling successes abound.[15]

Other organizations are developing other types of model-based DSS that will aid in the integrated use of knowledge resources. For example, ERP (Enterprise Resource Planning software) is designed to integrate all departments and functions across a company in a single computer system.[16] CRM (Customer Relationship Management) seeks to create a companywide business strategy designed to reduce costs and increase profitability by solidifying customer loyalty.[17] These and other models are the result of the search for the best means to use the firm's knowledge resources.

Database Management (DBM) and Online Analytical Processing (OLAP)

As noted earlier, technological advances have reinvigorated the field of modeling—as it has done to the entire field of business research. The concepts of modeling and database management have been around for decades, but it was not until the last half of the 1990s that the capabilities and expertise came together to make significant advancements in integrating data analysis tools with database management and acquisition. Several concepts of interest here are *data warehousing, data marts, OLAP,* and *data mining.*

Data warehousing is concerned with turning the masses of data that every organization has collected and is collecting into useful decision-making information for decision makers. *Data warehouses* feature large integrated, easily accessible databases with consistent encoding structures, usually collected from a variety of sources over an extended time period, organized by subject (customers, vendors, products) for the purposes of managerial decision making. Data-warehousing software includes advanced data extraction techniques that clean and transform data into a universally usable format. In addition, the system must be able to locate data that is often stored on multiple servers, mainframes, or on any combination of online storage devices. Data analysis, modeling, and delivery tools also are an integral part of the data warehouse.

Data marts are also being developed; these are often subsets of larger data warehouses in organizations. They are, in effect, miniature data warehouses, usually oriented to a specific department or other organizational unit; they are considerably smaller and less complex than the full data warehouse. The development of these integrated data acquisition, storage, and management devices has stimulated a huge amount of interest in the business community.

It has been estimated that more than 90% of large corporations worldwide have either built or are in the process of building data warehouses. Technology industry research firm META Group says the market for data-warehousing hardware and software, practically nonexistent in 1994, grew from $4.5 billion in 1996 to over $15 billion in 2000.

Large collections of data are generally not very useful unless one has the associated tools to analyze, summarize, and understand them. Here is where OLAP or Online Analytical Processing, comes into play. OLAP is simply a term for a variety of techniques, usually statistical in nature, to accomplish these objectives. Generally, a wide variety of statistical packages and software provide the basis for the analysis of the collected data. Excel, SAS, and SPSS are examples of commercially available software for just this purpose. These and other packages will be more fully discussed in Part 4 of this text. Software development in recent years has grown to facilitate ever increasing technological capabilities.

EXHIBIT 2.2

Example of Data-Warehousing and Data-Mining Techniques in Use

Reproduced by permission from Data Management Customer Solutions and Success Stories—Safeway, IBM website, http://www.software.ibm.com/data/solutions/customer/safeway/safeway2.htm. Copyright 1997 by International Business Machines.

IBM'S INTELLIGENT MINER HELPS SAFEWAY FOCUS ON THE INDIVIDUAL SHOPPER

The small, family-owned corner market is pretty much a thing of the past. There was a time when the neighborhood grocer recognized each one of his customers, called them by name. Responding to the individual needs of each shopper was relatively simple, because these were people he knew personally.

Today, large chains of supermarkets and convenience stores have replaced the smaller local grocery shops, offering consumers a wide variety of product choices and competitive prices. Things aren't so simple anymore. And the butcher probably doesn't even know your name.

But, understanding the individual customer is still the key, whatever the scope or size of the business. For Safeway plc, the problem looked like a statistical nightmare. How do you measure and respond to the complex, rapidly changing needs of more than six million customers each and every day, converting gigabytes of raw data into information that can be used to identify and target the buying habits of individual shoppers in everything from orange juice and cheese to the latest videos?

To the British public, Safeway is a household name almost synonymous with supermarkets. With its $10 billion in annual sales, 70,000 employees, and more than 410 stores, Safeway is the third largest grocery chain in the United Kingdom.

Recently, the company put together a strategic plan for growing its retail business and gaining competitive advantage in the future: Safeway 2000. According to Mike Winch, Safeway's Information Technology Director, the goal is to focus on more than products and storefronts. "Our ambition is to create a marketplace of one. We want to market our services to the individual. In fact, everything we do now is customer-centered. That's our business advantage."

To achieve this, technology has been an integral part of Safeway's strategy for reaching the individual customer. Realizing that managing information is a vital component in the success of a food retail chain, Safeway first started working with IBM some years ago to develop better information systems to support its growth in the marketplace.

Today, IBM's DB2 is the backbone of Safeway's Information Warehouse, the company's immense customer database, data analysis, and decision support program. Using the DB2 database has allowed Safeway to quickly spot regional variations in sales and pinpoint product trends. "It's this kind of information," says Mike Winch, "that helps us make the right marketing decisions. It could be said that information about products is as valuable as the products themselves."

But, the main thrust behind Safeway 2000's information strategy is to understand customer performance. "We now look at how we manage the business through our customers," explains Winch. "We have moved from managing information on 410 stores, with 25,000 product lines, to managing the individual needs of more than six million customers by looking at every item in their shopping basket."

It was obvious to Safeway's IT director that the complexity of managing nearly 30 gigabytes of data each day, warehoused in two data centers linked by satellite network to each point of sale, was a mind-boggling task in itself. The greater challenge, however, was to get behind the number crunching and understand the spending profile of the individual customer buried in the raw data. "You can have the best supermarket in the world, the best company, even the best products and staff," says Winch, "but that does not make one penny profit until someone buys something." Unless you use your information resources to effectively reach out to individual shoppers and target their needs, you can't compete.

(Continued)

EXHIBIT 2.2

Continued

Safeway was looking for the right kind of information technology to leverage the strategic value of its immense, highly dynamic customer database. Even more crucial was the requirement for a robust data mining tool capable of analyzing point-of-sale data on a day-to-day basis plus the detailed customer information gathered from the 6 million shoppers using Safeway Added Bonus Cards. These cards record each customer's purchases along with the shopper's name and address.

So Mike Winch turned to IBM for a solution. "This isn't a job for a statistician going off, looking at the numbers, and saying, 'Did you know?' Human analysts can't do a job of this magnitude. You need to get into data warehousing, and sophisticated data mining techniques." IBM's Intelligent Miner was the tool chosen to help the experts in Safeway's marketing, finance, and retail departments get a handle on the strategic information buried in its data assets. It was their key to understanding the customer's point of view.

"We aim to serve our customers well," comments Winch, "seeing each one of them as unique, meeting their individual needs. And that means knowing what products you sell to customers. We are the first food retailer in the UK to keep data on customer transactions down to the product level. Obviously, you have to understand that data, otherwise you cannot focus on customers effectively."

When you have to cope with a customer database capable of handling eight million transactions per week, you need all the help you can get. Intelligent Miner quickly became an essential component in Safeway's IT strategy. According to Mike Winch, "Our competitive advantage is linked to what we know. So we've embedded data mining in the very fabric of how we do our business on a day-to-day basis." Intelligent Miner's powerful algorithms, processing techniques, and analytical features have given Safeway's management team more flexibility and a deeper strategic understanding of customer requirements across the UK, which has translated into better business decisions.

Intelligent Miner is used to work the data stream continuously, segmenting Safeway's Information Warehouse into hundreds of customer characteristics that can be used to tailor individual mailings, analyze product performance, or forecast shopping patterns. "The value of the customer data we can continuously extract and analyze, using IBM's Intelligent Miner, is priceless," comments Winch.

The bottom line is that IBM's data mining toolkit, Intelligent Miner, has helped to create a comprehensive and comprehensible customer database that actually drives the growth of Safeway's business by focusing on the meaning behind the statistics. In Mike Winch's words: "It's a powerful tool that enables us to achieve our business goal—to create a 'market of one' tailored to the individual shopper."

One of the most important software groups to benefit from this growth is data-mining tools. Data mining is essentially a knowledge discovery process that extracts previously unknown, actionable information from very large databases or data warehouses.[18] Data-mining tools include a wide variety of analytical modeling techniques that help one gain insight into the underlying patterns in the data. These tools can include such analytical techniques as descriptive statistics, graphical plots, correlation, linear regression, discriminant analysis, factor analysis, and data visualization techniques, to name but a few of the more readily used modeling tools. Exhibit 2.2 provides an example of how Safeway plc, the third-largest grocery chain in the United Kingdom, used data-warehousing and data-mining techniques to improve its focus on the individual shopper.

Other, more integrated data-mining tools are now in the marketplace. SAS has developed the Enterprise Miner (http://www.sas.com), and IBM has developed the Intelligent Miner (http://www.software.ibm.com). IBM's Intelligent Miner enables users to recognize and classify correlation in their warehoused data by automatically performing such analyses as cluster analysis, predictive modeling, database segmentation, and classification, among other analyses, to build data-based models that will help a company improve its ability to react to markets. With this in mind, let us move to a discussion of how business research fits into KM.

BUSINESS RESEARCH (BR) AND KM

As one traverses through this chapter, one may ask, How does BR fit into this picture and why is it important to anyone anyway? Perhaps the best way to start this discussion is to refer to Figure 2.2 and tell you what BR is not. It is not KM, but BR, like IS and IT, is an important facilitating component of the KM process. As noted thus far in the text and as illustrated in this figure, one is seeing a blurring of fields of study traditionally distinct in the past. For example, research programs in the past were more likely to be ongoing or situationally significant investigations conducted and analyzed by research specialists, and information was subsequently communicated to managerial decision makers necessarily as a by-product of rather unconnected data files.

Similarly, prior to IT's explosive growth in the 1980s and 1990s, the business intelligence (BI) function was often framed by a separate group of individuals (often times planners or analysts), who would conduct ongoing scans and monitor the environment and competition to identify trends, opportunities, and threats for significant inputs to management decision makers. These scanners also tended to create reports and data (often not quantifiable or easily put into electronic format) that made data and information handling difficult at best.

Recipients of this information, the decision-making managers, were often not very concerned with the data, the analysis of this data, or the format of the data because it was unwieldy to handle, difficult to analyze, and out of their realm of expertise. As a result, specialization was the norm and there was little impetus given to increasing productivity gains from the intellectual assets of the firm because the inherent capabilities to do so were just not there.

Enter the IT explosion of the 1980s and the information age of the 1990s. The ability to create, collect, store, process, and assimilate knowledge into the fabric of the firm started to become a reality.[19] Content, methodologies, and use issues came together because of the need to create learning organizations comprised of people with a vast amount of information, knowledge, and wisdom at their disposal. Today, we see the beginnings of a seamless integration of information and knowledge into organizations of the future, but this will not be easy to fully accomplish.

Where does this leave the previous specialties of BI, BR, IS, and practicing functional managers? Well, the flattening of organizations and continuous cries for increased productivity will continue to force structural changes in how the information and KM function in the firm will be handled. Organizations will not allow intellectual assets to be underutilized in the face of huge investments in technology and the capability to more fully utilize intellectual assets. KM, BI, and IS functions will continue to be merged into the firm and individuals in the firm will have to take on increased responsibilities throughout the KM process. Insofar as BR is concerned, it has already blurred into BI, IS, KM, and managerial responsibilities. In other words, a basic knowledge of BR is a necessary skill required of operating managers and others if the full potential of KM is to be realized in the future.

SUMMARY

KM is emerging as a predominant paradigm in the first decade of this new century. This chapter attempts to define and illustrate the importance of KM in the context of ISM and BR. The basic transformation of data to information, knowledge, and wisdom is discussed in the context of today's rapidly evolving organizations. The field of KM is defined as the facilitating of how an organization creates, generates, stores, communicates, and uses its intellectual assets to accomplish its objectives. The interrelationships of major components of ISM are explored. Concepts such as KDD, DSS, data warehousing, data marts, OLAP, and data mining are discussed with regard to their contributions to the overall use of a firm's intellectual assets. Finally, BR's contribution to the KM field is outlined.

DISCUSSION QUESTIONS

1. What are the differences among data, information, analytical information, knowledge, and wisdom? What is necessary for data to evolve into wisdom?

2. Consult the web sites in Table 2.1 and derive a definition of KM that you think is appropriate. Why is KM so difficult to define?

3. Again, search the available KM web sites and identify a list of benefits that can be realized through the successful implementation of such

a system. Can the perfect KM system ever be designed? If so, why; if not, why not?

4. IS and IT have been said to be the essence of KM. How would you respond to such a statement? Give concrete arguments for your reasoning.

5. Examine Exhibit 2.1, "Vision and Strategy for Knowledge Management and IM/IT for Health Canada." Critique this vision and strategy after researching the current thinking about KM. What are its strengths and weaknesses?

6. Closely examine Figure 2.2, "The Fuzzy World of Knowledge Management (KM), Information Systems (IS), and Business Research (BR)." Is this an accurate portrayal of the interrelationships represented in this diagram? If not, why not? How would you modify or change it to reflect your views?

7. DSS and modeling are closely intertwined. How would you differentiate between the two?

8. Distinguish among data warehouses, data marts, data silos, and data files. What is the evolutionary structure of these data structures and what has enabled this evolution?

9. Choose IBM (http://www.ibm.com), SAS (http://www.sas.com), or SPSS (http://www.spss.com) and go to the web site. Using the information at the web site, write a brief synopsis of the capabilities of your chosen software in the areas of data warehousing and data mining. Be sure to be specific in your answers.

10. OLAP and techniques such as data mining have been used extensively in the knowledge-building process. Explain your knowledge of such techniques and give examples of how these techniques can be used in today's business environment.

11. Knowledge of BR is now becoming a necessity for managers and information specialists alike. Discuss the reasons for this phenomenon and relate to how it affects your future career path.

NOTES

1. This section draws heavily on "Knowledge Management—An Overview," online June 13, 2002, http://www.knowledgeboard.com/library/km_an_overview.pdf, downloaded September 16, 2003; and Anthea Stratigos, "Knowledge Management Meets Future Information Users," *Online* (January 2001), *25*:1, p. 65, http://infotrac.galegroup.com/itw/infomark/247/457/64736517w3/purl=rc1_ITOF_0_A68656986&dyn=11!mrk_one?sw_aep=orla57816, downloaded September 23, 2003.

2. For an interesting discussion of the history of KM, see Karl M. Wiig, "Knowledge Management: An Emerging Discipline Rooted in a Long History," 1999, http://www.krii.com/downloads/km_emerg_discipl.pdf, downloaded September 23, 2003.

3. "Knowledge Management—An Overview," op. cit. note 1.

4. See, for example, Megan Santosus and John Surmacz, "The ABCs of Knowledge Management," May 23, 2001, http://www.cio.com/research/edit/kmabcs.html, downloaded March 25, 2003, and Mutiran Alazmi and Mohamed Zairi, "Knowledge Management Critical Success Factors," *Total Quality Management* (2003), *14*:2, pp. 199–204.

5. "Vision and Strategy for Knowledge Management and IM/IT for Health Canada," December 1998, http://www.hc-sc.gc.ca/iacb-dgiac/km-gs/english/vision_en/chapterI.htm, downloaded September 23, 2003.

6. Leonard M. Jessup and Joseph S. Valacich, *Information Systems Foundations* (Indianapolis, IN: Que Education and Training, Macmillan Publishing, 1999).

7. Z. S. Demirdjian, "Marketing Research and Information Systems: The Unholy Separation of the Siamese Twins," *Journal of the Academy of Business, Cambridge* (September 2003), *3*:1/2, pp. 218–223.

8. "3Com: Networking Giant and Customers Profit with Knowledgebase Solution," Primus Case Study, April 2002, http://www.primus.com/customers/caseStudies/default.asp, downloaded October 1, 2003.

9. Lara Bayley, "Controls in Finance Whirlpool," *CA Magazine* (December 1993), pp. 66ff.

10. Edgar Rigdon, "Data Mining Gaines new Respectability," *Marketing News* (January 1997), p. 8; and Anthony O'Donnell, "Data Mining for Dollars," *Insurance & Technology* (October 2003), *28*:10, p. 9, downloaded from http://proquest.umi.com/pqdweb?index=1&did=000000411092411&SrchMode=1&sid=1&Fmt=4&VInst=PROD&VType=PQD&RQT=309&VName=PQD&TS=1064941683&clientId=20176, October 1, 2003.

11. See Steven D. Seilheimer, "Current State of Decision Support System and Expert System Technology," *Journal of Systems Management* (August 1988), pp. 14–19; and Efraim Turban and Jay Aronson, *Decision Support Systems and Intelligent Systems* (Upper Saddle River, NJ: Prentice Hall), 1998.

12. For an interesting article dealing with creating ESS, see Jim A. Van Weeldereen, "MEDESS; A Methodology for Designing Expert Support Systems," *Interfaces* (May–June 1993), pp. 51–61.

13. See "Data Warehouse," September 30, 2002, downloaded from http://www.cio.com/summaries/enterprise/data/index.html, December 17, 2002.

14. Jackie Heuter and William Swart, "An Integrated Labor-Management System for Taco Bell," *Interfaces* (January–February 1998), pp. 75–91.

15. See, for example, Ralph A. Oliva, "Use Modeling Power Tools to Improve Decisions," *Marketing News* (22 June 1998), pp. 7–8, and numerous case studies of IBM (http://www.ibm.com) and SAS (http://www.sas.com).

16. Christopher Koch, "The ABCS of ERP," February 7, 2002, downloaded from http://www.cio.com/research/erp/edit/erpbasics.html, October 1, 2003.

17. "What is CRM?," February 21, 2002, downloaded from http://www.destinationcrm.com/articles/default.asp?ArticleID=1747, September 30, 2003.

18. For good discussions on data mining and its success stories, see Shamsul Chowdhury, "Databases, Data Mining and Beyond," *Journal of American Academy of Business* (March 2003), *2*:2, pp. 576–581, and go to http://www.sas.com for success stories.

19. See, for example, Stratigos, op. cit. note 1, and the future of KM in the web sites listed in Table 2.1.

SUGGESTED READING

Understanding the Basics of KM

Alazmi, Mutiran, and Mohamed Zairi, "Knowledge Management Critical Success Factors," *Total Quality Management* (2003), *14*:2, pp. 199–204.

This article discusses a compilation of what has been found to be the critical success factors in adoption of KM framework. Provides an interesting list of factors.

"Knowledge Management—An Overview," online June 13, 2002, http://www.knowledgeboard.com/library/km_an_overview.pdf, downloaded September 16, 2003.

A very useful article for understanding the basics of present-day KM. Provides a good overview of both the past and future prospects of the field.

So What is KM?

Frappaolo, Carl, *Knowledge Management* (Oxford: Capstone Publishing Ltd., 2002).

This e-book provides an easy-to-read overview of KM, providing examples to illustrate the fine points. Both hardcopy and e-book.

Gamble, Paul R., John Blackwell, and Paul Gamble, *Knowledge Management* (London: Kogan Page Ltd., 2002).

A highly readable and accessible guide that explains the fundamentals of KM. A good introduction to the field.

Santosus, Megan, and John Surmacz, "The ABCs of Knowledge Management," May 23, 2001, http://www.cio.com/research/edit/kmabcs.html, downloaded March 25, 2003.

Provides a good basic understanding of knowledge management. A good ABC approach to start one's search.

Information Systems (IS) and KM

Stratigos, Anthea, "Knowledge Management Meets Future Information Users," *Online* (January 2001), 25:1, p. 65.

A succinct article that talks about how information systems users must interface with KM in the future. Notes the continued blurring of lines in the field.

"Data Warehouse," September 30, 2002, downloaded from http://www.cio.com/summaries/enterprise/data/index.html, December 17, 2002.

A good executive summary discussing the what, where, or why of data warehouses. An example is given to help illustrate this discussion.

Business Research and KM

Chowdhury, Shamsul, "Databases, Data Mining and Beyond," *Journal of American Academy of Business* (March 2003), 2:2, pp. 576–581.

This paper presents using data mining in data warehouses to build and extract DSS with statistical packages usually employed to analyze data in databases. Follows the integration of tasks into the KM function.

Demirdjian, Z. S., "Marketing Research and Information Systems: The Unholy Separation of the Siamese Twins," *Journal of the Academy of Business, Cambridge* (September 2003), 3:1/2, pp. 218–223.

The paper presents a model that integrates research and IS and recommends changes to the structure of IS curriculum.

3

Scientific Inquiry

Overview

Definitions

Observations, Facts

Concepts, Constructs, Definitions, Variables

Problems, Hypotheses, Laws

Theories, Models

Methods of Theory Construction

Model-Based Theory

Deductive Theory

Functional Theory

Inductive Theory

Relevance of Science in Business Research

Science Versus Nonscience

Scientific Method

State of the Art in Business Research

Levels of Scientific Endeavor

Where Do We Go from Here?

Summary

Discussion Questions

Notes

Suggested Reading

Every discipline maintains a specialized terminology to describe and communicate the essence of its subject area. For example, students in business quickly become familiar with such terms as *strategic planning, ROI, organizational climate, marginal utility,* and *market segmentation.* The field of research is no different. To study research, you, as a manager, must understand the fundamental terminology by which those in the field communicate. If you are going to specify and evaluate the research efforts to be performed, then you must be able to speak the language of the researcher.

The basic language of the researcher is founded in the philosophy of science. Although this might seem rather removed from decision making in the modern business organization, it is extremely relevant. Contemporary authors have stated the importance of the relationship as follows:

> At first glance the business practitioner or student might think that a discussion of science and knowledge is rather far afield from business research. The relationship among science, research, and effective business management is, however, surprisingly close in modern times. Although in the past businessmen could make adequate decisions on the basis of experience and hunch, the rapidly changing world of today makes this type of decision making obsolete. Businessmen must base decisions on understanding, on knowledge of *how* business variables interact, and on why they do so.[1]

If you accept these tenets, it would be difficult indeed to argue against obtaining a basic understanding of the nature and concepts of scientific inquiry. For this reason, the present chapter explores the basics of scientific inquiry.

The chapter begins with a discussion of the "building blocks" of science: critical concepts that you must understand to enter the analytical world of the researcher. Four methods of scientific theory construction are also discussed in the context of business research.

Once the fundamental concepts have been outlined, the relevance of science in business research is examined. The differences between non-science and science are discussed, to uncover the major differences inherent in each of the approaches to obtaining information. This discussion is followed by a brief presentation of the scientific method.

The final section in the chapter deals with the specific interrelationships between science and business research. The levels of development in any scientific endeavor are described and related to the state of the art in business research. Where do we go from here if significant advancements in the field are to be made? The question is examined through a realistic assessment of the current status of business research.

In order to define and illustrate the basic building blocks in scientific inquiry, a classic example will be taken from the literature in consumer behavior.[2] The example involves the relationship between purchase behavior and search behavior as predicted in a comprehensive theory of buyer behavior. Essentially, this theory attempts to understand buyer behavior through the use of scientific constructs. With this brief introduction, our building blocks will be analyzed.

Observations, Facts

Observations form the basis by which we recognize or note facts. They are our perceptions of reality. They are experiential in nature. Observations involve the act of noting some object or the occurrence of some phenomenon in our immediate environment.

Facts are those things or phenomena that we believe are true. Facts are generally consensual in nature, in that others who have observed the same phenomena agree to their existence. Facts may be collected in the business research process by[3]

1. Direct observation or sensing of natural phenomena or of experimental results

2. Direct inference from other data that are directly observed

3. Original documents

4. Reports and publications of fact-gathering agencies and researchers

5. Questioning of individuals

Observations and facts are the means by which science is grounded to reality. They provide the ultimate test for the value of any scientific inquiry.

Figure 3.1 presents a partial theoretical network relating purchase behavior and search behavior in an economic setting. The figure illustrates that facts are determined from reality through observations. More specifically, the concepts of "purchase" and "search for information" are inferred from facts through observations about reality.

Concepts, Constructs, Definitions, Variables

Concepts are the basic building blocks of scientific investigation.[4] They are creations of the human mind that are used in the classification and communication of the essence of some set of observations. *Concepts* can be defined as abstract ideas generalized from particular facts. Without concepts, there can be no theory.

Constructs are specific types of concepts that exist at higher levels of abstraction and are invented for some special theoretical purpose.

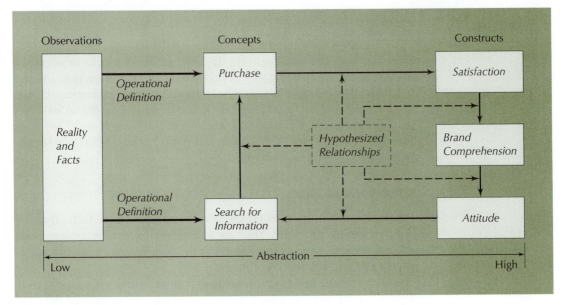

Observations   Concepts   Constructs

FIGURE 3.1

A Partial Theoretical
Network Relating
Purchase Behavior and
Search Behavior in an
Economic Setting

Source: A more complete
theoretical model is presented
in Gerald Zaltman, Christian
Pinson, and Reinhard
Angelmar, *Metatheory and
Consumer Research* (Hinsdale,
IL: Dryden Press, 1973), p. 81.

Generally, constructs are not directly tied to observations. Instead, they are inferred by some indirect method, such as paper and pencil questionnaires.

Referring to Figure 3.1, we can see the general relationships of our major building blocks thus far. The primary concepts of interest in our study are "purchase" and "search for information." These concepts are at relatively low levels of abstraction because they can be directly tied to observable facts; that is, the purchase of a product can be verified by sitting at the point of sale, and the search for information can be verified by such activities as asking for a salesperson's help, reading product information, and so on.

On the other hand, the constructs in our study are less directly observable. "Satisfaction," "brand comprehension," and "attitude" are all relatively unobservable by any direct means. Whereas we can empirically verify a purchase fairly easily, we cannot do so with the intangible constructs in our study. As researchers, we invent our constructs because the research literature suggests that some such phenomena exist in the human mind. However, we cannot cut open an individual's skull and physically extract an "attitude." We can only infer its existence from people's behavior, actions, and self-reports. These constructs are of value scientifically if they can help us explain some empirical relationship of interest to the researcher. However, it must be remembered that these are still concepts in the study. They are simply less observable than the concepts of "purchase" and "search for information."

Thus far, we have defined all our building blocks with the use of other concepts. This brings us to the distinction between two major types of definitions

| | Definitions | |
Concept/Construct	Constitutive	Operational
1. Purchase	The act of obtaining a good or service by paying money or its equivalent	The list of individuals who have signed a bill of sale for a GM auto in the past year
2. Satisfaction	The degree to which expectations are met in the performance of a product	The difference between an *a priori* and an *a posteriori* rating of performance expectations
3. Brand comprehension	The degree to which a particular product's attributes are understood	The proportion of a particular product's attributes recalled
4. Attitude	A learned predisposition to respond in a consistent manner	The summated rating received on an 8-item, 7-point bipolar scale
5. Search for information	A deliberate seeking of information, usually for the purposes of decision making	The summated rating received on a 5-item, 5-point Likert scale

that are necessary in scientific inquiry: constitutive and operational definitions.[5] *Constitutive definitions* define concepts using other concepts and constructs. They set the domain of interest for the concept of interest. Operational definitions put empirical meaning to constitutive definitions by specifying the means by which the concept or construct will be measured in reality. Specifically, *operational definitions* specify the procedures by which the concept will be measured or manipulated. For example, the constitutive and operational definitions for each of the five major concepts in Figure 3.1 are presented in Table 3.1.

It is difficult to talk about concepts and constructs in a research setting without referring to the term *variable*. A *variable* is simply a symbol or a concept that can assume any one of a set of values. The concept of variable is extremely important in measurement because it sets the limits on the range of values a measurement may take. Again using our example, we may specify that "purchase" is a dichotomous variable (i.e., capable of taking on one of two values, such as 0 for no purchase and 1 for purchase). "Search for information" is a polytomous variable (i.e., can take on multiple values, such as 1 for low search, 2 for medium search, and 3 for high search). "Satisfaction" is a continuous variable (i.e., capable of taking on an infinite number of values). The notion of variable is of great importance in the research process, as will be seen later on in the text.

Problems, Hypotheses, Laws

Managerial decision-making problems ultimately form the basis for all business research. *Managerial problems* can be defined as questions raised that are in need of a solution in a business setting. Although the variation in possible managerial problems is infinite, they all exhibit the following minimal and sufficient conditions for their existence:[6]

1. An individual who has a problem: the decision maker
2. An outcome that is desired by the decision maker (i.e., an objective)
3. At least two unequally efficient courses of action that have the same chance of yielding the same objective
4. A state of doubt in the decision maker as to which choice is best
5. An environment or context of the problem

An example will be used to illustrate these conditions.

Let us suppose that management at General Motors is contemplating developing an informational brochure for past buyers on the benefits of buying a GM product. Management is further considering whether or not to send the brochure to all past buyers or only some subset of them. Specifically, to be consistent with the corporate objective of selling their products in the most cost-effective manner possible, they want to know whether or not the brochures should be sent to all past buyers, only satisfied past buyers, or only dissatisfied past buyers. This scenario meets all the conditions identified earlier for a problem's existence: We have a decision maker who has an objective and is faced with a number of alternatives for which there is no clear-cut solution in a social setting.

From managerial problems come research problems. *Research problems* are specific restatements of managerial problems for the purpose of empirically investigating the latter. Research problems state perplexing managerial situations in a manner that is conducive to understanding the question that confronts the manager. The specific research problem in our example could be stated as follows.

What are the specific relationships between prior purchase of a product, satisfaction, and the search for additional information in a repurchase situation for GM products?

In essence, then, problems are questions about the relationships between one or more concepts that need to be answered.

Hypotheses are conjectural statements of the relationship between two or more variables that carry clear implications for testing the stated relations.[7] They are research tools to further define research problems. Hypotheses are tentative statements that are considered to be plausible given the available information. In our example, a rather complex hypothesis can be developed, explicitly linking the concepts and constructs defined thus far. This hypothesis can be stated as follows:

If a purchase of a GM product is followed by satisfaction, then brand comprehension will change in the direction of the new information, and consequently attitude will change. Furthermore, the more positive the attitude toward the GM product, the less the extent to which information concerning this product will be sought.[8]

The hypothesis as stated is developed because the *a priori* theoretical work (which will be addressed in the next section) suggests that the constructs are linked as just specified.

Once a hypothesis is verified by numerous empirical testings, the relationship can be considered a law. A *law* can be defined as a well-verified statement of relationship about an invariable association among variables. With this definition, we find that there is not an overabundance of "laws" in the field of business. Instead, we must be satisfied with what have been designated as "weak laws," which are not invariably true. This state of affairs applies to all the social sciences because of the seemingly unending complexity of the social setting.

Therefore, in our GM example, we would find limited support in the consumer behavior literature for the idea that any of the hypothesized relations have attained a law status. Instead, we would find a mixture of verifications and refutations of the hypothesized relations in Figure 3.1. It is only after extensive empirical testing that a relationship can be said to have attained a "lawlike" status.

Theories, Models

Facts, concepts, constructs, hypotheses, and laws do not float around aimlessly. What holds these components together in the scientist's world is theory. A *theory* can be defined as an interrelated set of statements of relationship whose purpose is to explain and predict. Abraham Kaplan succinctly states their purpose as follows:

> A theory is a way of making sense of a disturbing situation so as to allow us to most effectively bring to bear our repertoire of habits, and even more important, to modify habits or discard them altogether, replacing them [with] new ones as the situation demands . . . accordingly, theory will appear as the device for interpreting, criticizing, and unifying established laws, modifying them to fit data unanticipated in their formulation and guiding the enterprise of discovering new and more powerful generalizations.[9]

The establishment of theories is a primary goal of science because they are the means by which we explain and predict phenomena of interest. Theories are grounded in fact and supported by laws and other well-supported statements of relationship. The common misconception that theories are figments of our imagination and have no grounding in reality is utterly false. Scientific theories, by definition, must be empirically grounded because their ultimate purpose is the explanation and prediction of occurrences in reality.

Too often, one hears the comment, "This research is useless because it is too theoretical." On the contrary, good theoretical research helps direct the investigator in the search for understanding. It is inherently practical and applied because it highlights the important concepts and relationships in a problem situation. Examples of extremely useful theoretical structures abound in all areas of business research. Capital asset pricing theories help us understand finance; performance satisfaction theories help managers understand job performance factors; and consumer behavior theories help marketers understand buyer behavior. Without these theoretical structures, business researchers would have little guidance for future research efforts and managers would have little direction for making business decisions.

The final major concept to be addressed in this section is the model. A *model* is defined here as any highly formalized representation of a theoretical network, usually designed through the use of symbols or other such physical analogs. Models are used as representations of theoretical systems so that they can be tested, examined, and generally analyzed by those who create them. Models are simplified versions of phenomena that are of interest to the scientist.

Now back to our example. Previously, we specified the relationships and concepts identified in Figure 3.1 as a partial theoretical network. It is considered partial because it was derived from a larger theoretical framework in the consumer behavior literature. This larger theoretical framework is known as the Howard and Sheth Theory of Buyer Behavior and it is illustrated in Figure 3.2.[10] Basically, the theory is an interrelated network of perceptual and learning constructs whose designated purpose is to help explain economic behavior in the purchase situation. The theory is constructed from a mixture of "weak laws," empirical generalizations, and researcher ingenuity. The theory is relevant to our GM study because it specifies the major constructs of interest and their interrelationships.

The relationships of interest in our GM study are highlighted in Figure 3.2. As the model suggests, through a number of feedback loops and information flows, the precise partial theoretical construction to be tested is outlined in this generalized theory of buyer behavior. From the researcher's standpoint, this theory provides the framework for translating the situation into a researchable problem that can help solve management's dilemma in an effective and efficient manner. Note that concepts—and ultimately variables—are the key linkages between managers and researchers. Without agreement on the nature of the concepts under study, there can be little hope of solving management's dilemma. Another example of the importance of this is presented in Exhibit 3.1. In this example, *expectancy theory* is used in a human resource management context to help structure a management problem so that it becomes researchable.

These brief examples highlight some of the fundamental concepts and frameworks in scientific inquiry. Although these discussions were greatly simplified, it is hoped that you can begin to appreciate the usefulness of these tools in the field of business research. The next section of this chapter deals with another important topic of scientific inquiry, the process of theory construction.

FIGURE **3.2**

A Simplified Model of the Howard and Sheth Theory of Buyer Behavior

Source: From *The Theory of Buyer Behavior,* by J. Howard and J. Sheth, p. 30. Copyright © 1969 John Wiley & Sons, Inc. Reprinted by permission.

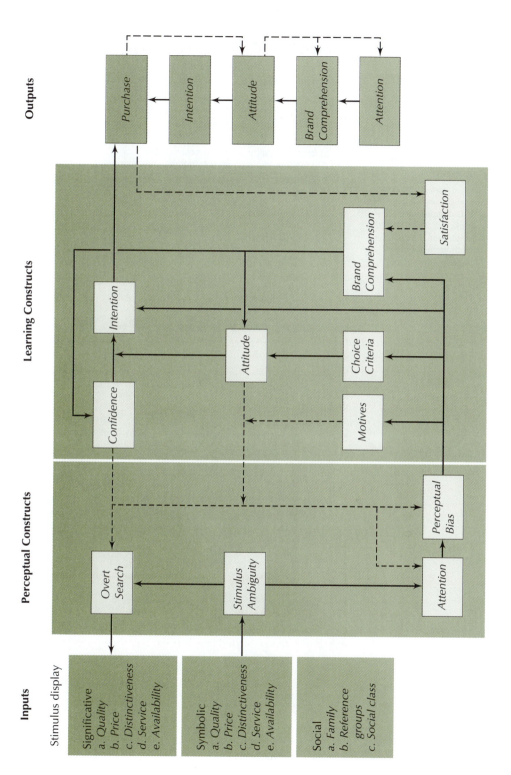

EXHIBIT 3.1

An Example of Using
Expectancy Theory in a
Human Resource
Management Context
to Translate a
Management Problem
into a Researchable
Problem

Adapted from *Personnel
Human Resource Management,*
by V. G. Scarpello and J.
Ledvinka, pp. 25–29, ©1988.
Reprinted with the permission
of South-Western College
Publishing, a division of
International Thomson
Publishing.

BRIEF DESCRIPTION OF THE THEORY

Expectancy theory attempts to explain why people expend effort on a task. It uses both personal and situational concepts in explaining human behavior.

Major Concepts	Constitutive Definition
(a) Expectancy	The perceived likelihood that an action will lead to successful performance.
(b) Instrumentality	The perceived likelihood that a successful performance will lead to desired outcomes.
(c) Valence	The value that a person places on the outcome.
(d) Effort	The amount of work expended to achieve a specific task.

GENERAL MODEL

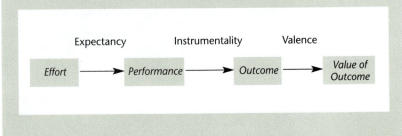

The model can be mathematically expressed as:

$$\text{Effort} = \text{Expectancy} \times \text{Instrumentality} \times \text{Valence}$$

EXAMPLE OF THE THEORY IN USE

Management wants to know if introducing an incentive bonus will increase worker productivity. In other words, the expectancy model in this situation would be:

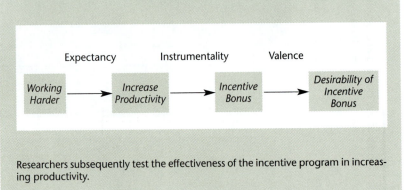

Researchers subsequently test the effectiveness of the incentive program in increasing productivity.

METHODS OF THEORY CONSTRUCTION

Given the central importance of theory development in any field of endeavor, it is desirable to examine the basic methods of theory construction in science.[11] We find a diversity of opinion in the literature concerning the "best" means to develop theory and thus understanding. The viewpoints range from the pure empiricist (i.e., that all theory must come directly from the data themselves) to the truly model-based position (i.e., we initially specify conceptual models without regard to the data itself and subsequently test their accuracy in reality). In business research, we can find examples in the literature of all major types of theoretical development. There have been successes and failures with each type. For present purposes, we will review the major methods of theory construction and withhold critical judgments of the methods.

Melvin Marx describes four major modes of theory construction that have received varying degrees of acceptance in the scientific disciplines. The modes differ primarily in terms of where the theory-building process starts in the basic "wheel of science" and what is the predominant form of logic used in developing the theory itself.

The wheel of science can be viewed as a circular arrangement, as presented in Figure 3.3. Depending on where one starts in the wheel, theories lead to the development of hypotheses, hypotheses then need data (obtained through observation) for testing purposes, which lead to empirical generalizations (statements of relationships or the further development of concepts and constructs), which subsequently lead to further theory development. A more detailed description of each of these methods of construction is given in the subsections that follow.

FIGURE **3.3**

Four Modes of Theory Construction and the Wheel of Science

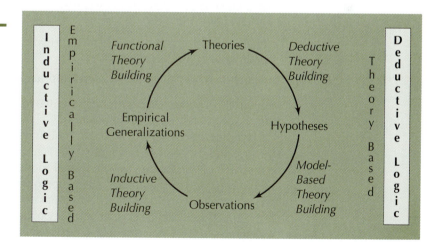

Model-Based Theory

As defined earlier, a model is any highly formalized representation of a theoretical network, usually designated through the use of symbols or other such physical analogs. The model is usually then expressed mathematically so that it can be manipulated for testing. Under this method of construction, the emphasis up front is on defining a conceptual network, then subjecting it to empirical testing. Essentially, substantive validity concerns (issues concerning the consistency and completeness of the network specification) are not of interest in the initial stages of theoretical development. The only concern in the initial stages in the model's testing is whether or not the model performs in the fashion desired by the researcher.

For example, if a researcher believes that information availability and computer literacy drive information use in decision making, then a preliminary model can be specified and tested. The researcher would merely have to develop sound measures of each of the three concepts (information availability, computer literacy, and information use in decision making), then test the hypothesized model.

Examples of this approach exist in all functional areas of business, but perhaps the best examples come from the field of economics.[12] Here, highly mathematical models are specified, then subjected to rigorous statistical testing to see if they perform in the hypothesized fashion. If the models perform, they are often refined and integrated into existing theoretical structures within the field. If they do not, they are discarded or modified for future testing.

Deductive Theory

Deductive theories are developed largely through the process of deduction. *Deduction* is a form of inference that derives its conclusions by reasoning through premises, which serve as its proof.[13] A simple example of deductive logic is as follows:

Premise: All decision makers have information needs.
Premise: Bob is a decision maker.
_____

Conclusion (or deduction): Bob has information needs.

Deduction is a widely used form of inferential logic; it is the basis by which many conclusions are drawn in business.

A *deductive theory* is one whose emphasis is distinctly on the conceptual structure and its substantive validity.[14] In Figure 3.3, the deductive theory is presented as one that is casually related to reality in its formulation but is stated precisely conceptually, then subsequently tested and modified. The deductive theory focuses on conceptual development prior to empirical testing. Although this method of theory building has been criticized for its general lack of reference to reality, it has led to useful results in a number of fields of interest.

An example of this type of theorizing is the Howard and Sheth Theory of Buyer Behavior presented in Figure 3.2. In essence, the theory was constructed using the available evidence in the literature. Since that time, empirical testing of the relationships contained within it has refined and developed the theory.

Functional Theory

Functional theories are those whose development is characterized by a continual interaction of conceptualizing and subsequent empirical testing. The major difference between deductive and functional theories is the degree to which conceptualization takes place in the early stages of the theory's development. With deductive theories, a generally grandiose scheme of conceptual relationships is formulated, and empirical testing is left for later stages of development. With functional theories, numerous smaller conceptual frameworks are usually constructed and tested until a more grandiose scheme can be built on empirical evidence. This method requires a constant interaction between theory and facts.

The examples of functional theory development in business are numerous.[15] One very good example in the finance literature is theory in the area of capital management.[16] Major theoretical advances in this field appear to have occurred over a number of years through a highly interactive method of alternating conceptualization and empirical testing.

Inductive Theory

The final major method of theory construction results in *inductive theory*. Inductive theory building is characterized by a strictly empirical approach to finding generalizations. Inductive reasoning relies on the repeated observation of reality, then the development of summary statements to explain and classify what is observed.

Although this method of theory construction has been criticized as being relatively unimaginative, it too has resulted in a substantial increase in our knowledge. The work of B. F. Skinner, which has been widely adopted in the organizational behavior literature, has led to a number of behavior modification programs that have been successful in economic organizations.

RELEVANCE OF SCIENCE IN BUSINESS RESEARCH

Up to this point, we have purposely avoided any thorough discussion of the nature of science and its relevance in business. This approach was used to enable us (1) to build the case for the need for research in the decision-making process and (2) to outline the basic terminology in the field of science.

With this accomplished, we now discuss the nature of science and its application in the business research setting.

Science Versus Nonscience

There is no universally accepted definition of science.[17] *Science* has been defined both as a body of knowledge and a method of inquiry. The body-of-knowledge definition suggests that science is a set of theories that describe and explain phenomena in a particular subject area. Although this definition is important in and of itself, we are not so concerned with this aspect in this text. Instead, we are more interested in the definition that defines a *science* as a systematic process of inquiry for solving decision-making problems. This is commonly referred to as the *scientific method*.

Here, again, little consensus can be reached as to the exact nature of the scientific method. There is agreement regarding the method's usefulness in generating knowledge and its systematic nature. Paul Rigby succinctly states this:

> The scientific method has been developed over the centuries for systematically increasing knowledge, and although it is certainly not the only means for increasing knowledge, it has proven to be one of the most effective. The great strides in human understanding and technological development taken by man during the last three to four hundred years were made possible primarily by scientific research. Modern industrial civilization is built upon the knowledge this method has produced.[18]

Science, then, can be considered as both a goal and a means to obtain that goal.

However, science is not the only means by which we collect information. Most of our information comes from nonscientific methods—that is, experience, semi- and unstructured observation, authority, and so on. It is not the purpose of this book to suggest that all information must be scientifically obtained to be of any value to the manager. To argue this would be folly indeed. Instead, one must argue for the use of scientific research *when it is of value and relevance to the decision maker.* This criterion is extremely important and will be dealt with in depth in Chapter 5.

The distinction between science and nonscience can be fuzzy.[19] Remember that in Chapter 1 the distinction was made between scientific business research and business research. This distinction was based on the testing of hypotheses. This formal distinction is an ideal the researcher should strive for in tackling research problems for the managerial decision maker. Neither the manager nor the researcher should be hung up on whether or not a study is labeled scientific. The fruitlessness of this line of thought is adequately demonstrated by the old question of whether management is an art or a science. It does not really matter. What does matter is that we maintain a clear perception of the information needs of management and how we best satisfy them. If scientific research is of *value and relevance,* do it. If not, do not do it, and obtain the information by the most cost-effective means available. Given this framework, let us examine the nature of the scientific method and the benefits we can obtain in business by using the method when appropriate.

Scientific Method

As mentioned earlier, there is no unanimous agreement as to the specific steps in the scientific method. However, there is consensus regarding the major characteristics of the scientific method. Briefly, these characteristics are as follows:[20]

> **The method is critical and analytical in nature.** This suggests an exacting and probing process to identify problems and the methods to arrive at their solutions.

> **The method is logical.** Logic is the method of scientific argument. The conclusions rationally follow from the available evidence.

> **The method is objective.** Objectivity implies that the results will be replicated by other scientists if the same study with the same conditions is performed.

> **The method is conceptual and theoretical in nature.** Science implies the development of conceptual and theoretical structures to guide and direct research efforts.

> **The method is empirical.** The method is ultimately grounded in reality.

> **The method is systematic.** This implies a procedure that is marked by thoroughness and a certain regularity.

As you can infer from its characteristics, the scientific method is designed to yield the best possible answers to the problems addressed. For this reason, the scientific method forms the basis for the business research process outlined in Chapter 1. This process is the primary organizational tool for the text.

STATE OF THE ART IN BUSINESS RESEARCH

Business research is categorized as applied social science research in most classificatory schemes. Social science research in general and applied business research in particular are thought of in the scientific literature as being at relatively low levels of development. When compared to the natural sciences, this is indeed difficult to dispute. The natural sciences, such as physics and chemistry, have developed intricate theoretical structures that have withstood the critical tests of empirical verification. Theoretical structures such as the law of gravity and the theory of relativity are now used extensively in the accurate prediction, explanation, and control of phenomena of interest to scientists. Business research has not yet been able to attain this level of precision in its endeavors.

Nevertheless, the practice of business research is currently experiencing radical change. An information explosion has resulted from the proliferation of microcomputers, the development of new data acquisition and communication technologies, and advanced methods of data analysis. Although it is difficult to obtain a definitive count, various estimates raise the level of worldwide Internet usage from approximately 40 million users at the end of the 1990s to 580 million in 2002, with projections for 2005 of approximately 1 billion active users.[21]

Given this explosion of usage and the continued commitment of organizations to fund interconnectivity among its customers and other stakeholders, change in the ways organizations conduct research will continue at a dizzying pace. Overall, these trends will help create the following changes in the business research industry.[22]

The research function will become increasingly woven into the fabric of the organization. The research function will become an "intelligence" or "information" function, integrated throughout the organization. Distributed multifunction technology (or, in essence, dispersed and interactive computing with many functions) is providing the means to decentralize decision making, while at the same time enhancing access to all types of information. This technology, coupled with continued economic pressures to downsize organizations, will increasingly lead to a blurring of the pure manager–researcher distinction. The net result will be that managers throughout the firm will be forced to become increasingly more interdisciplinary, particularly where the research and information technology functions are concerned.

The field will function more and more in real time. Totally interactive capabilities will lead to more timely and useful information. In other words, the collection, analysis, and reporting of results will become almost instantaneous as better data integration, collection, analysis, and reporting technologies become available to the researcher and manager. Real-time data collection systems are being integrated into management information systems with internal and external databases and with decision support systems and sophisticated data analysis capabilities.[23] Companies such as United Airlines, Connecticut Mutual Life Insurance, and Ford Motor Company have already implemented such systems and have recognized significant benefits in both service and productivity. The net effect of this will be timelier and more usable information.

New research techniques promise to make research settings more realistic and ultimately improve the quality of the research. The research environment is beginning to be enhanced by virtual reality.[24] Virtual reality simulations can put a respondent into a grocery store, a shopping mall, a tourist attraction, or any other environment without having the person leave the room. Virtual reality changes can be made in minutes. Imagine experiencing Disney World in the rain, in the sun, when it is hot, and when it is cold. This can be done quickly and easily in virtual reality. The net effect is more realistic research settings, with quicker and more controlled data collection.

Shared information and technology will lead to better theoretical networks. This change will lead to a better understanding of the dynamics of the business environment.[25] For example, until recently, knowledge of global business was largely anecdotal and often out of date. However, the development of worldwide computer-based information systems and the intense interest in globalization have stimulated interest in researching this important field. New theories are being developed, and old theories are being subjected to rigorous testing.

These and similar changes will come at an accelerated rate during this millennium. As a result, it is highly recommended that the astute manager and researcher keep abreast of recent developments in the research field. One method of keeping currency is regularly visiting the Web sites of some of the major research associations throughout the world. A partial listing of these associations and their Web sites is presented in Exhibit 3.2.

Finally, the growth in technology and information has created what one set of authors has referred to as "new gold mines and minefields."[26] The gold mines include the existence of thousands of publicly available online databases on almost every subject imaginable; the development of two-way video, voice, and data communications technology that allows almost instantaneous access to buying information; and the development of better and more sophisticated data analysis and modeling techniques. The minefields include the relative ease of access to and use of these powerful research tools, which may lead to overly simple descriptions of reality because of the apparent simplicity of producing summary results.

EXHIBIT 3.2

A Partial Listing of Major Research Associations and Their Web Sites

AAPOR	American Association for Public Opinion Research	http://www.aapor.org
AMA	American Marketing Association	http://www.ama.org
ARF	The Advertising Research Foundation	http://www.arfsite.org
AQR	Association for Qualitative Research	http://www.aqr.org.uk
CASRO	The Council of American Survey Research Organizations	http://www.casro.org
ESOMAR	European Society for Opinion and Marketing Research	http://www.esomar.org
MRS	The Market Research Society (UK)	http://www.marketresearch.org.uk
QRCA	Qualitative Research Consultants Association	http://www.qrca.org

FIGURE 3.4

Levels of Scientific
Research Endeavor

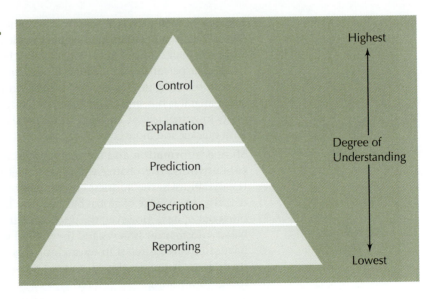

More specifically, it will become increasingly important in the future for managers to plan and direct the research function so that research helps managers make decisions, rather than research making decisions for managers. Research does not make decisions; managers do. The goal of management should be to use the new technologies in such a way that business research can evolve to higher levels of scientific understanding.

Levels of Scientific Endeavor

The levels of scientific endeavor are presented in Figure 3.4. In this figure, five stages are outlined. For our purposes, however, we will consider only the stages of description, prediction, explanation, and control in the realm of scientific research. Reporting, or the mere recording and presentation of facts or statistics, is not considered a scientific endeavor. Although this function is important to any type of research, it does not fall under the domain of scientific research (as defined in Chapter 1) because reporting lacks the critical analysis that is characteristic of scientific research. The levels of development can be described as follows:

> **Description is the simple identification of the major variables and their relationships in a problem situation.** For example, a study may be undertaken to describe the major characteristics of a "successful" first-line manager in the Caterpillar Tractor Company in Aurora, Illinois. A successful manager may be found to have a certain leadership style and personality type.

> **Prediction is achieved when the researcher can identify a variable or set of variables that is associated with the presence of**

some other variable. In our successful leader example, we can now state that there is an association between success, leadership style, and personality type. For instance, we may find that authoritarian individuals who have dogmatic personalities tend to be successful leaders in the Caterpillar Tractor Company in Aurora, Illinois.

Explanation is achieved when the researcher can answer why a certain phenomenon or relationship exists in reality. In the case of the successful leader, the researcher may now state, "Authoritarian individuals who have dogmatic personalities are successful leaders because the nature of the job and the type of worker employed require that the leader be a domineering, closed-minded boss who tells the employee what to do and when to do it."

Control exists when the researcher is able to manipulate one or more variables in a problem situation so as to effect the desired change in one or more variables associated with that situation. In our Caterpillar example, the researcher now selects a person who is an authoritarian with a dogmatic personality, places that individual in a first-line supervisory position, and obtains a successful leader. This level of scientific inquiry is the ultimate in any field because it reflects the highest degree of understanding.

As this brief treatise suggests, business managers inevitably want to obtain research information that will enable them to control specific problems facing the firm. Needless to say, there are few, if any, current research endeavors in which the manager can wind up being able to control a particular problem situation. The researcher cannot identify and control all the critical variables in the problem situation, and there is no way to measure those critical variables. The business researcher is not alone here, however. Social scientists have been plagued by these problems for a much longer time than our more recent cadre of business research professionals.

Where does this all leave us? Much current business research endeavor is still primarily descriptive, largely because of the social nature of the subject matter. The major research journals in accounting, economics, finance, management, and marketing all reflect a trend toward higher levels of scientific endeavor (largely prediction, but some explanation), but the field in general has a long way to go.[27]

The attempts at increased rigor in business research are not all positive. A number of scathing reviews have been directed at both the academic and professional business researcher.[28] These articles note a disturbing trend in much of the research literature. One author succinctly states,

Whether one does, sells, and/or buys consumer research, it stands to reason one should be able to critically evaluate and distinguish what is acceptable from that which is junk. However, judging from the papers which continue to be published in our most prestigious journals and from research reports which often form the basis for important marketing management and public policy decisions, it is all too apparent that *too large a portion of the consumer (including marketing) research literature is not worth the paper it is printed on or the time it takes to read.*[29]

This critique is directed particularly at consumer behavior research, but the same problems can be found in the other functional areas of business.[30] The major criticisms of research in these fields include minimal theoretical structures to guide research, inadequate or inappropriate research methodologies, lack of concern with measurement issues, the obsession with certain statistical fads in various fields, and misdirected research emphasis (i.e., studying trivial or irrelevant issues). All these criticisms are valid to a certain extent, but we must be careful not to throw the baby out with the bath water. Instead, we must assess where we are now and keep a clear perception of the goal of business research and the means to obtain that goal.

Where Do We Go from Here?

As stated in Chapter 1, the goal of business research is to provide the decision maker with relevant and useful information on which to base decisions. If researchers are performing their function well, they are providing managers with the type and quality of data needed to make sound decisions. This goal should not be forgotten as we enter the analytical realm of the researcher in the subsequent parts of this text.

To achieve this goal, the scientific method is applied to problems whenever possible. Note that not all problems can or should be addressed by scientific research. The problems that should be addressed are ones that are of sufficient concern and value to management to justify the expenditure for the research.

Further, management must start to become more involved in the research endeavor, to ensure that information is being identified and used to its fullest potential. Given ever-changing technologies, top managers must effectively manage the output of the research function and integrate it into the organization's management information system and the associated decision support system. Only the careful orchestration of data, information, computers, software, and human resources will result in the realization of the full potential of business research.

Researchers must strive to increase the rigor of their work and the level of exploration. Better samples, higher level research designs, and inferential statistics can all contribute to creating more valuable research information. The new technologies and information sources must be merged into the research process so that timely and cost-effective information can be made available to the appropriate decision maker. The research environment is changing rapidly, and innovative approaches to using technology to aid decision making will take the forefront in the next decade.

The chapters that follow outline the research process and the concerns of both managers and researchers for obtaining relevant and useful information. When reading through this text, develop a questioning attitude; this will help you as a manager to become aware of the many facets of the development of quality research. Coupled with a desire to solve managerial problems, this attitude will help us further the scientific advancement of business research and overcome the criticisms of past research discussed earlier.

SUMMARY

The relevance of scientific inquiry and its associated terminology in business research is examined in this chapter. Managers must have a basic understanding of the nature and concepts of scientific inquiry if they are to adequately perform their role as specifiers and evaluators of research.

The building blocks of science are defined and illustrated with an example. The major building blocks discussed are observations, facts, concepts, constructs, definitions, variables, problems, hypotheses, laws, theories, and models. The interrelationships of these components in scientific research are examined. Theories are described as the glue that holds the blocks together. Establishing theories is a primary goal of scientific endeavor because theories allow us to understand phenomena of interest.

Four methods of theory construction are then presented. The four methods result in model-based, deductive, functional, and inductive theories. All four types of theory construction are found in business research.

The final two sections of the chapter cover the relevance of science to business research and the current state of the art in our discipline. Science is shown to be relevant to the study of business because it is the only means to develop logically defensible solutions to decision-making problems. However, much business research is currently descriptive in nature and lacks the perspective needed to answer management's questions. A brief assessment of where we go from here ends the chapter.

DISCUSSION QUESTIONS

1. Why is the study of the nature of scientific inquiry essential for students of business research? What role does scientific inquiry play in the business research process?

2. What are the basic building blocks of scientific inquiry? Define them and discuss their interrelationships.

3. Define and contrast constitutive and operational definitions. Why is it so important to specifically define the relevant concepts in any study?

4. Define and contrast managerial problems and research problems. Give a specific example of a managerial problem and its associated research problem.

5. What role do theories play in business research? Give some examples to illustrate your point.

6. What are the major methods of theory construction? Give concrete examples illustrating each method of theory construction in a business context.

7. Discuss the current state of the art in business research. Compare and contrast business research to research in other social sciences and in the physical sciences.

8. Current technological advances are radically changing the way research will be conducted. Describe some of these technological advances and discuss their impact on the business research function.

9. What are the major levels of scientific endeavor? Define and discuss each level, giving examples of contemporary business research at each level.

10. It has been said that much of the business research that is conducted is of low quality. What implications does this have for the business manager?

11. What can managers do to assure themselves of high-quality research for decision-making purposes? Outline some specific actions that a manager can take to increase the probability that useful information will be obtained from the research function.

12. Identify a theoretical structure in your particular field of interest. Bring this theory to class and be ready to discuss the major concepts and constructs in the theoretical structure. What type of theory is it?

13. Using Exhibit 3.2 as a starting point, go to one of the Web sites of a major business research organization and develop a short two-page paper on what you find at the site. You are not restricted to the organizations in this exhibit. Be prepared to discuss your findings in class.

14. What is data mining? Find Web sites and information dealing with data mining and KDD. How have companies used this technology in the development of theory and decision making?

NOTES

1. Robert G. Murdick and Donald R. Cooper, *Business Research: Concepts and Guides* (Columbus, OH: Grid Publishing, 1982), p. 1.

2. This example is adapted from Gerald Zaltman, Christian Pinson, and Reinhard Angelmar, *Metatheory and Consumer Research* (Hinsdale, IL: Dryden Press, 1973), pp. 80–81.

3. Robert Murdick, *Business Research: Concepts and Practice* (Scranton, PA: International Textbook, 1969), p. 8.

4. For thorough discussions of the role of concepts in scientific inquiry, see John T. Doby, "Concepts and Theories," in John T. Doby, Ed., *An Introduction to Social Research,* 2nd ed. (New York: Appleton-Century-Crofts, 1967), pp. 31–49; and Carl G. Hempel, *Fundamentals of Concept Formation in Empirical Science* (Chicago: University of Chicago Press, 1952).

5. For a more thorough discussion of these definitions, see H. Margenau, *The Nature of Physical Reality* (New York: McGraw-Hill, 1950), pp. 232–242; W. Torgerson, *Theory and Methods of Scaling* (New York: Wiley, 1958), pp. 2–5; and Fred N. Kerlinger, *Foundations of Behavioral Research* (New York: Holt, Rinehart, and Winston, 1986), pp. 28–32.

6. Russell L. Ackoff, *Scientific Method: Optimizing Applied Research Decisions* (New York: Wiley, 1962), p. 30.

7. Kerlinger, op. cit. note 5, pp. 17–18.

8. Adapted from Zaltman, Pinson, and Angelmar, op. cit. note 2, p. 80.

9. Abraham Kaplan, *The Conduct of Inquiry* (Scranton, PA: Chandler, 1964), p. 295.

10. For a complete discussion of this theory, see John Howard and Jagdish Sheth, *The Theory of Buyer Behavior* (New York: Wiley, 1969).

11. This section is largely adapted from Melvin H. Marx, "The General Nature of Theory Construction," in Melvin H. Marx, Ed., *Theories in Contemporary Psychology* (New York: Macmillan, 1965), pp. 10–19.

12. For interesting reading on theory development by this and other means, see K. Sridhar Moorthy, "Theoretical Modeling in Marketing," *Journal of Marketing* (April 1993), 57:2, pp. 92–106; and Larry R. Smeltzer, "Emerging Questions and Research Paradigms in Business Communication Research," *Journal of Business Communication* (Spring 1993), 30:2, pp. 181–198.

13. For a complete discussion of the inferential process of deduction and a comparison with other models of scientific explanation, see Carl G. Hempel, *Aspects of Scientific Explanation* (New York: Free Press, 1965); Merrilee H. Salmon and Wesley C. Salmon, "Alternative Models of Scientific Explanation," *American Anthropologist* (1979), 81, pp. 61–74; and Earl Babbie, *The Practice of Social Research,* 6th ed. (Belmont, CA: Wadsworth, 1992), pp. 49–63.

14. Marx, op. cit. note 11, p. 15.

15. For an excellent review of functional theory development in modern organization theory, see Sherman Krupp, *Pattern in Organization Analysis* (New York: Holt, Rinehart, and Winston, 1961).

16. See Keith V. Smith, "State of the Art of Working Capital Management," *Financial Management* (Autumn 1973), pp. 50–55; and Consantin Zopoundis and Michael Doumpos, "Multi-criteria Decision Aid in Financial Decision Making: Methodologies and Literature Review," *Journal of Multi-criteria Decision Analysis* (July–August 2002), pp. 167–180.

17. See, for example, Roy G. Francis, "The Nature of Scientific Research," in John T. Doby, Ed., *An Introduction to Social Research,* 2nd ed. (New York: Appleton-Century-Crofts, 1967), pp. 12–15.

18. Paul Rigby, *Conceptual Foundations of Business Research* (New York: Wiley, 1965), p. 12.

19. For excellent discussions on the differences between science and nonscience, see Ackoff, op. cit. note 6, pp. 1–4; and Ernest Nagel, *The Structure of Science* (New York: Harcourt, Brace, and World, 1961), pp. 1–14.

20. See Francis, op. cit. note 17, pp. 12–15, for an extended discussion of these and other characteristics.

21. For example, see Ian P. Murphy, "Study Maps Course for Electronic Superhighway," *Marketing News* (17 January 1997), p. 14; and Graham T. T. Molitor, "Trends and Forecasts for the New Millennium," *The Futurist* (August–September 1998), pp. 53–59. For the latest numbers, see http://cyberatlas.internet.com or http://www.glreach.com/globstats

22. For discussions of possible changes, see John S. Mayo, "Telecommunications Technology and Services in the Year 2010," remarks made by the president of AT&T Bell Laboratories, delivered at the AT&T Bell Laboratories Technology Symposium, Toronto, Canada, 13 October 1993; Richard P. Derks, "Business Trends: Data Base Marketing," *Information Strategy: The Executive's Journal* (Winter 1994), pp. 34–37; Michael E. Whitman and Houston H. Carr, "The Impact of a Client/Server Architecture on Decision Support Systems," *Information Strategy: The Executive's Journal* (Winter 1994), pp. 12–22; and Marydee Ojala, "The Daze of Future Business Research," *Online* (January–February 1998), pp. 78–80.

23. See Derks, op. cit. note 22, pp. 34–35; Whitman and Carr, op. cit. note 22, pp. 21–22; and Ojala, op. cit. note 22, pp. 78–80.

24. See Mayo, op. cit. note 22, p. 5; R. R. Burke, B. Harlam, B. Kahn, and L. Lodish, "Comparing Dynamic Consumer Choice in Real and Computer-Simulated Environments," *Journal of*

Consumer Research (1992), *19*:1, pp. 71–82; and summary comments of a presentation to the University of Florida's Center for Retailing Education by R. R. Burke, reprinted in "The Virtual Store: A New Tool for Consumer Research," *Retailing Review* (Summer 1994), pp. 1–3.

25. For a discussion of some of these changes, see B. Ives, S. L. Jarvenpaa, and R. O. Mason, "Global Business Drivers: Aligning Information Technology to Global Business Strategy," *IBM Systems Journal* (1993) *32*:1, pp. 143–161; and Kamran Kashani, "Beware the Pitfalls of Global Marketing," *Harvard Business Review* (September/ October 1989), pp. 91–98.

26. See Leonard M. Lodish and David J. Reibstein, "New Gold Mines and Minefields in Marketing Research," *Harvard Business Review* (January–February 1986), pp. 168–182.

27. For example, see John Stevenson, "The State of the Art," *Direct Marketing* (August 1989), pp. 68–70; and John Nichols and Marsha Katz, "Research Methods and Reporting Practices in Organization Development: A Review and Some Guidelines," *Academy of Management Review* (1985), *10*:4, pp. 737–749.

28. See, for example, "Technique-Obsessed Marketing Researchers Can't Help Managers Make Tough Decisions," *Marketing News* (22 January 1982), *15*:15, pp. 1ff; Roman R. Andrus and James E. Reinmuth, "Avoiding Research Myopia in Marketing Analysis," *Business Horizons* (June 1979), *22*, pp. 55–58; Willard T. Carleton, "An Agenda for More Effective Research in Corporate Finance," *Financial Management* (Winter 1978), *7*:4, pp. 7–10; Jacob Jacoby, "Consumer Research: A State of the Art Review," *Journal of Marketing* (April 1978), pp. 87–96; "Quantification-Mania, Lack of Managerial Vision Has Slowed Progress of American Business," *Marketing News* (15 May 1981), *14*:23, part 2, p. 1ff; and Johan Arndt, "On Making Marketing Science More Scientific: Role of Orientations, Paradigms, Metaphors, and Puzzle Solving," *Journal of Marketing* (Summer 1985), pp. 11–23.

29. Jacoby, op. cit. note 28, p. 87.

30. See, for example, Nils Hakansson, "Empirical Research in Accounting, 1960–1970: An Appraisal," in Nicholas Dopuch and Lawrence Revsine, Eds., *Accounting Research 1960–1970: A Critical Evaluation* (Chicago: Center for International Education and Research in Accounting, 1973), pp. 137–173; and Eugene F. Fana, "Efficient Capital Markets: A Review of Theory and Empirical Work," *Journal of Finance* (May 1970).

SUGGESTED READING

Definitions

Kaplan, Abraham, *The Conduct of Inquiry* (San Francisco: Chandler, 1964).

> *Although sometimes tedious and overly scholarly, this is one of the most widely accepted standards of reference for logic and philosophy of science in the social sciences. It relates the conceptual world of science to its real counterpart: the world of scientific inquiry.*

Kerlinger, Fred, and H. B. Lee, *Foundations of Behavioral Research*, 4th ed. (Fort Worth, TX: Harcourt College Publishers, 2000).

> *The continuation of this classic text provides a good foundation for the building blocks discussed in this chapter. All definitions are presented and reviewed in a social science context.*

Zaltman, Gerald, C. R. A. Pinson, and R. Angelmar, *Metatheory and Consumer Research* (Hinsdale, IL: Dryden Press, 1973).

> *This text defines and discusses the use of concepts, hypotheses, and theories in advancing scientific inquiry and knowledge in the field of consumer research. Definitions and business examples are given throughout the book.*

Methods of Theory Construction

Babbie, Earl. *The Practice of Social Research*, 9th ed. (Belmont, CA: Wadsworth, 2001).

> *The author presents a down-to-earth presentation of theory construction and use. The differences in theory construction are explored and examples are provided to illustrate his points.*

Krupp, Sherman, *Pattern in Organizational Analysis* (New York: Holt, Rinehart, and Winston, 1961).

> *The author provides an excellent review of functional theory development in organizational theory. The consequences of this form of theory development are explored.*

Marx, Melvin H., "The General Nature of Theory Construction," in Melvin H. Marx, Ed., *Theories in Contemporary Psychology* (New York: Macmillan, 1965), pp. 10–19.

This excellent article presents the major terms of theory development. It expands on the information presented in this chapter.

Relevance of Science in Business Research

Dunnette, Marvin D., and Leaetta M. Hough, Eds., *Handbook of Industrial and Organizational Psychology,* Vol. 1 (Palo Alto, CA: Consulting Psychologists Press, 1994).

Good discussions of the relevance of science and theory. Focus is in the fields of industrial and organization psychology, but is relevant to a wide range of disciplines.

Libby, Robert, *Accounting and Information Processing: Theory and Applications* (Englewood Cliffs, NJ: Prentice-Hall, 1981).

This book reviews the basic relevance of the scientific process to the field of accounting. The author attempts to bridge the gap between the basic theory of human decision making and its application to accounting.

Smeltzer, Larry R., "Emerging Questions and Research Paradigms in Business Communication Research," *Journal of Business Communication* (Spring 1993), *30*:2, pp. 181–198.

This article is essentially the text of a lecture that described and evaluated research in the business communications field. The importance and relevance of systematic research are highlighted.

State of the Art in Business Research

Barker, Andy, Clive Nancarrow, and Nigel Spackman, "Informed Eclecticism: A Research Paradigm for the Twenty-First Century," *International Journal of Marketing Research* (Winter 2001) *43*:1, pp. 3–14.

A good discussion of past and future paradigms in both business and marketing research. Reflects a shift to informed eclecticism, which is highly consistent with the views in this text.

Mayo, John S., "Telecommunications Technology and Services in the Year 2010." Remarks made by the President of AT&T Bell Laboratories, delivered at the AT&T Bell Laboratories Technology Symposium, Toronto, Canada, 13 October 1993.

Mayo reviews major trends in information technology that will affect research services in the future. For the latest information, fax your request to AT&T Public Relations: 908-582-3800.

Ojala, Marydee, "The Daze of Future Business Research," *Online* (January–February 1998), pp. 78–80.

The author discusses some of the remarkable changes that have happened in the field of business research. Trends and recommendations for action are outlined in the article.

Zopoundis, Consantin, and Michael Doumpos, "Multi-criteria Decision Aid in Financial Decision Making: Methodologies and Literature Review," *Journal of Multi-criteria Decision Analysis* (July–August 2002), pp. 167–180.

A contemporary article that reviews emerging developments in the area of financial decision making. The literature review is extensive.

4

Secondary Data Collection in Business Inquiry

Overview

Secondary Data in Business Research

Uses of Secondary Data

Development of a Search Strategy

Data Retrieval Methods

Data Locations

Internet Searching

The Internet

Background

An Example—Researching Harley-Davidson on the Web

Managerial Considerations

Summary

Discussion Questions

Notes

Suggested Reading

OVERVIEW

This chapter is concerned with the use, collection, and evaluation of secondary data in business research. *Secondary data*—data that has been collected by others for another purpose—is important to us because it can save considerable time and effort in solving the research problem at hand. Researchers should always conduct a thorough secondary data search in the development and execution of any research program.

The discussion begins with a review of the role of secondary data in business research. First, the importance and uses of this type of data in decision making are highlighted. Then a generalized search strategy to obtain this type of data is presented. The search strategy includes the thought processes and activities that must be followed to obtain data relevant to the particular needs of researchers and/or managers.

Here, a distinction is made between manual and online data retrieval methods. Manual methods, although receiving less and less attention because of the explosion of the online information industry, are still critical to thorough and complete secondary data searches.

However, it is absolutely essential that anyone with serious information needs become familiar with the capabilities of online data retrieval methods. Once the major retrieval methods and specific data sources have been reviewed, the nature and background of the Internet is discussed. An example of Internet searching is then provided.

Finally, the important managerial considerations of secondary data are discussed. Here we deal with the primary question: Is the data of the appropriate type and quality to solve the problem at hand? If not, the researchers must then consider the collection of *primary data*—or the collection of data for a specific purpose from original sources.

SECONDARY DATA IN BUSINESS RESEARCH

The Ice Age, the Atomic Age, and other ages were all named for a particular phenomenon that dominated that era; following this pattern, we would have to call our present period the Information Age. The collection, generation, and dissemination of information are growing at a fantastic rate. One author has estimated that human knowledge doubled between 1850 and 1900.[1] A study by the National Science Foundation indicates that the number of published scientific and technical books skyrocketed from 3379 in 1960 to 14,442 in 1974.[2] In the decade between 1964 and 1974 alone, the number of books and periodicals published per scientist rose 63% and 19%, respectively. Since that time, it is estimated that scientific information has been doubling every five years, scientific knowledge every ten years, and scientific articles every four to five years.[3]

The net result is that there exists a tremendous amount of secondary data that is relevant to today's decision-making problems. If the data is applicable to our specific decision problem, it becomes secondary information. This information, if used wisely, can be of considerable importance, for two reasons:

Time savings. If there exists data that solves or lends insight into our research problems, then little primary research has to be conducted. Typically, secondary data takes less time to collect than does primary data.

Cost effectiveness. Similarly, secondary data collection in general is less costly than primary data collection. All things being equal, we would prefer to use secondary data *if it helps researchers to solve the research problem.*

Because resource constraints are always a problem for the researcher, it makes good sense to exhaust secondary data sources before proceeding to the active collection of primary data.

Uses of Secondary Data

Thorough secondary data searches should be made in the conduct of the business research process. The uses of this type of information can be conveniently arranged into four categories:

1. Problem recognition
2. Problem clarification
3. The formulation of feasible alternatives
4. The actual solution of the research problem

These four uses are described next.

Problem Recognition. A constant monitoring of secondary data can provide the impetus for problem recognition. As you will remember from Chapter 1, problem recognition is essential before any decision making can take place. The term *strategic environmental monitoring* means the searching for and processing of information about changes in a company's environment with the intent of planning future courses of action. The monitoring process usually entails a continual review of key environmental information sources.

One organization that specializes in providing secondary data to help spot trends in the defense industry is Forecast International/DMS of Newtown, Connecticut. It provides a variety of information in print, CD-ROM, and data diskettes, dealing with U.S. Defense Department budget forecasts, emerging technologies in the defense industry, and current and projected requirements of the world's armed forces, by product category. This data is frequently updated and provided to customers as a way to spot future product opportunities and problems.

United Way of America uses an extensive network of volunteer individuals from corporations, consulting firms, universities, and other associations, working with 50-plus "scanners" within United Way, to monitor secondary data. This monitoring allows United Way to identify problems and opportunities in the marketplace. The information is then used to develop programs and avoid potential problems in the accomplishment of its mission.[4]

Environmental monitoring can lead to the recognition of changes in a firm's strategic situation. This recognition may require subsequent modification of corporate strategies or may lead to the conduct of further research to understand the nature of the changes.

Problem Clarification. Secondary data can also be fruitfully used to help clarify the specific problem that an organization may be facing. Clarification usually means making the decision problem more researchable by delineating the components of the situation. For example, the fertilizer division of Chevron Chemical developed a customer information system (CIS) that created an internal bank of data on each customer.[5] When sales problems arose in the company, historical data and the salesperson's input were used to identify the causes of the sales slump. The system led to more satisfied and loyal customers.

J.C. Penney Company employs a comprehensive management information system (MIS) using secondary data to help define, select, and clarify problems that must be addressed by management. The research department monitors outside government and trade publications to understand the scope of problems they may be facing in the future. Subsequent primary research is then often used to bring closure to the decision situation.[6]

Another related use of secondary data is to help plan the study design and provide information to write the research proposal. A thorough library search and examination of previous study designs can yield information to help you design your study. Thus, you avoid the pitfalls that others have encountered in researching the same or similar problems. The experience of others, as documented in secondary sources, is almost always of help in developing an understanding of the problem at hand. It is up to the researcher to discover all known relevant information about a situation before collecting any primary data.

Formulation of Feasible Alternatives. Alternatives must exist before decision making can take place. Secondary data is usually very useful in generating viable alternatives to solve problems. Because of the multiplicity of data sources, research approaches, and managerial styles, a number of possibilities should be examined by the researcher.

For example, one online secondary data source that can help in the formulation of feasible alternatives in a variety of situations is the Census Bureau's (www.census.gov) computerized mapping system called TIGER

(Topologically Integrated Geographic Encoding and Referencing). TIGER contains files dealing with census geography (from city blocks up to national boundaries), postal geography (including street names, address ranges, and zip codes), map coordinates (in longitude and latitude), and other topological information. Roadnet, Inc., of Maryland uses TIGER with its own internal data to plan the routes of its fleets of delivery trucks. Many other applications of this program appear likely in the future.[7]

Similarly, when Buick changed its dated Park Avenue automobile into a sleeker sedan, it turned to a geodemographic mapping system to help analyze alternative markets to roll out the model. Secondary information based on census data helped Buick target selective markets, ultimately choosing a solution that increased Buick's sales by up to 20% in some markets.[8]

If secondary information is not used, possible alternatives (and potentially the best solution!) may be overlooked. Therefore, we must examine secondary sources to ensure that our research is as complete as possible.

Problem Solution. Besides being helpful in the definition and development of a problem, secondary data is often sufficient in and of itself to generate a problem solution. Many secondary sources in each of the functional areas of business provide insight and sometimes solutions to business problems.

For example, Nash Finch Company of Minneapolis, Minnesota, one of the nation's oldest and largest food wholesalers in the United States, used a combination of available secondary databases to help solve one of their site selection problems. Using a combination of U.S. census data, current demographic estimates and projections from Urban Decision Systems, and other secondary databases, Nash Finch developed an information system to help decide whether a particular site or town was a good market for a retail food store.[9]

Another example of using secondary databases to help an organization arrive at a solution to a problem is provided by Duke Power Company in Charlotte, North Carolina. Duke Power used VISION, a geodemographic segmentation system that assigns census block groups to one of 48 unique demographic segments. The system was used to develop a communication campaign to persuade current oil-burner owners to evaluate electric heat pumps for their homes. The VISION system enabled the company to develop a program, using readily available secondary information, that saved thousands of dollars in mailing costs.

As you can see, there are sound reasons for conducting secondary data searches in business inquiry. Exhibit 4.1 briefly describes the systematic effort and resources Ford Motor Company puts toward secondary data collection in its global operations. Secondary data can and should be used throughout the research process to help us solve managerial problems. With this in mind, we now turn to the question of how and where to find secondary data.

EXHIBIT 4.1

Secondary Data and
the Ford Motor
Company

Adapted from "The Cutting-
edge Libraries of the Ford
Motor Company," by T. Pack
and J. Pemberton, *Online*
(September/October 1998),
pp. 14–30. Reprinted by
permission.

FORD MOTOR COMPANY

The Ford Motor Company takes secondary data collection seriously, but its initiatives go far beyond the simple access and retrieval of worldwide secondary data. The heart of the organization is the Research Library and Information Services organization located in Dearborn, Michigan. Here, the RLIS staff maintains a collection of 30,000 books, about 600 journal subscriptions, and a large audiovisual library. The RLIS has a full-time staff of 20 (17 in the United States) and is directly responsible for three overseas libraries (one in England, two in Germany). The company uses both an intranet, which contains both Ford proprietary and external data, and an extranet, with several Web-based databases (such as the Community of Science for its patent database, Cambridge Scientific Abstracts for a number of environmental and materials databases, the VWWW database relating to the worldwide automotive industry, to name but a few).

The library staff is not only responsible for mere data searching but is also responsible for such activities as broadening the scope and content of the intranets and extranets, electronic publishing, data mining, and conducting "deep" searches to support managerial decision making. As a result, the RLIS is widely accepted throughout Ford because the services provided are useful throughout the company. This acceptance has resulted in more than 19,000 information requests being fulfilled in one year. Of these requests, 657 of them involved in-depth database searches. The shortest took three hours; the longest took three months, and included about $10,000 worth of online time.

DEVELOPMENT OF A SEARCH STRATEGY

The question of how to find secondary information can best be answered by developing a search strategy.[10] Figure 4.1 diagrams a generic search strategy to find information that can be applied in the uses previously outlined. The process begins with a statement of the data needed in a particular situation. This statement of need can be either very specific or very general, depending on the problem.

For example, the statement of need can be as concrete as "How many employees were there in the Xerox Corporation in 2004?" Or it can be as wide-ranging as "What general industry conditions affected the copier industry in the year 2003?" Obviously, the scope and complexity of information required are vastly different under the two statements of need. The search procedure outlined in Figure 4.1 can handle both requests because the major decision alternatives are presented in the diagram.

Data Retrieval Methods

After the data needs are identified, the researchers select the data retrieval method(s) to collect the desired information. The two major alternatives are manual and online retrieval methods. *Manual retrieval methods* are primarily physical search and collection methods of secondary data through

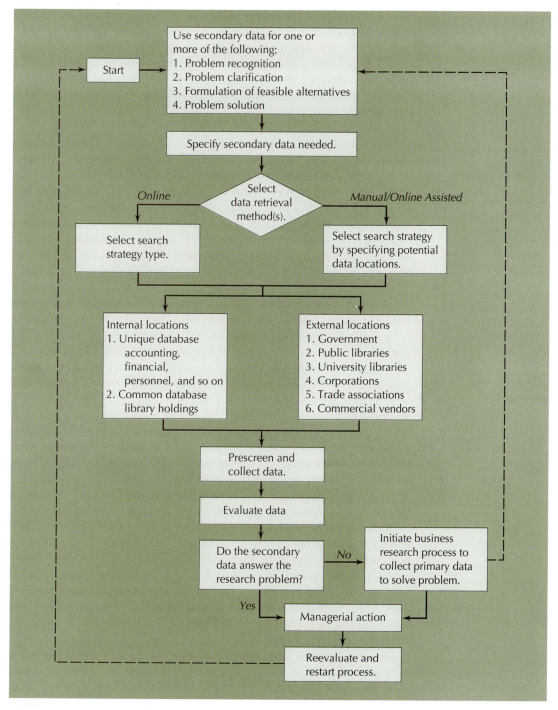

FIGURE 4.1

Secondary Information Search Strategy

the use of bibliographies, indexes, librarian referrals, and the like. Physical search methods include browsing in the library, searching reference texts, or systematically looking through hard copy reference materials for the desired data. Manual methods are often indispensable in finding relevant secondary information. Many organizations do not have comprehensive databases that can be accessed interactively. Therefore, researchers must become familiar with the basic means of finding data manually to satisfy information needs.

Once the specific data needs are expressed, the manual search can begin. The manual process usually leads through a maze of indexes, bibliographies, references, and all types of literature in search of relevant information. Perhaps the best way to organize the manual search is by the location of the data itself. We can systematically identify where data is expected to be and then begin our journey there.

Online retrieval methods primarily involve the use of computerized search and collection methods—that is, direct contact via the computer terminal between the research/manager and the secondary database, often through the Internet.

Although we do not want to get hung up in semantics, today most manual searches involve some contact with the computer (for example, a computerized cataloging system). In this text, we distinguish between manual and online methods by defining the former as one that uses hard copy directories, guides, indexes, reports, and the like to a significant degree to locate and collect secondary data. Of course, computerized searches also often lead to hard copy materials that are subsequently used for further searches and data collection. The point is, if one relies completely on either online or manual methods, he or she is likely to miss information that is either not online or is online but not yet in hard copy. No matter what retrieval method is selected, a specific search strategy for data collection must be designated. For manual methods, data locations need to be delineated and a search initiated. For online methods, the pertinent information retrieval services, Internet searching strategies, and/or databases must be identified.

Once the data needed is located, it must be collected and prescreened for relevance to the problem at hand. After this initial screening, the data must be evaluated in terms of its accuracy and applicability. Here, the primary question becomes: Does the secondary data collected answer the research problem? If it does, managerial action is taken, and the process begins anew. If the data does not satisfactorily answer the research problem, the decision must be made whether collecting primary data would be worthwhile. Managerial action then follows.

Again, it is worth noting that the researcher must choose a retrieval method after the data needs are specified. From the standpoint of thoroughness and efficiency, it is generally desirable to use some combination of manual and online methods to ensure that all pertinent information is gathered in the search process. However, from the standpoint of practicality, we often do not have the time or the resources to complete full-scale searches.

Therefore, we often choose one or the other of the two methods and make do with the information obtained. Usually the choice of method depends on the accessibility and resources available to the information searcher at the time of the study. A more complete discussion of these methods is presented in the following sections.

Data Locations

From a researcher's standpoint, secondary data locations can be classified as either internal or external to the firm. *Internal data* is secondary data available within the company, and *external data* is secondary data obtained from any other source. These two data locations can be further subdivided according to the type of data that is sought. Figure 4.1 outlines this breakdown, and the discussion that follows expands on this classification scheme.

Internal Locations. Internal locations can be categorized according to whether the information comes from the *unique database* or the *common database*. The distinction is whether the data is generally publicly available or not. The unique database typically includes accounting, financial, personnel, sales, and other information that is not readily available to external entities. This data can be of tremendous help in the detection and solution of problems within the organization.

The *common database* is composed of data that is not unique to the company but has been collected and stored within the organization for decision-making purposes. The data can be maintained either physically in a corporate library or electronically. The database usually exists as a result of a prior external secondary information search, which may not be oriented to a specific problem's solution. The databank often includes documents concerning competitive action, industry newsletters, periodicals, government documents, and other published data relevant to the firm's operations.

A major barrier to the use of these types of data in practice is that they are often not categorized or standardized. Ideally, the information should be compiled in a standardized form so that it can easily be understood and used by all decision makers in the firm.

As discussed in Chapter 2, most modern organizations are attempting to fully integrate their unique and common database information resources by creating data warehouses and sophisticated DSS. Here, not only are vast amounts of hard-copy data being converted into a digital format but a continuing new stream of oftentimes real-time data being input into ever increasing data reserves.

External Locations. The diversity of external secondary information sources is endless, and the amount of published data is staggering. Data of all types is published by government agencies, corporations, trade associations, universities, and commercial information services. The data is then stored in various locations, depending on its nature and type. As one would well expect, data is not always easy to find.

Luckily, however, a number of search aids can help the information seeker:

1. Research guides and bibliographies
2. Encyclopedias, dictionaries, and handbooks
3. Indexes
4. Statistical sources
5. References for individual companies
6. Other sources

As depicted in Figure 4.2, these information location aids can be used to identify data sources that may be helpful in the decision-making process. Space limitations preclude an exhaustive listing of such external search aids, but you are directed to Appendix A, "A Practitioner's Guide to Secondary Business Information Sources." This guide abstracts many of the important aids in the information search process and includes a specific section dealing with international business sources. However, this guide is by no means complete. Entire books are dedicated to identifying these information sources, and these should be consulted in any serious search. Three of these books are listed in the suggested readings at the end of this chapter.

Given the increased availability and search capabilities of the computer, it would be remiss if one did not include some reference to the incredible capabilities of the online search tools that exist in today's environment. Powerful *search engines* exist online that can aid in both manual and online searches. Exhibit 4.2 presents a sampling of these search engines. Here the distinction is made between basic search engines and *metacrawlers* that enable the researcher to use one site to access multiple search tools, thus combining the results of multiple search engines. The reader is directed to http://www.searchenginewatch.com for the latest developments on these powerful search tools.

A fast-growing field is commercially available databases in DVD CD-ROM format. Although this technology has wide acceptance in the marketplace, the DVD CD-ROM form of data collection is frequently not considered online because the user is not accessing remote databases through a time-sharing system. In addition, early projections of the growth of this technology have not been realized, largely because of the phenomenal growth of distributed processing and the Internet. Instead, the data is on a disc, played locally on a microcomputer or server. Table 4.1 presents a brief description of just a few of the databases available. Here again, good secondary sources of the available titles and producers/suppliers exist. *CD-ROMs in Print*, published by The Gale Group is one and is referenced at the end of Table 4.1.

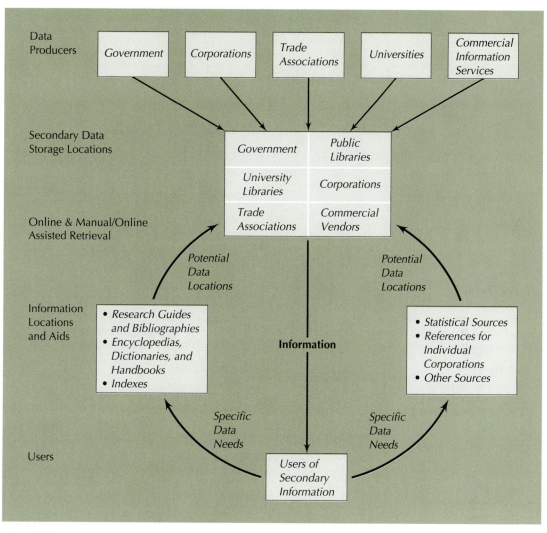

FIGURE 4.2

External Secondary
Information Search
Strategy

Internet Searching

An article written by Walter Kiechel III in *Fortune* magazine in 1980, "Everything You Always Wanted to Know May Soon Be On-Line," described the potential importance of online information searching to business. However, 25 years later, although giant strides have been made in online technology, significant amounts of information still (and probably will for a long time to come) remain offline. Numerous authors have been careful to note that online searching is not fully comprehensive yet and warn both the neophyte and experienced information seeker to recognize the limitations of what one writer calls a "huge construction zone."[12] Nevertheless, the overall growth of the information industry since 1980 has been no less than

SEARCH ENGINES

Crawls or "spiders" over the web searching for hits on the topics under inquiry.

Google

http://www.google.com
A crawler-based service with comprehensive coverage with great relevance.

AllTheWeb

http://www.alltheweb.com
A crawler-based service that is rated high against all search engines.

Yahoo

http://www.yahoo.com
The web's oldest directory, but recent changes have led it to using Google's crawler-based results.

MSN Search

http://search.msn.com
A blend of human-powered directory information and crawler coverage makes this a high quality search engine.

Ask Jeeves

http://www.askjeeves.com
First natural language search engine that has evolved into a crawler-based technology.

METACRAWLERS

Use multiple search engines for searching, then combines the results of multiple (meta) searches.

Vivisimo

http://vivisimo.com
Collects, organizes and categorizes results from multiple search engines.

EZ2WWW

http://www.ez2www.com
Provides results from such engines as AllTheWeb, AltaVista, Google and others.

Kartoo

http://www.kartoo.com
Visually presents results with the sites interconnected by keywords.

Ithaki

http://www.ithaki.net
Arguably the most global of metasearches. Available in 14 languages. Offers over 35 different categories for limiting searches.

Metacrawler

http://www.metacrawler.com
Provides results from at least 8 search engines. Offers a user-friendly interface.

INTERNET SEARCHING STRATEGIES

See http://www.rice.edu/Fondren/Netguides

EXHIBIT 4.2

Search Engines, Metacrawlers, and Tips

Adapted from Danny Sullivan, "Major Search Engines and Directories," April 29, 2003, and Chris Sherman, "Metacrawlers and Metasearch Engines," June 12, 2003, downloaded from http://www.searchengine.watch on Oct. 10, 2003. Check this web site for updates.

phenomenal. Furthermore, this growth is expected to continue in the near future as new information agendas develop.

The online industry underwent substantial change in the last five years—and it is still changing. Not only has the availability of information online exploded, but almost every aspect of obtaining and using information has evolved in recent years. Perhaps the biggest impetus for this change has been the growth of the Internet. The Internet and its associated technology is changing daily (a brief review and example of its capabilities will be given later in this chapter). You are strongly encouraged to explore this important information tool and keep up with the latest advances in this fast-changing field.

The arguments for the use of online retrieval services in business inquiry are convincing indeed. Essentially, the arguments can be enumerated as follows:

TABLE **4.1** A Selected List of CD-ROMs

Title	Description
Business Index on InfoTrac	Covers the most widely respected business publications including the *Wall Street Journal* and the business and finance sections of The *New York Times* and *Barron's*. Designed for use with the IAC's Business Collection, an automated micro-film retrieval system.
Business Periodicals OnDisc	Contains digitally scanned, full-text images of complete articles from business and management periodicals.
Econ/Stats I	Econ/Stats I contains eight databases: Consumer Price Index, Producer Price Index, Export-Import Price Index, Industrial Production Index, Money Stock, Selected Interest Rates, Industry Employment Hours & Earnings by State and Areas, and Capacity Utilization.
FEDSTATE™/1990 Census TIGER Files on CD-ROM	The TIGER (Topologically Integrated Geographic Encoding and Referencing) files are the geographic backbone of the 1990 census program. May be used to develop maps, to the county or city block level, for any location in the nation.
FINN-COIN	Contains company reports including detailed historical annual accounts and financial ratios of West German companies.
Predicasts F&S Index Plus Text	Covers worldwide business activities from over 1,000 trade, business, and government publications. Covers all manufacturing and service industries, focusing on companies, products, markets, and applied technology. One disc covers America, and the other covers international businesses.
Moody's Company Data™	Covers over 10,000 U.S. public companies, including all companies traded on the NYSE, AMEX, NASDAQ NMS, and NASDAQ exchanges, plus thousands of other OTC and regional exchange companies.
Moody's International Company Data	Contains full descriptions and unaltered financials in native currencies on approximately 7,500 non-U.S.-based companies in nearly 90 countries.

Source: CD-Roms in Print 1998. Copyright © 1997 Gale Research, 27500 Drake Road, Farmington Hills, MI 48331. This material originally appeared in *CD-ROMs in Print 1993*. Copyright © 1993 Meckler Publishing, 20 Ketcham Street, Westport, CT 06880. Reprinted with permission. See *CD-ROMs in Print*, 18th ed., © 2003, and 19th ed., © 2004, for the latest updates. The Gale Group, Inc. The Gale Group is part of the Thomson Corporation.

Time saving. Online searches are much faster than manual searches. The search procedures are usually extremely quick.

Thoroughness. Researchers can have more confidence that an important citation has not been overlooked in the search process.

Increased relevance. The experienced online researcher can selectively isolate the key concepts and terms to identify those citations and articles that are of direct relevance to the question at hand.

Cost effectiveness. The decreased search time and increased relevance of the materials identified in the process can result in a highly cost-effective procedure.

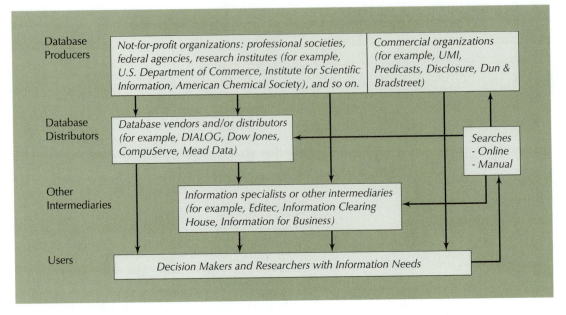

FIGURE **4.3**

Structure of the Online
Database Industry

These advantages have encouraged such organizations as Quaker Oats, Lucent Technologies, Exxon, and others to use online information retrieval to aid in decision making. For example, Ford Motor Company uses online searches to find patents, environmental information, and secondary information about the worldwide automobile industry, to name but a few of the applications of online database searching.

So as to better understand the process of getting specific information from online searches, we shall first describe the structure of the industry itself. Figure 4.3 diagrams the structure of the online database industry. Four levels in the information distribution structure are outlined: the database producers, the database distributors, other intermediaries, and the users of the information.

Where does the Internet come into play in this industry? Quite simply, the Internet is not a *source* of information; it is a *medium* or *tool* for finding information. As a result, the arrows in Figure 4.3 could represent the use of the Internet (or some other medium for that matter). The boxes in the figure represent the organizations, institutions, and/or individuals that produce, add value, or consume the information produced by the database industry.

Each of the four components is now briefly discussed, followed by summary comments on the industry.

Database producers are organizations that collect, assemble, or create data and subsequently store it in computer-accessible files. Both public and private sector organizations are producers of data. Currently, over 11,500 databases produced by over 3700 producers are available online. These databases include information on scientific

TABLE 4.2 Some Database Producers Relevant to Business Research

Databases	Description	Producer
BLS Series: Consumer Price Index, Labor Statistics, Producer Price Indexes http://stats.bls.gov	All aspects of labor, particularly employment and unemployment, economic growth, productivity, prices, wages, industrial relations, and occupational safety and health	U.S. Bureau of Labor Statistics, Washington, D.C. 20212 (202) 691-5200
SDC Platinum and Thomson One http://thomson.com/financial	A leading global provider of data, commentary, and analysis to investment bankers, M&A professionals, and venture capitalists	New York, NY 10007, (646) 822-2000
WilsonWeb, OmniFile, Book Review Digest http://www.hwwilson.com	Full text and abstracted business magazines in accounting, finance, management, and small business	H. W. Wilson Company Bronx, NY (800) 367-6770
ABI/INFORM, Business Dateline, ProQuest 5000™ http://www.umi.com	Topics covered include business, current affairs, science, education, social and consumer trends, opinion and analysis, personal profiles, arts and leisure, health and medicine, law and legislation, and sports	UMI, Ann Arbor, MI 48106 (800) 521-0600
Historical information, principles and standards, taxation, and others http://www.aicpa.org	A comprehensive library of authoritative accounting literature, corporate annual reports, annual reports for government units, and financial data	American Institute of Certified Public Accountants (AICPA), New York, NY 10036 (212) 596-6200
FIND/SVP Reports and Studies Index http://www.findsvp.com	Subjects include industries, business and finance, chemicals and plastics, computers and electronics, construction and machinery, consumer durables and nondurables, defense and security systems, energy, health care, media, retailing and consumer services, and other topics	FIND/SVP, New York, NY 10011 (800) 346-3787

Sources, Excerpted from *Computer-Readable Databases,* 8th ed., edited by Kathleen Young Marcaccio, copyright © 1992 by Gale Research Inc.; and from *Information Industry Directory,* 19th ed., vols. 1 and 2, edited by Bradley J. Morgan, copyright © 1971, 1974 by Anthony T. Kruzas, copyright © 1978, 1981, 1982, 1985, 1987, 1988, 1989, 1990, 1991 by Annette Novallo, copyright © 1995, 1998 by Gale Research Inc. Used by permission of the publisher. For the most up-to-date information, see the *Information Industry Directory* (annual) and *Gale Directory of Databases* (semiannual) by the Gale Group at http://www.gale.com

and technical journals, newspapers, magazines, books, patents, trade publications, corporate reports, economic indexes, statistical data, and others too numerous to mention. Table 4.2 presents a small sampling of databases that are particularly relevant to business research. The interested reader should consult specific source documents to understand the full range of information available to business researchers.[13]

TABLE 4.3 Database Distributors That Are Particularly Relevant to Business Research

Distributor	Descriptions of Services
DIALOG Information Services, Inc. http://support.dialog.com/products	DIALOG operates the world's largest online information retrieval service, providing clients with interactive access to more than 1200 databases in many subject areas including business, management, company, financial, industry, and product information.
Dow Jones & Company http://www.dowjones.com	The Dow Jones News/Retrieval service provides online access to up-to-the-minute business and financial information including stock market quotations, money market changes, actions by companies and government agencies, and labor negotiations.
EBSCO Information Services http://www.ebsco.com	EBSCO provides fully integrated information access to a variety of data sources. Bibliographic and full-text services are offered.
UGA Media http://www.ugamedia.com	UGA provides an international data service aimed at providing companies engaged in international trade useful tips and information. The organization provides a variety of databases, links, and information services to the b2b market.

Database distributors and/or vendors are organizations that subscribe to multiple databases made available by data producers and then market access time to a variety of users. At present, there are more than 9000 distributors of various sizes covering a wide variety of subject areas. Table 4.3 presents four of the larger database distributors, which provide a variety of services to the business professional. Again, you should consult particular online directories to obtain more information about the specifics of these distributors' services.

Other intermediaries in the online database industry include a wide range of information specialists and organizations, from the librarian who passes on the request for online information from a particular user, to specialized organizations such as Editec and Information Clearing House (ICH) that act as retailers of information services. For example, FIND/SVP publishes a catalog quarterly entitled *The Information Catalog,* offering "one-stop shopping for the most recent market and industry studies, company reports, and business and marketing reference sources."[14] In essence, this organization offers to search the literature for you and then provide you with the original source documents of particular studies, articles, and so on. In other words, they operate through database distributors and producers to gather information of the type specified by you.

The users of the information form the final link in the distributive structure. The users are any decision makers and/or researchers who make decisions from the data obtained from the retrieval process.

The users are the ultimate specifiers of information needs to the various organizations and entities in the distributive structure.

This four-component structure and the basic relationships diagrammed in Figure 4.3 may seem relatively clear-cut, but in reality there is considerable overlap. For example, such organizations as the Dow Jones News/Retrieval and the New York Times Information Bank fit into both the producer and distributor levels in the diagram. These organizations are not only primary generators of data, but also collectors of other organizations' information. Thus, you should become fully aware of the options and availability of various organizations' services in the secondary search process.

A related concern is the distinction between online and manual searches. As pointed out earlier, both manual and online searches are often conducted to identify the information desired. It is therefore to your advantage to become familiar with the major services available to the information seeker and the means to conduct both manual and online searches. However, the information technology field is so dynamic that often what is written today becomes obsolete tomorrow. For this reason, we do not attempt to describe everything that is and is not going on in the industry. It changes daily! What we do is provide some background on the development of, perhaps, the most important online tool today (the Internet) and then present an example of how it might be used to collect secondary information. At the same time, we set the stage for many of the online developments discussed in later chapters that are occurring in business research.

THE INTERNET[15]

"The Internet is, by far, the greatest and most significant achievement in the history of mankind."[16] With these words, Harley Hahn opens his Internet "how-to" text, *The Internet Complete Reference*. Hahn contrasts the Internet with such notable achievements as the great Egyptian pyramids and Michelangelo's *David*, and finds these lacking when compared with the Internet's implications for humanity. In that light, Hahn is probably correct in his view of the Internet. If our ability to communicate with each other is the cornerstone of our advancement, be it technological or sociological, then nothing that has existed over the course of history has had the potential to improve the human condition quite like the Internet. The Internet, unlike commercial online services such as DIALOG or LEXIS-NEXIS, is a communication *capability* that allows *people*—the emphasis here is on *people*—to communicate using computers and computer networks as the medium for that communication.

U.S. consumers have adopted the Internet faster than any previous communications technology, including television, radio, and the telephone.[17] It has already profoundly influenced the way the world does

INTERNET RESOURCES

CAUTION: Evaluate the information found on the Internet carefully. **The Internet does not have a quality control program.** Anyone can post information on the Web. While we have listed some full-text sources below, we cannot guarantee that the information given at the sites is accurate, complete or useful.

business, and yet its full potential has been far from realized. However, the "golden child" has its pitfalls. As critics have noted, the huge amounts of relatively unorganized data of often questionable (or even downright misleading) value can be a time and resource drain for the manager/researcher (see Exhibit 4.3). As a result, the information seeker must find ways to efficiently search and then subsequently evaluate the data obtained from the Internet. With this in mind, let us briefly review the background of the Internet.

Background

The roots of the Internet lie in a United States Department of Defense computer network created in the early 1970s. That network was referred to as the ARPANET. The ARPANET was originally intended to support the technology research and data-sharing efforts of the Defense Advanced Research Projects Agency (DARPA). Being an agency concerned with leading-edge technologies, DARPA (originally just ARPA) naturally wanted to be connected with other advanced research and technology organizations throughout the country. This network eventually was expanded to include universities. The ARPANET was a hodgepodge of separate computer networks linked together. Its purpose was to communicate ideas, concepts, opinions, and supporting data. That basic focus remains in the Internet. The Internet, which evolved from the ARPANET, has been a viable commercial entity only since about 1993, and the Internet of today is vastly different from what it was then.

In September 1993, the federal government proposed the construction of an advanced National Information Infrastructure (NII) to provide a seamless web of communication networks, computers, databases, and consumer electronics that would put vast amounts of information into the public and private sectors. This infrastructure was to unleash an information revolution that would enable U.S. firms to compete in the global economy and promote economic growth. Since that time, the Internet has begun to fulfill that promise, leading to significant international development and cooperation.

An interesting development is that the more the Internet advances, the faster it advances. According to the U.S. government, traffic on the Internet now doubles every 100 days, and that rate of increase is itself increasing. The Internet is growing so fast that any information written about it is out of date well before it can be published. So what is this Internet anyway?

Unlike the specialized online professional/research information services (such as DATASTAR, DIALOG, LEXIS®-NEXIS®) or consumer-oriented online services (such as America OnLine, CompuServe), the Internet is not controlled by anybody or any organization (neither governments nor commercial concerns, for example). In fact, many seek to keep government out of the Internet business in order to maintain its value as an avenue of open communication between the peoples of the world. The Internet remains a loose collection of smaller computer networks linked through common protocols. No one knows how big the Internet is or exactly all that it comprises. The Internet simply is—it exists.

Using the Internet is generally free. Gaining access to the Internet through an Internet service provider (ISP) may entail a fee, but generally these fees are very reasonable. Many ISPs allow unlimited access for a relatively small monthly fee. Some Internet online services charge a fee, but these services provide a list upfront of what they offer and what they charge. Generally, these are very reasonable too. As always, caveat emptor applies. This is especially true in an unregulated environment like the Internet.

Online commercial services, especially the professional/research information services, are pay-as-you-go, and users can run up substantial bills in seemingly short periods of time. These services generally have prefiltered their data and organized the information for quick retrieval, so strategic use of the service may mean shorter periods online. The key to using such services is to know exactly what you want before you log on. If you are not completely sure what you want, or want to share notes and seek research suggestions, you may well be better off using the Internet. Many, if not most, of the commercial online services can also be accessed through Internet connections. Their fees still apply, however.

An Example—Researching Harley-Davidson on the Web

One of the best ways to learn how to use the Internet is to use it. Unlike learning to fly an airplane, Internet mistakes are not fatal. So let's plunge in with an example. Let's suppose we are considering investing in the Harley-Davidson Motor Company and want to do a little research to see if such an investment would be potentially profitable. Essentially, we must determine what and how much we want to know (our data needs) about Harley-Davidson in order to make a sound investment decision.

As a starting point, we would like to know how Harley-Davidson has been performing financially over the past few years. What are Harley-Davidson's financial indicators, such as their current quoted stock price and price-to-earnings ratio? Is Harley-Davidson in any other markets besides motorcycles? This would seem important for a myriad of reasons, such as the possible smoothing effects such enterprises could have on the seasonal and cyclical effects of the motorcycle market. Do or could these other Harley-Davidson enterprises substantially enhance or detract from earnings, hence returns on investments? It would also seem prudent to see how select financial and investment services rate Harley-Davidson as an investment. Just how risky is investing in Harley-Davidson anyway?

Let's talk about the consumer. How does the consumer feel about Harley-Davidson's products? Are current sales the result of fads or transient "fashion" trends, or are Harley-Davidson's products insulated to some extent from such effects? Consumer motorcycle magazines could prove very helpful here. We could buy a few of these on the magazine stand. We could also review the online motorcycle magazines. Maybe we could look at the want ads and see how many used Harley-Davidson products are being offered for sale by their owners, and then compare that with Harley-Davidson production rates.

OK, let's get started. First, we have to gain access to the Internet. In this case, I used AT&T's WorldNet® Service. The default browser for AT&T is Microsoft's Internet Explorer. Just because this is the default browser doesn't mean I *have* to use it. If, for example, I wanted to use Netscape's Navigator, then all I would have to do is type "http://netscape.com" to go to Netscape's Web page, download the appropriate browser, and install it according to Netscape's instructions. In this case, I decided to use the default Microsoft browser. To begin, I figured that I should try to find the "home page" (Web site) for Harley-Davidson. Often, finding a company's home page is as simple as typing the Web pre-text, then the company name followed by a ".com" extension. Other times, you may have to use a search engine to find the home page. In this case, I typed "http://www.harley-davidson.com" and got the home page. Note that all the letters are lowercase. Most addresses, called URLs (uniform resource locators) use lowercase text, especially on the Web, but not always.

On the Harley-Davidson home page, I found a worldwide listing of markets. I clicked on the United States market to check the site for additional information. On the home page for Harley-Davidson USA (see Figure 4.4) was displayed all sorts of icons that I could enter to get further information. For example, by clicking on Products, I could review all the current models of Harley; learn what other enterprises Harley-Davidson is involved with (for example, the apparel industry) by clicking on another; learn about the corporate structure, corporate mission statement, current legal endeavors (such as the company's attempts to copyright certain aspects of the Harley-Davidson motorcycle and logos), and a host of other useful information by clicking on yet other icons. Clicking on the company icon put me into a realm where I could get the current stock quote via a link to the PR Newswire. The PR Newswire provides stock quotations for most companies traded on major stock markets. Included in this free PR Newswire service was a 180-day moving average chart of Harley-Davidson stock performance, market and company news briefs, and a market guide with key financial ratios. From the Harley-Davidson financial page, I also accessed the company's annual reports for the past three years. Clicking on another icon got me copies of Harley-Davidson's SEC filings (financial filings required by the Securities and Exchange Commission) for the past four years. This service was provided by a hypertext link to the SEC's EDGAR (Electronic Data Gathering, Analysis,

FIGURE 4.4

Harley-Davidson's
Home Page

Reprinted with permission of
Harley-Davidson.

and Retrieval) system online, an important source of SEC information for any United States–based company. Listed were such important SEC filings that Harley-Davidson has made over the past several years.

After I collected a rather comprehensive amount of information at Harley-Davidson's Web page, I decided to continue using Internet financial and investment services. Many of these are available on the Internet. Most of these services are commercial ventures that supply a limited amount of free service, along with enhanced paid subscription services. Because the amount of free financial information available at many of these sites is rather extensive, always check these first to see if the free services provide what you need. Also, if one site doesn't provide some particular information for free, another site might. The cost to you in searching all these sites is time; however, after you get to know the available sites and what they provide, you will know where to go in the future for the information you seek.

My next stop was Data Broadcasting Corporation's (DBC) Web site (http://www.dbc.com). DBC is a commercial concern that provides an extensive array of real-time market data to individual investors. This is the site that provided the access to the PR Newswire stock information that was linked through the Harley-Davidson home page. DBC provides links to many key financial services in addition to its own. After researching the available information at DBC, I decided to check what the Quote.Com company (http://www.quote.com) had on Harley-Davidson. From the Quote.Com site, links can be established with such financial and investor services as SEC's EDGAR online, Reuters, the Business Wire, Industry Watch, the Individual Investor, the Market Guide, Morningstar, S&P Stock Guide, and so forth. The list is quite comprehensive. However, sometimes the links provide limited access unless a fee is paid to Quote.Com. In several of those cases, I went directly to the respective sites, now that I knew what was available.

Next, I went to an online site commercial EDGAR support service (http://www.edgar.com). From here I could access a substantial amount of information for free. EDGAR.com also provided me a direct link to the Securities and Exchange Commission's (SEC) Web page (http://www.sec.gov). The SEC provides extensive investor and business support. Here I could directly access the SEC's EDGAR archives, along with other related documents. Spending a little time here should prove very beneficial. The SEC also lists complaints they have received on various companies, so look at these to see if there is any relevant information.

Now I wanted a little macroeconomic data to get a feel for what the economy may be doing in the future. The Department of Commerce's Bureau of Economic Analysis (BEA) seemed like a good place to start (http://www.bea.doc.gov). BEA provides extensive economic data and analysis results that should help indicate what the national economy is doing and what it could reasonably be expected to do in the near future. Other Commerce Department sites can augment BEA's information.

Consumer research was next. First, the online magazines seemed like the place to go. I also discovered that several independent articles are available (unpublished articles put on the Web by their authors as a service to fellow Internet navigators). Because I didn't know all the magazines and articles that were available, now was a good time to use some of the online directories and search engines.[18]

Starting with AltaVista, I entered "Harley-Davidson Motor Company" as my keyword input and initiated the search. It came back in a few seconds with 31,639 matches. This was far too many matches for me to review, so I narrowed the next search by adding "magazine" to my search. It still came back with a prohibitive number of matches. After a few iterations of altering keywords and refining the data searches, I eventually got the matches down to a few thousand. Because the most relevant matches were first, I started down the list reading the small reviews under the titles of the site matches. If a site looked as though it suited my research purposes, I linked to the site by placing the cursor on the title and clicking the mouse button. Among these sites were a few from other nations. For example, there was one from a Harley-Davidson dealer in Chile. As could be expected, the site documents were in Spanish. I wanted to read them in English, so I invoked AltaVista's translation service and, in a few seconds, had the documents translated into English.

Next, I decided to see what was available in the Yahoo directories. My search resulted in 9210 hits. Still way too many to process efficiently. I had time to read only a few. Hence, I had to select the most relevant based on the titles and reviews provided by Yahoo. Often, there comes a point in research where additional data becomes "me-too" data, which is data that simply tells you what you already know. Such data is limited in usefulness and often isn't worth the time or expense to procure it.

Finally, I decided to use my old standby, Lycos, and see what I came up with. I found a few online motorcycle magazines such as Motorcycle On-Line (http://www.motorcycle.com). Motorcycle On-Line provides an archive of motorcycle reviews, including many Harley-Davidson products. These reviews are valuable to the investor because, if Harley-Davidson is experiencing perceived quality problems, for example, this could have a significant impact on future sales, hence return on investments. I also found American Iron Magazine (http://www.americaniron.com), which publishes a traditional paper magazine dedicated to Harley-Davidsons and also supports a Web site.

Next, I wanted to talk with the owners of Harley-Davidson products. I wanted to know how they felt about the Harley-Davidsons they owned and if they would consider buying another one in the future. Yahoo has a good list of user discussion groups, so I reviewed Yahoo's directory for the appropriate groups. I found 82 matches for Harley-Davidson. Under those, I found three discussion groups that I actually tried. As expected, the discussions were fruitful.

After visiting all these sites and collecting data, the task turned to one of evaluating and assimilating. As noted earlier, all secondary data needs to be evaluated for its accuracy, relevance, and timeliness before it is used for decision-making purposes. In addition, several other managerial considerations need to be addressed in obtaining and using secondary data. These considerations are the topic of the next section of this chapter.

MANAGERIAL CONSIDERATIONS

Given the importance and value of secondary information, you should be concerned about the availability and use of this data in your organization. In particular, managers need to address three major concerns in the collection and use of secondary information.

1. The question of whether your organization should compile or buy its secondary information should be addressed periodically. As discussed earlier, there are many alternative sources from which to obtain secondary information. The information can be compiled and kept internally in a library (either online or off) or bought when needed from external information retailers or distributors.[19] The decision rests on a number of factors. The data should be compiled rather than bought if

1. The data will be proprietary and valuable
2. The organization will make intensive use of the data

3. The staff has expertise on the matter

4. The database is small enough to be internally manageable

5. The data could potentially be sold to others, thereby permitting the organization to recoup part of its costs

6. No satisfactory alternative is available

The data should be bought or leased if

1. The body of information is large and expensive to collect

2. The data will require frequent updating

3. The data is only needed on a sporadic basis

4. No competitive advantage will be lost by relying on an outside source

The manager must weight these considerations to determine the ultimate type and source of secondary information for decision making within the firm.

2. Require a thorough secondary information search to be conducted before any primary data is collected in a research project. This consideration will help ensure that costly primary research endeavors are not undertaken when appropriate secondary data can help answer the specific research problem. The prudent manager will exhaust the available secondary data before undertaking a research project. As we have already seen, secondary information can supplement the research process in all stages of a project's development. Thus, managers must insist that the research process include all relevant secondary data.

3. The nature and value of secondary data should be evaluated before it is used in a particular decision-making context. No matter what the source—manual or online, internal or external—secondary data should be evaluated for its appropriateness in each decision-making context. Secondary data was originally collected for purposes other than those for which we may choose to use it. Secondary data becomes secondary information to us when it is relevant and applicable to our decision problems. The data should be evaluated to see if it meets our specific needs.

This consideration is particularly true when it comes to using and comparing secondary data from international sources. The user of the data must constantly be aware of potential problems that might compromise the reliability of foreign secondary data sources. Exhibit 4.4 highlights some of the difficulties in obtaining and using such data.[20]

The primary criteria for evaluating secondary data are as follows:

Timeliness. Is the data timely? Is it up to date? Is it available when we need it? If it is not all available, or if it is outdated, the data should not be used for decision-making purposes.

Relevance. Is the data relevant to our particular needs? We may have timely data that does not apply to our decision-making needs. Relevance implies the ability of the data to satisfy the needs of the user.

EXHIBIT **4.4**

Difficulties in
Obtaining and Using
Foreign Secondary
Data

AVAILABILITY   In many developing nations, there are few secondary data sources. For example, Bolivia has not had a census since 1950 and Zaire since 1958. Generally, the availability of secondary data is roughly correlated with a country's level of economic development.

RELIABILITY   Governments are usually among the largest providers of secondary data. It is not unheard of for governments to underreport negative items, such as illiteracy rates and disease, and overreport such items as industrial production, market growth, and other variables that give a positive picture to the international community. Statistical data is often used for political purposes, such as obtaining assistance from the World Bank.

TIMELINESS   If data is available, it is frequently not timely or useful. Gathering data is a costly and time-consuming enterprise; many countries view data collection activities as a very low priority.

COMPARABILITY   Frequently, data cannot be meaningfully compared across countries because the use and classification systems are different. For example, in the soft drink and alcoholic beverage markets, forms and classifications of carbonated drinks, juices, and liquid concentrates can differ significantly. In Mediterranean cultures, for example, beer is considered a soft drink.

Accuracy. Are there sources of error in the data that affect its accuracy? As we have seen, sources of error may affect a study's validity and reliability. A primary difficulty in the use of secondary data is that often we know little about its accuracy. It is strongly suggested this data be evaluated according to (a) its completeness, (b) its source, and (c) the method of its collection.

Cost. How much does the data cost? Here, the primary question is whether the data is cost-effective for what you get. If secondary data is available, but not as relevant as you would like it to be, you may be better off conducting primary research at a higher cost. In other words, it may be more cost-effective.

Purpose. What is the point of view or purpose of the data? As noted earlier, the Internet is filled with .com, or commercial, Web sites whose purpose is often to persuade or sell some particular point of view. Who is the source? What is the site's purpose? These questions will help identify highly biased data that needs to be evaluated carefully in the decision-making process.

All secondary data should be evaluated against these criteria before the decision is made to collect and use it in the research process.[21]

The final concerns of management can be expressed as a set of questions (see Exhibit 4.5). Not all the questions are relevant for each and every research project, but they should be asked as necessary to ensure that secondary data collection is done appropriately in the organization. The effective use of secondary information can help researchers and managers in the performance of their respective tasks in the organization. However, it is up to the manager to

1. What are the key recurring secondary information needs within the organization? Should the data to satisfy these needs be internalized into the organization?
2. What are the information search procedures currently being used within the organization? Are these procedures producing the results desired?
3. Are online data retrieval methods currently used within the organization? If they are not being used, why not?
4. At the start of each research project, are thorough secondary information searches being conducted so as to avoid unnecessary primary data collection? If not, why not?
5. Is the secondary data collected being evaluated before it is used in the decision-making process?

ensure that the vast number of secondary information resources available are used productively within the firm.[22]

SUMMARY

Secondary data is data that has been collected by others for another purpose. Secondary data is important in research, because it can be fast and cost-effective when compared to the collection of primary data. As a result, researchers should conduct thorough secondary data searches before any primary data collection procedures are initiated.

There are four major uses of secondary information in the research process: (1) problem recognition, (2) problem clarification, (3) the formulation of feasible alternatives, and (4) the actual solution of the research problem.

How and where do we find secondary data? A generalized search strategy focuses on specifying the secondary data needed, selecting the data retrieval methods to collect this information, and evaluating the data to see if it is actually useful in the decision-making process (in other words, information).

The two major means of data retrieval described are the manual and online methods. The specific nuances of each retrieval method are outlined and discussed. "A Practitioner's Guide to Secondary Business Information Sources" is presented in Appendix A at the end of this text to help you in manual data retrieval. Online searching is described, and specific firms in the online database industry are outlined and reviewed.

The chapter continues with a brief discussion of the nature and impact the Internet has had on business research. An example of using the Internet to collect information on the Harley-Davidson Motor Company is given for illustrative purposes.

The chapter ends with the major managerial considerations in the collection and use of secondary data. Several major concerns are stated. Questions are developed for the analysis and review of secondary data collection's role in business inquiry.

DISCUSSION QUESTIONS

1. Secondary data is very important to the conduct of business research. Elaborate.

2. Elaborate on the general use of secondary data in business research.

3. The management team of a large retail chain would like to obtain data in selected areas to enhance the firm's competitive position. Can you help them out? The following are their areas of interest:

   a. General competitive facts

   b. Retail sales of clothing in San Francisco, California; Atlanta, Georgia; and Orlando, Florida

   c. Average retail clerk salaries

   d. Market potential of men's shirts in Atlanta, Georgia

   e. Corporate fitness programs

4. Describe a secondary data search strategy if you wanted a problem clarification concerning the reasons why employees are unhappy with the management practices of the firm.

5. Review the advantages and disadvantages of the online and the manual data retrieval methods.

6. Compare unique and common database information.

7. What type of database service would you use to obtain competitive intelligence data?

8. The general rule of thumb in research is to collect all possible secondary data before generating primary data. Can you elaborate on the reasons for this business research philosophy?

9. How would you evaluate secondary data gathered by your firm for research purposes? Is evaluation really necessary?

10. Conduct a secondary literature search for the concept of ethnocentrism. Use both online and manual searches if possible to develop a list of references that deal with the concept.

11. Using the Web, find examples of companies using secondary information to help them make decisions. Find four examples and be able to explain in your own words how the company benefited from the secondary information search.

12. Describe how Ford Motor Company uses a worldwide store of information in its decision making. Be specific, and bring examples to class from the Web (identify the URLs) to illustrate your points.

13. Choose any company and conduct a search like the one illustrated in this chapter using Harley-Davidson. Your goal is to determine whether or not to invest in the company of your choosing. Write up your results in five pages or less.

NOTES

1. Chris Mader, *Information Systems* (Chicago: Science Research Associates, 1979), p. 27.

2. Roger Christian, *The Electronic Library: Bibliographic Data Bases 1978–1979* (White Plains, NY: Knowledge Industry Publications, 1978), p. 2.

3. Graham T. T. Molitor, "Trends and Forecasts for the New Millennium," *The Futurist* (August–September 1998), p. 59.

4. See W. Wilkinson, "Getting and Using Information, the United Way," *Marketing Research* (September 1989), p. 6.

5. Henry Fersko-Weiss, "Tools of the Sales Trade," *Personal Computing* (August 1987), p. 81.

6. See Donald S. Tull and Del I. Hawkins, *Marketing Research* (New York: Macmillan, 1993), p. 28.

7. "Tiger Shows Its Teeth," *American Demographics* (June 1991), p. 23.

8. Michael J. Weiss, "Marketscapes: Mapping American Tastes," *Business Geographics* (October 1994), 2:6, p. 12.

9. Brian Numainville, "Easing a Company's Growing Pains," *Business Demographics* (February 1997), pp. 28–30.

10. Other search strategies have been proposed by Robert Berkman, *Find It Fast: How to Uncover Expert Information on Any Subject* (New York: Harper Collins, 2000); and David Stewart and Michael Kamins, *Secondary Research: Information Sources and Methods* (Newbury Park, CA: Sage Publications, 1993). For specific sources that help direct researchers to business information locations, see James Woy, Ed., *Encyclopedia of Business Information Sources,* 18th ed. (Detroit, MI: Gale Group, 2003); *Business Research Sources: A Reference Navigator* (New York: McGraw-Hill, 1999); Patrick F. Butler and Paula Berinstein, "The Top 10 Sources for Statistics," *Online* (March/April 1998), pp. 61–65.

11. Walter Kiechel III, "Everything You Always Wanted to Know May Soon Be On-Line," *Fortune* (5 May 1980), pp. 226–240.

12. See Herbert I. Schiller, "Electronic Highway to Where?" *National Forum* (Spring 1994), p. 19; Rick Boucher, "The Information Superhighway: Turning the Vision into Reality," *National Forum* (Spring 1994), pp. 16–18; Ian P. Murphy, "Study Maps Course for Electronic Superhighway," *Marketing News* (17 February 1997), p. 14; and David Curle, "Out-of-the-Way Sources of Market Research on the Web," *Online* (January/February 1998), pp. 63–68.

13. The reader can start by contacting the Information Industry Association, Washington, D.C.; the *NTIS Directory of Computerized Data Files, Software, and Related Technical Reports,* National Technical Information Service, Arlington, VA; and examining the six directories published by the Information Industry Group of the Gale Group. These directories include *The Information Industry* directory (annual) and *Gale Directory of Databases* (semiannual). Go to http://www.gale.com

14. See the *Information Catalog,* published bimonthly by FIND/SVP, 625 Avenue of the Americas, New York, NY 10011. (See www.findsvp.com.)

15. The author thanks Michael P. Hughes for his significant contributions to this section of the chapter.

16. Harley Hahn and Rick Stout, *The Internet Complete Reference* (Berkley–New York: Osborne, McGraw-Hill, 1994).

17. Peter Clemente, *State of the Net: The New Frontier* (New York: McGraw-Hill, 1998); and Harley Hahn, *Harley Hahn Internet Yellow Pages* (New York: McGraw-Hill, 2003).

18. Because search engine knowledge is changing so quickly, it is best to check online sources for the latest developments. Several such sources are http://www.searchengine.watch.com, http://www.search.page1.org, and http://www.brint.com, to name but a few. New sites appear regularly.

19. For a very good article dealing with choosing an online vendor, see George R. Plosker and Roger K. Summit, "Management of Vendor Services: How to Choose an Online Vendor," *Special Libraries,* vol. 71 (August 1980), pp. 354–357.

20. See discussions of these difficulties in Subhash C. Jain, *International Marketing*, 6th ed. (Mason, OH: Southwestern, 2001); and Warren J. Keegan and Mark C. Green, *Global Marketing*, 3rd ed. (Upper Saddle River, NJ: Prentice Hall, 2003).

21. See, for example, Nicole Auer, "Bibliography on Evaluating Internet Resources," 14 March 1998, http://refserver.lib.vt.edu/libinst/crit-THINK.HTM; and D. Scott Brandt, "Evaluating Information on the Internet," 28 January 1998, http://thorplus.lib.purdue.edu/~techman/evaluate.htm.

22. For interesting readings on how information technology can affect the firm, see Michael Scott Morton, Ed., *The Corporation of the 1990s: Information Technology and Organizational Transformation* (New York: Oxford University Press, 1991); and Thomas J. Allen and Michael Scott Morton, *Information Technology and the Corporation of the 1990s: Research Studies* (New York: Oxford University Press, 1994).

SUGGESTED READING

Secondary Data in Business Research

Berkman, Robert I., *Find It Fast,* 5th ed. (New York: Harper Collins, 2000).

A best-selling guide on how to uncover information on any subject. A great reference guide that supplements and extends this chapter.

Berinstein, Paula, "The Top 10 Sources for Statistics," *Online* (March/April 1998), pp. 61–65.

The author provides ten statistical sources that will answer many of the general questions people usually ask. Most are available free or at low cost on the Web.

Woy, James, Ed., *Encyclopedia of Business Information Sources,* 18th ed. (Gale Group Inc., 2003).

A bibliographic guide to more than 35,000 citations covering over 1,100 subjects of interest to business personnel. It is truly an encyclopedia of business information sources.

Development of a Search Strategy

Butler, F. Patrick, *Business Research Sources: A Reference Navigator* (New York: McGraw-Hill, 1999).

A comprehensive resource/guidebook for the conduct of business research. Covers both electronic and hard copy resources.

Schlein, Alan M., *Find It Online: The Complete Guide to Online Research,* 3rd ed. (Tempe, AZ: Facts on Demand Press, 2003).

An excellent guide to the workings and resources available on the Internet. The book outlines basic Internet terminology, outlines search strategy suggestions, and provides tools for researching the wide variety of information resources on the Web.

The Internet

Clemente, Peter, *State of the Net: The New Frontier* (New York: McGraw-Hill, 1998).

An in-depth examination of Internet users and trends, backed by primary research collected by FIND/SVP's Emerging Technology Research Group. Very informative.

Levine, John R., Margaret Levine Young, and Carol Baroudi, *The Internet for Dummies,* 9th ed. (New York: Wiley Publishing, 2003).

A basic guide to the Internet that is both easy to read and helpful for those starting serious searching online. Useful reading to recapture the basics of the Internet.

Mambretti, Joel, and Andrew Schmidt, *Next Generation Internet: Creating Advanced Networks and Services* (New York: John Wiley & Sons, 1999).

Provides a useful discussion on the changes coming to the Internet. More advanced than introduction to the Internet guides.

5

Problem and Proposal Development and Management

Overview

**Problem Identification
and Formulation**

Problem Identification

Problem Formulation

The Question of Value

Proposal Development

A Typology of Business-Related
Research Proposals

Structure of a Proposal

A Sample Research Proposal

Managerial Considerations

Proposal Evaluation Mechanisms

Other Control Mechanisms

Summary

Discussion Questions

Notes

Suggested Reading

**Appendix: Methods of Assessing
the Value of Research
Information**

Before any inquiry is begun, the researcher must have a problem to study and a plan of action to solve the problem. Generally, both the problem and the plan of action are specified in a research proposal. A *research proposal* is a formalized bid for research, usually written, specifying the problem to be studied and the plan of action to solve the stated problem. The proposal, if accepted, generally serves as the contract between the one requesting the investigation (usually the manager) and the proposer of the research project.

The proposal plays a critical role in the research process because (1) it specifies the problem to be addressed in the study and (2) it outlines the methods and procedures the researcher will use to solve the problem. From a managerial perspective, the proposal is of substantial interest because it outlines the information that will ultimately be made available for decision-making purposes.

However, we do have a chicken-and-egg problem when we study proposal development and evaluation before understanding the other components in the research process. This is because a fully developed research offering actually designates the research design, data collection, and analytical procedures. It is therefore difficult to examine the specifics of the proposal this early in the text. Nevertheless, its purpose and structure should be analyzed to solidify its role in the business process. With these issues in mind, you should attempt to grasp the basic issues in development and management of research that are discussed in this chapter.

The chapter begins with a discussion of problem identification and formulation. Problem identification is reviewed because it is actually the start of the research process and is essential to the development of the research proposal. If problems remain unidentified, there is no impetus for research. Problem formulation is a critical task because improper specification results in useless output from the research process, if the study is initiated. In addition, all participants in the research process must question the value of any investigation to be conducted. Planned primary research is an expensive proposition and should be viewed as an investment in the business at hand.

The focus of the chapter then switches to a detailed discussion of the need for research proposals to ensure quality information for managers. Then a typology of business-related proposals and their general structure are discussed. A sample research proposal is then presented for illustration purposes.

The chapter concludes with a discussion of the main managerial considerations in this critical communication stage of the research process. A framework for proposal evaluation is outlined, and major control mechanisms are identified.

Proper problem identification and formulation are essential if managers are to receive the information they need to make decisions. According to the American Marketing Association, "If any one step in a research project can be said to be more important than the others, problem definition is that step."[1] Although it may appear that problem formulation is the easiest of all the aspects of research, it is a difficult process. Let us examine the reasons why.

Problem Identification

Problem identification involves becoming aware that an undesirable or potentially undesirable situation exists for the firm.[2] Identifying some situations is simple (for example, high personnel turnover rates, discontent of the work force as evidenced by strike, declining profitability of a product line, selection of three alternative investment options). Other problems are of a more subtle nature, at least in the beginning. For instance, imagine how the leading manufacturers in the watch industry felt in the early 1970s when Texas Instruments introduced the digital watch, which sold like wildfire. Or imagine the dismay of the American automobile industry when the small, gas-efficient imports captured over 25% of the United States market in a matter of years. In both of these cases, the industry leaders were caught "sleeping." They did not monitor and anticipate major consumer, social, environmental, and technological shifts that were occurring in the markets they served. Here, problem identification was not so easy, and chaos resulted.

A secondary concern in identification is in management's initial specification, or misspecification, of the problem. One example of this is in the field of banking. The vice president of a midsized Southeastern bank was concerned about declining demand deposits at his institution. The manager felt that the institution's product line was inadequate, and he commissioned the in-house research staff to identify the product line most desired by consumers in its trading area. The study was developed and executed, and the results were given to the vice president. As a result of this study, several additional services were added: financial planning, consolidated statements, and more automated teller locations. Six months later, however, demand deposits were still falling, and the vice president accused the research staff of giving him "bum" information. Friction between the two offices then ensued.

Later that year, the research staff was routinely monitoring competitors' actions to see if any significant changes had been made in their strategies. They found that a pricing change had been made by several of their smaller competitors and subsequently adopted by some of the main competitors. The change lowered the price of banking services to a substantial segment

FIGURE **5.1**

The Research Problem
Development Process

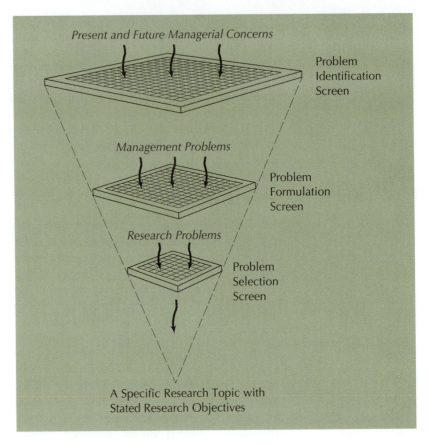

Present and Future Managerial Concerns

Problem
Identification
Screen

Management Problems

Problem
Formulation
Screen

Research Problems

Problem
Selection
Screen

A Specific Research Topic with
Stated Research Objectives

of consumers. When informed of the change, the bank's management adopted the new pricing arrangement. The slow decline of demand deposits was halted.

In this situation, the vice president had originally misspecified the problem to the research department. Rather than research on the makeup of the product line, the firm needed an assessment of the problem situation to identify the possible causes for the declining demand deposits.

Figure 5.1 diagrams the role of problem identification in the research problem development process. In a narrowing process, present and future managerial concerns are passed through a number of "screens" until a specific research topic with stated objectives is identified. The problem identification screen is viewed as a filter through which general management concerns are translated into managerial problems. In our banking example, the managerial concern of declining demand deposits was initially misinterpreted, resulting in a misspecification of a management problem: What product line should we be offering so the decline in demand deposits can be halted?

How do concerns turn into problems? Several factors may be involved:

Management awareness. An astute manager has wider experience and, in general, perceives more problems and/or opportunities than a less astute counterpart. In other words, the "aware" manager is more adept at identifying problems than either the neophyte or the poorly informed manager.

Urgency of concerns. In industries where little is changing, there is little perceived urgency to identify concerns that may become future management problems. Our mention of the watch industry exemplifies the opposite situation. If we are "up to our ears in alligators," managerial problems will abound because they all need attention now.

Research orientation. If an organization possesses a strong research and development orientation, future and present managerial concerns are more likely to be identified and converted into problem statements. Not all concerns are obvious. Exploratory research can identify and anticipate problems and opportunities that would otherwise go unnoticed.

As the narrowing in Figure 5.1 illustrates, not all concerns are converted into management problems, and not all management problems are converted into research problems. Quite simply, some concerns are not significant enough to need management attention. Others may never be perceived by management to be converted into problems.

Overall, effective problem identification requires a mixture of creativity, knowledge, experience, and sometimes sheer luck. Also, proper problem identification usually requires much communication between the manager and the researcher. As in our banking example, research is sometimes needed to diagnose problems that are particularly subtle and difficult to identify. Such inquiry sets the stage for further research to solve the problem that is isolated. There is no set procedure for proper problem identification, but Exhibit 5.1 poses some managerially oriented diagnostic questions to get you thinking about potential problems that may arise in your firm.

EXHIBIT 5.1

Some Possible Problem Identification Screening Questions

1. What is the current situation of the firm? Are there any undesirable situations that need attention?
2. Are there any conditions, processes, etc., that could be improved?
3. Are any problems foreseeable in the future that could affect the operations of the firm?
4. Are there any potential opportunities in the future that the organization may capitalize on?
5. Is the problem identified really a problem, or is it a symptom of another problem?
6. Does the identification of the problem follow from the available evidence?
7. Is research needed to identify the problem that underlies some undesirable situation?

Problem Formulation

Once a problem is properly identified, it must go through a "screen" that converts the general statement of management's problem into a researchable question that is capable of being investigated. If a specific problem is subjected to this step, management must consider it of sufficient priority and value to be formally investigated by the research process (which we will discuss in more depth later in the chapter). Again, not all management problems are translated into research problems. Some of the reasons for this are as follows:

Lack of value. In some cases, there exists a significant management problem, but the proposed research is not cost-effective. The research costs more than the benefits it is expected to produce.

Priorities. A problem is not of sufficient importance to require researching at the present time. For example, although the management of General Electric might like to know what makes production workers happy, this may not be so important as to interrupt its research activities monitoring the competition.

Researchability. As stated earlier, some problems are inherently unresearchable. In the pharmaceutical industry, management may be interested in testing the effectiveness of a newly developed cancer drug on humans. Not only would this be illegal; it would be unethical as well. For all intents and purposes, the subject would have to be considered unresearchable.

Limited resources. Many problems could be of sufficient priority and value to be researched, but the firm only has a limited pool of resources. Therefore, a number of problems do not get research attention simply because the research money or time is exhausted.

Problem formulation is important in the research process because it sets the direction and scope of the research endeavors that follow.[3] A properly defined research problem clearly states the context of the problem situation and the associated conditions that surround it. An example of a complete problem definition is presented in Exhibit 5.2. Major managerial concerns, the justification for problem identification, and statements of both the management and research questions are stated unequivocally.

The final stage of problem development is the actual selection of the question(s) to be studied. Here again, management's priorities and the perceived value of each research project determine which of the pool of research questions will ultimately be investigated. At this stage, the research problem is stated, and the objectives defining the goals of the investigation are agreed to by the user of the research and the researcher.

EXHIBIT **5.2**

An Example of a
Complete Problem
Definition in a Large
Computer Firm

SYMPTOM RECOGNITION (MANAGERIAL CONCERNS)

There is a growing discontent among computer programmers in the organization. Turnover has been increasing steadily over the past five years. Employees complain about inequitable salary structure.

PROBLEM IDENTIFICATION

1. Evaluation of internal and external data
   a. Monitoring of discontent and turnover information within the organization
   b. Any history of past discontent
   c. Literature suggesting similar problems in other computer firms
2. Isolation of problem area
   a. Management has no consistent salary allocation plan
   b. Exit interviews reveal discontent with salary system
   c. EEO has recently filed a complaint against the firm, alleging salary discrimination

MANAGERIAL PROBLEM STATEMENT

Is the present salary structure equitable?

RESEARCH PROBLEM STATEMENT

What are the main factors associated with salary levels for computer professionals in our organization?

Problem formulation is a step that needs a great deal of attention from managers and researchers alike. The old saying, "A problem well stated is half solved," is a little on the optimistic side, but there is some truth to it. An inadequately formulated problem is going to give you inadequate solutions at the end of the research process. Exhibit 5.3 lists a series of questions pertaining to the problem formulation and selection processes that will help promote critical thinking in the area of problem formulation.

The Question of Value

Once a problem is adequately identified and defined, an assessment of the value of the research is needed. Is the study of sufficient value (given management's best estimates of the situation) for it to be undertaken? In applied business research settings, the general principle for making this assessment is: The expected value of the information obtained from a particular study should exceed the costs of obtaining it. If the research endeavor fails this test of the value of its information, then management should seriously question the desirability of conducting the study.

However, some research may fail this initial assessment but still be instituted, because of the intrinsic value of the work itself. This is usually called

EXHIBIT 5.3

Problem Formulation
and Selection Screening
Questions

PROBLEM FORMULATION

1. Does the problem statement capture the essence of management's concern?
2. Is the problem specified correctly (i.e., is it the "real" problem)?
3. Is the problem stated unambiguously in terms of the specific variables and relationships of interest?
4. Is the scope of the problem clearly defined?
5. Does the statement of the managerial problem have clear implications for testing?
6. Is the problem statement free from personal bias?

PROBLEM SELECTION

1. Is the managerial problem of sufficient relevance and value to justify investigation?
2. Can data be obtained to solve the problem?
3. Do you have the capabilities and expertise to adequately research the problem?
4. Can the problem, as stated, be researched with the dollars and time available?
5. Are there other investigations that would have greater value to the firm?
6. What are the real reasons this problem was selected for study?

basic research. It can be defined as inquiry undertaken for the purpose of expanding the frontiers of knowledge. Most basic research is now conducted in universities.[4] Almost all research performed in business organizations must be cost-effective and of demonstrable value to the decision maker.

There are a number of explicit techniques to assess the value of research information (see the appendix at the end of this chapter). These formal methods all rely on management's estimates of such factors as costs, savings, profits, and probabilities of certain outcomes happening. As a result, most value assessments of research rely on an intuitive assessment of the costs and benefits of conducting or not conducting research.

For example, AT&T was recently contemplating a new product concept. It entailed the development of a personal information device that would provide customized news and information services. This product concept was market-tested with extensive primary research that ran into the hundreds of thousands of dollars. There was no explicit calculation of the costs and value of the research. Management felt that the company had to conduct the research before further extensive product development costs were incurred. Thus, the research was planned and conducted to answer the question of market acceptability.

This implicit value assessment appears to have become commonplace for two reasons. First, this approach is similar to what managers use in a wide variety of decision-making situations. They assess information from a wide variety of sources, then commit to the decision to do or not to do the research. Second, explicit methods involve considerable time, as well as uncertainty, in estimating the factors. On the other hand, it should be realized that the implicit approach is extremely prone to bias and misjudgment, based on the assumptions of the one doing the actual value assessment.

PROPOSAL DEVELOPMENT

The proposal is a formal means of communication between the requestor and the proposer of research. The formal written proposal is the culmination of manager–researcher interaction through which management concerns are converted into management problems and, ultimately, research problems. The proposal serves to define the scope of the investigation, the general analytical methods to be followed, and the anticipated results of the study. Both the manager and the researcher can derive substantial benefits from a well-written proposal.

From the manager's perspective, a well-written proposal can provide the following benefits.

Guarantees that the researcher understands the managerial problem. The document must clearly state the problem to be addressed and the results expected. If the researcher has misspecified or misunderstood the problem at hand, or is proposing research that will not give management the information it wants, the study can either be modified or stopped altogether, before valuable resources are spent on an investigation with little managerial value.

Acts as a control mechanism. When a research proposal is accepted, it establishes what the researcher will provide to management. In the case of outside research contractors, it serves as a contractual obligation. Therefore, management can use the proposal as a control mechanism, to make sure the information obtained from the study has been acquired by the method specified and is of the type promised.

Allows the manager to assess the proposed research methods. A preliminary assessment of the research methods can help ensure that the study will result in the information desired by the manager. Again, if the methods and techniques are inappropriate for management's purposes, the study can be modified before the resource expenditures are made.

Helps management judge the relative value and quality of the proposed investigation. There are really two areas where this is true. First, management may be forced by limited budgets to prioritize research projects addressing different problems. Proposals are typically important devices by which this is achieved. Second, an assessment of value and quality is also necessary when a particular research problem is going to be contracted to an outside research supplier. In this case, several suppliers may be bidding on the same project. The proposal is particularly useful to compare the relative merits of each supplier's bid.

Researchers also gain substantial benefits if their proposals are well thought out:

Ensures the researcher that the problem being investigated is the one management wants investigated. The sole justification for the research function in business is to provide information of value to managers for decision-making purposes. If the problem being investigated is not one management deems useful, then research is not performing its role and will cease to be a useful function from management's standpoint. The proposal clearly develops the research problem. If management does not perceive the problem to be of value, the proposal, in all likelihood, will be rejected.

Requires the researcher to think critically through the research process prior to the initiation of the study. One of the most common mistakes of inexperienced researchers is a failure to adequately think through the problem and the means to research it before the study is initiated. The research proposal forces critical thinking as to how the problem can be solved. It makes the investigator examine all stages in the research process to ensure that the study answers the problem in the most cost-effective way possible.

Provides the researcher with a plan of action. Thoroughly developed proposals are plans of action that outline the general purpose and strategy of the study. This can help focus the research on the stated problem.

States the agreement between the manager and researcher. Managers' expectations for research are often unrealistic. The proposal serves as a statement of what information will and will not be provided to management. Thus, the proposal should serve to reduce management dissatisfaction with the research function due to unrealistic expectations.

Given the substantial benefits that can be derived from the development of a well-thought-out proposal, managers must understand the types and structures of business-related proposals. For this reason, we now examine those issues.

A Typology of Business-Related Research Proposals

The diversity of research proposals is great.[5] No single type or format of research offering would be completely acceptable to all managers, let alone all researchers. Proposals vary in length and complexity, depending on the

FIGURE **5.2**

A Typology of Business-
Related Research
Proposals

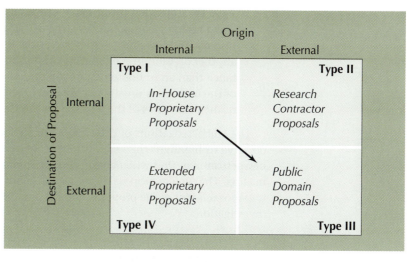

topic, the initiator of the research, and the audience. One useful way to study proposals is through the typology presented in Figure 5.2.

This typology classifies business-related research proposals into four major categories, based on origin (internal or external to the firm) and destination (internal or external to the firm). Proposal origin involves who is going to be the primary evaluator and recipient of the research output. The four types of offerings can be characterized as follows.

Type I: In-house proprietary proposals. These proposals originate inside the firm and are destined for use by management in the organization. Such proposals are usually the briefest and least complex of all the types. A short memo may be all that is required to initiate a research project in-house.

Type II: Research contractor proposals. A decision maker in an organization requests bids on a particular research project, or else an unsolicited proposal is received. In either case, the research bid originates external to the firm, but the results are to be used internally. These offerings are more detailed and longer than Type I. Type II and Type IV proposals are midrange: They strike a balance between the terse Type I and the extremely detailed Type III proposals. The arrow in Figure 5.2 indicates this increasing complexity.

Type III: Public domain proposals. These proposals are usually the longest and most detailed. A study that may be relevant to the organization is commissioned by an agency of the federal government with some outside research supplier (private or otherwise). Such proposals, and the results of the studies, are in the public domain. These offerings are often highly structured and may run hundreds of pages.

Type IV: Extended proprietary proposals. This type of proposal is also a midrange proposal. Here, the decision maker in the organization may find a request for research from a company affiliate (such as another division in the same organization). This offering is more detailed than an in-house proprietary proposal, because the requestor of the research is usually not as familiar with either the capabilities or the procedures of the research unit in the host organization.

The characterizations of these types are obviously generalities, but they do serve to highlight the wide variety of research proposals that the businessperson may find when attempting to obtain information for decision making. The type of proposal developed depends on the source of the offering, the recipient of the proposal, and the complexity of the problem under investigation.

Structure of a Proposal

The structure of the business research proposal also varies considerably.[6] Some organizations (particularly governmental agencies) demand highly structured and detailed proposals, with little latitude for variation. Other offerings are rather informal and may contain only the rudimentary elements necessary to be called a proposal.

The structure is closely related to the type of proposal being written. In-house proprietary offerings usually contain only minimal statements of the problem to be investigated and the benefits to be accrued, a brief presentation of the research strategy, and an estimate of the cost of the undertaking. On the other hand, public domain proposals are almost unbelievably detailed. These documents often require extensive literature reviews, letters of reference, line item budgets, and a meticulously stated research strategy.

Which is the best approach? Well, it all depends. Clearly the proposer must write at least to the minimum specifications of the requestor of the research. Beyond the minimum specifications, the structure should be determined by the type of study undertaken and the needs of the researcher. Ideally, the proposal should have enough detail to convey the essence of the problem, the research objectives, the general strategy and rationale to solve the problem, the results expected, and a statement of cost and time considerations.

Realistically speaking, however, most research proposals to business organizations are midrange proposals of intermediate complexity and detail. Exhibit 5.4 outlines the major structural components of a typical midrange proposal. Of course, the exact format of the proposal will vary depending on the situation, but this figure does summarize key categories of information that should be addressed in any midrange proposal. Fortunately, today there are many Web sites that provide information on

EXHIBIT **5.4**

Letter of transmittal. The letter of transmittal typically states the nature of the proposal and the main reasons the study should be conducted. The proposal writer usually invites any inquiries concerning additional information needed by the reader.

Title. The title should be short, descriptive, and imaginative.

Background. The background should briefly present the circumstances surrounding the immediate problem situation. In addition, the events leading to the development of the proposal should be reviewed. In some cases, depending on the type of proposal, the literature relating to the situation is presented here.

Statement of problem and research objectives. No matter what type of proposal is being prepared, the problem statement and the associated research objectives must be clearly stated.

Research strategy and methodology. This section presents the research methods and procedures to be used to accomplish the research objectives. Specifically, the steps to be taken to achieve the desired results are described in a logical format.

Nature of the final report. The proposal specifies the anticipated results of the study and the means by which they will be communicated to the individuals desiring the research.

Budget and schedule. The budget and the schedule for completing the research must accompany the proposal.

Background of consultants. If the proposal originated outside the organization requesting the information (Type II and Type IV proposals), the qualifications of the researchers should accompany the proposal (typically through a résumé).

the structuring and writing of research proposals. Exhibit 5.5 presents a sampling of these services.

A Sample Research Proposal

EXHIBIT **5.5**

To illustrate the key structural elements and format of a business research offering, an actual proposal for a feasibility study for a large land development is presented in Exhibit 5.6. The proposal is a midrange Type II proposal. It should be regarded as a sample rather than a definitive outline for all proposals.

http://www.proposalworks.com/	This Web site provides a wealth of information on proposal writing. The searchable proposal library contains many useful articles.
http://www.fedmarket.com/	A source specializing in writing proposals to the government. Provides tools and links to proposalworks.com for writing assistance.
http://www.proposalkit.com/	Provides information on Proposal Kit, Proposal Pack, and Proposal Pack Wizard, which are a software suite of business proposal and contract tools.
http://www.santcorp.com/	Also a provider of software tools and services to aid in the development of proposals. Training is also available.

EXHIBIT **5.6**

An Example of a
Midrange Type II
Research Proposal

E.S. & D. RESEARCH ASSOCIATES

Post Office Box 1169   Winter Park, Florida 32730

July 10, 2005

Mr. William Cate, General Manager
Project 530
P. O. Box 2611
Orlando, Florida 32816

Dear Mr. Cate:

Enclosed is our proposal for a feasibility study for Project 530. Although the proposal is general in nature at this time, we feel that it provides the detail necessary to assess the scope of activity as well as expected final output. Upon tentative authorization for such a study, we would be happy to provide a more detailed project plan for your approval.

The project will be conducted by professionals using the latest methodology and arriving at an objective series of recommendations for best utilization of the plant. In addition, we would reimburse any difference between projected and actual direct costs. You will note we have made no charges contingent upon percentage of development costs, and so on. In this way, we feel an objective and detached assessment can be made.

The proposed feasibility study would provide a comprehensive analysis of Project 530. It would satisfactorily serve as a document for investors as well as make recommendations for best utilization of the plat.

After you have had the opportunity to review the proposal, we will be glad to answer any specific questions or provide additional information.

Sincerely,

E.S. & D. Research Associates

A Proposal for a Feasibility Study for Project 530

Background

Central Florida has been one of the leading growth areas in the United States in the last three years. All indications are that the rate of growth will continue into the next decade, with the Orlando CMSA serving as the focal point for further development. Construction commitments approximating $1.75 billion ensure continued expansion. However, there are major factors that could have a negative effect upon continued economic progress. Among these are possible energy shortfalls, environmental concerns, and national economic conditions.

Because of previous overexpansion, combined with the oil embargo and an economic recession, many developments went into default during the 1973–1975 period. This previous experience has resulted in a more cautious optimism on the part of both developers and financial investors. This is reflected in the requirement

EXHIBIT **5.6**

Continued

for thoroughly researched feasibility studies by potential investors. In addition, with a rapidly changing growth situation, it is imperative that proposed developments reflect and take into account all factors that will have an impact upon the financial success of the project.

The size and scope of Project 530 is of such magnitude and diversity that an assessment of the feasibility of both the overall concept as well as the individual subcomponents must be carefully made. The proposed research would investigate and make recommendations addressing both the overall project, as now envisioned by the owners, as well as the feasibility of the various component subunits.

STATEMENT OF PROBLEM AND RESEARCH OBJECTIVES

The major research problem to be addressed is stated as follows:

Is Project 530 economically feasible as currently proposed? The specific research objectives of the study are

1. To assess the desirability of the proposed location
2. To develop a socioeconomic profile of the anticipated trading area for the various components of Project 530
3. To estimate the market potential for each of the major components of Project 530
4. To present recommendations as to the best use of the plat of land that is currently held for Project 530

RESEARCH STRATEGY AND METHODOLOGY

Major Areas of Investigation. Included in the proposed research are several major components. In addition to the overall socioeconomic factors pertinent to Project 530, the several subcomponents of the development will be analyzed as to their feasibility. Included would be

Site and access. An analysis of the site and access will be made, considering form qualities, transportation, infrastructure, competitive location, present (or proposed) barriers or conditions impinging on the proposed site.

Socioeconomic profile. A socioeconomic profile of the trade area will be presented. Included would be population projections, economic growth projections, competitive overview, housing inventory, and additional factors as described in the following pages.

Market potential for Project 530. Using secondary and primary data, the potential for major components of the project will be determined. Included among these would be townhouses, high-rise condominiums, an international hotel project, and shopping areas. The viability of each component, as well as that of the overall project, will be arrived at using market structure analysis as well as other commonly accepted market assessment techniques.

Specific target audiences would be identified for each component of the project. Identification would be based upon secondary as well as primary information.

Recommendations as to best use. On the basis of research results, recommendations as to best use of the property will be made. It is possible that the proposed project as presently envisioned will need to be modified as a result of the research findings.

Methodology. Three areas are of particular concern to the proper execution of the proposed study: the collection of secondary data, the collection of primary

(Continued)

EXHIBIT **5.6**

Continued

data, and the data analysis techniques that will be used in arriving at the final conclusions. Each of these concerns is briefly outlined so as to provide the client with an appreciation for the thoroughness of the proposed study.

Collection of Secondary Data. No research project should be undertaken without a thorough search of secondary data sources. These sources are valuable for this project for the following reasons:

1. They will solve some aspects of the project without having to engage in expensive primary research.
2. Secondary research costs are substantially lower than primary collection costs.
3. Obtaining secondary research data is considerably quicker than conducting primary research, thereby reducing the overall time frame for the project.
4. Obtaining secondary research data will help plan collection of primary data where the secondary sources are inadequate.

Therefore, in order to accomplish the stated objectives of the proposed study at minimum cost, an extensive secondary data search would be conducted.

The major types of information that will be secured from these secondary sources will include, but not be limited to, such items as:

Employment status and occupation	Rent (contract)
Apartments, units in structure	Second home
Components of population change	Population
Size of urbanized area of residence	Type of residence
Place of work	Value of residence
Income of family	Retail structure
Family size	Site and access
Industry	Occupancy/vacancy status
Means of transportation	

These data will further be analyzed as to the competitive situation and the overall potential for further economic development

Collection of Primary Data. After an extensive search of secondary information sources, primary data collection will begin. The primary data needed will include information about the needs and wants of potential markets for each of the successive stages of development for Project 530. The availability and usefulness of particular types of information in secondary sources will help dictate the necessary data to be collected by this means.

Primary data collection from each of the target markets for each stage of development will be designed to procure the data in the most effective and efficient manner possible. Mail surveys will be used to obtain information from specific target markets (i.e., professionals, potential lessees, and so on), to be isolated after an examination of the critical indicators in the Central Florida market. In addition, data will be collected through telephone and personal interviews with (a) key state and local planners and (b) decision makers who are familiar with the international hotel concept.

All questionnaires developed will use the latest psychometric scaling techniques to ensure the validity and reliability of the data collected. Specific attention will be given in the design stage to content and construct validity. Standard attitude scaling methods will be used whenever necessary to provide the best estimates of consumer preferences. All data will be thoroughly examined by the researchers to ensure its accuracy.

The sampling methods used will be those commonly accepted in contemporary research methodology. The samples selected will be clearly outlined and will provide

EXHIBIT 5.6

Continued

the client with estimates of the population parameters of interest. All procedures will be presented in detail in the final report for inspection by interested parties.

Data Analysis Techniques. Once the necessary data are collected, the data will be analyzed and summarized in a readable and easily interpretable form. The Statistical Package for the Social Sciences (SPSS®) will be used to summarize the data where needed. Here, it is anticipated that such statistical procedures as cross-tabs, analysis of variance, and some basic multivariate statistics would be used to analyze the data. Other statistical programs would be developed and used as the situation dictates.

All statistical manipulations of the data will follow commonly accepted research practices. The form of data presentation from these procedures would again be presented in an easily interpreted format. All statistical procedures will be performed by the computer to ensure accuracy and to minimize costs to the client.

NATURE OF THE FINAL REPORT

In preparing the final report, the consulting team will take into consideration the backgrounds and interests of the targeted audiences—financial institutions and investors. The resulting written presentation of findings will consist of a composite report serving segments of interest to the client. The technical aspects and basic findings would be presented in a simple and uncomplicated manner.

Furthermore, because the proposed study would take a significant amount of time and money, the consulting team would make an oral presentation of the major findings to the appropriate investors in the project. This form of presentation would emphasize both the recommendations and importance of the findings to the various targeted audiences. In addition, this presentation would allow the client to raise questions of concern to the researchers.

The following outline provides a brief description of the major sections to be presented in the final report.

I. Introduction
This section will provide the reader with information on the factors precipitating the study. The introduction will provide enough information so that the audience of interest could understand (a) why the study was undertaken and (b) important background information. Once these factors are described, the specific goals and objectives of the project will be discussed.

II. Management Summary
A brief and concise management summary would then be presented. Here, the major recommendations and findings contained in the study will be presented in a nontechnical and easily readable form. This summary will highlight the study without the full technical descriptions of the methodology or the extensive tables and charts that are essential in a complete presentation of the findings.

III. Acknowledgment of Study's Limitations
The readers of the report will be made aware of the study's limitations. Statements on the study's limitations are intended to identify particular inherent weaknesses that might affect the findings contained in the report. The consultants believe in the credibility of their work and, therefore, will be candid in describing such limitations.

IV. Description of Methodology
Once the objectives are explicitly outlined, telling the readers what the research intends to accomplish, the methodology section will outline how the study was executed. A complete description of the primary and secondary sources of information used will be included in this portion of the report. The major purpose of this section will be to describe the techniques used to

(Continued)

EXHIBIT 5.6

Continued

obtain the data. Not only will these techniques be fully described, but their use will be defended.

Because a sampling procedure will be used in primary data collection, the universe from which the sample is drawn will be described. The methods used to select both the size and members of the sample will also be outlined. Enough information will be included to enable readers to evaluate the accuracy and representativeness of any data collected. Finally, the methodology section will also contain a description of the statistical techniques used to analyze the relevant data.

V. Presentation of Findings

This section will present the information collected in the study and will comprise the majority of the report. The presentation, manipulation, and interpretation of the data will be integrated into a logical whole. The section will outline the rationale for particular analyses, the data relevant to it, and the actual results, and will then indicate what the result means to the client.

The presentation of the data analyses should provide a maximum of detail without being cluttered. Therefore, relevant tables, charts, and figures will be interspersed into the text of the report. These graphic aids will aid the reader in understanding the report's contents and give special emphasis to salient points of coverage.

VI. Conclusions and Recommendations

This section of the report will summarize all the key data and specifically relate them to the study's objectives. The conclusions to be drawn will be succinct statements flowing from the analysis of the results. There will be an objective interpretation of what the data really indicate. The recommendations section will translate the conclusions into specific actions to be undertaken. Alternative courses of action, along with their anticipated outcomes, will be outlined to the client.

VII. Appendix

This section will contain all data that are germane to the study's findings but that would disrupt the logical flow of information if presented in the body of the text. Such items will include in-depth descriptions of the sample design, the sample components, various statistical information, lengthy tables, copies of the questionnaires used in the study, and other components critical to the complete presentation of findings.

BUDGET AND SCHEDULE

Cost. The total cost for the feasibility study described in this proposal is $80,800. The total budgeted breakdown of costs is presented in the following schedule.

Item	Cost
Direct costs (materials and supplies):	$ 5,000*
Computer time, sample lists, telephone, postage, printing, travel, maps, aerial photographs, questionnaires, publication costs, other supplies and materials	
Direct costs (other personnel)	5,000*
Research assistants, clerical, data analysis	
Consulting fee	70,800
59 workdays @ $1200/day	
Total cost	$80,800

*If direct costs are actually lower than stated, the difference will be refunded to the payor.

EXHIBIT **5.6**

Continued

Terms. Twenty-five percent will be due upon initiation of contract. An additional 25 percent will be due at the end of three months. Remainder to be paid upon delivery of final report.

Time Frame. The time frame necessary for completion of this study will be approximately six months from the starting date. Periodic reports (monthly) will be provided to the clients as to the progress being made on the study.

BACKGROUND OF CONSULTANTS

The consultants have had a broad base of consulting experience in both the public and private sectors. Individual vitas are attached for the client's inspection.

MANAGERIAL CONSIDERATIONS

Although management may believe its job to be largely finished after a proposal is developed, this belief is quite erroneous. Management must take an active role in the research endeavor to ensure that the study is conducted as planned and delivers the results desired. This role is described succinctly as follows:

> If research is to be fruitful, it must be a daily concern of top management, and if it is to succeed, those who conduct the research must have a solid understanding of the true needs of their company, of its industry, and of its customers. Unless the top management of a company and its research staff are in the most constant and intimate kind of communication, research will amount to little and can be a great drain on any company.[7]

This manifests itself in a very important management function—control. *Management control* is the regulation and coordination of business activities by managers. The business activity in this instance is the research function. Here, managers must regulate the research process to obtain the needed information for decision making. In essence, some means of communication must be developed and maintained between managers and researchers throughout the research undertaking.

Control also means that standards are set prior to the project's beginning and are periodically monitored to make sure they are being achieved. When management is actively involved in proposal development and its subsequent evaluation and acceptance, there is agreement among the parties as to the resources, methods, and output that a particular project will encompass. Fortunately, there are tools useful in this task: research proposal evaluation mechanisms and other types of control mechanisms, including Gantt charts, program evaluation and review techniques (PERT), and the budget itself.

The importance of these tools in the control process is indicated by the fact that many requests for proposals (RFPs) require that one or more of these items be attached to the submitted proposal.[8] Fortunately, these and other project management tools are readily available in standard software packages. The interested reader is strongly encouraged to become familiar with the use of project management software.[9] An overview of the basics of each of these control mechanisms is presented next.

Proposal Evaluation Mechanisms

All business research proposals are subjected, either implicitly or explicitly, to some sort of evaluative process during the course of their development.[10] Usually, proposals are submitted to middle or top management for review before approval is given to conduct the investigation. This review can be elementary, as in a cursory examination of the major components of the offering. Or a highly structured evaluation procedure can be used to assess the quality and value of the research bid to the users within the organization.

The degree of formality in the evaluation process is usually associated with the amount of structure requested in a proposal's development. More specifically, if a requestor of research demands a highly structured bid, the evaluation process is usually similarly formal. Therefore, in our typology of business-related proposals, we would expect to find the least rigorous evaluations conducted on in-house proprietary offerings and the most rigorous reviews on public domain proposals.

Informal evaluation procedures may consist of an evaluator reading a proposal and ranking it against others submitted for funding. More formal procedures usually require the identification and selection of evaluative criteria, weighting of these criteria, and the determination of some numerical value representing the score or grade the proposal receives. As an illustration, Figure 5.3 shows a sample evaluative framework developed to grade the Project 530 proposal in Exhibit 5.6 of this chapter.

Let us suppose you are the general manager for Project 530. You have recently put out for bid a feasibility study that is supposed to answer the research problem and achieve the objectives stated in the proposal in Exhibit 5.6. You receive a number of proposals, including the one prepared by E.S. & D. Research Associates. You are now faced with the task of evaluating the various bids and recommending a research contractor to the board of directors. What do you do?

A good first step would be to develop a list of criteria on which you want to evaluate each of the proposals. The criteria should be closely related to the factors that are considered important to the generation of timely and usable decision-making information. These factors were identified in Exhibit 5.4, which outlined the major structural components of a midrange business proposal. They are the factors that will help the evaluator determine the quality and relative value of a particular research offer.

Given the general components, you would now want to develop an expanded evaluative instrument to rate each proposal. Without getting into the details of instrument design, a possible evaluative framework is laid out in Figure 5.3. Essentially, this framework is constructed around five major areas that are important in determining the ultimate quality of the information received from the study.

The next step you would probably take is to read each proposal thoroughly, then rate each one on the criteria identified in the evaluative framework. For instance, after reading the E.S. & D. proposal, you may rate item 1a as only fair

Institution	Principal Investigator	Proposal Number
Proposal Title		

Evaluative Dimension	Check (√) where applicable					General comments and reasons for the rating given
	Weak	Poor	Fair	Good	Excellent	

1. Problem Development
  a. The background of the problem is fully developed.
  b. The conditions leading to a statement of the problem are specified.
  c. The proposal demonstrates the researcher(s) is familiar with the complexities of the problem.						

2. Research Strategy and Methods
  a. The research strategy to solve the problem is presented in a succinct fashion.
  b. A justification is presented for the strategy selected.
  c. The research design is appropriate.
  d. The sampling design is appropriate.
  e. The data collection procedures are appropriate.
  f. The data analysis proposed is appropriate.						

3. Expected Research Results
  a. The information to be received is outlined.
  b. The proposed output satisfied the research objectives.						

4. Budget and Schedule
  a. The time frame is realistic given the objectives of the study.
  b. The budget is appropriate given the objectives of the study.
  c. The value of the research is sufficient to justify the expenditures of corporate resources.						

5. Background of Researchers a. The qualifications and experience of the researcher(s) indicate he or she is capable of undertaking the study.						
6. Overall Evaluative Score						

Please provide any additional comments or remarks that are relevant to this proposal.

Reviewer's Name	Reviewer's Signature	Date

FIGURE 5.3

A Sample Evaluative Framework for a Business Research Proposal

because the background of the problem is not as fully developed as you thought it would be. Each successive item would be rated according to your judgment of how well the proposal met the criteria set forth.

Once all the items have been rated for each proposal, you would be faced with the overall evaluation. Although you may want to weight each component differently to reflect its relative importance to you, let us say each item is of equal value, for simplicity's sake. Scoring might go from 1 to 5 (for weak to excellent). The overall evaluative score would thus be found by adding up the individual item scores as indicated at the top of the evaluation instrument. For example, if item 1a received a fair evaluation, it would receive a score of 3. Similarly, if item 1b received a good evaluation, the score assigned would be a 4. You would then add up all the item scores to arrive at an overall evaluative score (ranging from 15 to 75). This rating could be compared to the other proposals' scores for selection purposes.

This relatively simple example gives you an idea of one possible method of proposal evaluation. The important ideas to get from this brief discussion of evaluation are as follows:

1. All proposals go through some type of evaluation process.
2. The type of evaluation is usually associated with the type of proposal requested.
3. The evaluator should develop a logical evaluation procedure for choosing among alternative proposals.
4. For all intents and purposes, the evaluation of the proposal sets the standards for controlling the research project.

With these issues in mind, we now move to other management control mechanisms in the area of proposal development and management.

Other Control Mechanisms

Gantt Charts. A Gantt chart is designed to provide management with a basic means of comparing planned to actual performance.[11] Henry Lawrence Gantt's creation is the forerunner of most performance review devices in existence today. The Gantt chart forces the researcher to logically organize and schedule the major activities of the research project. Figure 5.4 is a Gantt chart for "A Proposal for a Feasibility Study for Project 530." This tool is useful from a control perspective because management can easily assess whether the study is progressing as planned.

Program Evaluation and Review Techniques (PERT). PERT is a slightly more sophisticated scheduling and control device. This tool requires that the researcher identify all the major tasks and make estimates of the time required to finish each task. These activities are then arranged in a logical network to reflect the order in which they have to be completed. The wide acceptance of PERT in research endeavors is evident in the relevant literature.[12]

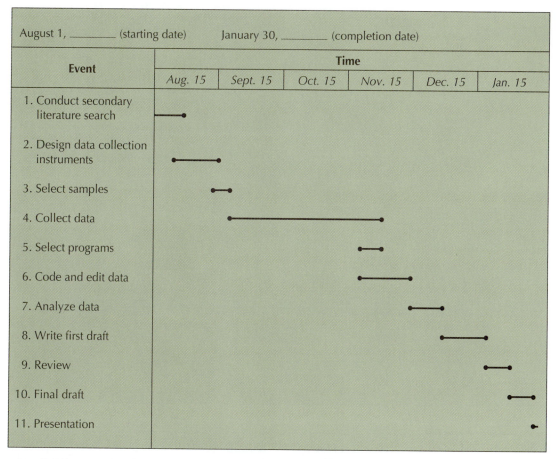

August 1, _____ (starting date)   January 30, _____ (completion date)

Event	Time					
	Aug. 15	Sept. 15	Oct. 15	Nov. 15	Dec. 15	Jan. 15
1. Conduct secondary literature search						
2. Design data collection instruments						
3. Select samples						
4. Collect data						
5. Select programs						
6. Code and edit data						
7. Analyze data						
8. Write first draft						
9. Review						
10. Final draft						
11. Presentation						

FIGURE 5.4

A Gantt Chart for Project 530 Proposal

It is neither possible nor within the scope of this text to review the details of PERT. However, an extremely simplified PERT chart is presented in Figure 5.5 for the Project 530 proposal. You are encouraged to do further reading in this area to develop your planning and scheduling skills.

Budgets. The budget is perhaps the single greatest control mechanism at management's disposal. Budgets determine the amount and allocation of resources to research projects and tasks. The process of budgeting and its implications for control are well discussed in other texts, so we will not elaborate here. Again, the reader is encouraged to do further study in this area.[13]

The importance of the involvement of management at this stage in the research process cannot be overstated. Several key managerial questions are presented in Exhibit 5.7. These questions are designed to stress the fact that managers cannot hope to get the information they desire unless they actively involve themselves in problem and proposal development.

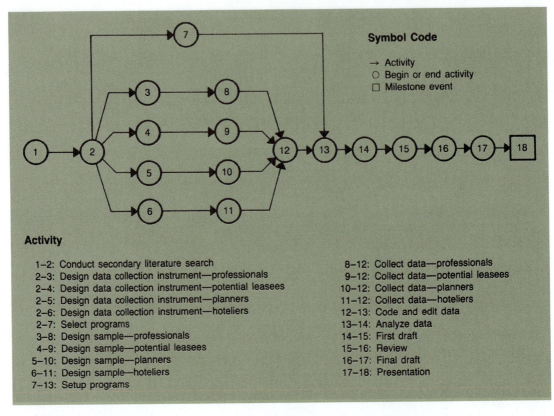

Symbol Code

→ Activity
○ Begin or end activity
□ Milestone event

Activity

1–2: Conduct secondary literature search
2–3: Design data collection instrument—professionals
2–4: Design data collection instrument—potential leasees
2–5: Design data collection instrument—planners
2–6: Design data collection instrument—hoteliers
2–7: Select programs
3–8: Design sample—professionals
4–9: Design sample—potential leasees
5–10: Design sample—planners
6–11: Design sample—hoteliers
7–13: Setup programs

8–12: Collect data—professionals
9–12: Collect data—potential leasees
10–12: Collect data—planners
11–12: Collect data—hoteliers
12–13: Code and edit data
13–14: Analyze data
14–15: First draft
15–16: Review
16–17: Final draft
17–18: Presentation

FIGURE **5.5**

A Simplified PERT Chart for Project 530 Proposal

It is also appropriate at this point to note that integrated software is making the development and control of the research process much easier. For example, using Microsoft® Office software, management has access to both proposal development tools (Word as the word processor and PowerPoint® for oral presentation slides) and management tools (Project for creating Gantt charts and planning networks and Excel as a spreadsheet-based budgeting and analysis package). We can expect continued improvements in integrated software.

EXHIBIT **5.7**

Key Managerial Questions Pertaining to Problem and Proposal Development and Evaluation

1. Is the problem identified the proper one? Is it misspecified?
2. Is the managerial concern being addressed correctly in the research problem statement? If it is not, what needs to be done to correct the situation?
3. Is the research problem of enough value to commit scarce resources to it? Are other problems more pressing?
4. Does the research proposal clearly state the problem, the expected results and benefits, and enough methodological detail to ensure that the research will produce the information desired by management?
5. Are the appropriate control mechanisms incorporated to ensure that the research goes as planned and obtains the information desired by management?

SUMMARY

The chapter begins by examining the topics of problem identification and formulation, an integral part of research proposal development. The problem development process can be viewed as a series of screening procedures that serve to refine managerial concerns and narrow the focus of management to a set of specific research topics that are of value and relevance to management. A set of diagnostic questions is presented to aid in this process.

Once researchable problems are targeted for investigation, the question of value is addressed. There are explicit and implicit means for value assessment. Implicit assessments are most common because (a) the process follows managers' customary decision-making approaches and (b) resource constraints and estimation problems are inherent in the explicit methods.

The discussion then moves to the structure and nature of business-related research proposals. The benefits of proposals to researchers and managers are identified. A typology of proposals is outlined, and a sample research proposal is presented. The chapter ends with a discussion of managerial considerations with regard to proposal development.

Once researchable problems are targeted for investigation, an assessment of their value must be made. In the appendix to this chapter, four methods of analyzing value are presented: Bayesian decision analysis (examined in greatest detail because of its thoroughness and theoretical appeal), the simple savings method, return on investment, and the present value method.

DISCUSSION QUESTIONS

1. What is a research proposal? What role does it play in the research process?

2. Why should a manager specify the development of well-thought-out proposals before a research project is undertaken? Similarly, why is it desirable from a researcher's standpoint to have developed a proposal before initiating any study?

3. Problem identification and definition has been said to be the most important step in the business research process. Discuss the validity of this statement.

4. Problem development is a very difficult task. Discuss this statement in the context of business research and give specific examples to support your points.

5. What is the research problem development process? Discuss the major screens in this process and provide an example to illustrate your discussion.

6. Not all management problems are translated into research problems. Discuss the reasons for this fact.

7. What are the major methods used to assess the value of research information? Compare and contrast these methods, indicating the strengths and weaknesses of each valuation method.

8. What can you do as a manager to aid the identification of managerial concerns that will be of researchable value for the firm?

9. What are some of the major factors that affect the conversion of management concerns to management problems? Are there any mitigating circumstances that would affect the importance of these factors in different industries?

10. Identify the major structural components of a midrange business research proposal. Discuss why these components are necessary from a managerial standpoint.

11. If you were a manager in charge of evaluating research proposals, what information would you require for evaluation purposes? State all types of information required and justify their inclusion to your boss.

12. What does managerial control have to do with the research function? Why is it that managerial control is essential in the conduct of good business research?

13. What are some of the major tools that can be used by managers in the control of the research function? Discuss the relative merits of these tools with regard to management's control over the resources of the firm.

14. Why should managers become involved in the research proposal stage of the research process? Identify some consequences if the manager is not involved at this stage.

15. Using whatever online tools you choose, identify some of the major software programs that can be helpful in problem and proposal development and management. Be sure to discuss the sources used to find this software and be prepared to outline how each software program would be helpful in the proposal development process.

NOTES

1. *Problem Definition,* Marketing Research Techniques, Series No. 2 (Chicago: American Marketing Association, 1958), p. 5.

2. For other significant discussions of this topic, see John W. Buckley, Marlene H. Buckley, and Hung-Fu Chiang, *Research Methodology & Business Decisions* (New York: National Association of Accountants, 1979), pp. 13–19; Russell L. Ackoff, *Scientific Method: Optimizing Applied Research Decisions* (New York: Wiley, 1962), pp. 67–76; and Donald R. Cooper and Pamela S. Schindler, *Business Research Methods* (New York: Irwin/McGraw Hill, 1998), pp. 58–68.

3. See Forrest S. Carter, "Decision Structuring to Reduce Management–Research Conflicts," *MSU Business Topics* (Spring 1981), pp. 40–46; Robert G.

Murdick and Donald R. Cooper, *Business Research: Concepts and Guides* (Columbus, OH: Grid Publishing, 1982), pp. 90–102; and Donald R. Cooper and C. William Emory, *Business Research Methods* (Chicago: Irwin, 1995), pp. 55–59, for good discussions of the problem formulation process.

4. For a good discussion of the status of basic research in the corporate environment, see "Bell Labs: Threatened Star of U.S. Research," *Business Week* (5 July 1982), pp. 46–52.

5. For another discussion of the types of proposals, see Murdick and Cooper, op. cit. note 3, pp. 105–126.

6. For examples of structure, see R. Moyer, E. Stevens, and R. Switzer, *The Research and Report Handbook: For Business, Industry, and Government* (New York: Wiley, 1981), pp. 73–77; and Lawrence F. Locke, Waneen Wyrick Spiduso, and Stephen J. Silverman, *Proposals That Work: A Guide to Planning Dissertations and Grant Proposals*, 4th ed. (Thousand Oaks, CA: Sage, 2000).

7. J. L. Miller, "The Real Challenge of Research," *Business Horizons*, vol. 1 (Winter 1958), p. 59.

8. RFP (request for proposal) is an acronym used largely by federal government agencies and departments. An RFP is basically a formalized request for bids and proposals to perform some service for the agency.

9. Two project management software packages that have received wide acceptance are Microsoft Project and Harvard Project Manager. The interested reader is encouraged to review these and other packages to judge their versatility and capability.

10. For some examples of evaluative frameworks for research proposals, see David R. Krathwohl, *How to Prepare a Research Proposal*, 3rd ed. (Syracuse, NY: Syracuse University Press, 1988), pp. 1–5; Stephen Isaac and William B. Michael, *Handbook in Research and Evaluation* (Editions Pub, 1995); Murdick and Cooper, op. cit. note 3, pp. 125–126; and Louis P. Plebani, Jr., and Hemant K. Jain, "Evaluating Research Proposals with Group Techniques," *Research Management* (November 1981), 24:6, pp. 34–38.

11. For a discussion of Gantt charts, see Project Management Institute, *A Guide to the Project Management Body of Knowledge (PMBOK GUIDE)—2000 Edition*, (Project Management Institute, 2000) and Harold Kerzner, *Project Management: A Systems Approach to Planning, Scheduling, and Controlling*, 8th ed. (John Wiley & Sons, 2003).

12. For more complete discussions of PERT networks, see Wayne L. Winston, *Operations Research: Applications and Algorithms*, 4th ed. (Belmont, CA: Duxbury Press, 2004); and J. Wiest and F. Levy, *A Management Guide to PERT/CPM* (Englewood Cliffs, NJ: Prentice-Hall, 2006).

13. See Charles T. Horngren, George Foster, and Srikant Datar, *Cost Accounting: A Managerial Emphasis*, 11th ed. (Englewood Cliffs, NJ: Prentice-Hall, 2002); and Robert S. Kaplan, *Advanced Management Accounting*, 3rd ed. (Englewood Cliffs, NJ: Prentice-Hall, 1998).

SUGGESTED READING

Problem Identification and Formulation

Cooper, Donald R., and Pamela Schindler, *Business Research Methods*, 8th ed. (New York: McGraw-Hill, 2003).

Chapter 3 presents a useful discussion of the problem identification and formulation process. Useful as a supplement to the process proposed in this text.

Punch, Keith F., *Developing Effective Research Proposals* (Thousand Oaks, CA: Sage Publications, 2000).

Provides a useful discussion of research questions and problems. The book is useful in the larger regard in terms of overall proposal development.

Proposal Development

Johnson-Sheehan, Richard, and Sam Dragga, *Writing Proposals*, (Pearson Longman, 2001).

This book offers a systematic approach to writing proposals. It also offers Internet strategies for aiding in the proposal development process.

Pfeiffer, William Sanborn, and Charles H. Keller, *Proposal Writing: The Art of Friendly and Winning Persuasion* (Upper Saddle River, NJ: Prentice-Hall, 1999).

This book provides readers with an overview of proposal writing with a discussion of the issues that influence the development of effective proposals. A good overview of the proposal writing field.

Porter-Roth, Bud, *Proposal Development: How to Respond & Win the Bid*, 3rd ed. (Central Point, OR; PSI Research-Oasis Press, 1998).

The purpose of this book is to provide a practical guide to writing and winning proposals. Useful information can be gleaned in the proposal development area.

METHODS OF ASSESSING THE VALUE OF RESEARCH INFORMATION

A number of explicit methods to assess the value of research are readily available to the manager.[1] These techniques all rely on management estimates of factors such as profits, costs, and probabilities. Therefore, the projections of value produced by the techniques depend on the accuracy of the data management provides for the analysis. This may seem like a severe limitation of the tool's usefulness, but it is the best we have available at this time. At the least, the methods of valuation force managers to make explicit their assumptions in a particular decision-making setting, and they help managers in making decisions that are not consistent with their own judgments and experience.

The major methods of valuation are Bayesian decision analysis, the simple savings method, return on investment, and the present value method. A brief presentation of each of these methods follows.

Bayesian Decision Analysis

Bayesian decision analysis is perhaps the best means of determining the value of information before initiating a research project.[2] Bayesian analysis provides the manager with a formal way to evaluate research as an information-supplying and cost-incurring activity.[3] The output of the procedure is a dollar value, representing the maximum amount the manager would be willing to pay for research to reduce uncertainty in the decision-making situation. This dollar amount is derived from an assessment of the dollars to be saved or gained from making a correct decision. To illustrate this approach to information valuation, let us look at a simple decision example in the financial setting.

Suppose the financial manager for Century Drill & Tool Company is contemplating an investment of corporate funds into a highly speculative business. The manager must make a decision to invest or not invest in the next two months. A study designed to determine the success of the business endeavor is being considered.

The manager believes that the condition of the economy in the coming year will largely determine the business's success. The manager therefore reviews the available economic forecasts and estimates that there is a 40% chance that the economy will prosper (growth rate greater than or equal to 10%) and a 60% chance that the economy will falter (growth rate less than 10%). Using these judgments, the financial manager calculates the estimated profitability of investing in the business if either economic consequence occurs. The payoff for not investing under either condition is simply zero because no investment was made to obtain a return. This type of problem—where we have only two alternatives with one of the actions having zero payoffs—is known as a *venture analysis* problem.[4]

TABLE 5A.1 Conditional Payoff Matrix for Two Alternative Investment Strategies

| | States of Nature (S_i) | | | |
| | S_1 = Optimistic Economic Outlook ($\geq$10%) | | S_2 = Pessimistic Economic Outlook (<10%) | |
Alternatives	$p(S_1)$	Payoff	$p(S_2)$	Payoff
A_1 = Invest	0.4	$600,000	0.6	−$250,000
A_2 = Do not invest	0.4	0	0.6	0

This information is summarized in Table 5A.1. Such a table is called a *conditional payoff matrix* because it shows the anticipated payoffs with each alternative, given a particular economic consequence. In this table, the economic forecasts are designated states of nature (S_1 and S_2). They reflect the uncontrollable, or uncertain, aspect of the decision-making environment. The alternatives are simply the choices the manager has to make in this decision situation: invest or do not invest. The probabilities ($p(S_1)$ and $p(S_2)$) are simply management's estimates of the probable occurrence of each economic situation in the coming year. For instance, if the manager chooses A_1 (invest in the business), and S_1 (economic growth of 10% or more) occurs (there is a 40% chance of this happening), then Century Drill stands to make $600,000 in profit.

Now, further suppose that the study being considered to determine the success of the proposed business endeavor could give us perfect information. In other words, we would know precisely if S_1 or S_2 was going to occur, and we would make our decision appropriately. In this case, if we knew S_1 was going to occur, we would choose A_1, the investment option, because we would get a payoff of $600,000 (as opposed to $0). Alternatively, if we knew S_2 was going to occur, we would choose A_2 and not invest in the project ($0 is greater than −$250,000). Now given these estimates, we can calculate the *expected monetary value* (EMV) of the business venture with perfect information, as follows:

EMV of the business venture

with perfect information = 0.4($600,000) + 0.6(0)

Expected value of decision under certainty = $240,000

Similarly, the EMV of the business venture with the currently available information can be calculated as follows:

EMV of the business venture

with present information[5] = 0.4($600,000) + 0.6(−$250,000)

= $240,000 − $150,000

Expected value of optimal action = $90,000

Therefore, the EMV of perfect information (EMVPI) can simply be defined as the EMV of the business venture with perfect information minus the EMV of the business venture with present information, or

EMVPI = Expected value of decision under certainty
          – expected value of optimal action

EMVPI = $240,000 – $90,000
       = $150,000

In other words, $150,000 is the maximum dollar amount an expected value decision maker would be willing to pay for perfect information.[6] In the case of our financial manager, this dollar figure sets an upper limit on the cost of the research to obtain additional information for decision-making purposes. If the proposed cost of the research is $50,000, our manager should not eliminate the possibility of doing the research. Instead, the manager should assess the predictive accuracy of the proposed study and then calculate the EMV of imperfect information (EMVII). EMVII takes into account the fact that errors are inherent in any study, so the value of information is less than EMVPI. The problem now becomes the assessment of the dollar value of such errors, so that they can be subtracted from EMVPI to give us a realistic estimate of EMVII.

From basic statistics, we find that two basic types of errors are possible in our simple scenario: Type I and Type II errors. If the research we commissioned indicated that S_1 would occur (we will designate this by R_1), but S_2 turned out to be the true state of affairs, this would be a Type I error. We will symbolically represent the conditional probability of this error as $p(R_1/S_2)$ and call this our Type I error. Type I errors are usually designated as α; therefore, $\alpha = p(R_1/S_2)$.

The other error possible is a Type II error. A Type II error, in our example, occurs when the research indicates that S_2 would occur (we will designate this as R_2), but S_1 turned out to be the true state of affairs. Our conditional probability designation for a Type II error is then $p(R_2/S_1)$. Because β is usually the symbol for a Type II error, we write $\beta = p(R_2/S_1)$.

Given this framework, we can now calculate the expected costs of error. The expected cost of making each type of error can be defined as[7]

Expected cost of Type I error = $\alpha \times p(S_2) \times V_2$

Expected cost of Type II error = $\beta \times p(S_1) \times V_1$

where V_2 and V_1 are the absolute values of the expected payoffs for the investment alternative under S_2 and S_1. Now let us suppose our financial manager believes the research that can be purchased for $50,000 will possess a predictive accuracy of around 75%. This means that there is a 25% chance that the research will give the manager a wrong answer, or

$$p(R_1/S_2) = p(R_2/S_1) = 0.25$$
$$\alpha = \beta = 0.25$$

Therefore, the

Expected costs of a Type I error $= \alpha \times p(S_2) \times |V_2|$

$$= (0.25) \times (0.6) \times |-250,000|$$
$$= \$37,500$$

and

Expected costs of a Type II error $= \beta \times p(S_1) \times |V_1|$

$$= (0.25) \times (0.4) \times |600,000|$$
$$= \$60,000$$

With this information, we now have all the components needed to calculate the EMVII. As stated earlier, EMVII is simply EMVPI minus the cost of the associated errors found in the study. Specifically,

$$\text{EMVII} = \text{EMVPI} - \alpha p(S_2)|V_2| - \beta p(S_1)|V_1|$$
$$= \$150,000 - \$37,500 - \$60,000$$
$$= \$52,500$$

Remembering that the cost of the research was $50,000, the value of the research is only slightly higher ($2500) than its cost. The manager is put in a situation of deciding whether the research should be purchased given the time constraints inherent in the decision and other factors that may not be quantifiable. In the cases where the EMVII is substantially greater (or less) than the cost of the research, the decision to accept (or reject) the research project is much easier for the decision maker.

This example of Bayesian decision analysis is quite simple, but the logic of the procedure holds for more complex decision situations. The primary advantage of Bayesian analysis is that it forces managers to make explicit their assumptions about the outcomes of their actions and subsequently provides them with a realistic value for the information, given their subjective judgments.

Simple Savings Method

As its name suggests, this method of information evaluation is simple. Here, management is required to make judgments about the relative costs of making a wrong decision under various alternatives. For example, our financial manager thought that if a wrong decision was made to invest in the business venture, the firm could stand to lose $250,000. Furthermore, let us suppose the manager believes the following information to be true:

	Chance of a Correct Decision	Estimated Cost of a Mistake
Use known facts.	60%	$0.4 \times \$250,000 = \$100,000$
Do research on the investment.	80%	$0.2 \times \$250,000 = \$50,000$

If this is the case, the manager should be willing to pay up to $50,000 ($100,000 – $50,000) to obtain research to increase the chances of making a correct decision.

This method is extremely basic and does not provide as much information as the Bayesian method does, but it is very simple and presents a first approximation of the value of proposed research.

Return on Investment

The return on investment (ROI) approach views research as an investment that brings some return to the firm. Oscar Mayer & Company has used this method to evaluate its research undertakings in the past.[8] The approach uses the following formulation in calculating research's ROI.

$$\text{Return on investment} = \frac{\text{worth of findings} \times \text{error factor}}{\text{cost of research}}$$

where

Worth of findings = the dollar contribution resulting from research project undertaken

Error factor = management's estimate of the percentage of wrong decisions made without research

Annual research budget = the amount spent on research to obtain the dollar contribution above

For example, our finance manager is again faced with the question of whether or not to commission a research project. Here, the manager must once more make some judgments as to the relative worth of the project and the "hit" rate on past management decisions in this area. Suppose our manager feels that 60% of the decisions made without research in this decision area were correct in the past. Furthermore, the $50,000 research project is expected to generate an additional $500,000 worth of contribution. The ROI for the project is as follows:

$$\text{Return on investment} = \frac{\$500,000 \times (1 - 0.6)}{\$50,000}$$

$$= 400\%$$

In this case, the research project appears to be a worthwhile investment.

Present Value Method

The present value approach to valuation also treats research expenditures as an investment. The method is an intuitively more pleasing formulation than either the return on investment method or the simple savings method. This technique discounts the incremental cash flows by the marginal cost of capital over the lifetime of the research project. The model is defined as[9]

$$\underset{\text{Current}}{(R_0 - C_0)} + \underset{\text{First year}}{\frac{(R_1 - C_1)}{1 + K}} + \underset{\text{Second year}}{\frac{(R_2 - C_2)}{(1 + K)^2}} + \cdots + \underset{\text{Final year}}{\frac{(R_n - C_n)}{(1 + K)^n}}$$

where

$R_0 \ldots R_n$ are annual cash flows attributable to the research

$C_0 \ldots C_n$ are the annual cash expenditures for the research

K is the firm's marginal cost of capital

This method of evaluation requires realistic assessments of continuous cash flows and expenditures directly attributable to the research program.

For example, suppose our financial manager is interested in assessing the value of a research project that is designed to give a go or no-go decision on a potential new product for the firm. Further suppose that management estimates (a) the firm's marginal cost of capital to be 15%; (b) the cost of the research project to be $15,000, all incurred in the first year; and (c) the net annual incremental cash inflows over the estimated three-year life of the product to be $20,000, $30,000, and $15,000, respectively. With this information, the present value of the research expenditure can be calculated as follows:

$$(\$20,000 - \$15,000) + \frac{(\$30,000 - 0)}{1 + 0.15} + \frac{(\$15,000 - 0)}{(1 + 0.15)^2}$$

$$= \$5000 + \$26,087 + \$11,342 = \underline{\$42,429}$$

In other words, over the life of the project, the valuation of the firm will be increased by $42,429 using a required rate of return of 15%.

NOTES

1. The following section draws heavily from James H. Myers and A. Coskun Samli, "Management Control of Marketing Research," *Journal of Marketing Research* (August 1969), 6, pp. 267–277.

2. The Bayesian discussion is adapted from Donald S. Tull and Del I. Hawkins, *Marketing Research: Meaning, Measurement and Method*, 6th ed. (New York: Macmillan, 1993), pp. 817–827.

3. For a more thorough discussion of Bayesian decision theory and its applications, see Paul E. Green and Donald S. Tull, *Research for Marketing Decisions*, 4th ed. (Englewood Cliffs, NJ: Prentice-Hall, 1978), pp. 34–62.

4. Tull and Hawkins, op. cit. note 2, Appendix A. *Note:* More complicated venture analysis problems are possible.

5. Another way of approaching these calculations is to use the absolute value of the payoff A_1S_1, $|-\$250,000|$, which reflects the fact that a rational manager would not spend more than what the expected loss would be, given no additional information. Then the EMV of the business venture with present information is $0.4(\$600,000) - 0.6 |-\$250,000| = \$90,000$.

6. There are other types of decision makers. See Green and Tull, op. cit. note 3, pp. 26–31.

7. Tull and Hawkins, op. cit. note 2, Appendix A, "Expected Value Analysis."

8. See Dik Warren Twedt, "What Is the 'Return on Investment' in Marketing Research," *Journal of Marketing* (January 1966), 30, pp. 62–63; and Steven M. Heyl, "Decision Matrix Points the Way to Better Research ROI," *Marketing News* (15 September 1997), pp. 18 and 30.

9. Myers and Samli, op. cit. note 1, p. 269.

6

Fundamentals
of Research Design

Overview

The Nature of Research Design

Error Reduction Through Design

Potential Sources of Error in the
Research Design Process

Managerial Strategies for Dealing
with Error

Major Types of Designs

Ex Post Facto Designs

Experimental Designs

Specific Design Configurations

Validity Concerns

Specific Designs

Online Research Design Issues

Managerial Considerations

Summary

Discussion Questions

Notes

Suggested Reading

**Appendix: Advanced
Experimental Designs**

OVERVIEW

The previous chapter outlined the general structure and purpose of the research proposal. There it was stated that one of the salient components of a proposal is the specification of the research design to solve the particular problem at hand. Furthermore, *research design* was defined as the structure of the research project to solve a particular problem. Users of research must be concerned with these issues because an inadequate design will yield information that is incapable of answering the research question under study. Therefore, this chapter focuses on the fundamentals of research design.

The chapter begins with an expanded view of the nature and purpose of design. From the standpoint of the manager, research design is the means by which valuable information is obtained to make decisions. From the perspective of the researcher, it is the structure of the research project (used as a plan to minimize errors inherent in the research process) that ensures that the appropriate data is collected for decision-making purposes.

After the general nature of design is explained, the problem of error reduction is explored in more depth. The major potential sources of error are identified and illustrated using a pertinent business research example. Then managerial strategies for dealing with error are presented.

The discussion then moves to the two main types of designs that are available to the researcher. These types, ex post facto and experimental, are distinguished from each other by the relative degree of control of the important variables in the study. This is noteworthy because the control of variables affects the kinds of conclusions that can be drawn from the investigation.

Once the designs are presented, the discussion moves to online research design issues. Here the impact and role of the Internet on the research design process is explored.

The chapter ends with the major considerations in design that management should review. Specifically, several principles of design are expounded, and questions are posed for the interested manager.

THE NATURE OF RESEARCH DESIGN

Research design can be thought of as the road map for researchers. It is the means by which investigators plan the collection of data to answer a pertinent research question. Like travelers who plot their trips in advance, good researchers outline their goals and the means to achieve them in the research design. Research design can be thought of as an active process leading to the specification of the structure to solve a particular research question. As mentioned in the previous chapter, the formal specification of research design is an integral part of the research proposal.

Simply stated, the primary purpose of research design is to guide researchers in their quest to solve problems under study. The design process can be thought of as a series of decisions: what concepts will be studied, how these concepts will be measured, what approach will be used to study the problem, who will be studied, how the data will be collected and analyzed, and, ultimately, how the information that was collected will be presented to solve the problem. Researchers are faced with a multitude of decisions concerning the means and methods to achieve the objective of the study. The process is further complicated by the fact that there is no definitive means by which to choose the best design.

To illustrate, suppose we are researchers for a large consumer products manufacturer that is contemplating establishing an incentive system for decreasing absenteeism among its production employees. The primary research problem to be addressed is stated as follows: What incentive system should our company institute to help reduce production worker absenteeism? The researchers now must select the methods to solve the stated problem. The decisions to be made are those that are designated design considerations, commonly referred to as the research methodology.

For example, one of the first decisions facing us is what type of research design should be used to study the problem. Should we specify several incentive plans, institute them in the workplace, then look at their ability to reduce absenteeism (experimental design)? Or should we ask the workers what incentive system they would like to see implemented, design a system, then implement the system (ex post facto design)? As we will see later in this chapter, each of these major design alternatives has advantages and disadvantages that researchers must weigh to provide the best possible information to management. The outcome of this decision affects what concepts will be studied (measurement decisions), how they will be measured (scaling decisions), who will be studied (sampling decisions), how the data will be obtained (data collection decisions), and how it will be analyzed (analytical decisions).

From the standpoint of the researchers, design decisions require a series of tradeoffs due to the constraints of the problem situation. Rarely does the researcher have unlimited time and resources to execute a research project. Instead, the research team is restricted by some dollar resource constraint, some time frame, or some other parameter that serves to limit the type of design chosen to solve the research problem. The question that is most often asked is, How can we best solve the stated research problem with the available resources, to provide the information that management needs to make decisions?

There is no one procedure we can follow that will lead to the single best design for a particular question. In fact, if we were to select three researchers, put them in a room, and ask them how to address our absenteeism problem, we could easily get three different approaches. Each approach would probably have some merits; these would have to be evaluated against the disadvantages inherent in the proposed approach to arrive at the best possible design, given the problem situation and resource constraints.

Generally, much research experience and a thorough knowledge of the important design components are needed to evaluate alternative design configurations.

ERROR REDUCTION THROUGH DESIGN

As stated previously, the primary purpose of design is to guide the investigator in answering the research problem. This is accomplished through the careful construction of the research design so that the results obtained are as error-free as possible. On the surface, this goal would appear to be relatively simple to achieve, but it is actually very complex because of the numerous sources of potential error that may affect the results of any investigation.

Potential Sources of Error in the Research Design Process

Many types and sources of errors can affect the results of business research. These errors can be classified in a number of ways, depending on one's perspective.[1] For the purposes of this text, we will divide the potential sources of error into four major categories. As outlined in Table 6.1, these categories are planning, collection, analytical, and reporting errors. This breakdown of errors is useful because it classifies research errors on the basis of when they occur in the research undertaking. After the types of errors are described, managerial strategies for dealing with them are presented.

Planning Errors. Let's again assume we are undertaking a research project regarding which incentive system our company should institute to help reduce production worker absenteeism. As outlined in Table 6.1, the first class of errors that may affect the quality of research results are *planning errors*. These errors are reflected in the setup of the design to collect information. In our incentive study example, as in any other example we could have chosen, the types of error that can occur in this category are almost boundless.

First and foremost, a critical error arises from the ill-structuring or misspecification of the problem. As noted in Chapter 5, the misspecification of the research problem results in information that cannot solve management's problems. To illustrate, suppose management's real concern is increasing worker productivity rather than ensuring a worker's presence or absence on any particular day. In other words, management is actually concerned with getting the employees to work more productively when they are on the job, rather than getting them to show up for work more often. This problem misspecification is a major source of error that will affect the usefulness of the research results.

Planning errors are also associated with the selection of an inappropriate research design. The research design process is a highly interactive stepwise

TABLE **6.1** Potential Sources of Error in the Research Design Process

Sources of Error	Planning Errors	Collection Errors	Analytical Errors	Reporting Errors
Description	Planning errors are those that deal with the setup of the design to collect information. Think of planning errors as sources of misinformation caused by inappropriate structure when the research is undertaken.	Collection errors are those associated with the collection of information to answer the research problem. These sources of misinformation arise during the actual collection of data if the study is generally well-structured (or perhaps they compound errors if the study is ill structured).	Analytical errors are those that are due to the inappropriate analysis of the data collected. Think of these errors as involving the actual manipulation of the data.	Reporting errors are those that are due to the incorrect interpretation of the results of the study. Reporting errors generally have to do with reading erroneous meanings into the relationships and numbers identified from data analysis.
Major factors that create error	The research problem is misspecified or ill structured in the first place. The research design is inappropriate and poorly thought out before the research is undertaken.	The measurement procedures are invalid and unreliable. The data collected are not representative of the individuals or organizations to whom the results are to be generalized. The data collection methods result in inaccurate and/or misleading data.	The analytical techniques are applied inappropriately to the data available. The analytical techniques are not appropriate given the type of data collected.	Inadequate researcher familiarity with the subject area under study. The relationships in the study are not grasped fully by the researchers. The deliberate distortion of facts due to some vested interest or pet project.
Primary managerial strategies to reduce research errors	Develop well-thought-out proposals that clearly specify the methods, cost, and value of the research to be undertaken.	Carefully and thoughtfully execute the specified research design.	Justify the analytical procedures used in manipulating and summarizing the data.	Have unbiased and knowledgeable reviewers examine the results of the study.

procedure, in which decisions in earlier stages affect the outcomes in later stages. For example, suppose we choose to study our misspecified problem with an inappropriate design containing poorly constructed measurement instruments and an inadequate sample to address the research question at hand. The errors built into the research before it is even begun ensure that the results of the study will be meaningless. Careful planning and critical review of the research proposal can minimize planning errors.

Collection Errors. *Collection errors* arise during the actual collection of data. These types of errors can serve to magnify errors that are a result of inadequate planning. For example, if we go ahead and study absenteeism instead of productivity, and then measure this concept inappropriately, we could further compound our errors. The research results would then be totally useless, given management's original concern.

However, for the moment, let us assume that the study is reasonably well planned—that is, defined appropriately and structured correctly. In this case, planning errors are considered minimal (we can never eliminate all errors, as discussed later). The researchers must now collect data of an acceptable quality to answer the original research problem. To minimize error, researchers focus on the following concerns:

1. The measurement procedures are of acceptable quality.
2. The data collected is representative of the population we are studying.
3. The data collection methods are yielding accurate data.

All of these concerns are related to important issues that we must deal with in research design. Specifically, we must ensure that the data collected is of the right type and quality to answer the research question. If the data is not collected appropriately, then even the best planning cannot yield information that is truly valuable to management. Measurement, sampling, and data collection are important issues that will be dealt with in later chapters.

Analytical Errors. It is clear that errors in the research process are extremely problematic because of their compounding nature. *Analytical errors* can arise from the way in which the researchers choose to analyze the data. Substantially different results may be derived from the same data, depending on what analytical techniques are used. Thus, researchers must be familiar with the statistical techniques that are available to them to summarize the data.

Later chapters in this book survey some of the statistical tools available to researchers to analyze data. These chapters review some of the assumptions and limitations in the use of selected statistical techniques. The goal is to familiarize the reader with the proper usage of these tools. Expert statistical advice on the application and limitations of the tools should always be sought, so as to use analytical methods properly in any investigation.

Reporting Errors. The final category of research errors arises from the incorrect interpretation of the results of the study. *Reporting errors* generally have to do with putting erroneous meaning into the relationships and numbers identified at the data analysis stage of research. Such errors can be either intentional or unintentional.

Although much has been written about intentional errors (i.e., research fraud and the deliberate distortion of research results), little is known about unintentional errors because they are more difficult to detect. Intentional errors are ethical concerns, discussed in the chapter on ethics. Unintentional errors usually result from ignorance or honest mistake. Critical and unbiased reviews can help minimize such errors. But one must wonder how much of the seemingly contradictory research results in the literature derives from reporting errors.[2]

Managerial Strategies for Dealing with Error

As we have seen, there are many potential threats to valuable decision research: all errors that may ultimately distort the results of a particular project. From the manager's perspective, biased or distorted information is of truly limited value if the goal is to obtain useful information for decision making. Of course, a manager may request research to verify his or her preconceived notion about what alternative is the correct one to follow. But these are not legitimate research endeavors; their purpose is not to generate information. In this book, we assume that the request for information is legitimate—that information is needed for decision making. It is our problem to generate the best available information given our resource constraints.

In order to accomplish the goal, we must approach the research task as we do any other managerial problem. Basically, we must plan our work, work our plan, and ensure that the goals we set are actually achieved. To this end, management can reduce the major sources of error in the research process by using the strategies listed in Table 6.1. These strategies can help ensure that the results obtained are of the quality and type desired. Briefly, the major strategies are as follows:

> **Planning error reduction strategy.** Requires the development of well-thought-out proposals that clearly specify the methods, cost, and value of the research to be undertaken. In essence, this strategy forces researchers to clearly state the research problem and the means by which they are going to solve it (design). This strategy can be ensured by having the proposals evaluated by impartial and research-competent individuals and committees.
>
> **Collection error reduction strategy.** Requires the careful and thoughtful execution of the specified research design outlined in the proposal. This strategy keeps the researchers oriented to the plan and focuses their efforts on the research problem identified. One means by which this strategy can be monitored is through periodic

progress reports, checking that the procedures used are following generally accepted research practices.

Analytical error reduction strategy. Requires a justification of the analytical procedures used in summarizing and manipulating the data. A justification of the procedures to be used requires the researcher to think through the implications of the analysis. In other words, be sure to ask why a particular analytical technique is being used to answer the research problem.

Reporting error reduction strategy. Requires unbiased and knowledgeable reviewers to examine the results of the study. Professional researchers are accustomed to having their work subjected to public review and criticism. This is a fundamental characteristic of all scientific research. The researchers should not be afraid to subject the fruits of their work to review and to explain and defend their conclusions.

These managerial strategies for error reduction are admittedly broad, but they help ensure the usefulness of the research when it is completed. However, the specific sources of error are nearly endless. The reviewer and user of research must be able to identify the factors that detract from the accuracy of the results. For this reason, the rest of this chapter (and the rest of the text for that matter) deals with these issues. *Keep in mind that design is the ultimate means by which the researcher controls error and obtains accurate and useful information.* Therefore, both the manager and the researcher must understand the basic components of design.

MAJOR TYPES OF DESIGNS

One of the first considerations in design is what general approach will be used to solve the problem. As stated earlier, there are two basic types of design: ex post facto and experimental.[3] The distinction between these types largely has to do with the researchers' control over the independent variable(s) chosen in the study.

Chapter 3 defined a variable as a symbol or a concept that can assume any one of a set of values. An *independent variable* in a study is a presumed cause of the *dependent variable,* the presumed effect.[4] The independent variable produces a change in the dependent variable. (See Exhibit 6.1, A Note on Causality.) The independent variable is the one (or more, in the case of multivariate models) that the researcher believes precedes and affects the dependent variable. In other words, the changes in the dependent variable are what we try to predict, understand, or explain by using the independent variable. For example, if we return to the incentive-system absenteeism problem, the independent variable would be the incentive system because it is presumed to affect absenteeism—the dependent variable.

EXHIBIT **6.1**

A Note on Causality

The notions of cause and causation are often problematic for the researcher because of the vagueness of these terms. Bertrand Russell has been so bold as to state, " . . . the word 'cause' is so inextricably bound up with misleading associations as to make its complete extrusion from the philosophical vocabulary desirable."* However, the terms are an integral part of our vocabulary, and they are used frequently by all types of people. For this reason, we will attempt to clarify these notions in this exhibit.

There are two basic meanings of cause and causation in the literature: determinant cause and probabilistic cause.† A *determinant cause* can be defined as any happening that is a necessary and sufficient cause of another happening that occurs at a later time. A *necessary condition* implies that the causal factor (antecedent variable) must be present if a predicted change is to occur in the dependent variable of interest. A *sufficient condition* implies that the sheer presence of the causal factor is all that is needed to create the change in the dependent variable. As one can well imagine, business researchers cannot talk in terms of deterministic causes, because of the complexity of the environment in which they operate. Instead, we talk of probabilistic causes.

A *probabilistic cause* can be defined as any happening that is necessary but is not sufficient for a subsequent happening. In other words, the causal factor's presence is needed in order to effect a change in the subsequent happening, but it is not sufficient in and of itself to cause that change. This brings in the concept of probability—the more a particular causal factor's presence contributes to creating change in the dependent variable, the higher the probability that the change will take place. Given the environment in which business researchers operate, this notion of cause is the most appropriate one. When we refer to cause and causality in this text, probabilistic causes will be meant.

Three kinds of evidence are used to demonstrate the existence of causality. If the researchers can demonstrate evidence of the three, they can have a high degree of confidence that a causal factor has been identified. The three types of evidence are:

1. **Concomitant variation.** The degree to which two variables are associated. They must covary if a causal relationship is to exist.

2. **Time order of occurrence.** The causal factor must precede the dependent variable if causality is to be established.

3. **Absence of other causal factors.** The most difficult of all the evidence to demonstrate: that no other variable or variables could possibly cause the change in the dependent variable. As we will see later in this chapter, experimental designs are the only ones that can adequately demonstrate this type of evidence.

* Bertrand Russell, "On the Notion of Cause, with Applications to the Free Will Problem," in H. Feigl and M. Brodbeck, Eds., *Readings in the Philosophy of Science* (New York: Appleton, 1953), p. 387.

† The definition of these causes follows those in R. L. Ackoff, Shiv Gupta, and J. S. Minas, *Scientific Method: Optimizing Applied Research Decisions* (New York: Wiley, 1962), Chapter 1.

The selection of the independent and dependent variables should flow logically from a well-thought-out problem statement.[5]

At this time, it is also appropriate to introduce moderating and intervening variables.[6] A *moderating variable* is one that has a strong effect on an independent–dependent variable relationship. For example, research

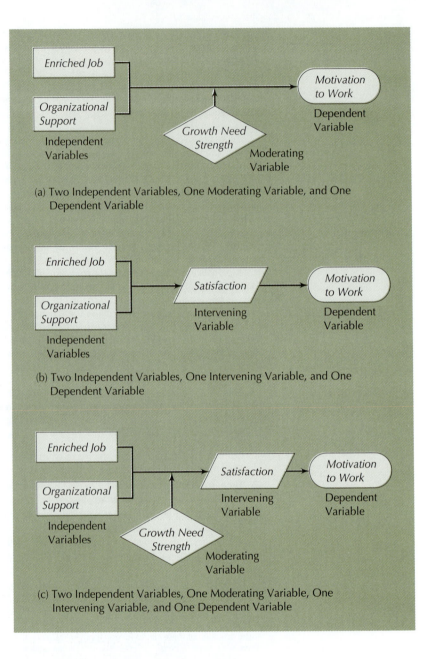

FIGURE 6.1

Schematic Diagrams
Illustrating
Moderating and
Intervening Variables

Source: Adapted from Uma
Sekaran, *Research Methods for
Managers* (New York: Wiley,
1984), pp. 50–58.

(a) Two Independent Variables, One Moderating Variable, and One
    Dependent Variable

(b) Two Independent Variables, One Intervening Variable, and One
    Dependent Variable

(c) Two Independent Variables, One Moderating Variable, One
    Intervening Variable, and One Dependent Variable

suggests that variance in employees' motivation to work (the dependent
variable) can be largely explained by the two independent variables of
enriched jobs and organizational support. However, research also indicates that this relationship only holds for those who are high in growth
need strength. In this case, growth need strength would be identified as a
moderating variable. This theoretical network is illustrated in Figure 6.1(a).

On the other hand, an *intervening variable* is one that emerges as a function of the independent variable(s) operating in a situation and helps to explain the influence of the independent variable(s) on the dependent variable(s). For example, if the independent variables, enriched jobs and organizational support, make more satisfied employees, who are then more motivated to work (dependent variable), then employee satisfaction would be designated an intervening variable. The independent variables (enriched jobs and organizational support) create satisfaction (an intervening variable) that, in turn, increases the motivation to work (the dependent variable). This network is shown in Figure 6.1(b).

Independent, intervening, moderating, and dependent variables can be present in the same theoretical structure. The possibilities are many. Extending the motivation to work example, rather complex models can result, as illustrated in Figure 6.1(c).

Here, it is extremely important to emphasize that *a relationship does not imply causality.* Just because satisfaction is related to motivation to work, one cannot say that satisfaction *causes* motivation to work. Other plausible explanations have not been ruled out, and the time order of occurrence has not been established. It may be a *spurious correlation*—a relationship that happens by chance alone and has no real meaning. Many relationships exist in our world, but they may or may not involve cause and effect. Exhibit 6.2 gives examples.

Once the variables of interest are identified in the study, the researchers must choose what type of design to use to solve the problem. In ex post facto

EXHIBIT 6.2

Causality, Proof, and Leaping Out the Window

All too often, people (including researchers!) who look at studies that have found a statistical relationship between a presumed cause (independent variable) and a presumed effect (dependent variable) are too eager to declare, "See, this proves it." Two examples illustrate this point.

Example 1 A study is conducted in a company that correlates pay scales with performance levels. A highly positive and significant relationship between the variables is found to exist in the study. The proponent of higher pay scales in the company says, "As this study proves, higher pay leads to higher performance."

Example 2 A study is designed to assess the relationships of various fruit drinks' product attributes (i.e., acidity, sugar and clarity levels, and size of container) to the sales volume of each type of drink. Sugar levels are found to have an extremely high positive correlation with the sales of fruit drinks. At the next product development meeting, the product manager states, "As our product study proves, we need to put more sugar in our fruit drinks to increase sales volume."

These two examples illustrate leaping out the window: making statements with too little supporting information. In both situations, there are other plausible explanations for the identified relationships. Can you argue effectively against statements such as these?

designs, the researcher does not attempt to manipulate the independent variable because it is not manipulable for some reason or another.[7] In *experimental designs,* on the other hand, the researcher manipulates or in some way controls the independent variable and then measures the effect on the dependent variable of interest. Table 6.2 compares these two types of design.

When examining Table 6.2, there are several important items to keep in mind. First, the table merely summarizes the major types of designs and their primary characteristics; the complexity of most research studies is not shown. Many designs in business research combine two or more of the approaches listed in Table 6.2. For example, field studies, surveys, and field experiments can be used together in the same research project, and other multiple design configurations are possible.

The second major item of interest in the table is the arrangement of design types by (1) the primary goal of the study and (2) the degree of understanding of the problem situation. The table suggests that field studies are typically associated with exploratory studies, where the researchers have a relatively low degree of understanding of the problem. Of course, other design subtypes can be used in that type of situation if the problem dictates. It is entirely possible that a combination of several design subtypes could be used to help solve a descriptive problem with which the researchers have a basic familiarity. Let's use our incentive-absenteeism problem to illustrate the differences between the various types of designs, while describing the characteristics of each.

Ex Post Facto Designs

Field Studies. Suppose the researchers assigned to the project feel they have very limited knowledge of the relationship between incentive systems and absenteeism, let alone what types of incentive systems have been successfully used in the past. In this case, the researchers would probably use some type of field study to explore the problem situation in some depth. *Field studies* usually combine literature searches, experience surveys, and single- or multiple-case studies in which the researchers attempt to identify important variables and their relationships in a particular problem situation. Critical, nonintrusive observation is frequently used in field studies as a method of collecting data. Field studies are ex post facto designs because the researchers make no attempt to control or manipulate incentive systems. They systematically study what other people's and organizations' experiences are with the concepts under study.

The researchers may start by searching the literature for firms that have successfully implemented incentive systems. (This is called a literature search.) Or they may start by asking a number of knowledgeable managers what their experiences are with absenteeism-reducing mechanisms. (This is called an experience survey.) Alternatively, or in conjunction with these other approaches, the researchers may prefer to closely examine one or several firms' experiences with incentive systems. (These are called single- or

TABLE **6.2** A Comparison of the Major Types of Research Designs

Design Type	Ex Post Facto Designs		Experimental Designs	
Distinguishing Factor	No manipulation of independent variable(s)		Manipulation of independent variable(s)	
Primary Goal of Study	*Exploratory*	*Descriptive/predictive*	*Causal (explanatory)*	
Degree of Understanding	Low —————————————————————————— High			
Subtypes	Field Studies	Surveys	Field Experiments	Laboratory Experiments
Description	Combination of literature searches, experience surveys, and single- or multiple-case studies, where the researcher attempts to identify significant variables and their relationships in a problem situation. Generally leads to a more rigorous study at a later date.	Usually use standardized questionnaire. Probably the most widely used type of design in business research. Also probably the best method for studying and describing large populations.	Research studies in a realistic setting, where the researcher intervenes and manipulates independent variable(s). Characterized by a high degree of reality, but afford the researcher much less control than a laboratory experiment.	Studies in an artificial setting, where the researcher intervenes and manipulates independent variable(s). Researcher can control most sources of error. Drawback: carried out in artificial settings.
Examples Using Incentive-Absenteeism Problem	The researchers search the literature looking for firms who have successfully implemented incentive systems. They then pick one of two organizations and intensively study the successes and failures of that organization in designing and implementing a system.	The researchers search the literature and find a number of firms that tried to develop incentive systems to reduce absenteeism. They prepare a paper-and-pencil questionnaire to be sent to the general managers of those firms, with the intention of identifying the characteristics of successful incentive systems.	Researchers identify three possible incentive systems to reduce absenteeism. They randomly assign an incentive system to every department within their organization. The project goes on for two months. The researchers review the performance (in terms of reduced absenteeism) of each incentive system at the end of two months.	Researchers identify three possible incentive systems to reduce absenteeism. The researchers randomly select individuals in the organization to come to the personnel office to take part in a study. The workers are randomly assigned to four groups. One type of incentive is randomly assigned to the worker groups (one of which is a control group). The entire situation is highly controlled by the researchers. The incentive systems' impact on absenteeism is reviewed at the end of two months.

multiple-case studies.) These endeavors are usually *exploratory research,* in which the primary goal is to increase the researchers' understanding of the nature of the problem. Exploratory studies are generally followed by a more rigorous study at a later date, when the situation is better understood by the researchers.

Japanese corporations have a strong belief in the use of field studies to help in understanding context-specific problems.[8] For example, when Canon was looking for a new distribution strategy for U.S. camera sales, it sent three managers to U.S. retail outlets to talk with store owners and managers to identify problems and opportunities for Canon in introducing its AE-1 camera. Other companies such as Sony, Matsushita, and Toyota also tend to rely heavily on field studies in developing strategies for market entry and channel maintenance.

A field study's primary strength is that it is conducted in a realistic setting. The researcher can closely observe the interrelationship of the variables under study without the artificiality usually present in the other types of research design. However, the field study does have its weaknesses:

1. The lack of control by the researcher because of the field study's ex post facto nature

2. The fact that the field study is generally hard to interpret because of the noise in the study setting

3. The danger that the field study can become infeasible because of cost and time considerations

Surveys. The second subtype of ex post facto design is the *survey,* which uses some sort of questionnaire to describe and/or predict some phenomenon. Surveys differ from observation studies in that they require interaction with the respondent (by paper-and-pencil questions, personal interviews, online interaction, etc.). Surveys are probably the most widely used type of design in business research endeavors because they allow researchers to study and describe large populations fairly quickly at a relatively low cost. In addition, surveys are extremely versatile; they can be adapted to almost any research setting. Generally, however, the researchers must possess at least a rudimentary understanding of the problem at hand before a survey can be designed.

Most surveys have as a central objective a search for relationships among variables.[9] Surveys have been used successfully to help test hypotheses, evaluate programs, describe populations, build models of human behavior, develop useful measurement scales, and make other methodological improvements in business research. Surveys usually use a well-constructed or standardized questionnaire to collect data from the relevant unit of analysis under study (usually an individual). However, it is important to note that the researcher does not intervene in an attempt to control the independent variable, but only seeks to measure the relative level of the relevant concepts isolated for study. To illustrate this point, let us return to our incentive-absenteeism problem.

Now suppose the researchers are vaguely familiar with the use of incentive systems to reduce absenteeism in a manufacturing setting. They are acquainted with other firms that have used this technique to decrease absenteeism in their organizations. They review the literature and find a number of firms that have tried to develop incentive systems to reduce absenteeism. This review helps them construct an online survey to be sent to the general managers of firms that have implemented, successfully or unsuccessfully, such behavior modification programs. The initial goal of the study is to describe the characteristics of those firms and incentive systems that have successfully reduced absenteeism. Studies designed to describe the major characteristics of some problem situation constitute *descriptive research*.

Experimental Designs

Field Experiments. As noted earlier, experiments are those designs in which the researcher manipulates or in some way controls the independent variable(s) under study. The two major subtypes of experiments are commonly designated field and laboratory experiments. The primary distinguishing factor between these two types is the degree of realism in the research setting.[10] For our purposes, we will define a *field experiment* as a research study in a realistic setting, where the researcher intervenes and manipulates some independent variable(s) and controls the situation as carefully as the conditions permit.

Again, let us use the absenteeism problem for an example of a field study in business research. Suppose the researchers have identified three possible incentive systems that appear to have worked successfully:

1. A lottery with monetary reward
2. A lottery with compensatory time-off reward
3. A lottery with a material prize, such as an automobile

Now the researchers request permission from management to go into the organization and randomly assign one incentive system to each of three departments that are similarly configured in terms of personnel characteristics. A fourth is assigned no incentive plan. In other words, the researchers are manipulating the independent variable (incentive system) to see its effect on the dependent variable (absenteeism). After two months, the investigators compare the average weekly absenteeism rates for each of the three treatments (the three incentive systems) and the status quo group (which is commonly called a *control group*) to see if there are statistically significant differences.

This example suggests the main characteristics of field experiments: (1) The investigator manipulates or controls some independent variable; (2) the study is conducted in a realistic setting; and (3) the experimenter attempts to control the research setting in some fashion.[11] The primary strengths of a field experiment are that (1) the study situation is usually highly realistic, allowing the independent variable's effect to be accurately assessed, and (2) compared with ex post facto research, a stronger inference can be made about the relationship between the variables under study. The inference is stronger because the experimenter actually manipulated the variable of interest and then measured its effect on the dependent variable.

The field experiment is not without its weaknesses, however. The realistic setting, although a strength, can also be a weakness because of the limitations it may present to the researcher. Specifically, it may be impossible to manipulate an independent variable or to randomize treatments in real life. (This point will be discussed in depth later on in the text.) For example, if departmental managers in our example are being evaluated on departmental performance, they may not want to receive the control group status. Thus, it may not be completely feasible to manipulate the independent variable in a manner that is desirable from a research standpoint.

Laboratory Experiments. The alternative to field experiments is laboratory experiments. *Laboratory experiments* are conducted in an artificial setting, where the researcher intervenes and manipulates some independent variable(s) in a highly controlled way. These designs allow the researchers to control most major aspects of research design error; they are the most scientifically rigorous of all the designs specified thus far. The primary goal of laboratory (as well as field) experiments is to specify the exact relationship between two or more variables in the problem situation.

Suppose researchers want to test the effectiveness of the three incentive systems identified earlier using a laboratory experiment. The researchers would randomly select individuals in the organization to come to the personnel office to take part in a study. These individuals are then put in separate facilities and given new tasks. This time, workers are randomly assigned to the four groups. Incentive systems (treatments) are randomly assigned to three groups; the fourth, the control group, receives no incentive system. The work situation is closely monitored by the researchers. Again, the absenteeism rate is monitored and reviewed at the end of two months.

As this scenario indicates, the laboratory experiment is more highly controlled by the researchers than any other of the design alternatives. The primary strength of the laboratory experiment is that the investigators have almost complete control of the study situation. The previously mentioned designs, although generally high in realism, all suffer from the fact that extensive "noise" can greatly affect the outcome of the study. Researchers can never be quite sure if changes in the absenteeism rate, if found in the various groups, were due to the incentive system or some other uncontrollable variable such as a high unemployment rate, a bad economy, or the season. This point is less true for field experiments than the ex post facto designs. But the strongest conclusions could be drawn from the laboratory experiment because the investigators were able to minimize the possibility that other causal factors had a chance to operate.

However, all is not perfect if the researchers choose to study a problem with this design. The artificiality of the research setting may actually generate changes in the dependent variable that might not happen in the real world. For example, let's say a laboratory experiment shows a relationship between a money lottery incentive system and decreased absenteeism. If this system is implemented and put into the actual work environment, it may prove a miserable failure because peer pressure will not allow the program to work on the shop floor.

Second, the artificiality of the experimental process may actually change the behavior of the individuals in the experiment. Perhaps the most widely known case has to do with the findings of the Hawthorne Studies, where it has been argued that the experimentation procedures themselves changed the behavior of the workers, not the treatments of more humanistic management techniques.[12]

The manager and the researcher must become aware of the primary strengths and weaknesses of each design alternative and then select the one, or the combination, that will best answer the research problem at hand. Choosing a design to solve a problem is not a simple process, and the decision must be weighed in light of the goals of the research, the resource constraints of the project, and the overall feasibility of the design in the particular situation. Our discussion will now turn to specific design configurations and their nuances in the business research setting.

SPECIFIC DESIGN CONFIGURATIONS

Thus far, we have discussed the general nature and purpose of design and outlined the basic design structures available to researchers. This section of the chapter builds on this framework by outlining some general classes of variables that affect a study's internal and external validity, and then describes a number of specific design configurations commonly used in business research.[13] Validity is examined first because of the importance of the degree of confidence researchers and managers can have in the results of the study. Validity has to do with limiting research errors so that the results are accurate and usable when delivered.

Validity Concerns

Internal Validity. *Internal validity* can be defined as the degree of confidence that the results are true given the study situation. Stated in another fashion, internal validity is strongest when any alternative interpretations of the study's results are ruled out. Internal validity is essential if a study is to be meaningful to managers. Without it, there can be no confidence that the relationships identified in the investigation are really justifiable. Both researchers and managers should be concerned that a study possesses internal validity.

The difficulty comes in assessing internal validity because it is never fully measurable. There are simply too many sources of invalidity for the researcher to measure, or even identify, in the business research environment. The best the researcher can hope for is that the major threats to validity are isolated and at least recognized in the study's design.

It should now be evident that in all the design subtypes outlined in Table 6.2, one of the primary goals of the researcher was to control some of the threats to internal validity. Specifically, field studies are generally the weakest in terms of control of extraneous variables that can threaten validity, whereas laboratory experiments are generally the strongest. We will now list some classes of variables that may affect a study's internal validity. Their influence on specific research designs will be discussed later in the chapter.

History. Events outside the investigation may affect the results of the study. For example, a high unemployment rate during the study period of our incentive-absenteeism field study would affect normal absenteeism rates.

Maturation. Changes occur in respondents over time, such as aging or just plain fatigue. For example, the employees under the incentive systems get bored with the program.

Testing. Some effects can be attributed to the process of being tested, which can sensitize and bias a respondent's action. For example, the employees realize that the incentive system program is actually only an experiment, so they change their work habits because they are being watched.

Instrumentation. There can be changes in the measurement instrument during the study. For example, workers are counted absent if they do not show up by 8:00 A.M. during the first month of the test, but the time is changed to 11:00 A.M. during the second month.

Selection. The selection procedure can have an artificial effect on the results of the study. For example, the departments in our organization are taken intact. It is not known whether there will be different responses to incentive systems among machinists and clerical workers.

Mortality. Respondents may be lost from the study. For example, a number of workers in one department quit while the study is being conducted.

These are by no means all the threats to internal validity, but they do represent the major classes of variables that have been known to affect a study's results. We will now turn to the concept of external validity.

External Validity. *External validity* can be defined as the degree to which the study's results can be generalized across populations, settings, and other similar conditions. Ideally, the manager would like to maximize both the internal and external validity of a study. However, there must be compromises in both forms of validity due to considerations of practicality and/or resource constraints. It should be noted, however, that one cannot have external validity without evidence of internal validity. That is, we must be confident that the results are in fact due to what the study says they are before we can generalize to other populations, settings, and so on.

Just as in the case of internal validity, there are a number of sources of external invalidity. Again, these sources of invalidity are often very difficult to detect, let alone measure. This makes it all the more important to be aware of these potential sources of error in design. Three major classes of variables are presented here.

> **Testing interaction.** Because of artificial effects created by testing respondents, conclusions may not be reliably generalized to other situations. For example, the respondents in our incentive study reduce their absenteeism rate because they know they are being watched by management.

> **Selection interaction.** The effect that the type of respondents has on a study's results may limit its generalizability. For example, the testing of the incentive system in our organization may not be generalizable to other organizations because our workers are not like workers in other industries.

> **Setting interaction.** The artificial effects that are created by the specific setting of the study may not be replicable in other situations. For example, the incentive systems are tested in a specific company, in a specific city, in a specific period, with a unique group of managers. This setting may not be generalizable to any other situation.

We now turn to a discussion of some specific types of design configurations. This discussion will focus on the format of the designs and their potential weaknesses and strengths in terms of the two validity types.

Specific Designs

First, we present a notational system to represent the major events in various types of studies. This system will be used to discuss both ex post facto and experimental designs. The system is as follows:

> X represents the exposure of a test group to an experimental treatment (independent variable). Then we will measure the effects of the treatment on the dependent variable of interest.

> O refers to a measurement or observation of the dependent variable being taken on some individual, group, or object.

> R means that individuals or groups have been selected and assigned at random for the study's purposes. Randomization, as will be seen in Chapter 9, is the all-purpose procedure for statistically eliminating

bias in a study's design. Generally, this procedure is a safeguard against differences (which might lessen the validity of inferences about the effect of the treatment) between experimental groups. We are seeking equivalence among the groups.

Left-to-right placement of X's and O's indicates a sequence of events. When X's and O's are in a given row, this means that the treatments and measurements occur on the same individuals and/or groups. When there are multiple groups or cases in the study, the vertical positioning of symbols is designated to mean the simultaneous occurrence of the activities they represent.

With this notational system, we are now ready to examine some basic research designs. Table 6.3 summarizes the major points of discussion in this section. Note that this table, like any attempt to summarize a complex phenomenon, oversimplifies the strengths and weaknesses of each type of design. The researcher must examine the application of each design to a particular problem to truly see its strengths and weaknesses. The more basic designs are discussed in this section; some more advanced experimental designs are covered in an appendix at the end of this chapter.

Ex Post Facto. Ex post facto designs are relatively easy to picture using our notational system because they involve no treatment manipulation whatsoever. Both field studies and surveys are represented by an O or O's, depending upon the nature of the study. For example, if we were to select two firms, one with an incentive system and another without one, to do an in-depth field study on the incentive-absenteeism problem, the design could be represented this way:

$$O_1$$
$$\hspace{4cm} \text{(a)}$$
$$O_2$$

where O_1 is the observation on the first firm under study and O_2 is the measurement on the second firm.

Similarly, surveys are represented in the same fashion. For example, if we were to randomly sample 200 firms from a population of 1000 firms that have implemented incentive systems for reducing absenteeism, and then send a questionnaire to the general managers asking about each firm's experience with its program, the design configuration would look like this:[14]

$$\text{(R)} \quad O_1 \hspace{4cm} \text{(b)}$$

where O_1 represents the survey (observation) on the 200 randomly selected (R) general managers of the chosen firms.

Both of these design configurations are known as *cross-sectional* designs because the measurements on the variable(s) of interest are taken at one point in time. In essence, cross-sectional designs provide the manager with a snapshot of variables at one instant in time. The alternative to cross-sectional studies are longitudinal designs, which measure the same sample or population

TABLE **6.3** Sources of Invalidity for Basic Research Designs*

Basic Designs	Internal						External		
	History	Maturation	Testing	Instrumen-tation	Selection	Mortality	Testing Interaction	Selection Interaction	Setting Interaction
Ex Post Facto designs									
O	−	−		−	−	?		−	?
Experimental Designs									
Preexperimental designs									
1. One-shot case study X O	−	−			−	−		−	?
2. One-group pretest–posttest design O_1 X O_2	−	−	−	−	+	?	−	+	?
3. Static-group comparison design X O_1 / O_2	+	?	+	+	−	−	−	−	?
True experimental designs									
4. Posttest-only control group design (R) X O_1 (R) O_2	+	+	+	+	+	+	+	?	?
5. Pretest–posttest control group design (R) O_1 X O_2 (R) O_3 O_4	+	+	+	+	+	+	−	?	?
6. Solomon four-group design (R) O_1 X O_2 (R) O_3 O_4 (R) X O_5 (R) O_6	+	+	+	+	+	+	+	?	?

Sources of Invalidity

*In the tables, a minus sign indicates a definite weakness, a plus sign indicates that the factor is controlled, a question mark indicates a potential problem, and a blank indicates that the factor is not relevant.

Source: Donald T. Campbell and Julian C. Stanley: *Experimental and Quasi-Experimental Designs for Research*, copyright © 1963, Houghton Mifflin Company, used by permission.

repeatedly (at least twice) through time. For example, if we survey the same random sample of firms now, six months from now, and one year from now on the incentive problem, we would in essence have an ex post facto longitudinal design of the following configuration:

$$(R) \quad O_1 \quad O_2 \quad O_3 \qquad\qquad\qquad (c)$$

where O_1, O_2, and O_3 are the three observations on the same firms at the three points in time.

As pointed out earlier, field studies are typically intensive studies of a small number of cases, whereas surveys are generally less intensive investigations of a larger number of respondents and/or organizations. However, these types of designs share many of the same problems from a validity standpoint because of their ex post facto nature. First, because of the lack of control of variables by the researcher in ex post facto designs, history is a major validity problem. As noted in Exhibit 6.1, one of the primary criteria for showing cause-and-effect relationships is the absence of other possible causal factors. This cannot be done in after-the-fact studies. This is probably the biggest weakness of these types of investigations.

Depending on the particular configuration of the study, the five other threats to internal validity and the three primary threats to external validity are indicated in Table 6.3. In ex post facto designs, it is extremely difficult, if not impossible, to measure the extent to which these threats have affected a study's conclusions. Researchers must carefully plan their study design so that, given reasonable conditions, the results obtained are as error-free as possible.

How do researchers do this? They carefully define their research problems. They set specific objectives and narrow the scope of their research. They design their measurement instruments carefully and select their samples meticulously. They develop a data collection strategy and conscientiously implement this plan. In essence, they follow the guidelines of sound research by concerning themselves with the research issues presented in the remainder of this text. We now turn to the highly controlled experimental designs.

In our discussion of experimental designs, a distinction will be made between preexperimental (often called quasi-experimental) and true experimental designs. This distinction centers on the concept of control. We first describe three basic preexperimental designs for comparison purposes.

One-Shot Case Study. This design is the simplest and weakest of all the experimental designs. In fact, many researchers suggest that the one-shot case study is so weak that its usefulness in an experimental setting is questionable indeed. The design involves only one treatment, one group, and one measurement:

$$X \quad O \qquad\qquad\qquad (1)$$

The only difference between this and ex post facto studies is that the researchers intervene and manipulate some variable and then take a measurement. For example, researchers put in an incentive system in the organization and measure the absenteeism rate the following month.

As might be expected, this design suffers from all the major sources of invalidity. The sources of invalidity can all be substantial and are not measurable. Therefore, one-shot case studies are to be avoided if at all possible.

One-Group Pretest–Posttest Design. This design, although it is relatively weak, is often used in business research. Notationally, it can be represented as follows:

$$O_1 \quad X \quad O_2 \tag{2}$$

It can also be extended as a longitudinal design, as follows:

$$O_1 \quad O_2 \quad O_3 \quad X \quad O_4 \quad O_5 \quad O_6 \tag{2a}$$

This extended design also is known as a *time series design* because multiple measurements are taken from the same group over time.

This design is better than (1) because it allows a comparison between pre- and posttreatment periods. For example, if researchers measured absenteeism one month before (O_1) the implementation of an incentive system and one month after (O_2), a comparison could be made to see if the treatment worked. Design (2a) is slightly better; we could average, let us say, three months' observations before and after the treatment.

The weaknesses of designs (2) and (2a) include history, maturation, testing, instrumentation, and severe problems with external validity because of the nonrandom selection of the treatment group. The problems with testing, maturation, and history are particularly acute because of the multiple measurements over time on the same testing units. The researcher focuses on the difference ($O_2 - O_1$) between the two measurements. In other words, did something occur to the dependent variable (absenteeism) after the treatment (incentive system) was implemented?

Static-Group Comparison. This research configuration is a slight modification of design (2). Two groups are chosen for study: One receives the treatment, and the other does not (control group). Diagrammatically, it can be represented as follows:

$$\begin{array}{l} X \quad O_1 \\ \quad O_2 \end{array} \tag{3}$$

As in all the preexperimental designs, the groups may not be randomly selected. They may exist already intact. As a result, there is always a question whether the groups were in fact equal at the outset. However, the very existence of two groups allows at least a cursory examination of the treatment's effect without the internal validity threats of testing, instrumentation, and, to a lesser extent, history. Selection may also be a major problem because of the lack of randomization.

Posttest-Only Control Group Design. The simplest of the true experimental designs can be diagramed as follows:

$$(R) \quad X \quad O_1$$
$$(R) \qquad \quad O_2 \tag{4}$$

It is exactly the same as design (3), but the respondents are randomly chosen and assigned to either the treatment or the control group. This design meets the criteria of a true experiment: manipulation of at least one variable and randomization of group and treatment selection.

The design performs fairly well (as do all true experiments) on all dimensions of internal validity. The primary strength of the design is ensured by the randomization of groups (the rationale for this will be seen in Chapter 9). The degree of external validity is largely a question mark; it depends on the conditions of the experiment. The relevant measurement of interest, as in design (3), is the difference between O_1 and O_2.

Pretest–Posttest Control Group Design. The next true experimental design follows the format of design (2), except that a control group is added and the selection of groups is random. The layout of this design can be expressed as follows:

$$(R) \quad O_1 \quad X \quad O_2$$
$$(R) \quad O_3 \qquad \quad O_4 \tag{5}$$

The design is useful because it controls most sources of invalidity through the use of a pretest, a control group, and randomization. History is controlled as long as the pre- and postmeasurements are taken relatively close in time. Maturation and testing are also controlled because of the control group and the pretests and posttests. The problem of external validity is again a question mark because of the extensive control and testing by the researcher. However, this design is often used in television advertising campaign tests using separate cable facilities.

The main measure of interest using this design can be expressed as follows:

$$(O_2 - O_1) - (O_4 - O_3)$$

This expression allows the factors of history, maturation, testing, and instrumentation $(O_4 - O_3)$ to be subtracted from the measurement $(O_2 - O_1)$ that represents the treatment and all other effects. Thus, this design is relatively strong in its control of sources of invalidity.

Solomon Four-Group Design. The last basic experimental design to be examined is a combination of designs (4) and (5). Diagrammatically, design

(6) looks like this:

$$
\begin{array}{llll}
(R) & O_1 & X & O_2 \\
(R) & O_3 & & O_4 \\
(R) & & X & O_5 \\
(R) & & & O_6
\end{array}
\tag{6}
$$

This design has been called the "ideal model for controlled experiments" because it explicitly controls for all sources of internal invalidity outlined in Table 6.3.[15] This design allows the researcher to isolate most of the major sources of error because of its four-group, six-measurement format.

Although this design is the ideal, it is far from the most used study configuration in business research. In fact, one can find few instances of its application in the available literature because of practicality and cost considerations. As stated earlier, researchers in business are faced with limited resources. It is rare indeed that the cost of securing the two additional groups and taking two to four additional measurements would be justified by the value of information obtained from such a venture. However, this design should be studied to understand what sources of error we are trying to control and to realize that there are means to control them if the setting so dictates.

ONLINE RESEARCH DESIGN ISSUES

Since the start of the millennium, online research has surged into prominence. *Online research* is defined in this text as Internet-facilitated design, collection, analysis, and/or reporting of systematically collected information for decision-making purposes. This definition reflects the complexity and diversity of online research issues. This online environment now permeates the research design process, and the developments in this field must be followed closely. Online influences in the design process are discussed throughout the subsequent chapters in this text. However, it is important to note that *online research is an enabling technology and **not** the research design process.*

The rapid growth and acceptance of online research capabilities has been no less than phenomenal in recent years. For example, one source estimated that the online market research revenues in the United States alone grew from $92.7 million in 1999 to $333.5 million in 2001 and was projected to be close to $500 million in 2002 (a 500% plus growth in just a five-year time frame).[16] Companies such as Procter & Gamble and General Mills have further indicated that they are amenable to a gradual move of all research to the Internet and have demonstrated this willingness by investing directly in MarketTools (http://www.markettools.com), a company said to be focusing "on becoming one of the leaders in Internet-based, technology-enabled, full-service, standardized primary market research."[17] All the major research suppliers have, or are now in the process of cementing, the capabilities of providing self-serve, assisted, and/or full-service online research capabilities.[18]

This movement to online research is centered on four basic reasons:

- *Speed*—Online data collection is much faster than traditional offline collection methods of mail, telephone, or personal interviews. Projects that have taken weeks or months to complete are now completed in days.

- *Affordability*—The difference between the costs of conducting online research compared to the traditional offline methods appears to be substantial. The potential for self-service, or in-house research administration and analysis, is enhanced because of the tremendous breakthroughs in instrument design, data collection, and analysis.

- *Ease of Use*—Traditional offline methods take time to set up, implement, and analyze and are not typically "user friendly" to researchers or managers alike. Online studies are inherently more flexible, which makes them easier to use.

- *Promise of Higher-Quality Data*—Researchers are beginning to promise higher-quality data from online sources. The basic arguments for this position center on the validity of the data itself and higher response rates in data collection.

Although these points are all positive, a substantial number of problems exist that can potentially slow the growth in online research in the near future. The level of Internet connectivity in the international marketplace is one such problem. Other problems relate to user capabilities and sampling efficiency that will be discussed in later chapters. Many of the glowing promises for online research remain to be "proven," much less realized.[19] However, the initial results are indeed promising. Offline research approaches are alive, healthy, and will persist in this world for years to come.

With these issues in mind, it is critical to recognize online research is a medium for deploying effective research designs, interacting with respondents, and analyzing and communicating information. The research problem and objectives drive research design, not a technology that can facilitate design. In all cases, after the problem and appropriate design type have been determined, one should ask how online technology could aid in the solution of the problem at had. The following chapters integrate online technologies into the essence of the design process.

MANAGERIAL CONSIDERATIONS

It is difficult for most managers to thoroughly grasp the intricacies of selecting and applying specific designs to problem situations. This is particularly true when evaluating the use of a specific design in an international context. Exhibit 6.3 presents some of the factors that would have an impact on the selection of a design in an international setting. Information-seeking managers

EXHIBIT **6.3**

International
Considerations in
Research Design

The same types of research design are used in international and domestic research. The primary difference is how the designs are applied in the foreign setting. Some designs may be impractical or have to be modified to accommodate the differences. Some of the factors that create these differences are language, culture, literacy rates, and the research infrastructure.

Language. Language and translation problems abound in international research. Often, words do not translate neatly, or there are not functional equivalents for the same concepts. The problem is exacerbated in multilingual societies, where several languages or dialects are spoken. Much more exploratory research must be done before a study is initiated.

Culture. Culture plays a major role in design because different countries present different problems in the design and collection of data. For example, a tradition of courtesy in Japan makes people generally very cooperative to research inquiry. However, this same courtesy can lead to an overly optimistic picture because of people's desire to please the interviewer. In Muslim countries, women are not allowed to converse with male interviewers, and women are not allowed to be interviewers. This forces changes in the selection of a research design.

Literacy. There are great differences in literacy throughout the world. For example, the illiteracy rate in Somalia is 93.9%, in Saudi Arabia 75.4%, and in the Yemen Arab Republic 91.4%. Where these levels of illiteracy exist, written questionnaires are obviously useless. The research design must accommodate this factor to ensure that the information collected is meaningful.

Infrastructure. The research infrastructure places limits on what designs are feasible. In many countries, the telephone and postal systems are unreliable and Internet penetration is limited. In addition, the research capabilities of organizations in developing countries are often inadequate or nonexistent. As a result, certain designs, although desirable from a methodological perspective, may not be feasible.

must be aware of the general research design alternatives available to them. This awareness will help decision makers get the information they require.

The general managerial considerations in design can be summarized using a statement of concerns and a list of questions that should be considered by the researchers undertaking a project. The concerns are as follows:

There is no single correct design for a research problem. As stated earlier, multiple designs are possible for the same research problem. Managers should not seek one design to definitely answer all their questions. It simply does not exist.

Design research to answer the research problem. This may appear to be obvious, but it is a genuine management concern. It is easy to get caught up in a "method myopia," for one reason or another. If a survey satisfactorily answered one management problem, that does not mean it is the best procedure to use to answer an entirely different problem. Similarly, just because Internet research possibilities exist, do not fit the design to the technology. The technology should be adopted if it fits the problem and design at hand.

EXHIBIT 6.4

Key Managerial
Questions Pertaining to
Research Design

1. Which design will best answer the research question?
2. Will the design produce information that is timely and cost-effective for management decision making?
3. Does the design adequately control the main sources of error inherent in the research process?
4. How does the design explicitly control the major threats to internal and external validity?
5. Has the researcher clearly thought through the details of the research design so as to avoid major modifications and unanticipated cost overruns after the project is initiated?
6. Why is the design chosen by the researcher the best one to answer management's problem?
7. How do online research technologies fit into the research design process? Is the design driven by the research problem or by the technology?

All research design represents a compromise. The design chosen is usually a compromise among a number of factors. Cost considerations, practicality, and the information requested require compromises in design. The design that provides the best information may not be practical, for one reason or another. The best means to approach this issue is to choose the design that will result in the highest expected payoff to management in the given situation.

A research design is not a framework to be followed blindly, without deviation. Once a design is chosen to study a particular problem, it should be viewed as a road map. If you map out your trip but then find a bridge washed out, you take a detour to get to your destination. Also in research, situations may dictate that the original research design must be modified. However, researchers should be required to justify their "detours" to management with well-thought-out arguments.

These general concerns should be remembered by managers, but they should also ask more specific questions regarding design issues. Pertinent questions such as those listed in Exhibit 6.4 will help ensure that the research will generate the information required for decision-making purposes.

SUMMARY

Research design is defined as the structure of the research project to solve a particular problem. The primary purpose of design is to guide researchers in their quest to solve problems under study. Design considerations entail decisions concerning what concepts will be studied, what approach will be used to study the problem, who or what will be studied, how the data will be collected and analyzed, and how the information collected will be presented to the manager. Typically, the research design is formally stated in a well-thought-out proposal.

Research design is one of the primary means by which the researcher ensures that unwanted error in the data collected is controlled. The chapter identifies four possible sources of error in the research design process: planning, collecting, analytical, and reporting errors. Four strategies by which research managers can control these errors are subsequently identified.

The discussion then moves to an analysis of the two main types of research design: ex post facto and experimental. The two types differ mainly in the researcher's relative control of the independent variables in the study. Field studies and surveys, as well as field and laboratory experiments, are then described and highlighted with examples.

Specific design configurations are then presented. Seven design configurations are diagramed using a notational system and then analyzed according to their potential for controlling six types of internal invalidity and three types of external invalidity.

Once the designs are presented, the discussion moves to online research design issues. Here the impact and role of the Internet on the research design process are explored.

The chapter ends with four major managerial considerations and seven key questions pertaining to the research design issues. This framework sets the stage for the subsequent chapters, which deal with other issues in the design area: measurement, questionnaire design and scale construction, sampling, and data analysis procedures.

DISCUSSION QUESTIONS

1. What is research design? Why is research design important to the manager? To the researcher?

2. Why is the study of research design so difficult? Give concrete examples to support your answer.

3. What are the major types of errors that research design attempts to minimize? Define these; then propose researcher–managerial strategies that can be used to reduce these errors.

4. What are the two major types of design that can be used to solve business research problems? Discuss the weaknesses and strengths of these two design types.

5. Discuss the difference among field studies, surveys, field experiments, and laboratory experiments. Give examples of how each one of these designs could be used to study organizational climate and its effect on job satisfaction in a business setting.

6. What is causality? Why is this concept such a problematic concern to the business researcher?

7. What are the major types of validity that one must be concerned with in research design? Discuss the major threats to these types of validity and what the researcher can do to control these threats.

8. From a managerial perspective, discuss some of the basic concerns management should have about research design. How should management become involved with the research process?

9. Discuss the differences among independent, dependent, intervening, and moderating variables. What effect does each of these variables have on research design?

10. Suppose you are devising a study to evaluate various sales force incentive plans. Your boss feels that three different plans should be tested. In addition, your boss feels that the important variables to be examined are change in sales, change in motivation, change in attitudes toward the job, and change in cold-call sales. First, diagrammatically outline a design for this study. What are the independent and dependent variables in this design? What type of design is this? What are the advantages and disadvantages of the design you have chosen? What other designs could be used to solve this problem?

11. The concepts of variance and error are central to the process of research design. Given this statement, discuss the following:

    a. Define variance and discuss its relevance to design issues.

    b. Discuss the concepts of internal and external validity and relate these issues to the concepts of variance and error in research design.

    c. Is it ever possible to have error-free research? Defend your position.

12. Discuss some of the factors that would be important in the design of multinational research studies. How would these factors affect the selection of a specific research design?

13. Using the Internet and a search engine of your choice, identify some business research firms (in any functional area) that design and conduct research. First, identify these firms, and then note what types of research are their specialties.

14. Using the Web, find articles discussing some of the difficulties of implementing research designs in a global setting. Be sure to identify the sources of these articles and summarize the important issues inherent in conducting global research.

NOTES

1. For example, see the list of errors identified by Donald S. Tull and Del I. Hawkins, *Marketing Research: Measurement and Method*, 6th ed. (New York: Macmillan, 1990), pp. 67–72; Rex V. Brown, "Evaluation of Total Survey Error," *Journal of Marketing Research* (May 1967), 4, pp. 117–127;

James Hulbert and Donald Lehmann, "Reducing Error in Question and Scale Design: A Conceptual Framework," *Decision Sciences* (1975), *6*, pp. 166–173; and Henry Assael and John Keon, "Non-sampling vs. Sampling Errors in Survey Research," *Journal of Marketing* (Spring 1982), pp. 114–122.

2. See, for example, Marc J. Wallace, "Methodology, Research Practice, and Progress in Personnel and Industrial Relations," *The Academy of Management Review* (January 1983), *8*:2, pp. 6–13; Michael Withey, Richard L. Daft, and William H. Cooper, "Measures of Perrow's Work Unit Technology: An Empirical Assessment of a New Scale," *Academy of Management Journal* (1983), *26*:1, pp. 45–63; and Jacob Jacoby, "Consumer Research: A State of the Art Review," *Journal of Marketing* (April 1978), pp. 87–96, for discussions of contradictory research results and plausible explanations for this state of affairs.

3. Although we choose to use this classification of design types, which is widely accepted (see, for example, Fred N. Kerlinger, *Foundations of Behavioral Research,* 4th ed. (Fort Worth, TX: Harcourt College, 2000)). There are other classifications. See, for example, Stephen Isaac and William B. Michael, *Handbook in Research and Evaluation* (San Diego: Robert Knapp, 1972), pp. 13–30; David Aaker and George S. Day, *Marketing Research* (New York: Wiley, 1983), pp. 49–52.

4. Kerlinger, op. cit. note 3, p. 35.

5. See Roger E. Kirk, *Experimental Design: Procedures for the Behavioral Sciences* (Belmont, CA: Brooks/Cole Publishing, 1968), pp. 4–7, for a good discussion of the selection of independent and dependent variables in an experimental setting.

6. This section is adapted from Uma Sekaran, *Research Methods for Managers* (New York: Wiley, 1984), pp. 50–59.

7. This definition follows the one used by Kerlinger, op. cit. note 3.

8. See Johnny K. Johansson and Ikujiro Nonaka, "Market Research the Japanese Way," *Harvard Business Review* (May–June 1987), pp. 16–22.

9. John A. Sonquist and William C. Dunkelberg, *Survey and Opinion Research* (Englewood Cliffs, NJ: Prentice-Hall, 1977), p. 2.

10. For two relevant examples of business research using a field experiment, see Gregory Upah, "Product Complexity Effects on Information Source Preference by Retail Buyers," *Journal of Business Research* (1983), *11*, pp. 107–126; and Lawrence A. Crosby and James R. Taylor, "Effects of Consumer Information and Education on Cognition and Choice," *Journal of Consumer Research* (June 1981), *8*, pp. 43–56.

11. Control in this instance relates to variance control, particularly error variance. As will be discussed in later chapters, researchers can control error variance through sampling, measurement, and so on. In this case, error variance was controlled in one sense through randomization of experimental treatments to groups.

12. See Alex Carey, "The Hawthorne Studies," *American Sociological Review* (June 1967), *32*, pp. 403–417.

13. Much of this section is based on the work reported in Donald T. Campbell, "Factors Relevant to the Validity of Experiments in Social Settings," *Psychological Bulletin* (1957), *54*, pp. 297–312; and later expanded upon in Donald T. Campbell and Julian C. Stanley, *Experimental and Quasi-Experimental Designs for Research* (New York: Houghton Mifflin, 1990).

14. The concepts, rationale, and purpose of sampling and randomization will be further explained in Chapter 9, Sampling Design.

15. S. L. Payne, "The Ideal Model for Controlled Experiments," *Public Opinion Quarterly* (Fall 1951), pp. 552–557.

16. See *Insight Express*, "How to Develop Online Surveys that Work," September 2002, http://wp.bitpipe.com/resource/org_1027448665_672/Online_Service.pdf, downloaded October 29, 2003; and *Inside Research* (January 2002) *13*:1.

17. "MarketTools Strengthens MR, Takes Aim at Big MR End Users," *Research Business Report,* June 2002, http://www.markettools.com/press/MarketTools_June2002.pdf, downloaded October 30, 2003.

18. See, for example, "SPSS and AOL Announce a New Agreement," October 15, 2003, http://www.spss.com/press/template_view.cfm?pr_id=625, downloaded October 30, 2003; and explore the Web sites of other major research suppliers: http://www.npd.com/about/about_consumer.html, http://www.nfow.com/comprehensive.asp, and http://www.harrisinteractive.com/solutions/vert_servicebureau.asp.

19. See *Insight Express*, "Proven: Online Research's Time has Come," May 2002, http://www.insightexpress.com/proven/whitepaper.pdf, downloaded October 30, 2003.

SUGGESTED READING

The Nature of Research Design

Babbie, Earl, *The Practice of Social Research,* 9th ed. (Belmont, CA: Wadsworth, 2001).

> *Chapter 4 in Babbie's text provides an excellent overview of the nature of research design as it applies to the social sciences. The book as a whole is a good supplement to this text.*

Miller, Delbert, *Handbook of Research Design and Social Measurement,* 6th ed. (Newbury Park, CA: Sage, 2002).

> *A useful reference for all aspects of the research design issue. The handbook is a very good reference for anyone dealing with design issues.*

Error Reduction Through Design

Hulbert, James, and Donald Lehman, "Reducing Error in Question and Scale Design: A Conceptual Framework," *Decision Sciences* (1975), *6*, pp. 166–173.

> *This article presents an easy-to-understand conceptual framework for reducing error in one aspect of research design. It is "must" reading for the serious methodologist.*

Katzer, Jeffrey, Kenneth H. Cook, and Wayne Crouch, *Evaluating Information: A Guide for Users of Social Science Research,* 4th ed. (New York: McGraw-Hill, 1997).

> *This paperback presents a very good analysis of the role of bias and error in the research design process. It is well worth reading for all students and users of research.*

Major Types of Designs

Berger, Paul D., and Robert E. Maurer, *Experimental Design* (Belmont, CA: Wadsworth, 2002).

> *Provides a good discussion about experimental design and its application to management problems. Attempts to minimize mathematical detail and focuses on the application and understanding of experimental designs.*

Cox, Brenda, David Binder, and B. N. Chinnappa, Eds., *Business Survey Methods* (New York: John Wiley & Sons, 1995).

> *Describes current methods, technologies, and innovative research approaches to business surveys. A compendium of invited papers from internationally recognized researchers.*

Rea, Louis M., and Richard A. Parker, *Designing and Conducting Survey Research: A Comprehensive Guide,* 2nd ed. (San Francisco: Jossey-Bass, 1997).

> *A practical guide for the design of survey research. Useful for professionals and students alike. Covers a wide range of topics of interest to those involved in the survey process.*

Specific Design Configurations

Campbell, O. T., and J. C. Stanley, *Experimental and Quasi-Experimental Designs for Research* (New York: Houghton Mifflin, 1990).

> *A "must" for any researcher's bookshelf. This is the indispensable classic on research design. The strong point of the book is its emphasis on error reduction in design.*

Kirk, Roger E., *Experimental Design: Procedures for the Behavioral Sciences,* 3rd ed. (Monterey, CA: Brooks/Cole, 1995).

> *This is one of the better texts on experimental designs. It deals with the more advanced designs in greater detail than is possible in this text.*

Online Research Design Issues

> *The best readings in this area are on the Web because the field is so dynamic today. As one professional researcher involved in the business told me, we operate "on the fly." As a result, the issues are constantly changing. Start by examining some of the Web sites identified in notes 16 through 19.*

ADVANCED EXPERIMENTAL DESIGNS

The use of experimental designs allows the researcher to manipulate, or in some way control, independent variable(s) and then measure their effect(s) on the dependent variable(s) of interest. For simplicity, we limited the discussion in the chapter to the conceptual underpinnings of design rather than attempting to deal with the complexity of the more advanced design configurations that are actually used. However, all that is required to perform a true experiment correctly is a single-level experimental variable (an independent variable), a dependent variable, and the random selection of an experimental group and a control group.

For example, if our intent is to determine whether perks or incentives will increase the happiness of our executives, we could randomly assign mid-level executives to two groups. On day one, we premeasure the happiness of both groups. We then give the experimental group a new desk, chair, and a personal computer; the control group receives nothing. At the end of day two, we postmeasure the happiness of the two groups. This is a simple example in the form of design 5 from Table 6.3, using one independent variable (incentives) and one dependent variable (happiness).

More often, however, management is interested in studying more than one level of an independent variable in order to determine an optimal level for that variable. In our example, we might want to see what combinations of incentives would produce the greatest happiness. Similarly, management might want to combine the impact of two or more variables and measure their combined effect on an outcome variable. In addition, one must always consider the control of one or more of the nuisance variables that may bias the outcome of an experiment. In any event, all of these considerations fall into the realm of advanced experimental design.[1] The use of advanced designs in no way contradicts our previous discussion of the basics of design. It simply gives us vehicles for optimally implementing our design considerations to solve management problems. The designs are simply more efficient.

Advanced experimental design emphasizes two approaches to reducing potential sources of bias in an experiment: experimental control and statistical control of nuisance variables. *Nuisance variables* are variables that may be important and have an impact on the relationships being tested but are not of primary concern in the study. *Experimental control* occurs when the researchers control nuisance variables by research design. Usually, this is accomplished by one or more of the following three means: (1) randomly assigning subjects to treatments, (2) holding the nuisance variable constant across all subjects, or (3) including the variable as one of the treatments in an experimental design. On the other hand, *statistical control* is the use of regression procedures (or, more specifically, covariance analysis) to remove the effects of a nuisance variable statistically from a particular study. Both types of control emphasize the reduction of error so that the researcher may more confidently state the effects of the independent variable on the dependent variable. Statistical control has an advantage: The experimenter can remove potential sources of bias that are difficult or impossible to eliminate by experimental control. Both types rely on the concept of randomization to achieve the objectives of the design.

Because space limitations make it impossible to cover every advanced design here, we briefly discuss five general design types: the completely randomized design, the randomized block design, the Latin square design, the factorial design, and the analysis of covariance experiment. Our discussion will focus on the impact on a dependent variable of one or more independent variables or levels thereof. But these principles of design are easily transferred to an experiment with several dependent variables.[2]

Completely Randomized Design

The most easily understood formal experimental design (in the class of complete block designs) is the completely randomized design.[3] In this design, we are concerned with the impact of a single variable, usually set at several levels, on a dependent variable of interest. Randomization is used to create equivalent groups in this design. The researcher randomly assigns the unit of analysis to one or more groups, then randomly assigns treatments to groups.[4]

Suppose management is interested in the effect of three different styles of superior-to-subordinate instructions on the comprehension of a particular job. After randomization, one group of 25 subordinates is given verbal instructions concerning the job; the second group of 25 is given written instructions; and the third group of 25 is given no instructions at all. All the subordinates are allowed to ponder the situation for ten minutes. They are then given an objective test as to how well they comprehend the job to be undertaken. The design is laid out in Table 6A.1. The purpose of the design is to determine which instruction level (independent variable) creates better comprehension (dependent variable) of the approach to the job. In other words, the primary research question asked is: Does the manipulation of the variable instruction have an impact on comprehension of the approach to the job?

Associated with this and every experimental design is a model that can help analyze the sources of variability in the design. The better the model can represent this variability, the better the researcher can evaluate the effects of the treatment. We discuss this class of analytic models in Chapter 14. Commonly referred to as *linear* models, they encompass the analysis of variance and regression methods of data analysis.

Randomized Block Design

In the completely randomized design, we allowed individual differences to remain unchecked. Obviously, we cannot expect that the three groups will be equivalent on all subject characteristics. A way to help the system work better is to select the subjects from the most homogeneous

TABLE 6A.1 Completely Randomized Design

Instructions	Treatment Levels		
	Experimental Groups		Control Group
	a_1 (Verbal)	a_2 (Written)	a_3 (No Specific)
	$x_{1,1}$	$x_{1,2}$	$x_{1,3}$
	$x_{2,1}$	$x_{2,2}$	$x_{2,3}$
	$x_{3,1}$	$x_{3,2}$	$x_{3,3}$
	.	.	.
	.	.	.
	.	.	.
	$x_{25,1}$	$x_{25,2}$	$x_{25,3}$
Treatment means	$\overline{X}_{.1}$	$\overline{X}_{.2}$	$\overline{X}_{.3}$

where x_{ij} = ith individual comprehension score in the jth treatment.

Grand mean = $\overline{X}_{..}$

population possible. If we are able to identify some of the important sources of individual differences beforehand, we can select a homogeneous group of subjects to be used in the experiment. Usually this allows us to generalize the results of the experiment better than with the completely randomized design.[5]

For example, suppose the 75 subordinates are taken from five different departments. We assume that workers from the same department can be expected to be more homogeneous with respect to certain characteristics (i.e., cohesion) than workers from different departments. Differences among the departments can be regarded as a nuisance or extraneous variable that can be experimentally controlled through the use of a randomized block design.

The randomized block design is formed by randomly assigning the subordinate workers within each department to the three experimental conditions. The layout of this design is depicted in Table 6A.2. We can regard this design as consisting of five independent experiments—the first containing workers of the one department, the second containing workers of the second, and so on. In each case, the employees within each block are randomly assigned to the three treatment levels of instructions. First, the workers were grouped by department, and then they were randomly assigned to treatments by departments (blocks).

Essentially, this design would give us information similar to the completely randomized design. It also allows us to examine effects due to departments (blocks) and the interaction effect of instructions by departments. In essence, a randomized block design can be more powerful than the completely randomized design if the blocking accounts for considerable variability in job comprehension. However, the cost (both in time and money) to match the units of analysis may not be worth the additional power of the design.[6]

Instructions: Blocks (Departments)	Treatment Levels			Block Means
	Experimental Groups		Control Group	
	a_1 (Verbal)	a_2 (Written)	a_3 (No Specific)	
b_1	5 (workers)	5 (workers)	5 (workers)	$\overline{X}_{1.}$
b_2	5 (workers)	5 (workers)	5 (workers)	$\overline{X}_{2.}$
b_3	5 (workers)	5 (workers)	5 (workers)	$\overline{X}_{3.}$
b_4	5 (workers)	5 (workers)	5 (workers)	$\overline{X}_{4.}$
b_5	5 (workers)	5 (workers)	5 (workers)	$\overline{X}_{5.}$
Treatment means	$\overline{X}_{.1}$	$\overline{X}_{.2}$	$\overline{X}_{.3}$	

5 workers per cell, each dependent measure is x_{ijk}

where

i represents worker,
j represents blocks (departments), and
k represents treatment levels (instructions).

The Latin Square Design

The Latin square optimizes experimental control by including the nuisance variable or variables as one of the treatments in the design. The Latin square is used when the researcher wants to simultaneously control two nuisance variables.[7] The designated levels of these variables are assigned to the rows and columns of the Latin square. In our randomized block example, the workers are equated on the basis of departmental cohesion. Workers in the same department may have homogeneity in terms of ability levels, because they were picked according to the same selection standards.

Because job approach based on comprehension is the dependent variable, we might also want to control the extraneous variable, ability. To do this, we first must determine a criterion for measuring the ability of the workers. We then randomly select three workers from each department, one from each of the three designated ability levels (high, medium, and low). As illustrated in Table 6A.3, departments are crossed by abilities to form a 3×3 table. The three treatment levels a_k (the three instruction treatments) are randomly assigned to the nine cells so that any row and column intersection will contain the treatment level only once.

The fact that you need an equal number of blocks for each nuisance variable forces the table to be square. The design is exceptional for studies involving fixed units of analysis, such as departments, organizations, stores, and other entities. The design also lets the researcher minimize sample size by allowing the same units of analysis to react to all different levels of the independent variable. Generally, this design is more powerful

TABLE 6A.3 Latin Square (3 × 3)

Block (Departments)	Ability of Workers			Block Means
	c_1 High	c_2 Moderate	c_3 Low	
b_1	$(a_1)x_{1,1,1}$	$(a_2)x_{1,2,2}$	$(a_3)x_{1,3,3}$	$\overline{X}_{1..}$
b_2	$(a_2)x_{2,1,2}$	$(a_3)x_{2,2,3}$	$(a_1)x_{2,3,1}$	$\overline{X}_{2..}$
b_3	$(a_3)x_{3,1,3}$	$(a_1)x_{3,2,1}$	$(a_2)x_{3,3,2}$	$\overline{X}_{3..}$
	$\overline{X}_{.1.}$	$\overline{X}_{.2.}$	$\overline{X}_{.3.}$	

where x_{ijk} = comprehension score of the ith department on the jth ability given the kth treatment level (instruction type);

$$\text{Ability means} = \overline{X}_{.1.} \overline{X}_{.2.} \overline{X}_{.3.}$$
$$\text{Departmental means} = \overline{X}_{1..} \overline{X}_{2..} \overline{X}_{3..}$$
$$\text{Grand mean} = \overline{X}_{...}$$
$$\text{Treatment level means} = a_1 = (x_{1,1,1} + x_{3,2,1} + x_{2,3,1})/3 = \overline{X}_{..1}$$
$$a_2 = (x_{2,1,2} + x_{1,2,2} + x_{3,3,2})/3 = \overline{X}_{..2}$$
$$a_3 = (x_{3,1,3} - x_{2,2,3} + x_{1,3,3})/3 = \overline{X}_{..3}$$

than the ones previously mentioned. However, it is limited to only two extraneous variables at a time. More than two nuisance variables must be handled by a Greco-Latin square design.[8]

Another weakness of the Latin square design is that it is not possible to assess the interaction of the control variables (or the control variables and the independent variable). Sometimes, we wish to test two or more variables at the same time to determine whether the variables interact or act independently to produce an observed effect on the dependent variable. The design that can handle this problem is called a *factorial design*.

Factorial Design

When a researcher uses a factorial design, the emphasis is on the evaluation of the combined effects of two or more treatments on some dependent variable(s).[9] This type of design can be completely randomized or randomized block; we will examine only the completely randomized style.

To illustrate this type of design, let us take an example concerning magazine advertising. Suppose we have reason to believe that a magazine's advertising format will influence the comprehension of the basic points of the ad. Two independent variables that might be of interest are the length of the printed lines and the color contrast in the printed letters on the page. Assume we choose three often-used line lengths (5, 7, and 12 inches) and three different contrast levels (low, medium, and high). We can now

TABLE **6A.4** Factorial Design: Completely Randomized

	Contrast Levels		
	b_1 Low	b_2 Medium	b_3 High
Line Length: a_1, 5 in. a_2, 7 in. a_3, 12 in.	$x_{1,1,1}$	$x_{2,2,1}$	$x_{3,3,1}$
A Treatment Means	$\overline{X}_{1..}$	$\overline{X}_{2..}$	$\overline{X}_{3..}$
B Treatment Means	$\overline{X}_{.1.}$	$\overline{X}_{.2.}$	$\overline{X}_{.3.}$

where, for example (all cells contain data):

$x_{1,1,1}$ = comprehension score of an individual given both treatments of 5-inch line length and low contrast levels

$x_{3,3,1}$ = comprehension score of an individual given both treatments of 12-inch line length and high contrast levels

Grand mean = $\overline{X}_{...}$

manipulate the two variables simultaneously in a 3×3 factorial design, as illustrated in Table 6A.4. This would give us nine unique treatment groups. We would randomly assign subjects to the treatment groups or cells of the design, expose them to the combined treatments, and measure their comprehension levels for the ad.

The factorial design is an economical one. It gives the same information as two or more single-factor randomized designs, with the added feature of an estimate of the interactive effects of the independent variables. The interaction is an important aspect of research and will be discussed at length in Chapter 14. We can also resort to higher level factorial designs to control important and unwanted variability.[10] Such higher order factorials are beyond the scope of this book, so if they need to be incorporated into your research, you should seek additional information.

Analysis of Covariance

Analysis of covariance experiments are randomized designs that exercise statistical control rather than experimental control over nuisance variables. Such designs are referred to as natural, quasi-, or field experiments (discussed in Chapter 6). The analysis of these experiments makes it possible to use the combined techniques of analysis of variance and regression methods to remove the effects of nuisance variables.[11] Analysis of covariance can be used in conjunction with any well-thought-out experimental design. To use statistical control, however, note that very strict assumptions must be satisfied, so it may prove untenable in many business research applications.[12]

NOTES

1. See, for example, Roger E. Kirk, *Experimental Design: Procedures for the Behavioral Sciences,* 3rd ed. (Belmont, CA: Brooks/Cole, 1995); B. J. Winer, *Statistical Principles in Experimental Design* (New York: McGraw-Hill, 1971); and G. Keppel, *Design and Analysis: A Researcher's Handbook* (Englewood Cliffs, NJ: Prentice-Hall, 1973).

2. See Winer, op. cit. note 1, pp. 233–240; or Jeremy D. Finn, *A General Model for Multivariate Analysis* (New York: Holt, Rinehart, and Winston, 1977).

3. Most texts in design divide randomized designs into building block subsystems. These are complete block, incomplete block, factorial, and covariance designs. For a complete explanation of this subsystem, see Kirk, op. cit. note 1.

4. See, for example, Ralph Katz, "The Effects of Group Longevity on Project Communication and Importance," *Administrative Science Quarterly* (1982), *27,* pp. 81–104.

5. See, for example, P. J. McClure and E. J. West, "Sales Effect of a New Counter Display," *Journal of Advertising Research* (March 1969), *9,* pp. 29–34.

6. *Power* refers to the probability of rejecting the null hypothesis when an alternative hypothesis is true. It is a statistical notion of detecting real differences in the data, to be discussed in Chapter 12.

7. See, for example, William E. Cox, Jr., "An Experimental Study of Promotional Behavior in the Industrial Distributor Market," in Raymond M. Haas, Ed., *Science, Technology and Marketing* (Chicago: American Marketing Association, 1966), pp. 578–586.

8. See Seymour Banks, *Experimentation in Marketing* (New York: McGraw-Hill, 1965), pp. 170–175.

9. See, for example, David Caldwell and Charles O'Reilly III, "Responses to Failure: The Effects of Choice and Responsibility on Impression," *Academy of Management Journal* (1982), *25:*1, pp. 121–136.

10. See, for example, H. Bruce Lammers, Laura Leibowitz, George Seymour, and Judith Hennessey, "Humour and Cognitive Responses to Advertising Stimuli: A Trace Consolidation Approach," *Journal of Business Research* (1983), *11,* pp. 173–185.

11. See, for example, Michael P. Allen and Sharon K. Panian, "Power Performance and Succession in the Large Corporation," *Administrative Science Quarterly* (December 1982), pp. 538–547.

12. Kirk, op. cit. note 1, pp. 457–459; and Elazar J. Pedhazur, *Multiple Regression in Behavioral Research* (New York: Holt, Rinehart, and Winston, 1982), Chapter 13.

7

Foundations of Measurement

Overview

The Nature of Measurement

Components of Measurement

The Measurement Process

Levels of Measurement

Nominal

Ordinal

Interval

Ratio

Evaluation of Measurement Scales

Validity

Reliability

Managerial Considerations

Summary

Discussion Questions

Notes

Suggested Reading

OVERVIEW

Once a research problem is specified and a particular type of design is chosen to solve the problem, the task of measurement must be undertaken. *Measurement* can be defined as a rule for the assignment of numerals (numbers) to aspects of objects, persons, states, and events.[1] Measurement is of concern to managers and researchers alike because it is the means by which reality is represented in the researcher's analytical world. If the representations of reality are inaccurate, the results of the research are bound to be wholly inadequate from the decision maker's perspective. Simply stated, the researcher and manager receive from the study no more than they initially put into it. As elsewhere, GIGO (garbage in, garbage out).

The theory of measurement should therefore be of concern to everyone associated with the research process. The theory of measurement deals with the delineation of rules and procedures designed to increase the probability that goings-on in the world of concepts will correspond to goings-on in the world of reality.[2] For example, if the researcher is concerned with measuring the job satisfaction of employees of a particular firm, measurement's task would be to ensure that the level of job satisfaction is actually assessed by the measurement scale. The delineation of measurement rules and procedures is the subject of this chapter.

The chapter begins with an in-depth discussion of the measurement process. The nature of measurement is clarified, and the process is explicated. Measurement is clearly positioned in the research process. Then the topic of level of measurement is addressed. Level of measurement involves the specific rules for assigning numbers to characteristics and the resultant properties of scales. The four levels of measurement are outlined and examined.

After the foundations of measurement are presented, the discussion moves to the evaluation of measurement procedures. Evaluation is necessary to ensure that mapping procedures result in high-quality measures—in other words, valid and reliable measures. Sources of variation in measures are examined and related to the concepts of validity and reliability. The chapter ends with an examination of managerial considerations in measurement.

THE NATURE OF MEASUREMENT

The goal of measurement is to translate the characteristics and properties of empirical events into a form that can be analyzed by the researcher. Measurement is the procedure that is used to symbolically represent aspects of reality in the researcher's analytical world. Proper measurement ensures that the research being conducted is grounded in real-life happenings that are of interest to decision makers. Thus, the correct execution of this procedure is critical to obtaining usable information for the manager.

Measurement is concerned with the assignment of numbers to empirical events according to a set of rules. The actual procedure by which this is done is the *measurement process:* the investigation of the underlying characteristics of the empirical events of interest and the assignment of numbers to represent these characteristics. Measurement devices are numerous and diverse, but we know that three components are necessary for measurement to take place:

1. Observable empirical events
2. The use of numbers to represent these events
3. A set of mapping rules[3]

Components of Measurement

Empirical events can be simply defined as a set of observable characteristics of an object, individual, or group. The term *observable* implies that one can perceive (or at least infer) that an object, individual, or group possesses a particular characteristic (perhaps to some degree). For example, if we were to study the relationship between the sex of administrators and the job satisfaction of their subordinates, it would be necessary to first identify the unit of analysis. The *unit of analysis* can be defined as the primary empirical object, individual, or group under investigation. In this case, the units of analysis are the individual administrators and subordinates. Once the primary empirical entity is specified, the researcher can identify the characteristics (concepts) that are of immediate concern—in this case, the sex of the administrator and the job satisfaction level of the subordinate. These are the concepts to be measured.

The second component in measurement is the use of numbers to represent empirical events. *Numbers* are numerals or other symbols that are used to identify or designate. Here, we are concerned with the use of numeric designations to give meaning to the characteristic(s) of interest. Numbers, in and of themselves, are meaningless. The researcher, after a thorough study of the nature of the phenomenon, assigns meaning to the numbers, largely through the specification of the appropriate level of measurement. The closer the correspondence of these representations to reality, the greater the probability that the study will have meaning to the manager. Levels of measurement will be treated in greater depth later in this chapter.

The final essential component in all measurement is a set of mapping rules, often referred to as *coding* by researchers (see Chapter 12). *Mapping rules* are statements that dictate the assignment of numbers to empirical events. These rules ultimately describe the characteristics that we are measuring. The rules are developed by the researcher for the purposes of the study. This may seem like a rather arbitrary procedure, but it is in the best interests of the researcher and manager alike that the measurement

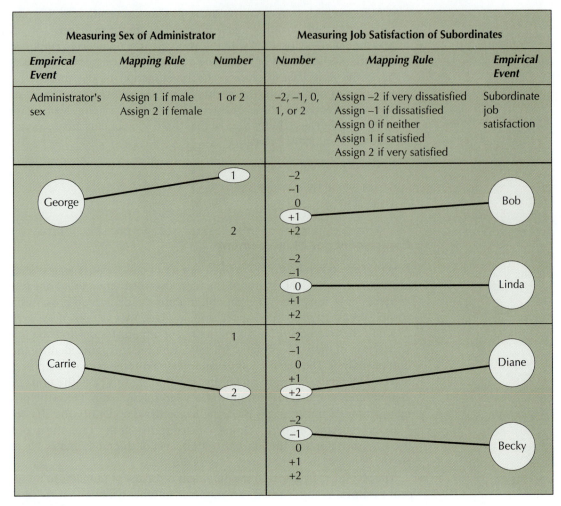

Measuring Sex of Administrator			Measuring Job Satisfaction of Subordinates		
Empirical Event	Mapping Rule	Number	Number	Mapping Rule	Empirical Event
Administrator's sex	Assign 1 if male Assign 2 if female	1 or 2	−2, −1, 0, 1, or 2	Assign −2 if very dissatisfied Assign −1 if dissatisfied Assign 0 if neither Assign 1 if satisfied Assign 2 if very satisfied	Subordinate job satisfaction

FIGURE 7.1

An Example of the Relationships Among the Three Components of Measurement

process ultimately results in symbolic representations that correspond as closely as possible to reality.

In order to illustrate the three major components necessary for any measurement, Figure 7.1 diagrams the relationships among empirical events, numbers, and mapping rules in our example concerning the relationship between sex of administrators and job satisfaction of subordinates. In a very simplified example, suppose we have two administrators, each overseeing two subordinates. We already stated that our primary unit of analysis is the individual. We first identify the two administrators (George and Carrie) and the four subordinates (Bob, Linda, Diane, and Becky). We then assess the empirical events of interest (administrator's sex and subordinate's job satisfaction) for each individual and assign numbers according to the mapping rules developed by the researcher.

FIGURE 7.2

The Measurement
Process

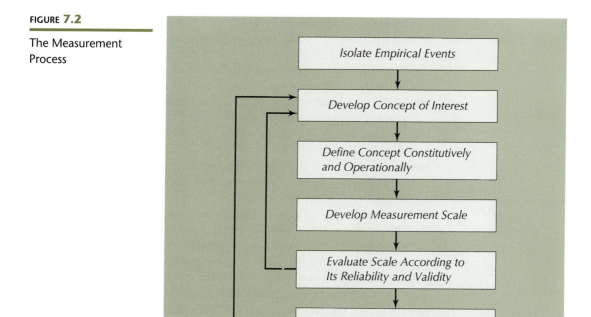

The Measurement Process

The example outlined in Figure 7.1 merely illustrates the major components necessary for any measurement to occur. It does not cover any of the complexities or nuances of the measurement process. The measurement process can be described as a set of interrelated steps that begins with the isolation of empirical events of interest and ends with the actual measurement of the characteristics under study. Figure 7.2 is a schematic diagram of the measurement process.

The process is initiated when the researcher isolates empirical events that are in need of measurement. As stated in Chapter 3, theory usually dictates what is to be measured. Typically, this activity is a direct consequence of problem identification and formulation. In essence, empirical events are summarily stated in the form of concepts and/or constructs and are related through a statement of the problem.

The next step in the process is precise definition of the concepts identified. The two types of definitions used for this purpose are constitutive and operational. It is important to note the centrality of these definitions in the measurement process pictured in Figure 7.2. If concepts are not explicitly defined, it is highly unlikely that useful measurement scales can be developed. Without these definitions, how do we know we are measuring what we want to measure?

Constitutive definitions define concepts with other concepts and constructs setting the domain for the concept of interest. If a concept is correctly defined constitutively, it should be readily distinguishable from any other concept. This notion of distinguishability is the crux of good measurement and, ultimately, good research. Vague constitutive definitions only serve to hinder the interpretability and credibility of the research.

Once the constitutive definition is decided on, an *operational definition* must be reached, to accurately reflect the essence of the former. Operational definitions detail the mapping rules and means by which the variable will be measured in reality. The definition states the procedures that a researcher must follow to assign some number to the concept being measured.

Up to this point, we have treated the measurement process as relatively clear-cut. However, this is normally not the case in actual research endeavors. Generally, researchers are faced with a variety of different theory-based (or, for that matter, researcher-based) constitutive and operational definitions from which to choose. For example, let us take the concept of job performance. The concept may mean the successful/unsuccessful outcome of a specific task to one researcher, whereas to another it represents the worker's reactions to the consequences of completing a specific job. The researcher and manager must agree on the essence of the concept (constitutive definition) so that both parties mean the same thing when they discuss job performance. If the manager's and the researcher's definitions are different, the research is doomed from the start.

Once agreement is reached on the conceptual definition of a concept, the researcher must choose between alternative operational definitions. For example, if the constitutive definition of job performance is the degree to which an employee adequately accomplishes the tasks of his or her position, this concept could be operationally defined in a number of ways. The definition could include the proportion of available workdays the employee was present, the quantity of output, the quality of output as measured by the number of mistakes, or even the lack of tardiness.

Obviously, there are many different potential definitions for any one concept. This is particularly true when dealing with the vast array of abstract, imprecise concepts used in business inquiry. This diversity is both a blessing and a nightmare to the researcher. It is positive in that the researcher can choose the definitions that fit most appropriately with the stated problem. On the downside, a definitive statement of the relationship between any two concepts becomes almost impossible.[4]

After the definitions are properly stated, the assignment of numbers must be made. The goal is to assign numerals so that the properties of the numbers are paralleled by the properties of the events to be measured.[5] This task is accomplished when the researcher thoroughly understands the empirical nature of the events to be measured, then translates this knowledge into the development of the appropriate measurement scale that displays similar properties. A *measurement scale* is a device that is used to assign numbers to aspects of objects and events.[6]

The development of measurement scales is of extreme importance in business research. The rest of this chapter and much of the following one focus on the development of measurement scales. This chapter concentrates on the delineation of the general properties (commonly referred to as *levels of measurement*) and the necessary traits of good measurement scales (the issues of validity and reliability). Chapter 8 outlines some specific types of scales (such as Likert and Osgood's semantic differential) and discusses how they should be constructed to ensure proper instrument design.

LEVELS OF MEASUREMENT

Scales of measurement vary in complexity. *Simple scales* are one-item devices that are used to measure some characteristic. An example of a simple scale of a dichotomous nature is as follows:

1. Are you: ❑ Male _____   ❑ Female _____?

Complex scales are multi-item devices that are used to measure some characteristic. An example of a complex scale used to measure the energy consciousness of an individual is as follows:

Energy Consciousness Scale

	Strongly Disagree	Disagree	Neither Disagree nor Agree	Agree	Strongly Agree
1. Energy problems are of primary importance at this time.	1	2	3	4	5
2. A person who saves energy is an asset to society.	1	2	3	4	5
3. I believe there is a critical energy shortage.	1	2	3	4	5
4. I feel I can make a contribution to society by conserving energy.	1	2	3	4	5
5. I believe people who waste energy are foolish.	1	2	3	4	5

Although the complexity of measurement devices varies tremendously, all scales possess the properties of at least one of the four levels of measurement. This is a critical consideration because the level of measurement of the variables in the study is critical in deciding which type of statistic to apply to analyzing the relationships or differences in the data. This text stresses a thorough understanding of levels of measurement and their impact on the selection of analytical techniques.

Most discussions of the levels of measurement can be traced back to S. S. Stevens.[7] Stevens classified scales of measurement into four major levels: nominal, ordinal, interval, and ratio, as shown in Table 7.1. In his treatise, Stevens discussed:

1. The mathematical transformations that could be performed on each level
2. The summary statistics that are appropriate to each level
3. The precise relationship between level of measurement and the use of statistical techniques

These points are generally well accepted in practice today.[8] However, many psychometricians disagree with the strict interpretation of point 3 in Stevens's original formulation. This issue will be addressed in the following discussion.

Nominal

Nominal measurement is used to classify objects, individuals, or groups. Many of the frequently studied variables in business inquiry are nominal in nature. Essentially, nominal scales use numbers only as labels to identify or categorize objects or events. Some examples are sex, religious preference, occupation, and geographic area of residence. When using nominal measurement scales, we divide our measure into mutually exclusive and collectively exhaustive subsets (unless they are naturally so constructed).

For example, suppose we are conducting a study to determine the relationship between sex of an individual and communication ability. Assume that we obtain 100 responses to an in-house company survey. Because one concept under study in our investigation is gender, we ask the simple-scale question, "Are you: Male _____ Female _____?" We are ultimately going to classify our respondents into two subsets for analysis: male and female.

As we will see later, one of the goals of the measurement process is to create variables that we can analyze or manipulate in our study design. In our gender example, we have created a *binomial* or *dichotomous* variable that is a 0/1, absence or presence type of variate.[9] Keep in mind that in measurement we try to assign symbols or numerals in our mapping procedure to measure how much of the characteristic the object, group, or individual possesses. The most common approach is to assign 0 for a male or 1 for a female, or vice versa. We could have used M and F or even 11 and 12 to designate the categories of the variable. *The important point to remember is that nominally scaled numbers only act as labels for the classification or category.* Therefore, we cannot perform arithmetic operations on these numbers because they only indicate the presence or absence of some characteristic.

TABLE 7.1 The Four Major Levels of Measurement

Level	Description*	Basic Empirical Operations	Typical Usage	Typical Statistics	
				Descriptive	Inferential
Nominal	Uses numerals to identify objects, individuals, events, or groups.	Determination of equality/inequality	Classification	Percentages/mode	Nonparametric
Ordinal	In addition to identification, the numerals provide information about the relative amount of some characteristic possessed by an event, object, and so on.	Determination of greater or less	Rankings/ratings	Median (mean and variance)[†]	Nonparametric (parametric)[†]
Interval	Possesses all the properties of nominal and ordinal scales; in addition, the intervals between consecutive points are equal.	Determination of equality of intervals	Preferred measure of complex concepts/constructs	Mean/variance	Parametric
Ratio	Incorporates all the properties of nominal, ordinal, and interval scales, plus an absolute zero point.	Determination of equality of ratios	When precision instruments are available	Geometric mean/ harmonic mean	Parametric

*Because higher levels of measurement contain all the properties of lower levels, we can convert higher level scales into lower level ones (i.e., ratio to interval or ordinal or nominal; or interval to ordinal or nominal; or ordinal to nominal).

[†]See Fred N. Kerlinger, *Foundations of Behavioral Research* (New York: Holt, Rinehart, and Winston, 1973), p. 441; and Jum C. Nunnally, *Psychometric Theory*, 2nd ed. (New York: McGraw-Hill, 1978), pp. 43–50, for discussions of the use of parametric statistics in ordinal scales.

Source: Adapted with permission from S. S. Stevens, "On the Theory of Scales of Measurement," *Science* (7 June 1946), *103*, pp. 677–680. Copyright © 1946 by the American Association for the Advancement of Science.

For example, if we assigned employee numbers for identification purposes in our study of sex and communication ability, doing the arithmetic operations of adding, subtracting, multiplying, and/or dividing these nominal scales would be meaningless. What does it mean to add employee number 10 to employee number 30? Or what does it mean if we multiply employee number 50 times employee number 6? It is meaningless from a practical standpoint.

Our second concept, communication ability, has interesting connotations as a nominal measurement. In reality, communication ability in individuals is perhaps better reflected as a continuous phenomenon (i.e., a matter of degree) that could be represented by a higher level of measurement (perhaps interval). However, for our purposes, we are only concerned with three categories—high, medium, and low.[10] Scale conversion from a higher level of measurement to a lower level is very common in business inquiry, for a number of reasons, which we will discuss in the next chapter.

Ordinal

Ordinal measurement provides information about the relative amount of some trait possessed by an object or individual.[11] This level of measurement contains all the information of a nominal scale plus some relative means of ordering. It provides information on whether an object possesses more or less of a characteristic, but not how much more or less.

Ordinal measurement has three major uses in business inquiry:

The ranking of items. This use is the least controversial of the three. A numbering system is used to reflect the relative ordering of phenomena. For example, if we have a respondent rank five stocks as to their perceived future earnings (where a 1 is assigned to the top future earner, and so on, until all stocks receive a number between 1 and 5), this results in an ordinal ranking scale. If accurate, the scale represents the relative perceived future earnings of the stocks. But it does not provide the researcher with information about the amount of difference between each stock.

The rating of a characteristic. This use of the ordinal scale is the most controversial. The researcher assigns numbers to reflect the relative ratings of a series of statements, then uses these numbers to interpret relative differences. For example, suppose we develop a simple scale to assess students' perceptions of the business research course, then assign a number rating the value of that statement. The scale is illustrated below. The number in parentheses represents the numerical value assigned to an individual's response to the question.

Business Research is a terrific course. (Check one.)

   (6) I definitely agree.   (3) I moderately disagree.

_(5)_ I generally agree. _(2)_ I generally disagree.

_(4)_ I moderately agree. _(1)_ I definitely disagree.

If the researcher assumes interval scale measurement (i.e., that the intervals between the numbers are indeed equal), then the more powerful statistical descriptors and analytical techniques can be used to analyze the data.

The creation of other complex scales and indices. This use of ordinal scales has also been roundly criticized by a number of statisticians. However, this criticism has not stopped the widespread use of complex scales and indices in business. Basically, these scales are collections of items that are rated by respondents and then summated by researchers to arrive at the measure of some complex phenomenon. A prime example of this usage is the overall evaluation score for proposals, which is shown in Figure 5.3. In fact, most performance evaluations are of this general type.

What is the most appropriate use of ordinal scales? There are proponents and opponents of each of the uses given. We take the view that if ranking procedures are used, they are strictly ordinal in nature and should be treated that way. On the other hand, complex scales and indices of an ordinal nature are useful in certain situations and allow us to use the more powerful parametric statistical techniques. However, it must be recognized that these scales do not provide absolute information regarding how much more or less of the characteristic various objects or events possess. We side with the view of Fred Kerlinger, who states, "The best procedure would seem to be to treat ordinal measurements as though they were interval measurements but to be constantly on the alert to the possibility of gross inequality of measurement."[12]

A final note on ordinal measurement scales pertains to the assumed origin of the scale. Up to now we have discussed ordinal scales with arbitrary origins. In some instances, one of the classes of this type of scale may logically be assigned the number 0 to represent zero amount of the property under consideration.[13] This is referred to as an ordinal scale with a natural origin. In research, we usually assign the numerical value of 0 to the category that never occurs or to a neutral point.

Interval

Interval scales possess all the properties of nominal and ordinal scales, plus the added property that the intervals between the points on the scale are equal.[14] Most researchers would like all of their scales to possess interval properties because such scales can be used to distinguish how much more of a trait one individual or object might have than another. To illustrate the basic properties of this level of measurement, let us take a simple example.

Suppose I measure both your body temperature and mine with a Fahrenheit thermometer. I am taking intervally scaled measurements because there are equal distances between degree gradations and because the zero point on the thermometer is arbitrary. I find your temperature to be 92 degrees and mine to be 90 degrees. I now can subtract 90 from 92 and determine that your temperature is 2 degrees more than mine. I can also add 92 and 90 together and divide by 2 and determine that our average body temperature is 91 degrees. Because interval scales can be thoroughly manipulated mathematically and produce continuous variables with known distributions, we can use the more powerful parametric statistical procedures on the variables. In addition, we can also use nonparametric analyses if so desired.

Ratio

Ratio measurements retain all the properties of the first three measurement scales, with the added characteristic of an empirical absolute zero point. At the absolute zero point, there is a complete absence of the characteristic being measured. Ratio scales are very common in the physical sciences. In business research, they are rarer because of the fuzzy nature of many of the concepts and constructs studied.

However, we do find some important ratio measurements in business inquiry. Obvious examples are sales, costs, any monetary amounts, and any counts of physical objects (inventory, customers, and so on). In such cases, all statistical descriptive measures and inferential techniques are applicable. However, it must be noted that the use of a ratio scale produces little advantage in analysis over an interval measure.

EVALUATION OF MEASUREMENT SCALES

Once the variables of interest have been identified and defined conceptually, a specific type of scale must be selected. To a large degree, the choice is determined by the underlying properties of the concept to be studied and the researcher's anticipated use of the variable in the data analysis stage of the research process. In order to select the proper type of scale, the researcher must choose a device that accurately and consistently measures what it is supposed to measure and achieves the research objectives of the study. This process entails the evaluation of the adequacy of the measurement device.

As Figure 7.2 showed, the measurement process is a stepwise procedure in which a scale is developed and subsequently evaluated by the researcher. However, we have the same chicken-and-egg problem that applied to proposal evaluation in Chapter 5. In practice, the researcher consciously thinks ahead to the desirable properties of the measurement device and how the variable will be used in data analysis before the scale is developed or selected. Furthermore, it is difficult to discuss the particulars of the different types of measurement scales until we have at least a basic understanding of how they will be evaluated for adequacy. So we will defer the discussion of the types of scales to the next chapter and instead deal directly with the chief concerns of scale evaluation in the following pages.

The first of these concerns—choosing a measure that accurately and consistently measures what it is supposed to measure—relates to validity and reliability in measurement. This is by far the most important: to ensure that the scales will produce information that is of relevance to the decision maker. This chapter concentrates on considerations in the assessment of validity and reliability.

The second concern—the researcher's anticipated use of the variable in the data analysis stage of the research process—is also difficult to discuss without an appropriate knowledge of the types of statistical techniques that can be used to analyze the data. Some aspects of this relationship were mentioned in discussing level of measurement earlier in this chapter, but comprehensive coverage of this issue will be provided in the data analysis chapters later in the text. Let us examine the concepts of validity and reliability as they pertain to measurement.

Validity

A measurement scale is valid if it does what it is supposed to do and measures what it is supposed to measure.[15] If a scale is not valid, it is of little use to the researcher because it is not measuring or doing what it is supposed to be doing. Validity has to do with one type of error variation, as discussed in the preceding chapter. There are three basic types of validity in measurement that most researchers must be concerned with.[16] These validity concerns are outlined in Table 7.2 and are discussed next.

Content Validity. This validity type has to do with the degree to which the scale items represent the domain of the concept under study. The assessment of content validity is not a simple matter for complex concepts. It is difficult, if not impossible, to enumerate all the dimensions of the concept. For example, the concept of organizational climate is multidimensional; it can be described by literally hundreds of attributes. The problem is to find a procedure that taps the critical dimensions of the variable being measured. We suggest the following procedures.

1. **Conduct an exhaustive search of the literature for all possible items to be included in the scale.** Enumerate these dimensions and put them in a scaling format similar to the one you anticipate using in the study.

2. **Solicit expert opinions on the inclusion of items.** Find experts in the field and ask for suggestions as to any additions or deletions to the scale.

TABLE 7.2 Three Basic Types of Validity in Measurement

Types of Validity	Definitions
1. Content validity	The degree to which the scale items represent the domain or universe of the concept under study
2. Construct validity	The degree to which the measurement scale represents and acts like the concept being measured
a. Convergent validity	The degree of association between two different measurement scales that purport to measure essentially the same concept
b. Discriminant validity	The degree to which the measurement scale may be differentiated from other scales purporting to measure different concepts
3. Criterion-related validity	The degree to which the scale under study is able to predict a variable that is designated a criterion
a. Predictive validity	The extent to which a future level of some criterion variable can be predicted by a current measure on the scale of interest
b. Concurrent validity	The extent to which some criterion variable measured at about the same time as the variable of interest can be predicted by the scale under investigation

3. **Pretest the scale on a set of respondents similar to the population to be studied.** Encourage suggestions and criticisms as to the contents and/or wording of the scale.

4. **Modify as necessary.** Suggestions 2 and 3 should be used to modify the device to ensure Jum Nunnally's criteria for content validity: the adequacy with which important content has been sampled, and the adequacy with which the content has been cast in the form of test items.[17]

Although this procedure cannot guarantee content validity, it can give the researcher and the manager a reasonable degree of confidence.

Construct Validity. Construct validity is the degree to which the scale represents and acts like the concept being measured. There are two aspects involved in the assessment of construct validity: one primarily theoretical in nature and the second primarily statistical.

The theoretical aspect calls for some justification of the concept itself. For example, to study the construct of "readiness to go on vacation," one could search the literature and find that the variable has been used fruitfully in the travel literature. This fact gives the researcher some evidence of the first aspect of construct validity. Or if the concept appears useful in explaining the existence of some phenomenon, its usefulness lends some credence to its existence.

The second concern is usually treated statistically. Two approaches are commonly used in evaluating construct validity.[18] The first approach involves the assessment of convergent validity. *Convergent validity* is defined as the degree of association between two different measurements that purport to measure essentially the same concept. For example, suppose we identify the construct of "readiness to go on vacation" and develop a scale to measure it. We could determine convergent validity by correlating our instrument with one that was designed to measure "readiness to relax." Intuitively, these two constructs should display a high degree of association. If they do, they can be said to exhibit convergent validity and thus some degree of construct validity.

The second approach to demonstrating construct validity is the evaluation of discriminant validity. *Discriminant validity* is largely the opposite of convergent validity. It is the degree to which our measurement scale can be differentiated from other scales purporting to measure different concepts. Again using our vacation example, if we correlated our "readiness to go on vacation" scale with a "readiness to go to work" scale, we would expect to find a high degree of negative correlation. If we did not find this, it would suggest that our measurement device is really not measuring what it is supposed to be measuring.

The principal technique for establishing construct validity using the concepts of convergent and discriminant validity is the multitrait-multimethod matrix.[19] Essentially, this technique uses an intercorrelation matrix of a number of different measurement scales to verify construct validity. Another approach that is often used is a sophisticated statistical procedure known as factor analysis. This technique is discussed further in Chapter 15, which deals with advanced data analysis techniques.

Criterion-Related Validity. This form of validity has to do with the degree to which the scale under study is able to predict a variable that is designated a criterion. This type of meaning assessment is often used in business. Again, there are two general subtypes that are usually examined: predictive and concurrent validity.

Predictive validity is the extent to which a future level of some criterion variable can be predicted by a current measurement on the scale of interest. Here, the emphasis is primarily on the criterion (predicted) variable rather than the measured variable.[20] For example, we might use our scale of "readiness to go on vacation" to predict the number of days of vacation by an individual in the next six months. Similarly, *concurrent validity* is largely criterion-oriented, with the only major difference being the time dimension. With concurrent validity, the measure of the predictor and criterion variables is made at about the same point in time.

Although this type of validity assessment is used in business research, it has been subjected to criticism. The greatest difficulty appears to be in the selection of an appropriate criterion for validation purposes. A number of serious questions can be posed. What criterion should be used to validate the scale of interest? What criterion should be used to test the predictive validity of the scale? Is the criterion selected an adequate variable for validation purposes? In essence, these questions bring us back to the all-important

FIGURE **7.3**

Illustrations of Possible
Reliability/Validity
Situations in
Measurement

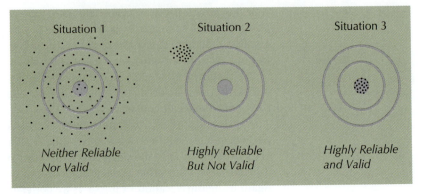

issues of construct and content validity. The evaluation of these two types of validity is generally of more importance than the assessment of criterion-related validity.

Reliability

Reliability refers to the consistency and stability of a score from a measurement scale. For a scale to be valid, it must also be reliable. Reliability is differentiated from validity in that the former relates to consistency whereas the latter relates to accuracy. In order to illustrate this difference, we can use the analogy of a marksman shooting a rifle at a target. Figure 7.3 demonstrates the different situations that could result when the marksman fires the rifle. In situation 1, the bullets hit all over the target, with very little consistency. In measurement terms, we would say the instrument is not very reliable. Moreover, if we do not have a reliable instrument, it can in no way be valid because of the gross and unknown errors created by the measurement device itself. We cannot be sure of validity unless we demonstrate reliability.

Situation 2 illustrates another important fact. Here, the bullets are tightly clustered, but they are far from the bull's-eye. This represents a highly reliable scale that is not valid. In other words, the scale is consistently measuring something, but it is not the concept we are attempting to measure. It is important to evaluate validity and reliability together to ensure that usable measures are developed. Usable measures are of the type demonstrated in situation 3 of our analogy—highly reliable and valid.

Numerous techniques are available for assessing the reliability of a measurement scale. They can largely be summarized under three general methods: the test-retest, alternative forms, and internal consistency.[21] The three approaches to estimating reliability all attempt to determine how much systematic, or true, variance exists in various tests of the measurement scale. The usual means of estimating the amount of systematic variance in a measure takes the form of a correlational exercise, where the scores from the measurement scale are correlated with the scores from some variation or replication of that scale. If this correlation is high, one can infer that the scale is consistent and will yield the same, or nearly the same, results over repeated administrations. In other words, the scale can account for a very high degree of systematic variance relative to error variance. Exhibit 7.1, "The Foundations of Reliability," develops these notions.

When attempting to assess the reliability of a measure (no matter what method is used), the value we examine is generally called the *coefficient of reliability*. This value can range from 0 to 1.0, with 1.0 perfectly reliable and 0 perfectly unreliable. A perfectly reliable measure is seldom attainable, but one should strive for the best measures possible. There are no rules available for what constitutes a reliable measure. However, one author suggests the following minimum standards, based on how the measure is being used by the researcher, for the coefficient of reliability in the development of behavioral measures.[22]

0.7 is used for exploratory research.

0.8 is used for basic research.

0.9 or better is used in applied settings where important decisions will be made with respect to specific test scores.

The only real difference in the three approaches to estimating reliability is in what the scores of the test will be correlated against. With this difference in mind, we will now discuss the three methods of reliability assessment.

Test-Retest Method. In order to estimate reliability with the test-retest method, we must administer the same scale to the same set of individuals at two different times. Normally, the measurement device is either administered immediately or delayed for some time. In either case, the two sets of scores are correlated with each other to obtain an estimate of reliability. Although it is useful, this method has been criticized for the following reasons.

1. Different results may be obtained, depending on the amount of time elapsed between testing.

2. There may be changes in the subject from day to day. Thus, we would expect lower reliability over time.

3. In addition, if there is any change between the test-retest scores, we cannot be sure if the change was real or due to the unreliability of the instrument.

4. Finally, mathematically the correlation between the test-retest scores is only partially dependent on the nature of the changes in the instruments. Specifically, there is a confounding of the coefficient of reliability because it contains the correlation of each item

EXHIBIT **7.1**

The Foundations of
Reliability

Usually, the discussion of reliability begins with the simple notion that any observed scale score can be divided into two parts.[*] In terms of any observed score, these components are

$$X_o = X_t + X_e \qquad (1)$$

where

X_o = observed score

X_t = true score

X_e = error score

The true score is never known. However, it is usually assumed to be the mean score of administering the same scale to the same individual repeatedly. The error term X_e can increase or decrease X_o due to any number of sources of measurement error (vague instructions, ambiguous questions, and so on). Reliability assessment is essentially an attempt by the researcher to make X_e as small as possible so that X_t is the vast majority of X_o.

Now, if it is assumed that the errors in a set of measures are distributed independently and randomly, we can also show the variance of an observed score as containing two parts:

$$V_o = V_t + V_e \qquad \text{or} \qquad \sigma_o^2 = \sigma_t^2 + \sigma_e^2 \qquad (2)$$

where

V_o, σ_o^2 = observed scale variance

V_t, σ_t^2 = true variance or the actual differences on the concept of interest (commonly known as between-groups or systematic variance)

V_e, σ_e^2 = error variance

Given this formulation, one can readily see that if V_e is large relative to V_t, then V_o would be highly suspect, or unreliable. Thus, the more systematic, or true, variance is present as a part of total variance, the more reliable the measure is.

More specifically, a coefficient of reliability can be thought of as the percentage of total variance that is of the systematic type. Symbolically, this can be represented as follows:

$$\text{Reliability} = \frac{V_t \,(\text{true variance})}{V_o \,(\text{observed variance})} \qquad (3)$$

where

$$V_t = V_o - V_e \qquad (4)$$

Thus, reliability can alternately be stated as

$$\text{Reliability} = 1 - \left(\frac{V_e}{V_o} \right) \qquad (5)$$

or

$$\text{Reliability} = \frac{V_o - V_e}{V_o} \qquad (6)$$

As we can readily see in equations (5) and (6), as V_e gets larger, reliability decreases.

EXHIBIT **7.1**

(Continued)

We can further manipulate equation (5) to define the standard error of measurement as:

$$\sigma_m = \sigma_o \sqrt{1 - \text{reliability}}$$

where

σ_m = standard error of measurement ⟶ (7)

These two equations (the coefficient of reliability estimate (5) and the standard error of measurement (7)) represent the two most common ways in which the reliability of a set of measures is expressed.

---

* This useful framework is largely explained by Fred Kerlinger, *Foundations of Behavioral Research* (New York: Holt, Rinehart, and Winston, 1973), Chapter 26; and E. J. Mason and W. J. Bramble, *Understanding and Conducting Research* (New York: McGraw-Hill, 1978), Chapter 10.

with itself. This fact could produce a high test-retest correlation and a corresponding inflated estimate of reliability.

Although there are difficulties with this method of reliability assessment, it is still used in certain circumstances. Generally, however, it is recommended that a two-week time period be the norm between the test and the retest situations. Also, if possible, this estimate should be used in conjunction with estimates of reliability generated by the internal consistency method.

Alternative Forms Method. Alternative forms techniques of estimating reliability assess the equivalency of content of sets of items. The techniques usually involve the administration of equivalent scales to the same individuals, with or without a time interval.[23] Most common is the use of a time interval of about two weeks, much like the test-retest method. The important difference between the test-retest and the alternative forms methods is in the testing instrument itself. In the former, the same scale is used; in the latter, a different but equivalent form is used.

The biggest problem with these methods of assessment stems from the equivalent-forms requirement. It is usually very difficult and costly to develop truly equivalent forms of the same scale. One can never be fully sure whether the unreliability of a scale is due to an unreliable measure or a nonequivalent form. This estimation procedure is appealing to some, but using the generally robust coefficient alpha would seem to be the more prudent course of action.[24]

Internal Consistency Method. The internal consistency method assesses the homogeneity of a set of items. The basic rationale for such reliability assessments is that items in a scale should behave similarly. The specific techniques included in this method are the split-half technique, Cronbach-Alpha, and Kuder and Richardson's KR-20.

The *split-half technique* is accomplished by splitting a multi-item scale in half and then correlating the results of the scores in the first half with those in the second half. Usually the items in the scale are randomly assigned to one half or the other. The problem with this estimate of reliability is that the estimate of the coefficient of reliability is wholly dependent on the manner in which the items are split.

An internal consistency technique that overcomes this shortcoming is the *Cronbach-Alpha*.[25] In essence, this technique computes the mean reliability coefficient estimates for all possible ways of splitting a set of items in half. This presumably results in a better estimate of reliability. The Cronbach-Alpha is perhaps the most widely used method of reliability assessment in business research. This is because of the aforementioned problems in the test-retest method, and because cost and time constraints often make alternative forms assessment not desirable.

One limitation in the use of the Cronbach-Alpha is its assumption that the items in the scale are at least interval in nature. An alternative method employing the same notions was developed for use with dichotomous, or nominally scaled, items. This is called the *KR-20 formula*.[26] Although the computational forms of this technique can become rather complex with multi-item scales, its use is facilitated by many statistical software packages on the market today.

MANAGERIAL CONSIDERATIONS

There are a number of significant reasons why management should take an active interest in how measurement scales are developed. Specifically, the major reasons are as follows:

Without good measurement, the results of the study are bound to be of little value to the manager. Although good measurement cannot guarantee valuable results, it is necessary for obtaining valuable information. In other words, appropriate measurement procedure is a necessary, but not sufficient, condition in generating useful research results.

The way in which concepts or variables are scaled actually sets limits on how the results of the study can be later analyzed. As mentioned earlier, the measurement process is highly interactive. For example, a researcher chooses to measure sales of an organization with a nominal measurement scale such as the following:

The sales of your organization are (check one)

_____ $1 million or less

_____ more than $1 million

Later, the analysis of that variable will be limited to a specific type of statistical technique. More importantly, the manager cannot go back and ask for other analyses that may prove to be more fruitful from a decision-making standpoint (other, more detailed breakdowns of sales volume, or specific relational statements between input variables and dollar sales).

As more and more global and international research is being undertaken in the field of business research, the opportunities for poor measurement because of misinterpretation, misunderstanding, or just plain carelessness is growing. Numerous examples of misinterpretation and careless translation blunders serve to remind both manager and researcher alike that doing research in the international business setting is difficult indeed.[27] There are enough errors in all other aspects of the research process to make a strong case for carefully developing one's measurement devices.

Once the measurement scales are set and the data is collected, the researcher cannot (without incurring a great deal of expense and time) go back and collect additional information. This is significant to managers because this stage in the research process provides the last opportunity for management to review the research project before significant resources are expended on data collection.

The next logical question is how managers should get involved in this stage of the research process. To address this question, we suggest a number of guidelines that users (managers) of information can require of the doers (researchers) before the study is undertaken.

Require clearly defined constitutive concepts. Clearly defined concepts will ensure that the researcher and the manager are talking about the same thing. Our example of job satisfaction highlighted the importance of coming to an agreement on critical concepts in the study.

Require clearly defined operational definitions that specify the procedures of measurement. It was noted that there are many possible ways to operationalize a single concept. It is critical that the measurement procedure be appropriate to management's needs.

Require multi-item (complex) scales whenever the situation and resources permit. Multi-item scales are desirable for two reasons:

a. It is very difficult, if not impossible, to measure complex concepts and constructs with a single-item scale.

b. The individual items in complex scales tend to have counterbalancing errors, thus increasing reliability.

Require the assessment of validity and reliability, and pretest whenever possible. The assessment of validity and reliability just makes plain good sense. Invalid and unreliable measures are only capable of generating invalid and unreliable results. Pretesting a scale helps ensure that these two goals are achieved.

Require the researcher to show how the scales will be analyzed and how this analysis will answer your managerial question(s). This forces

1. What are the key concepts/constructs that need to be measured?
2. Are these concepts the critical ones needed to answer your managerial problems?
3. Have the concepts been adequately defined constitutively?
4. Do the operational definitions of the concepts adequately define how the variables will be measured?
5. What level of measurement is being used in the scale developed? Is this level appropriate for the desired analysis of your data?
6. Is a complex scale being used to measure the concept of interest? If not, why?
7. How reliable and valid is the scale? Has the scale been pretested?
8. Will the data generated from the measures developed accomplish the objectives of the study and answer the research question(s)?

the researcher to think ahead, to make sure that the data being collected will be of the appropriate type and quality to solve your problems. Without this forethought, the data collected may not be of value in your decision making.

This framework can be restated in the form of a number of key managerial questions pertaining to the management process. These questions, presented in Exhibit 7.2, reflect the main issues that management should be concerned about in the measurement process.

SUMMARY

Measurement can be defined as a rule for the assignment of numerals (numbers) to aspects of objects and events. The procedure by which measurement is accomplished is the measurement process—a set of interrelated steps that begins with the isolation of empirical events of interest and ends with the actual measurement of the characteristics under study. The goal of measurement is to translate the characteristics and properties of empirical events into a form that can be analyzed by the researcher.

The development of measurement scales is of extreme importance in business research because the value of the results depends on the quality of the scales themselves. This chapter reviews the four major levels of measurement and the necessary traits of good measurement scales. A discussion of the specific types of scales is deferred to Chapter 8.

The four major levels of measurement identified are nominal, ordinal, interval, and ratio scales. The principal characteristics of these types of scales are highlighted and reviewed.

The discussion then moves to the evaluation of measurement scales. In order for researchers to select the proper type of scale for a particular purpose, they must possess a thorough understanding of the concepts and variables under study. This understanding leads to (1) the development of scales to measure the variables and (2) the subsequent concern with the proper measurement and use of the variables. The section closes by reviewing the main components of validity and reliability.

The chapter ends with a discussion of managerial considerations in measurement. Four reasons for management's concern are outlined and a five-point guideline for management's involvement in the measurement process is proposed. A list of key questions pertaining to the measurement process is also provided.

DISCUSSION QUESTIONS

1. What is measurement? Discuss its importance in the context of managers receiving information from business research for decision making.

2. What is the measurement process? Outline the major components in this process and discuss their importance to the business research process.

3. What are levels of measurement? Identify and discuss the differences between the four major levels of measurement.

4. Discuss what is meant by the evaluation of measurement and its importance in business research. Be specific and give examples to support your points.

5. What major types of validity must researchers be concerned with in measurement? Define and discuss the differences in these validity types.

6. Discuss the difference between validity and reliability. How are the two concepts related?

7. What are some of the major methods to assess reliability? Discuss the relative merits and weaknesses of each approach.

8. Identify the major concerns managers should have in terms of measurement. What are some of the means managers can use in the measurement process so that they receive the information they desire?

9. Conduct a secondary data search on the concept of validity of measurement. What articles did you find and how do they relate to this chapter? Be specific and be sure to describe your search process and the topics as discussed in this chapter.

10. Using the Internet, search the literature to find four examples of measurement scales that have been used in business research. Are the measures assessed in terms of their validity and reliability? Evaluate these scales in the context of this chapter.

NOTES

1. S. S. Stevens, "Mathematics, Measurement and Psychophysics," in S. S. Stevens, Ed., *Handbook of Experimental Psychology* (New York: Wiley, 1966), p. 22.

2. Leonard S. Kogan, "Principles of Measurement," in Norman A. Polansky, Ed., *Social Work Research* (Chicago: University of Chicago Press, 1960), pp. 87–105.

3. For in-depth discussions of these three components, see Harry Upshaw, "Attitude Measurement," in Hubert M. Blalock, Jr. and Ann B. Blalock, Eds., *Methodology for Social Research* (New York: McGraw-Hill, 1968), Chapter 3; Warren S. Torgenson, *Theory and Methods of Scaling* (New York: Wiley, 1958), Chapters 1 and 2; Fred N. Kerlinger, *Foundations of Behavioral Research* (New York: Holt, Rinehart, and Winston, 1973), Chapter 25; or Theodore J. Mock and Hugh D. Grove, *Measurement, Accounting and Organizational Information* (New York: Wiley, 1979), Chapter 1.

4. For excellent discussions of the examples used here, see either the job performance-satisfaction or the leadership-effectiveness literature in most management texts.

5. Donald S. Tull and Del I. Hawkins, *Marketing Research: Measurement and Method*, 6th ed. (New York: Macmillan, 1993), pp. 298–300.

6. Mock and Grove, op. cit. note 3, p. 20.

7. Originally discussed by S. S. Stevens, "On the Theory of Scales of Measurement," *Science* (1946), *103*, pp. 677–680; subsequently refined in Stevens, op. cit. note 1.

8. Although most social scientists agree with the fourfold classification, there are dissenting views. See, for example, Kogan, op. cit. note 2, pp. 89–91, for a sixfold classification.

9. As we will discuss later, the distribution a variate possesses has a lot to do with the way it is sampled as well as how we wish to analyze it. It is assumed at this point that the student is familiar with the binomial, multinomial, and normal distributions. When a variable takes on two or more characteristics (such as our communication ability variable), the distribution is usually assumed to be binomial (possesses the characteristics or not) or multinomial (possesses three or more characteristics). A simplified discussion of these two distributions can be found in Leslie Kish, *Survey Sampling* (New York: Wiley, 1965), pp. 45–47.

10. At first, this scale conversion (from an interval to a nominal scale) may seem strange, but it is performed frequently due to management's requests and the objectives of the research. Perhaps the best examples of such conversion are in the market segmentation literature in the field of marketing.

11. E. J. Mason and W. J. Bramble, *Understanding and Conducting Research* (New York: McGraw-Hill, 1978), p. 138.

12. Kerlinger, op. cit. note 3, p. 441. Also see Edgar F. Bogatta and George Bohrnstedt, "Level of Measurement: Once Over Again," *Sociological Methods & Research* (November 1980), *9:2*, pp. 147–160.

13. Torgenson, op. cit. note 3, p. 19.

14. Mason and Bramble, op. cit. note 11, p. 138.

15. For an excellent discussion of validity, see Jum C. Nunnally, *Psychometric Theory*, 2nd ed. (New York: McGraw-Hill, 1978), pp. 86–113.

16. George W. Bohrnstedt, "Reliability and Validity Assessment in Attitude Measurement," in Gene F. Summers, Ed., *Attitude Measurement* (Chicago: Rand McNally, 1970), pp. 91–98.

17. Nunnally, op. cit. note 15, pp. 91–94.

18. See Donald T. Campbell and Donald W. Fiske, "Convergent and Discriminant Validation by the Multitrait-Multimethod Matrix," *Psychological Bulletin* (1959), *56*, pp. 81–105.

19. Ibid.

20. Kerlinger, op. cit. note 3, pp. 459–460.

21. For other excellent discussions on reliability, see Kerlinger, op. cit. note 3, Chapter 26; and J. Paul Peter, "Reliability: A Review of Psychometric Basics and Recent Marketing Practices," *Journal of Marketing Research* (February 1979), *16*, pp. 6–17.

22. Nunnally, op. cit. note 15, pp. 245–246.

23. Kogan, op. cit. note 2, p. 97.

24. Nunnally, op. cit. note 15, p. 234.

25. The Cronbach-Alpha can be found in L. J. Cronbach, "Coefficient Alpha and the Internal Structure of Tests," *Psychometrika* (September 1951), *16*, pp. 297–334.

26. G. Kuder and M. Richardson, "The Theory of Estimation of Test Reliability," *Psychometrika* (September 1937), *2*, pp. 151–160.

27. See David A. Ricks, *Blunders in International Business* (Cambridge, MA: Blackwell, 1997).

SUGGESTED READING

The Nature of Measurement

Blalock, H. M., *Conceptualization and Measurement in the Social Sciences* (Beverly Hills, CA: Sage, 1982).

The author uses a verbal discussion of measurement tenets. For the nonquantitative student, this book provides an excellent overview of the problems and procedures for valid and reliable measurement.

Ghiselli, Edwin E., John P. Campbell, and Shelton Zedeck, *Measurement Theory for the Behavioral Sciences* (San Francisco: W. H. Freeman, 1981).

A well-written introduction to measurement. The reader needs a limited background in descriptive statistics. Don't let the title fool you; it is oriented to the research practitioner.

Nunnally, J. C., *Psychometric Theory,* 2nd ed. (New York: McGraw-Hill, 1978).

A comprehensive, analytical view of measurement. Strong emphasis on estimating the errors in measurement using statistical methods. A "must" for the researcher's bookshelf.

Levels of Measurement

Kogan, Leonard S., "Principles of Measurement," in Norman Polansky, Ed., *Social Work Research* (Chicago: University of Chicago Press, 1960), pp. 87–105.

Stevens's framework is the most widely accepted, but Kogan presents a sixfold classification that makes interesting reading.

Stevens, S. S., "On the Theory of Scales and Measurement," *Science* (1946), *103,* pp. 677–680.

This is the seminal article on levels of measurement. This article forms the basis for most of the discussion in this section.

Evaluation of Measurement Scales

Cote, Joseph A., and M. Ronald Buckley, "Measurement Error and Theory Testing in Consumer Research: An Illustration of the Importance of Construct Validation," *Journal of Consumer Research* (March 1988), *14,* pp. 579–582.

This article presents an illustration of the impact of measurement error on theory testing in the field of consumer behavior.

DeVellis, Robert F., *Scale Development: Theory and Applications* (Newbury Park, CA: Sage, 1993).

A short book that discusses developing valid and reliable scales where none exist.

Glick, William H., "Conceptualizing and Measuring Organizational and Psychological Climate: Pitfalls in Multilevel Research," *Academy of Management Review* (1985), *10*:3, pp. 600–616.

This article is an excellent review of the evaluation of measurement scales of organizational and psychological research in the field of management.

Peter, J. Paul, "Reliability: A Review of Psychometric Basics and Recent Marketing Practices," *Journal of Marketing Research* (February 1979), *16,* pp. 6–17.

This article examines the issue of reliability in the marketing literature. It succinctly reviews methods of reliability assessment.

8

Scaling and Instrument Design

Overview

The Nature of Instrument Design

Scale Development

Item Phrasing

Response Formats

Frequently Used Scaling Techniques

Instrument Design

Scale Sequencing and Layout

Online Design Aids

Pretesting and Correcting

Managerial Considerations

Summary

Discussion Questions

Notes

Suggested Reading

I n the last chapter, measurement and scale development were introduced and discussed. This chapter expands this framework by examining the scaling topic in more depth and subsequently relating it to the broader subject of instrument design. *Instrument design* is defined as the formal construction of the data collection device (usually a questionnaire) to obtain the data needed to solve the research problem.

The chapter begins by examining the nature of instrument design. The measurement process is again related to scale development and, ultimately, the study instrument itself. The complexity of designing good measurement instruments is highlighted.

The discussion then moves to the intricacies of scale development. Issues of interest include the nuances of item wording and response format explication. Guidelines for proper scale development are presented, and several scaling techniques are briefly reviewed in order to familiarize you with some commonly used measurement devices.

Next, actual instrument design is analyzed. The determination of item sequencing and layout is examined first. Basic guidelines are presented to help sensitize you to the factors that may affect the collection of unbiased data. The discussion then turns to online design and data collection aids that use artificial intelligence and researchers' experience to help in the design process. This section ends with the topics of pretesting and correcting the instrument, the last steps before the actual data is collected. The chapter concludes with managerial considerations in instrument design.

THE NATURE OF INSTRUMENT DESIGN

Instrument design is very much an art. Ask items in one way, and the answers reflect what the public is thinking. But phrase them just slightly differently, and the results can be entirely different. A panel of experts from the National Research Council concluded that survey designers simply did not know much about how respondents answered items.[1] No wonder it is difficult to construct totally valid and reliable questionnaires.

There will probably never be a definitive means to develop the perfect data collection instrument;[2] too many variables affect the construction and use of an instrument in a study situation. Nevertheless, new developments in the field and general guidelines can be useful in the design of any measurement instrument. These topics are the primary concern of this chapter.

One of the primary difficulties in instrument design is determining what to include and exclude. Researchers often lose focus of what they are trying to accomplish in the research design. For example, a study is to answer the question, Are pay level and perceived job satisfaction significantly related to worker performance in General Motors? The instrument design process should be driven by the three major concepts in the study: pay level, perceived job satisfaction, and worker performance. All other variables (IQ, perceptions of the economy, and so on), although perhaps interesting and easy to collect, are extraneous to the study at hand. We must stay focused on the relevant concepts and variables if we are to design effective measurement instruments.

As you remember from the previous chapter, measurement is the means by which researchers represent and manipulate concepts in the analytical world. An instrument is the primary device by which this task is accomplished. Items make up scales, which in turn make up instruments. Thus, we must fully understand the concept of scaling so that we can construct scales to assemble instruments that are valid and reliable (the basic measures of scale adequacy). The measurement process as it relates to scale development and instrument design is the focus of the following pages.

The measurement process as outlined in Figure 7.2 is expanded in Figure 8.1 to highlight the salient points that will be covered in the

FIGURE **8.1**

The Measurement
Process As It Relates to
Instrument Design

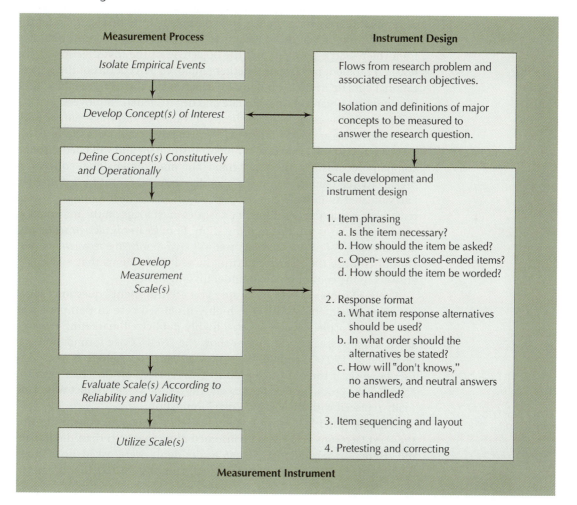

Measurement Process	Instrument Design
Isolate Empirical Events	Flows from research problem and associated research objectives.
Develop Concept(s) of Interest	Isolation and definitions of major concepts to be measured to answer the research question.
Define Concept(s) Constitutively and Operationally	

Develop Measurement Scale(s)	Scale development and instrument design
	
	1. Item phrasing
	  a. Is the item necessary?
	  b. How should the item be asked?
	  c. Open- versus closed-ended items?
	  d. How should the item be worded?
	
	2. Response format
	  a. What item response alternatives should be used?
	  b. In what order should the alternatives be stated?
	  c. How will "don't knows," no answers, and neutral answers be handled?

Evaluate Scale(s) According to Reliability and Validity	3. Item sequencing and layout
Utilize Scale(s)	4. Pretesting and correcting

Measurement Instrument

following pages. The subject of scale development is dealt with in detail first, followed by an examination of the broader issues of instrument design.

SCALE DEVELOPMENT

The basics of scale development were discussed in the measurement chapter. There, a distinction was made between simple and complex scales, with the primary distinguishing factor being one-item or multiple-item scales. It was also stated that complex scales are usually preferred over simple scales because of the multidimensionality of most phenomena studied in the business setting and the need for multi-item scales for the enhancement of reliability and validity. We now take a more detailed look at scale development.

In the development of any measurement scale, there are a number of key factors to consider. As is indicated in Figure 8.1, the first factor is the determination of what variables and concepts need to be measured for the study. The determination of what is to be measured should flow naturally from the research problem and the objectives of the investigation. If the research problem is to assess whether employee perceptions of the organizational climate in the firm differ by department, then the variables to be measured are organizational climate and departmental employees.

Of course, the complexity of these two variables and the resultant measurement scales are vastly different. Departmental employees can be easily assessed by a simple scale in a self-report instrument and validated through company records. On the other hand, the multidimensional concept of organizational climate requires a complex scale that must be constructed and validated as illustrated in the previous chapter. The important point to remember is that the development of scales must focus on the information needed to answer the research question. *In other words, design scales and instruments to collect only the data you will need to solve the research problem.* Do not confuse the issue by putting in additional items that won't give answers to the central problem under investigation.

A related problem is to make sure you ask the "right" questions about the concepts in the study. One highly discussed case in point was the Coke–"New" Coke debacle. In this highly publicized marketing research snafu, Coke's research did indeed show the "New" Coke formulation was preferred by 190,000 people. What they did not ask, though, is would the user be upset if the old Coke product was replaced on the shelf with the "New" Coke product.[3] Needless to say, the "New" Coke product was eventually replaced with the Classic Coke formulation.

Similarly, a market research study made the erroneous conclusion that the West Germans and French consumed more spaghetti than the Italians did. Here, the problem was that respondents were asked how much branded and packaged spaghetti they ate, ignoring the fact that the heavy user Italians usually bought their spaghetti in bulk (not in branded and packaged boxes).[4] We need to be extremely clear in stating exactly what concepts we want to collect data on and how we are going to ensure that this data is what we think it is.

Once the associated concepts are identified for study, attention shifts to the actual development of the measurement scales. The major factors to consider include item phrasing and determining the type of response format.

Item Phrasing

Developing the measurement scale is largely a matter of wording items to assess the concept being studied. A primary consideration in the actual phrasing of items is the method by which the scales are administered to the study respondents. The main methods of administration are the more traditional offline (personal, telephone, and mail) and online (Internet facilitated). If nonpersonal methods of survey administration are used, the accompanying items and instructions must be generally much more succinct, mainly because there is no interaction between the researcher and respondent. The growth of online, interactive data collection on the Internet, e-mails, and intranets within companies have created the opportunity for more communication between remote researchers and their respondents. This opens up entirely new possibilities for data collection. More will be said about this phenomenon later in the text.

In all methods of administration, the items should be clearly stated, unambiguous, and easily understood, so as to ensure the most error-free data possible. To illustrate one potential layout, Exhibit 8.1 presents a sample instrument that was used to assess the viability of a new sports-related concept named the River End Golf and Entertainment Center. Note that this is a traditional paper-and-pencil question that requires respondents to check their responses to a series of questions. The data must be coded and then transferred to an electronic medium for analysis and summarization. As one will see later in this chapter, online design aids are facilitating the construction, collection, and analysis of the data collected, easing the time required for respondent and researcher alike. It must be remembered however, that regardless of the design and delivery method chosen, the basic concepts of scale and instrument design still apply. Please examine the layout of this questionnaire as the guidelines for scale development and instrument design are reviewed.

In constructing items for scales, researchers should ask themselves a series of questions, as follows.

Is the Item Necessary? An item should be asked only if it is needed to solve the research problem. Unnecessary items serve only to confuse potential respondents. The critical test for inclusion of any item in a scale is whether or not it is essential to the accomplishment of the study's goals.

EXHIBIT **8.1**

A Sample Instrument

RIVER END GOLF AND ENTERTAINMENT CENTER

The *Golf and Entertainment Center* at River End is a new sports related concept that will include a variety of golf and entertainment related activities. The center would have a distinctive golf focus and would include entertainment and learning activities for both the family and golfer. It is anticipated the Center would include a miniature golf course, a driving range, a putting course, a golf instruction center, a nine hole pitch and putt, a nine hole executive golf course, a nationally recognized golf and recreational retailer and other entertainment related activities. This Center would be located on property adjacent to the Mall at River End.

1. Do you golf? ❑ No ❑ Yes, if so, how many rounds per year? _____

2. Based on what you know now, how interested would you be in visiting this golf center if it was located adjacent to the Mall at River End? (*Check one box*)
   ❑ Extremely interested
   ❑ Very interested
   ❑ Somewhat interested
   ❑ Not very interested
   ❑ Not at all interested

3. How likely are you to visit this attraction based on the above description? (*Check one box*)
   ❑ Definitely will visit
   ❑ Probably will visit
   ❑ Might or might not visit
   ❑ Probably will not visit
   IF ❑ Definitely will not visit—**STOP**, *Thank you for your time and consideration, otherwise, please continue.*

River End expects to have a number of basic options for this center. Please think about each option, then answer the questions that follow.

Option 1: Adventure Golf—A themed, 18 hole miniature golf course with synthetic and natural areas which includes decorative landscaping, waterfalls, rock formations and other amenities. Intended for speedy play and total family enjoyment. Priced at $6.00 per person.

4a. Based on the $6.00 price, how would you rate this option's value for the money? (*Check one box*)
   ❑ Excellent value
   ❑ Very good value
   ❑ Good value
   ❑ Fair value
   ❑ Poor value

4b. Based on the $6.00 price, how likely are you to try this option? (*Check one box*)
   ❑ Definitely will try
   ❑ Probably will try
   ❑ Might or might not try
   ❑ Probably will not try
   ❑ Definitely will not try

4c. Based on the $6.00 price, do you feel the price is fair?
   ❑ Yes ❑ No, *then what would you expect as a fair price?* _____Price

EXHIBIT **8.1**

(Continued)

Option 2: Driving Range—A double ended driving range with natural and synthetic hitting areas for all skill levels of golfers. Includes 30 all weather, lighted hitting and landing areas with natural target greens. Simulated rough and fairway areas with quality golf balls. Price would be $7.00 for 75 balls.

5a. Based on the $7.00 price, how would you rate this option's value for the money? (*Check one box*)
- ❏ Excellent value
- ❏ Very good value
- ❏ Good value
- ❏ Fair value
- ❏ Poor value

5b. Based on the $7.00 price, how likely are you to try this option? (*Check one box*)
- ❏ Definitely will try
- ❏ Probably will try
- ❏ Might or might not try
- ❏ Probably will not try
- ❏ Definitely will not try

5c. Based on the $7.00 price, do you feel the price is fair?
- ❏ Yes ❏ No, *then what would you expect as a fair price?* _____ Price

Option 3: Putting Course—A professionally designed 18 hole challenge putting course with natural grass playing characteristics. Surrounded by natural landscaping, the course is designed to put the player into realistic putting situations with lengths of holes varying from 30 to 100 feet. Price is $8.00.

6a. Based on the $8.00 price, how would you rate this option's value for the money? (*Check one box*)
- ❏ Excellent value
- ❏ Very good value
- ❏ Good value
- ❏ Fair value
- ❏ Poor value

6b. Based on the $8.00 price, how likely are you to try this option? (*Check one box*)
- ❏ Definitely will try
- ❏ Probably will try
- ❏ Might or might not try
- ❏ Probably will not try
- ❏ Definitely will not try

6c. Based on the $8.00 price, do you feel the price is fair?
- ❏ Yes ❏ No, *then what would you expect as a fair price?* _____ Price

6d. Would you prefer ❏ natural grass or ❏ artificial turf on this challenge course?

Option 4: Nine Hole Pitch and Putt—A nine hole challenge course with shots ranging from thirty (30) to ninety (90) yards. Multiple tee positions will be available to simulate realistic approach shots that a golfer could experience during a round of golf. Price: $9.00 per round.

(Continued)

EXHIBIT **8.1**

(Continued)

7a. Based on the $9.00 price, how would you rate this option's value for the money? (*Check one box*)

❑ Excellent value

❑ Very good value

❑ Good value

❑ Fair value

❑ Poor value

7b. Based on the $9.00 price, how likely are you to try this option? (*Check one box*)

❑ Definitely will try

❑ Probably will try

❑ Might or might not try

❑ Probably will not try

❑ Definitely will not try

7c. Based on the $9.00 price, do you feel the price is fair?

❑ Yes ❑ No, *then what would you expect as a fair price?* _____ Price

Option 5: Nine Hole Executive Golf Course—A professionally designed and challenging 9 hole executive course with bunkers, water traps, and all amenities. Designed expressly for enjoyment and to improve one's golfing game. Price: $14 for 18 holes.

8a. Based on the $14 price, how would you rate this option's value for the money? (*Check one box*)

❑ Excellent value

❑ Very good value

❑ Good value

❑ Fair value

❑ Poor value

8b. Based on the $14 price, how likely are you to try this option? (*Check one box*)

❑ Definitely will try

❑ Probably will try

❑ Might or might not try

❑ Probably will not try

❑ Definitely will not try

8c. Based on the $14 price, do you feel the price is fair?

❑ Yes ❑ No, *then what would you expect as a fair price?* _____ Price

Option 6: Learning Center—A golf academy and learning center will provide individual instruction with certified PGA and LPGA instructors. The center will have state of the art computer assisted video instruction, workshops, club fitting, and other fundamentally driven instructional aids. Price: $45 for 45 minutes of instruction.

9a. Based on the $45 price, how would you rate this option's value for the money? (*Check one box*)

❑ Excellent value

❑ Very good value

EXHIBIT **8.1**

(Continued)

❏ Good value
❏ Fair value
❏ Poor value

9b. Based on the $45 price, how likely are you to try this option? (*Check one box*)
❏ Definitely will try
❏ Probably will try
❏ Might or might not try
❏ Probably will not try
❏ Definitely will not try

9c. Based on the $45 price, do you feel the price is fair?
❏ Yes ❏ No, *then what would you expect as a fair price?* _____ Price

9d. How important would the following endorsements be in your determination of choosing to go to a golfing academy such as this?

	Extremely	Somewhat	Not at All
Golf Digest Pitch and Putt Center	❏	❏	❏
A Professional Teaching Pro	❏	❏	❏
A Touring Pro	❏	❏	❏
The National Golfing Institute	❏	❏	❏

10. Would you be willing to pay a daily fee of $45 to have access to all the golfing facilities (learning academy excluded) ❏ Yes If No, ❏ what would be a fair price? _____

11. Would you be willing to pay a membership fee of $250 for unlimited use of the golfing facilities for a year? (academy excluded) ❏ Yes If No, ❏ what would be a fair price? _____

> Listed below are a number of features that could be included in the *Golf and Entertainment Center.* Please rate the importance of these features to you personally.

	Extremely	Somewhat	Not at All
12. a. mini golf	❏	❏	❏
b. driving range	❏	❏	❏
c. lighted driving range	❏	❏	❏
d. all weather, heated driving range	❏	❏	❏
e. double tiered driving range	❏	❏	❏
f. pitch and putt	❏	❏	❏
g. golf academy	❏	❏	❏
h. putting course	❏	❏	❏
i. 9 hole executive course	❏	❏	❏

> Finally, can you please answer a few demographic questions for classificatory purposes only? We thank you for your time and consideration.

13. How many minutes driving time from Mall at River End, do you live? _____
Work?_____

14. Please indicate your sex and age. ❏ Male ❏ Female Age? _____

How Will the Item Be Asked? This touches on a host of related issues that deal with the general form of the item and the way in which it is presented to the respondent. The way in which the item is asked may actually affect what type of response is obtained. Different people attribute different meanings to the same word, and all individuals have a different frame of experience when reading and interpreting items. Therefore, items must be carefully designed and then pretested so that the meaning that is intended by the researcher is actually conveyed to the respondent.

Another concern involves asking sensitive or potentially embarrassing questions.[5] For example, questions about age, income, corporate policy, and the like are frequently considered off-limits by potential respondents. Some explanation of how the data will be used, and a promise of anonymity, can help overcome this reluctance. Phrasing the question in terms of others' feelings, rather than the feelings of the actual respondent, can also be helpful.

Should the Item Be Open- or Closed-Ended? The decision to use an open- or a closed-ended item is based on how well the researcher understands the issue under study, the issue itself, and the type of research being undertaken. For example, in a study of the factors related to employee turnover in a retail merchandising firm called Salesmark, it would be unwise to assume we know all the contributing factors in this situation. We know little about the company-specific factors, and the research is largely exploratory. In this case, open-ended items would be more appropriate than closed-ended ones.

Open-ended items are those that respondents have total freedom to answer as they please. In closed-ended items, the number of responses is limited by the researcher. If the researcher knows little of the precise nature of the subject under study, open-ended items are often the only alternative if important data is not to be overlooked in the inquiry process.

Open-ended items possess several inherent advantages and disadvantages.[6] One advantage is that respondents can choose to say anything they want, without being restricted by the researcher's preconceived ideas. Researchers often use this type of question in early stages of the research process to ensure that later structured questionnaires adequately represent the feelings of respondents.

The disadvantages of open-ended items are the primary advantages of closed-ended questions. The answers to open-ended questions are difficult to code, time-consuming to analyze, and overall less efficient to handle than the closed-ended alternatives. Furthermore, open-ended questions are more difficult to handle in the data collection process, especially from a personal interviewing standpoint.

How Should the Item Be Worded? The basic rule is to keep the items as simple as possible. Considerations such as educational level of the respondent, background, language concerns, and a multitude of other factors affect the interpretation and understanding of the measurement instrument. Many people have difficulty reading even basic material, let alone highly involved instruments thought up by researchers to answer specific questions. In many national and international studies, multilanguage instruments are necessary to accomplish the objectives. The wording should remain simple and avoid ambiguous phrases and questions whenever possible.[7]

For example, in a recent multinational study of entrepreneurship in the United States, South Africa, and Portugal, the problems of wording had to be resolved. First, the basic instrument was developed in English in the United States. Then the English version was tested both here and in South Africa, for interpretation and meaning, by pretests and by expert review. The instrument went through a translation into Portuguese by a local expert; then it was backtranslated by another independent expert, to ensure that the meaning was equivalent to the original version. A pretest was then conducted to ensure readability. This process made the wording and meaning appropriate for all samples involved.[8]

Other points to consider in wording relate to the degree of item specificity and the use of double-barreled questions. In general, more specific questions are preferred to general ones, to minimize ambiguity. Double-barreled questions are those that call for two responses. For example, "How satisfied are you with your current pay and current job description?" would be double-barreled, because it asks for two responses. Such questions should be avoided because they generally create confusion among respondents.

Response Formats

Closely related to item phrasing is the determination of response formats. Three questions must be dealt with:

What response alternatives should be used? The response alternatives available in scale design are the open-ended and closed-ended (dichotomous and multiple-choice) responses. The response alternatives selected for use should reflect the purposes of the investigation and the researchers' degree of understanding. In exploratory research, open-ended questions are frequently used because the objective is to find out as much as possible about the problem situation. Conversely, as more is known, the tendency is to move to more structured response alternatives.

In what order should the alternatives be stated? Research has found that the order in which the response alternatives are presented can affect the results.[9] Specifically, if the alternatives are lengthy and complex, or if respondents do not feel strongly about a subject, the alternative presented last will be chosen more often than the others. If this bias is suspected, a procedure known as the *split-ballot technique* can be used to counteract it. Separate instruments are developed, rotating the order of the response alternatives so that each alternative appears in the final position an equal number of times.

How will "don't know," "no answer," and "neutral answer" responses be handled? There is no easy solution.[10] Many researchers prefer to force individuals into making a choice, eliminating neutral or "don't know" categories. For example, examine the following two alternative response formats.

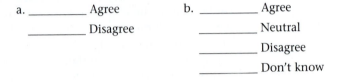

a. _____ Agree

_____ Disagree

b. _____ Agree

_____ Neutral

_____ Disagree

_____ Don't know

Obviously, alternative (b) is the more exacting of the two because of the inclusion of a middle point (neutral) and a "don't know" category.

Research suggests that including these two categories will generally increase the overall number of responses. Approximately 10% to 20% of the answers move to the neutral category.[11] It is argued that the forced-choice alternative makes people lean in one direction or another on an issue, even though some people legitimately have neutral opinions on some issues. Thus, forcing them to choose distorts the true state of affairs. The final choice in this matter depends on the researchers' preferences and the objectives of the research.

Frequently Used Scaling Techniques

As indicated in the previous chapter, a number of scaling techniques are commonly used in business research. To familiarize you with them, this chapter presents an overview of some of the major scaling techniques. The techniques presented are outlined in Table 8.1.

Two classes of scales are discussed: rating scales and attitude scales. In a *rating scale,* the rater evaluates a single dimension of a person, object, or other phenomenon at a point along a continuum or in a category; then a numerical value is assigned to that point or category. *Attitude scales* measure an individual's attitude toward any object or phenomenon. Rating is an integral part of attitude scaling, but attitude scales are generally more complex, multi-item scales.[12] Attitude scales are combinations of rating scales, designed to measure some or all aspects of a respondent's feeling toward some behavior or object. Which technique is chosen depends largely on the nature of the variable being measured and the researcher's preferences.

Graphic Rating Scales. These scales are relatively simple to use.[13] The researcher identifies the characteristic that needs to be measured; the

TABLE 8.1 Some Commonly Used Scaling Methods in Business Research

Scaling Technique	Description	Example
1. Rating scales	Raters evaluate a person, object, or other phenomenon at a point along a continuum or in a category. A numerical value is then assigned to this point or category.	Examples of various rating scales are presented below.
a. Graphic rating scales	Raters mark, or indicate in another fashion, how they feel on a graphic scale of some sort. A common graphic scale is the thermometer chart, shown in the example at the right.	On a scale of 0 to 100, please indicate how you would grade the movie you just saw. 100 ─ Very best 50 ─ Indifferent 0 ─ Very worst Record grade _____
b. Itemized rating scales	Raters select one of a limited number of categories that are ordered in some fashion. The number of categories is usually between 2 and 11. The itemized scale at the right is a 3-point scale.	How interested would you be in purchasing an electric footwarmer? ☐ Very interested ☐ Somewhat interested ☐ Not interested
c. Comparative rating scales	Raters judge a person, object, or other phenomenon against some standard or some other person, object, or other phenomenon. The scale can take a variety of forms. One comparative rating scale is the rank-order scale, shown at right.	Please rank the following automobiles in terms of their desirability to you. Assign 1 to the most desirable auto, 2 to the next, 3 to the next, etc. ____ BMW ____ Mercedes 450SL ____ Corvette ____ Jaguar
2. Attitude scales	Any one of a variety of scales that measure an individual's predisposition toward any person, object, or other phenomenon. These scales differ from rating scales in that they are generally more complex, multi-item scales.	Examples of various attitude scales are presented below.
a. Likert scale	Respondent indicates degree of agreement or disagreement with a variety of statements about some attitude, object, person, or event. Usually the scales contain 5 or 7 points. The scales are summed across statements to get the attitude score.	1. I have somewhat old-fashioned tastes and habits. Strongly disagree 1 Disagree 2 Neither agree nor disagree 3 Agree 4 Strongly agree 5
b. Semantic differential	Respondent rates an attitude object on a number of 5- or 7-point bipolar adjectives or phrases. The selection of adjectives or phrases is based on the object, person, or event.	Please rate Coors beer on the following dimensions. Bitter : __ : __ : __ : __ : __ : __ : __ : Not bitter Sweet : __ : __ : __ : __ : __ : __ : __ : Sour

For in-depth discussions of these and other scaling techniques, see Warren S. Torgerson, *Theory and Methods of Scaling* (New York: Wiley, 1967); Allen Edwards, *Techniques of Attitude Scale Construction* (New York: Appleton-Century-Crofts, 1957); Charles Osgood, George Suci, and Percy Tannenbaum, *The Measurement of Meaning* (Champaign, IL: University of Illinois Press, 1957); and Gene F. Summers, Ed., *Attitude Measurement* (Chicago: Rand McNally, 1970).

appropriate wording of the item and response formats are determined; and graphical presentation is made, reflecting these decisions. Table 8.1 illustrates one form of graphic rating scale, the thermometer chart. A graphic scale is often used because respondents can easily relate to the visual nature of the representation.

Itemized Rating Scales. These scales are perhaps the most frequently used in business research because of their simplicity and adaptability to almost any type of measurement situation. The rater is usually forced to select one of a limited number of choices (usually between 2 and 11) on a continuum applicable to the phenomenon being measured. Table 8.1 provides one illustration of this type of scale. Also, the questionnaires presented in this chapter use rating scales to varying degrees.

Comparative Rating Scales. Graphic and itemized rating scales are non-comparative: The rater makes a judgment without reference to another person, object, and so on. On the other hand, comparative rating scales make explicit reference to some norm or objective for comparison purposes. In Table 8.1, the *rank-order scale* asks the rater/respondent to compare four types of autos to each other, then rank them according to their desirability. Because these scales are rank-ordered, they are ordinal measurement devices.

Likert Scale. Frequently called the *method of summated ratings,* the Likert scale is also a widely accepted and adopted technique.[14] Using Likert scales, the respondent indicates the amount of agreement or disagreement with a variety of statements about some attitude or object. The scale is highly reliable when it comes to the ordering of people with regard to a particular attitude.[15]

Semantic Differential. This scale is usually composed of a number of bipolar adjectives or phrases on which the respondent must evaluate an attitude object.[16] The procedure is widely used in image studies, where competitive comparisons are often made. Note that both Likert scales and semantic differential scales are usually five- or seven-point scales, so a midpoint is an available option for the respondent. However, six-point and other even-numbered scales are sometimes used by researchers who want to eliminate "fence sitting" at the midpoint. Usually, the number of scaling points is based on researcher preference and the nature of the study. Table 8.1 includes a classical seven-point semantic differential scale using the characteristics of bitterness and sweetness.

Other Scaling Methods. The scaling approaches indicated here are by no means the only ones available to the business researcher. Thurstone's Equal-Appearing Interval Scale, Guttman's Scale-Analyzing, and the Staple Scales are but a few of the many variants that have been around for some time.[17] These techniques can be useful in a variety of study situations, but we leave the investigation of these alternatives to the interested reader.

In addition, more advanced techniques such as multidimensional scaling (MDS) and conjoint analysis have been gaining popularity in the business research field.[18] Because these more advanced techniques are closely related to data analysis, an introductory discussion will be presented in the chapter on advanced multivariate data analysis.

Before we end our discussion of scale construction, it should be noted that numerous standardized scales have been tested for validity and reliability.[19] Researchers should thoroughly explore the possibility of using these scales, rather than reinventing the wheel each and every time a measurement problem comes up.

INSTRUMENT DESIGN

Once the major decisions have been made as to what variables need to be measured and what scales are going to be used to measure them, we must put the entire package together. This is called *instrument design*. As discussed previously, the researcher must be aware of many factors in scale and instrument design. Everything from the type of question to the wording must be conscientiously examined to ensure that the data collected will be of the proper type and quality to answer the research question.

As stated earlier, the process of design is very much an art, but there are a number of guidelines and new technology aids to help researchers in constructing sound measurement instruments. These guidelines and aids relate to questions concerning scale sequencing and design, the advent of online design aids (both on and off the Web), and recommendations about pretesting and correcting.

Scale Sequencing and Layout

The presentation and organization of the data collection instrument can be crucial to a study's success. Most important are the sequencing of scales and the presentation of the measurement device in an interesting and easily understood fashion. There are no steadfast rules for this important facet of design, but some rules of thumb can help.

> **Start the questionnaire with simple and interesting questions.** If the start of the questionnaire is complex and hard to understand, there is a very high probability of the respondent not completing the interview. There should be relatively easy and general questions up front, to get the respondent involved in the questioning process. Once the respondent is involved, the questions can become more specific and difficult if the situation dictates.

Provide clear and easy-to-read instructions, with transitions if necessary. Instructions should be provided to explain the purpose of the questions, with directions on how to complete the scales, if necessary. The instructions should describe and explain clearly what the respondent is expected to do. If there are abrupt changes in the types of scales in the measurement instrument, the instructions should direct the respondent to the change in response format.

Ask sensitive and classificatory information last. Classificatory information is used to classify respondents or other study units. Common examples of classificatory information are age, income, sex, household size, and so on. Similarly, sensitive information should be asked last because of the possibility of scaring respondents away before they answer the rest of the questions in the study. For example, if the first question asked a person's income, some people might see this as an invasion of privacy and refuse to answer the rest of the questionnaire. Therefore, place these questions last so that if respondents do refuse to answer past that point, it will not affect the answers already given.

Lay out the questionnaire so that it is easy to read and follows the flow of the questioning process. This recommendation is often overlooked. Eye-appealing and well-laid-out questionnaires make it easy for the respondent to fill out the questions correctly and completely. Cluttered surveys with awkward designs are difficult to complete. Anything that helps make completing the questionnaire easier for the respondent will most likely help the return and completion rate for the study.

Online Design Aids

Tremendous strides have been made in the field of technological aids that help in instrument design and other stages of the research process. Several of these (like the growth of the Internet, advances in online databases and search mechanisms, and integrated software development) were discussed earlier. Here, we concentrate our discussion on some of those developments that directly affect the whole issue of scale and instrument design. However, keep in mind that these developments are interlinked with advance in the areas of sample selection, actual data collection, subsequent data analysis, and, ultimately, data presentation and delivery, as we will see later in the text.

These aids are generally designed to help researchers do some combination of the following:

- Construct questionnaires by suggesting the form, structure, and wording for items, given the parameters provided by the researcher
- Randomize and administer items so as to help eliminate bias due to item order and position

- Provide options for the collection of data to help accomplish the researcher's objectives and goals for the study
- Provide other add-on services such as speeding data collection and aiding in data analysis and/or its presentation

It is important to remember that, despite the benefits these aids provide, significant researcher judgment and input is still needed to ensure proper instrument design. Ultimately, it is the researcher's call as to whether the instrument will accomplish the objectives of the study at hand.

Perhaps the best way to illustrate these aids is to provide an overview of some of the Web sites and organizations that are leading the way in the development of online design aids for researchers.[20] Exhibit 8.2 outlines several organizations, identifies their associated Web sites, and gives brief descriptions of their activities in these and related areas. As can be seen from the exhibit, the firms provide a variety of services, including "how-to" wizards, preconstructed questionnaire templates, libraries of sample questions, online or manually administered formats of the same questionnaire, and a variety of add-on collection, analyzing, and presentation options of the data obtained from the study.

To further illustrate some of the capabilities of these online services, it is useful to examine one of these sites in more depth. We chose MarketTools' Zoomerang because it is a highly flexible self-service product

EXHIBIT 8.2

A Sampling of Web Sites and Companies Providing Online Instrument Design Aids

Zoomerang http://www.zoomerang.com	MarketTools' Zoomerang provides Web-hosted technology for the development and conduct of online survey research. Survey templates aid in the design process. zSample and analysis capabilities make this an integrated survey solution.
Perseus Development http//:www.perseusdevelopment.com	Using SurveySolutions for the Web software, you can design professional surveys quickly, distribute them via the Web or e-mail, collect responses automatically, analyze survey results, and create presentations of survey results instantly.
SPSS http://www.spss.com	SPSS Data Entry Builder can be a stand-alone or network-based solution to creating surveys and online data collection forms. Works seamlessly with other SPSS products to create an integrated development, collection, analysis, and presentation software package.
Socratic Technologies http://www.sotech.com	A full-service firm that produces software and conducts research. Their software is Web-enabled and includes libraries of questions and design help.
Decision Analyst http://www.decisionanalyst.com	With Decision Analyst's American Opinion Online panel of approximately 200,000 people, surveys on a variety of topics can be conducted over the World Wide Web. Online development of survey instrument is provided.
WebSurveyor http://www.websurveyor.com	WebSurveyor provides a local PC-installed client that interacts with hosted central services. Used to help design and implement online surveys. One must recruit or have one's own sample.

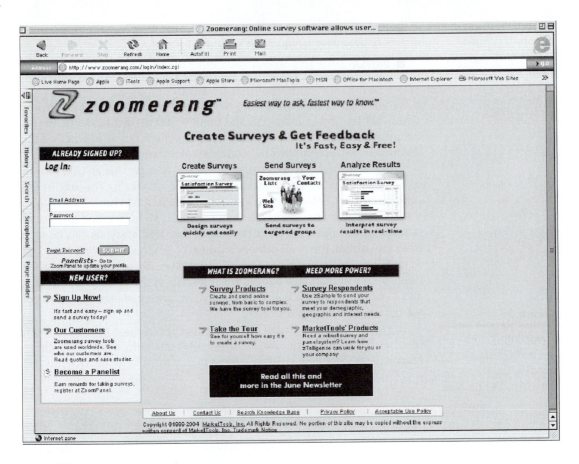

FIGURE 8.2

Home Page for zoomerang.com from MarketTools

Reprinted with permission.

that demonstrates many of the current features that are now available. Figure 8.2 presents the home page for Zoomerang (http://www.zoomerang.com). It must be noted that MarketTools also offers full-service research capabilities, but here we are primarily concentrating on the self-service end of the spectrum.

As noted on the home page, one can create, deploy, and analyze surveys using the Zoomerang product. One can develop one's own custom online survey or explore and modify many of the survey templates that are available online, as presented in Exhibit 8.3. In addition to these capabilities are the possibilities of enabling skip, or branch, logic in the asking of questions, including images in the survey instrument, and many other specialized design features. This flexibility has encouraged organizations such as General Mills, Procter & Gamble, and People Soft to use this product. The interested individual is strongly encouraged to log on to Zoomerang's Web site and explore the capabilities of this survey development tool.

EXHIBIT **8.3**

A Brief Customer Satisfaction Survey Template from Zoomerang

zoomerang

support logout

home | new survey | my surveys | email list | account info

Design theme: Graphite Gray

Customer Satisfaction (Brief)

Edit Title Area

To add a survey title and upload an image, click the "Edit Title Area" button.

1 How long have you used [ENTER PRODUCT/SERVICE NAME]?

- Less than 1 month
- 1 to 6 months
- 6 months to 1 year
- 1 to 3 years
- Over 3 years
- Never used

Edit Insert Delete

2 How often do you use [PRODUCT/SERVICE]?

- Once/week or more
- 2 to 3 times/month
- Once/month
- Less than once/month
- Never used

Edit Insert Delete

3 Please rate your OVERALL satisfaction with [PRODUCT/SERVICE]?

Extremely Unsatisfied	Unsatisfied	Satisfied	Extremely Satisfied
1	2	3	4

Edit Insert Delete

4 Compared to available [PRODUCT/SERVICE CATEGORY] alternatives, [PRODUCT/SERVICE] is . . .

- Much better
- Somewhat better
- About the same
- Somewhat worse
- Much worse
- Don't know
- Never used

Edit Insert Delete

5 Will you use/purchase [PRODUCT/SERVICE] in the next 6 months?

- Absolutely
- Very likely

Edit Insert Delete

(Continued)

EXHIBIT **8.3**

(Continued)

○ Unlikely
○ Absolutely not
○ Not sure

⊙ Edit
⊕ Insert
⊖ Delete

6 Would you recommend [PRODUCT OR SERVICE] to others?

○ Absolutely
○ Very likely
○ Unlikely
○ Absolutely not
○ Not sure

⊙ Edit
⊕ Insert
⊖ Delete

7 Please rate your level of SATISFACTION with each of the following:

	1	2	3	4	
	Extremely Unsatisfied	Unsatisfied	Satisfied	Extremely Satisfied	
Product/service quality	1	2	3	4	N/A
Value for the price	1	2	3	4	N/A
Purchase experience	1	2	3	4	N/A
Installation or first use experience	1	2	3	4	N/A
Usage experience	1	2	3	4	N/A
After purchase service	1	2	3	4	N/A
Repeat purchase experience	1	2	3	4	N/A

⊙ Edit
⊕ Insert
⊖ Delete

8 Is there anything you would like to tell [COMPANY] about its [PRODUCT/SERVICE] that was not already asked in the survey?

⊙ Edit
⊕ Insert
⊖ Delete

Thank you for your feedback.

⊙ Edit
⊕ Insert
⊖ Delete
⊕ Insert

Page 1

Pretesting and Correcting

Once the instrument is completed in draft form, it should be subjected to a pretest.[21] Ideally, a pretest should be conducted on a set of respondents who are similar to the final study respondents. For example, a questionnaire designed to assess top executive opinions of current business conditions should not be pretested on a group of college students. It is not desirable because executives have different frames of reference from college students, and this would most likely affect the meaning they attach to various questions in the scales.

Pretesting should be considered essential in the design of any instrument. Such pretests often identify problems in wording, questionnaire format, and other areas that have a profound impact on the validity of the findings. Once problem areas are located, the researcher can make the necessary corrections to ensure the highest quality data possible.

To summarize, the process of scale development and instrument design is definitely an art. It takes a lot of patience and experience to develop sound data collection instruments that are reliable and valid. No list could ever include all the decisions researchers must face in the instrument design process, but Exhibit 8.4 gives 21 guidelines to help direct you in the design of good measurement instruments. Use these guidelines with common sense in your quest for useful measurement instruments.

EXHIBIT **8.4**

Some Guidelines for Instrument and Scale Design

Source: Stanley Payne, *The Art of Asking Questions.* Copyright © 1951, renewed 1979, by Princeton University Press. Excerpt, pp. 228–237, reprinted by permission of Princeton University Press. For a complete listing of the 100 questions Payne proposes, see this source.

1. Clearly understand the research question yourself before you attempt to develop scale questions.
2. Phrase the questions so they are easily understood by potential respondents.
3. Fit the type of question (open-ended, dichotomous, multichotomous) to the degree of understanding of the desired respondents (i.e., if the opinion is vague, then use open-ended questions; if it is clear-cut with several alternatives, use multiple-choice questions).
4. Never take anything for granted. Consider all assumptions implicit in the question.
5. Choose the scale questions which will best answer your research question, but always ask if open-ended questions are the best means to obtain those answers.
6. With dichotomous and multiple-choice questions, attempt to make the answers mutually exclusive and exhaustive. If this is not possible, provide answers which can handle multiple or both answers.
7. Make provisions to handle "Don't Know" and "No Opinion" answers in the scales.
8. Avoid double-barreled questions where two or more issues are addressed in the same question. Always try to put only one issue in a question.
9. Provide adequate, easy-to-read, and understandable instructions for respondents.
10. Do not talk down to respondents. No one likes their intelligence insulted.

(Continued)

EXHIBIT **8.4**

(Continued)

11. Use good grammar in wording questions, but do not appear to be overly formal.
12. Avoid overly long, complex questions. Keep it simple.
13. Use simple words if they express the meaning you want conveyed.
14. Avoid specialized language. Trade jargon is all right for a specialized audience, but it is wholly unacceptable for the general public.
15. Use examples carefully in the introduction of questions. The introduction of examples may divert a respondent's attention from the main issue.
16. Highlight or underscore important words which should be emphasized.
17. Eliminate unnecessary redundancy in questions and answers.
18. Hold sensitive and difficult questions and issues until the later part of the instrument.
19. Be considerate of the respondent's time and privacy.
20. Always pretest before collecting data with any instrument.
21. Always say thank you at the end of each questioning session.

MANAGERIAL CONSIDERATIONS

There are five primary managerial concerns with regard to instrument design. Four of these concerns have to do with the quality of the data that will be produced by the instrument. The fifth involves the image the instrument projects to the particular public under study.

Instrument design is an art that requires substantial time and effort to ensure high-quality data. Management must understand that high-quality, valid data collection instruments usually cannot be developed overnight. To obtain high-quality instruments, substantial time and effort must go into the development and pretesting of the devices to ensure they are doing what they are supposed to be doing. Management must have an appreciation for the process and support the research function with the appropriate resources to obtain quality data. International research studies require special attention when it comes to scale and instrument design. Managers should require significant justification of the development of instruments to be used in international and multinational settings. Exhibit 8.5 presents some considerations in developing useful scales and instruments for these settings.

Poorly designed instruments lead to low-quality data. The only thing worse than no research is bad research. Sloppy research, particularly in the area of instrument design, leads to low-quality data: large amounts of unknown errors in the data itself. If the time and resources cannot be justified to develop high-quality instruments, then the investigation should not be conducted.

Researchers should be required to justify the inclusion of scales in an instrument by demonstrating how the data will help solve the research question. A basic tenet in instrument design is to keep it simple.

EXHIBIT **8.5**

Issues in the Design of
International
Measurement Scales
and Instruments

Source: Adapted from Susan P.
Douglas and C. Samuel Craig,
"Marketing Research in the
International Environment," in
Ingo Walter and Tracy Murray,
Eds., *Handbook of International
Management* (New York:
Wiley, 1988), pp. 5.19–5.28.
Reprinted by permission of
John Wiley & Sons, Inc.

Translation	When conducting multinational or multilingual studies, translation of the instrument is critical. Both verbal and nonverbal stimuli must be considered to ensure conceptual equivalence of meaning. *Back-translation* is often used to ensure equivalence: The instrument is translated into the native language by one person, then back-translated into the original language by someone else. Colors, words, and symbolism of meaning all need to be examined from the respondents' perspective. For example, white in Japan is a color of mourning; green in Malaysia symbolizes danger and, in some other Asian countries, infidelity.
Scale Construction	Scales must often be creatively designed to accommodate literacy levels and cultural differences. Frequently, graphic scales have to be used in developing countries to make sure respondents understand. In addition, "show cards" or pictures frequently accompany the questions to enhance the meaning for respondents. Meanings must be conceptually equivalent to ensure comparability across languages. This often means that the scale items are modified somewhat so that the meaning is constant across nations or language groups.
Topic Bias	Cultures differ in their willingness to provide certain information. For example, willingness to talk about sex or provide information on personal issues varies considerably. Yea-saying and nay-saying responses can be related to cultural background. Respondents from Asia tend to provide socially positive responses to "help" the researcher.
Interpretation	Procedures to avoid bias are necessary in the interpretation of responses, particularly open-ended responses. Researchers and managers must be careful not to use their native culture as the reference in interpreting the results of an international investigation. Multiple interpreters should review the responses and back-check the results to see if there are similar interpretations of the data from the measurement instrument.

Unnecessary questions only serve to clutter and confuse the primary issues under investigation. All scales and measured variables should play an integral part in solving the research question. If they are unnecessary, they should be eliminated from the questionnaire.

Consider using the latest online instrument design aids in the developmental process. Giant strides are being made in research because of numerous new technical developments relating to online data collection. Seriously consider the pitfalls and benefits of these technologies. Above all, remember that your results will only be as good as the researchers and managers that oversee the research process.

EXHIBIT **8.6**

Key Managerial
Questions Pertaining to
Instrument Design

1. Is the instrument free of awkward sentences, poor grammar, and other flaws in style that make it difficult to comprehend?
2. Are adequate instructions provided with the instrument to help the respondents answer the items?
3. Will the instrument, as designed, aid in collecting the data needed to solve the research problem?
4. Do the researchers have a clear plan of how they will use every item contained in the instrument to answer the research question?
5. Has the instrument been adequately pretested to ensure that it is unambiguous and easy to follow?
6. If the instrument is to be administered under your company's logo, does it project the desired image of your firm?

The instrument is a reflection of the image of the firm. Therefore, it should be evaluated for the image it projects to its respondents.
This last concern has the least to do with the conduct of good research. Most business research involves people because business is a social phenomenon. Contact with the respondents should be evaluated to ensure that it portrays the image desired by management. Measurement instruments happen to be a major contact between the firm and its public. Thus, to ensure that the best image possible is projected from the research function, questionnaires and all other forms of public contact should be of professional quality.

Through the monitoring of these five concerns, management can help protect its interest in the generation of timely and relevant data for decision-making purposes. To help guide management in accomplishing this objective, we present in Exhibit 8.6 a number of key managerial questions pertaining to instrument design.

SUMMARY

Instrument design is defined as the formal construction of a data collection device (usually a questionnaire) to obtain the information needed to solve a research problem. The chapter focuses on the primary concerns in the development of valid and reliable scales and data collection instruments. This process is very much an art; there is no definitive means that can be used to ensure the development of good questionnaires.

A data collection instrument is composed of scales, which in turn are composed of items. Much of the chapter focuses on the development of good scales of measurement. Specifically, the topics of item phrasing and response formats are examined. Suggestions for proper item and response development are highlighted and reviewed. In addition, some frequently used rating and attitude scaling techniques are presented.

The final section of the chapter dealt with the design of the instrument itself. The topics of scale sequencing and layout were discussed. The discussion then moved to an overview of the fast-developing field of online instrument design aids. Web sites and organizations were highlighted for their developments in this area. The important stage of pretesting and correcting any instrument before it is actually used was then addressed. The last part of the chapter noted some of the salient managerial considerations given today's complex and changing world.

DISCUSSION QUESTIONS

1. What is meant by the term *instrument design*? Discuss the interrelationships of instrument design with the measurement process.

2. Discuss some of the basic concerns of item phrasing. Provide examples to illustrate these concerns.

3. What should the researcher be concerned with when deciding upon which response format to use? What ultimately dictates the response format used in any study?

4. What are some of the more frequently used scaling techniques in business research? Give an example of each technique.

5. Outline some of the major rules of thumb in scale sequencing and layout. What should be the basic concern when constructing any instrument?

6. Why is pretesting so essential in scale development and instrument design? Discuss some of the possible consequences if an instrument is not pretested before it is administered to your potential respondents.

7. Outline and discuss the major managerial concerns in instrument design. How can managers become involved in this process?

8. Identify five major cultural differences in a country of your choice, and discuss what impact these differences would have in the design and use of measurement instruments. Be specific.

9. Design a questionnaire that could be used to evaluate the usefulness of various courses taken during your college career. Be creative, and make any assumptions necessary to design this survey instrument.

10. Using the Web sites (or others you can find) outlined in Exhibit 8.2, select a company providing online design aids and write a brief report discussing the online products of the vendor. Be sure to discuss all relevant aspects of the vendor's online services.

11. Go to zoomerang.com (http://www.zoomerang.com) and use the free trial offer to design a survey of your choice. Print your survey and be prepared to discuss your experience using this online service.

NOTES

1. See "Excuse Me, What's the Pollsters' Big Problem?" *Business Week* (16 February 1987), p. 108.

2. For texts dealing with the subjects of scale and instrument design, see Stanley Payne, *The Art of Asking Questions* (Princeton, NJ: Princeton University Press, 1951); Warren Torgerson, *Theory and Methods of Scaling* (New York: Wiley, 1958); A. N. Oppenheim, *Questionnaire Design and Attitude Measurement* (New York: Basic Books, 1966); and Robert F. DeVellis, *Scale Development: Theory and Applications* (Newbury Park, CA: Sage, 1991).

3. See Ronald Alsop, "Coke's Flip-Flop Underscores Risks of Consumer Taste Tests," *Wall Street Journal* (18 July 1985), p. 27.

4. See David A. Ricks, *Blunders in International Business* (Cambridge, MA: Blackwell, 1997).

5. See M. D. Geurts, R. R. Andrus, and J. Reinmuth, "Researching Shoplifting and Other Deviant Consumer Behavior Using the Randomized Response Research Design," *Journal of Retailing* (Winter 1975–76), pp. 43–48; and Guy Begin and Michel Boivin, "A Comparison of Data Gathered on Sensitive Questions Via Direct Questionnaire, Randomized Response Technique, and a Projective Method," *Psychological Reports* (1980), *47*, pp. 743–750, for discussions about alternative methods and problems in collecting sensitive data.

6. For discussions on the open- and closed-ended question, see Howard Schuman and Stanley Presser, "The Open and Closed Question," *American Sociological Review* (October 1979), *44*, pp. 692–712; and B. S. Dohrenwend, "Some Effects of Open and Closed Questions," *Human Organizations* (Fall 1970), pp. 416–422.

7. Again, the classic reference dealing with the subject is Payne, op. cit. note 2. Other useful discussions can be found in Gilbert Churchill, *Marketing Research,* 4th ed. (Chicago: Dryden Press, 1987), Chapter 7; and Delbert Miller, *Handbook of Research Design and Social Measurement* (New York: David McKay, 1970), pp. 76–86.

8. See M. H. Morris, D. L. Davis, and J. W. Allen, "Fostering Corporate Entrepreneurship: Cross-Cultural Comparisons of the Importance of Individualism Versus Collectivism," *Journal of International Business Studies* (First Quarter, 1994), pp. 65–89.

9. M. Parten, *Surveys, Polls and Samples* (New York: Harper and Row, 1950), pp. 210–211.

10. For good discussions on these issues, see Hans Zeisel, *Say It with Figures,* 5th ed. (New York: Harper and Row, 1968), pp. 40–58; Stanley Presser and Howard Schuman, "The Measurement of a Middle Position in Attitude Surveys," *Public Opinion Quarterly,* (1980), *44*, pp. 70–85; and Rebecca F. Guy and Melissa Norwell, "The Neutral Point on a Likert Scale," *Journal of Psychology* (1977), *95*, pp. 199–204.

11. Presser and Schuman, op. cit. note 10, p. 83.

12. For an excellent discussion of rating and attitude scales, see Claire Selltiz, Lawrence S. Wrightsman, and Stuart Cook, *Research Methods in Social Relations* (New York: Holt, Rinehart, and Winston, 1976), Chapter 12.

13. See J. P. Guilford, *Psychometric Methods* (New York: McGraw-Hill, 1954), for a discussion of graphical scales.

14. For discussions of Likert scales, see Rensis Likert, "A Technique for the Measurement of Attitudes," *Archives of Psychology* (1932), no. 140; Miller, op. cit. note 7, pp. 92–93; and Allen Edwards, *Techniques of Attitude Scale Construction* (New York: Appleton-Century-Crofts, 1957), pp. 149–170.

15. Miller, op. cit. note 7, p. 92.

16. See Charles Osgood, Charles Suci, and Percy Tannenbaum, *The Measurement of Meaning* (Champaign, IL: University of Illinois Press, 1957), for a complete discussion of the semantic differential. Also, John Dickerson and Gerald Albaum, "A Method for Developing Tailormade Semantic Differentials for Specific Marketing Content Areas," *Journal of Marketing Research* (February 1977), *14*, pp. 87–91.

17. See Warren Torgerson, *Theory and Methods of Scaling* (New York: Wiley, 1967), for discussions of these and other methods.

18. For discussions of MDS and conjoint measurement, see P. E. Green and V. R. Rao, *Applied Multidimensional Scaling: A Comparison of Approaches and Algorithms* (New York: Holt, Rinehart, and Winston, 1972); P. E. Green, F. J. Carmone, and S. M. Smith, *Multidimensional Scaling* (Boston: Allyn and Bacon, 1989); and J. F. Hair, R. E. Anderson, and R. L. Tatham, *Multivariate Data Analysis* (New York: Macmillan, 1987).

19. For example, useful scales can be found in John P. Robinson, Robert Athanasian, and Kendra Head, *Measures of Occupational Attitudes and Occupational Characteristics* (Survey Research Center, Institute for Social Research, University of Michigan, Ann Arbor, 1969); John P. Robinson and Phillip Shaver, *Measures of Social Psychological Attitudes* (Survey

Research Center, Institute for Social Research, University of Michigan, Ann Arbor, 1973); Marvin Shaw and Jack Wright, *Scales for the Measurement of Attitudes* (New York: McGraw-Hill, 1967); Delbert C. Miller, *Handbook of Research Design and Social Measurement* (Newbury Park, CA: Sage, 1991); Gordon Bruner and Paul Hansel, *Marketing Scales Handbook* (Chicago: American Marketing Association, 1992); and William O. Bearden and Richard G. Netemeyer, *Handbook of Marketing Scales* (Thousand Oaks, CA: Sage, 1998).

20. The Web sites listed in Exhibit 8.2 are just a sampling. Many others are available; for example, see http://www.harrisinteractive.com, http://www.insightexpress.com, http://www.hostmysurvey.com, to name a few.

21. For an interesting article on pretesting, see Nina Reynolds, Adamantios Diamantopoulos, and Bodo Schlegelmilch, "Pretesting in Questionnaire Design: A Review of the Literature and Suggestions for Further Research," *Journal of the Marketing Research Society* (April 1993), pp. 171–182.

SUGGESTED READING

The Nature of Instrument Design

Foddy, William, *Constructing Questions for Interviews and Questionnaires: Theory and Practice in Social Research* (Cambridge, UK: Cambridge University Press, 2003).

A very useful paperback that discusses the problems associated with constructing questions for data collection purposes. A useful read for beginner and practitioner alike.

Oppenheim, A. N., *Questionnaire Design and Attitude Measurement* (New York: Basic Books, 1966).

This classic book is essential reading for anyone interested in the topic of instrument design. The author gives a very good overview of the problems in questionnaire design.

Scale Development

Bearden, W. O., and R. G. Netemeyer, *Handbook of Marketing Scales,* 2nd ed. (Thousand Oaks, CA: Sage, 1998).

A compilation of multi-item, self-report measures frequently used in consumer behavior and marketing research. Helpful in not reinventing the wheel.

Miller, Delbert C., and Neil J. Salkind, *Handbook of Research Design and Social Measurement,* 6th ed. (Thousand Oaks, CA: Sage, 2002).

This reference is a classic in scale development in the social sciences. Treats a wide variety of issues covered in Chapters 7 and 8 in this text.

Instrument Design

Dillman, Don A., *Mail and Internet Surveys,* 2nd ed. (New York: John Wiley, 2000).

This book presents the Tailored Design Method for the design and application of self-administered surveys in a changing environment. A good read for those concerned with the whole issue of instrument design.

Reynolds, N., A. Diamantopoulos, and B. Schlegelmilch, "Pretesting in Questionnaire Design: A Review of the Literature and Suggestions for Further Research," *Journal of the Marketing Research Society* (April 1993), pp. 171–182.

An interesting article that reviews the literature on pretesting. The appendix in the article outlines recommendations for items to be pretested.

See the Web sites in Exhibit 8.2 and others for contemporary discussions of this issue.

9

Sampling Design

Overview

The Nature of Sampling
Terminology
The Rationale for Sampling
The Sampling Process

An Introduction to the Philosophy of Sampling

Sample Designs
Sample Design Choice Considerations
Probability Designs
Nonprobability Designs

Practical Considerations in Sampling

Incidence and Response Rates
Internationalization of the Marketplace

Online Sampling Design Issues

Managerial Considerations

Summary

Discussion Questions

Notes

Suggested Reading

Appendix: A Computational Example Illustrating the Properties of the Central Limit Theorem

The research design process must include the selection of whom or what to study—the sample selection decision. In essence, the sampling decision requires you to answer key questions that affect the quality and generalizability of the results obtained from a study. This chapter deals with the salient issues related to sampling.

The chapter begins with a discussion of the terminology and nature of the sampling process. Significant terms are defined and explained in a nontechnical format. This terminology is then used (1) to explain the rationale of sampling in business research, (2) to develop the sampling process, and (3) to describe the basic types of sampling designs that are available to researchers.

The overall focus of the chapter is to establish a basic understanding of the underlying nature of sampling, so that various sampling designs can be understood and evaluated. The later sections of the chapter outline some of the sampling designs available to the researcher. The advantages and disadvantages of each design are presented so that a preliminary evaluation of their applicability to a specific problem can be undertaken. Practical considerations in sampling are then discussed. Important managerial considerations in sampling are outlined in the last section, and a computational example is presented in the chapter appendix.

THE NATURE OF SAMPLING

Sampling possesses its own specialized terminology.[1] Before discussing the subject matter, we present some of the basic definitions in this area. Then the rationale and the process of sampling will be examined.

Terminology

Element. An *element* is the unit from which the necessary data is collected. This data forms the basis of analysis, from which conclusions are drawn and research problems are solved. An element can be a unit of analysis (UOA), because business research may go beyond collecting data solely from individual respondents or interviewees. A unit of analysis can be an organization (a retail, wholesale, or manufacturing firm), inanimate objects (an automobile taken from a production line), or the most common element of survey research—the individual.

Population. A *population* is defined as the complete set of unit analysis that are under investigation. The population is the finite (closed) or infinite (exhaustive) set of units of analysis that could be included in a study.

For example, if a team of researchers is asked to study the work habits of midlevel executives of a company with six divisions located in different parts of the country during the month of May, the population they would be sampling from is a finite and closed population consisting of the midlevel executives within the six divisions during a specific time frame. These midlevel executives also represent the units of analysis in the study.

If the researchers were given the broader task of studying the work habits of midlevel executives of organizations throughout the world, for all practical purposes this would entail sampling from an infinite population of units of analysis. Whether the researchers have specified a finite or an infinite population of UOAs, they will usually choose a smaller set of elements to study: the *sampled UOA*, or simply the *study sample*.

Sampling Units. *Sampling units* are nonoverlapping elements (UOAs) from a population.[2] Sampling units can be individuals, households, city blocks, census tracts, departments, companies, or any other logical unit that is relevant to the study at hand. The study sample is a subset of UOAs chosen for study. As we will see later in this chapter, various procedures can be used to select the study sample from a population, ranging from a simple one-stage sample selection to more complex multistage processes. A sampling unit can be an individual element (UOA) or a set of elements (UOAs).[3]

For example, a midlevel executive in one organization is an element in the previously discussed work habit example. In a simple one-stage sampling procedure, the sampled unit would simply be the executive (UOA). However, it may be more convenient, take time, and be less costly to use a more complex procedure: selecting departments of midlevel executives (UOAs) first, then selecting executives from within these departments to arrive at our study sample. This would be a two-stage sampling procedure, where the departments are the primary sampling units and the individual executives the secondary sampling units. We describe this process in more detail in a later section.

Sampling Frame. A *sampling frame* is a physical representation of objects, individuals, groups, and so on, important to the development of the final study sample. It is the actual list of sampling units at any stage in the selection procedure. In our simple example, the sampling frame is a listing of all midlevel executives. In our complex example, the sampling frames consist of all the departments and all the individual executives within these departments.

A sampling frame should ideally be a complete enumeration of all elements of the population as specified by the research problem. However, in reality, what generally happens is that sampling frames already in existence may define the population.[4] Again referring to our work habit example, the researchers might use midlevel executives who belong to the American Management Association as their sampling frame to study work habits because there is a readily available list of members and their addresses.

Clearly, not all midlevel executives belong to the AMA, so the population may be modified for the study's purposes.

Another problem relating to the sampling frame arises when there is no listing of the elements to be sampled. A case in point would be the sample of midlevel executives attending a professional meeting on a particular day. If no prior list of attendees is available, there is no actual frame from which to sample. In this situation, the sample selection process follows a set of rules (for example, choose every fifth manager registering for the meeting on that day), rather than selecting from a fixed sampling frame.

The sampling frame decision is an important one. Most sampling frames suffer from inadequate identification of all the elements in the population as specified by the researchers.[5] For example, telephone directories are used as a sampling frame for many studies. However, as you know, directories do not include the substantial number of households with unlisted numbers. With telephone interviewing, a procedure has been developed to overcome the problem. It is called *random-digit dialing*. This procedure gives a greater percentage of existing phone numbers and essentially has all of the properties of a probability sample, which will be discussed later in this chapter.[6] The sampling frame must be evaluated carefully to make sure it is representative of the population designated for study.

Sample. The subset of UOAs chosen for study from a population is called the *study sample* or merely a *sample*. To be useful to the researcher, the sample must reflect an approximation of the characteristics in the population that it is supposed to represent. A good sample will permit generalization of the findings to the population with a certain level of statistical confidence. An efficient sample combines the generalization qualities of a good sample with the added benefits of cost minimization in design and execution.

Parameter/Statistic. The objective of sampling is to make inferences about a specified population of interest based on the information contained in the sample. What we infer from the sample information are estimates of certain population parameters. *Parameters* are summary descriptors of a given variable in the population; a *statistic* is the summary descriptor of a given variable in the sample.[7] Keep in mind that sample statistics are used to make estimates of population parameters.

Sampling Errors. Good research must reduce errors in order to maximize information content and provide generalizable results. Good sampling designs adhere to this premise. Generally, there are two classes of sampling errors: procedural errors of sampling (the biases in the sampling procedure itself) and imprecision associated with using statistics to estimate parameters.[8] Both classes of sampling error can play havoc with research results. However, using sampling theory and probability samples (through randomization), the researcher can estimate the degree of imprecision that might be associated with any sampling design.

Statistical and Sample Efficiency. *Statistical efficiency* is a measure of like-sized sample designs that evaluates which design produces the smaller standard error of the estimate.[9] *Sample efficiency* is a characteristic in sampling that emphasizes high precision and low per-unit cost for every fixed unit of precision.

To illustrate, suppose we draw two samples, one using a simple random sampling procedure and the other a cluster sampling procedure (to be discussed later). Our objective is to measure an interval-level variable from the sample so as to estimate its corresponding population parameter. All things being equal, the simple random sample would be a statistically more efficient design because it would produce a lower standard error of the estimate than the cluster sample, given the same sample size. Therefore, to obtain the same level of precision (standard error estimate), a cluster sample would have to be larger.

On the other hand, cluster samples are generally significantly less expensive to obtain than simple random samples because they take less time and fewer resources. Look at two possible scenarios in our example:

> **Scenario 1.** It costs $10,000 to conduct a simple random sample of 500 people with a precision level of 0.10, *or $1000 per 0.01 unit of precision.*

> **Scenario 2.** It costs $7500 to conduct a cluster sample of 750 people with a precision level of 0.10, *or $750 per 0.01 unit of precision.*

In this case, it is obvious that the simple random sample is the more statistically efficient design, but the cluster sample possesses greater sample efficiency. These considerations significantly affect the researcher's choice of a sampling plan.

Sampling Plan. A *sampling plan* is the formal specification of the methods and procedures to be used to identify the sample chosen for study. The sampling plan is a formal statement of the sampling process for a particular study; it defines the strategy and critical events in the selection of a sample.

The Rationale for Sampling

We can now discuss the rationale for sampling in business research. As mentioned earlier, a primary purpose of sample statistics is to infer the characteristics of some population. Sometimes it is possible and desirable to measure each and every element in the population on some characteristic of interest. Such complete measurement of the population is known as a *census*. Although a census has the advantage of completeness, some very important considerations make sampling an attractive alternative for business researchers:

> **Resource constraints.** Most business researchers are forced to deal with resource constraints, including the all-important factors of cost

and time. Well-selected samples can be less costly and provide very reliable results.

Accuracy. Well-drawn samples can provide the researcher with pre-specified margins of error. As we will see later in the chapter, the use of probability samples makes it easier to estimate the degree of sampling error in a study.

Destructive measurement. Very often, measurement can be a destructive process. For example, if a firm producing tires had to test the air-holding capabilities of each tire by blowing every tire up until it exploded, there would be no product to sell. The ability to sample and make inferences about the population is very important in such cases.

The Sampling Process

The *sampling process* can be thought of as a series of steps,[10] summarized in Table 9.1. Essentially, these seven steps designate critical activities in the selection of any sample.

Step 1: Select the Population. The process begins by defining the population that we are interested in studying. A properly defined population

TABLE **9.1** Steps in the Sampling Process

Step	Description
1. Select the *population* relative to the research problem and design.	The explicit designation of elements of concern in the study. A proper definition usually includes four components: elements, sampling units, extent, and time.
2. Select what *sampling units* are appropriate in the population.	Designation of the appropriate units for sampling. It may be one element (a midlevel executive) or multiple elements (a division in a corporation).
3. Select a *sampling frame*.	The means of physically representing the population—for example, the company telephone book, the membership roster of the AMA, a complete mailing list, and so on
4. Select a *sample design*.	The method by which a sample is ultimately selected. There are probability and nonprobability type designs.
5. Select the *size of sample* needed to accomplish research objectives.	The selection of the number of people or objects to study in the population.
6. Select a *sampling plan*.	The development of the specific procedures by which the sample will be chosen.
7. Select the *sample*.	The actual activities that are performed in the process of selecting a sample. The sampling plan is implemented.

explicitly designates all elements of concern. This definition usually includes four components: elements, sampling units, extent, and time. (Extent is simply the range of conditions or situations to which the population under study is restricted.)

For example, in our midlevel executive example, the following population might be selected for study:

Element	All midlevel executives
Sampling unit	in the Century Drill & Tool Company
Extent	in Chicago, Illinois,
Time	during the week of June 15, 2005

Alternatively, a broader definition such as the following could be accepted:

Element	All midlevel executives
Sampling unit	belonging to the American Management Association
Extent	in the United States
Time	on June 15, 2005

Obviously, the generalizability of the results obtained from the study of these two populations would be vastly different.

Step 2: Select Sampling Units. Sampling units are the units of analysis from which the sample is drawn. The complexity of the research and ultimately the sample design chosen often dictate the sampling units to be selected. In our definitions, the sampling unit was readily identifiable, but in complex sample designs, one or more sampling units may be selected. More will be said about such designs later in the chapter.

Step 3: Select a Sampling Frame. Selection of the sampling frame is a critical step in this process because if the sampling frame chosen does not adequately represent the population selected in step 1, the generalizability of the results of the study will be questionable. Suppose we use the sampling frame of all midlevel executives who attended the American Management Association's annual conference to represent the population of all midlevel executives in the AMA. Because only a portion of the midlevel executives belonging to the AMA would attend this conference, we would run the risk of biasing our results in some unknown direction. The cardinal rule is to select the sampling frame that best represents the population defined, given the resource constraints of the study.

Step 4: Select a Sample Design. The *sample design* is the method or approach that is used to select the units of analysis for study. There are about as many different types of sample designs as there are research designs available to the researcher. Each design has its particular advantages and disadvantages; a later section of this chapter will fully describe the major types of sample designs and their respective merits.

Step 5: Select Size of Sample. The sample size depends on a number of factors, including the following:[11]

> **Homogeneity of sampling units.** Sometimes in a specified population, the sampling units are more similar with regard to certain traits. For example, a population consisting of very small retail fishing tackle stores is more homogeneous with respect to smallness and the sale of fishing tackle than a population containing small retail stores and large department stores carrying some fishing tackle. Generally, the more alike the sampling units in the population are, the smaller the sample will need to be to estimate the population's parameters of interest.
>
> **Confidence.** Even in a specified, known population, we are never certain of the true population parameter values. We would like to estimate the true value without examining every element in the population. *Confidence* is the degree to which the researchers want to be sure they are in fact estimating the true population parameter. The degree of confidence is specified as a probability. For example, the researcher might want to be 95% confident of estimating the true number of small fishing tackle stores in a certain area that are carrying a particular type of fishing lure. This is the probability of capturing the true population parameter in replicated study samples, or the confidence interval. All things being equal, more confidence requires a larger sample size.
>
> **Precision.** How close should the estimate be to the true population parameter? In our example, the researcher might want to come within one or two stores of the actual number that are carrying the fishing lure. The researcher has specified the precision of the estimate. Statistically, this is the size of the standard error of the estimate. All things being equal, greater precision requires a larger sample size.
>
> **Statistical power.** This term refers to the researcher's ability to recognize a relationship when it in fact exists. A researcher does not like to make a costly mistake. For example, to determine if there are differences between the number of fishing lures sold by small fishing tackle stores in the city of Los Angeles and those in coastal Los Angeles county, a sampling plan could help test a hypothesis of no difference. If the test indicates that there is no difference but there really is, the researcher will make a costly wrong decision based on this information. The probability of making a correct decision to reject the null (no difference) hypothesis when in fact it is false is called the *power* of the test. In general, the more power desired by the researcher, the greater the sample size needed.

Analytical procedures. The type of analytical procedure chosen for data analysis can also affect sample size selection. Certain minimum data requirements must often be met before a particular procedure can be appropriately used in a study. More will be said about this in the analytical chapters of the text.

Costs, time, and personnel. Here again, we meet with the cold, hard facts. It has been said that resource constraints are the major obstacle to doing good research.[12] Generally, to obtain sample efficiency, researchers try to maximize the information obtained relative to cost. Sample size determination requires a balancing of statistical efficiency, sampling efficiency, and practicality from the researcher's viewpoint.

These are the primary factors initially, but many practical considerations enter into the process of reaching the final study sample size. Frequently, multiple procedures and influences are integrated into the decision to ensure that the study's objectives will be met.[13] In this chapter, after the major sample designs are discussed and several examples of initial sample size determination are given, we examine practical considerations in sampling.

Step 6: Select a Sampling Plan. The sampling plan specifies the procedures and methods to obtain the desired sample. Besides the explicit statement of steps 1 through 5 of the sampling process, it includes a statement of how and when the sampling procedures will be conducted. The sampling plan, if designed correctly, guides the researcher in the selection of the study sample so that potential errors in the sampling process are minimized.

Step 7: Select the Sample. The final step in the process is actually selecting the sample. The specific units of analysis are enumerated and designated for the next step in the research process, data collection.

AN INTRODUCTION TO THE PHILOSOPHY OF SAMPLING

To clarify the rationale and basis for sampling, a brief introduction to the subject is presented here. Many excellent books have been written on this complex subject.[14] The goals of this exposition are modest: (1) to acquaint you with the underlying philosophy of sampling and (2) to show how and why samples can be used to estimate the parameters of a population of interest.

In most populations in business research settings, there is very little element similarity (i.e., homogeneity of characteristics). For example, in studying the attitudes of consumers toward the purchase of electric blankets, we would in all likelihood find very negative and very positive attitudes. Therefore, if we are to use some subset (a sample) of a population to estimate the overall attitudes toward the purchase of electric blankets, we must use procedures that will lead to accurate and unbiased estimates of that particular characteristic. Fortunately for both researchers and managers, certain sampling techniques allow us to generate good samples from which to do this.

What are good samples? A good sample has the following general characteristics.[15]

1. It enables the researchers to make decisions concerning what sample size to take to obtain the answers desired.[16]

2. It identifies the chance, or probability, that any primary unit of analysis will be included in the final study sample (although the probability may be different for different UOAs).

3. It enables the researchers to quantify the accuracy and imprecision (errors) in choosing a sample, rather than taking a complete canvas of the population (a census).

4. It enables the researchers to quantify the degree of confidence that can be placed in population estimates made from sample statistics.

What we are talking about here is the development of probability samples. As the next section will discuss in more detail, the two major types of sample designs are probability and nonprobability. In *probability designs*, each element in the population has a known, nonzero chance of being selected for inclusion in the study sample. In *nonprobability designs*, the chance of each element being selected is not known. Only probability samples have all the characteristics of good samples just outlined. The discussion of sampling in our example applies only to probability samples; the purposes and characteristics of nonprobability samples will be discussed later.

The characteristics of probability samples are derived from the foundations of statistical theory and random variables. The probability distributions for random variables possess a variety of shapes, but many of the population estimators (statistics) possess one of three basic theoretical frequency distributions: the binomial, multinomial, and normal distributions.[17] As you will recall from Chapter 7, there are four levels of measurement, and these generally correlate with the three distributions. Nominal and ordinal measurements usually show binomial or multinomial distributions; the normal distribution often applies to interval and ratio measurements.

In a random probability sample, each member of the population has a known chance of being chosen for the sample, and the selection of one member of the population does not affect the chances of any other member being selected. If such a sample is taken from a population, we can say that the sample is *representative* of the population.[18] This characteristic is fundamental to the study of *inferential statistics*—the science of estimating population parameters from sample statistics. Inferential statistics, with which all business researchers must be concerned, is based on the assumptions of taking random samples from specified populations of interest. In essence, random sampling provides the unbiased mixing and ultimate selection of the units of analysis to be studied.

Especially in random sample sizes over 30, many estimators possess the theoretical frequency distribution that is approximately bell-shaped, commonly called the *normal curve*.[19] This frequency distribution is particularly important because it provides the underlying basis for many of the inferences made by researchers. The Central Limit Theorem demonstrates the importance of this theoretical distribution to statistical inference and business research.

Essentially, the Central Limit Theorem states that if we take a sufficiently large sample (n) from a population (N) and then measure a particular characteristic of the sample, the sampling distribution of the means of that characteristic will be nearly normally distributed. Furthermore, the mean of this normal distribution will equal the population mean (μ).[20] The variance will equal σ^2/n, where σ^2 equals the population variance. The square root of σ^2/n is known as the *standard error of the mean* ($\sigma_{\bar{x}}$) and is defined as follows:

$$\sigma_{\bar{x}} = \frac{\sigma}{\sqrt{n}}$$

What is important in all of this is that we can use the Central Limit Theorem in conjunction with sample estimates to make specific statements about the population of interest. A complete development of the rationale behind this theorem is beyond the scope of this text, but a simple example of the theorem and its properties is presented in the appendix at the end of this chapter. You are encouraged to review this appendix if you are unfamiliar with the basics of the Central Limit Theorem.

SAMPLE DESIGNS

The *sample design* is the method used to select the units of analysis for study. Such methods can be classified in a variety of ways; the most usual breakdown is into probability and nonprobability sampling designs. The previous section on sampling theory related explicitly to probability designs. Using probability designs, the researcher can generally estimate the sampling errors and ultimately express these in confidence intervals for managers. This section now examines considerations in sample design selection and some useful sampling designs.

Sample Design Choice Considerations

What design type to use in a particular study depends on a number of factors. First and foremost, the type selected must be consistent with the objectives of the study. If the manager desires specific error estimates for population parameters, some type of probability sampling design must be

TABLE 9.2 Sample Design Choice Considerations

	Design Type	
Consideration	**Probability**	**Nonprobability**
1. Cost	More costly	Less costly
2. Accuracy	More accurate	Less accurate
3. Time	More time	Less time
4. Acceptance of results	Universal acceptance	Reasonable acceptance
5. Generalizability of results	Good	Poor

used because nonprobability designs are incapable of providing such estimates. If the objective of the study is only exploratory, with little concern for the generalization of results to the population, then a nonprobability sampling design may be all that is needed.

Other considerations that must also be addressed are presented in Table 9.2. Researchers and managers alike must give attention to these basic considerations because they affect the overall cost and the quality of the results that will be obtained from the study. The considerations are cost, accuracy, time, acceptance of results, and generalizability of results.

Table 9.2 presents the relative ratings of probability and nonprobability samples on these five criteria. Of course, this table is a simplification, because several types of design are grouped under each broad classification, but it does serve to highlight general tendencies. For instance, probability samples tend to be more costly and more accurate, tend to take more time, are more universally accepted, and are usually better in terms of the generalizability of results than are nonprobability samples.

Why so? Probability samples are usually more costly because they require complete enumeration of the population, and then we must locate the particular study units of analysis. They are more accurate because of their known sampling distributions. They take more time because they are generally more exact procedures. They are generally accepted because of their known sampling properties. Also, the generalizability of results is good because probabilistic sampling usually gives conclusions that are replicable, given similar study conditions.

As just outlined, the two main types of sampling designs are probability and nonprobability. Table 9.3 presents some of the more widely used design subtypes under each classification. Rather than elaborating on the subtleties in all the possible design subtypes, this section merely presents an overview of each subtype. Following this overview is an example of the application of the sampling process to a particular type of problem, using a simple random sample design. This example will give you a taste of the actual decisions the researcher must make to select the study sample. Let us begin our look at some of the basic design types, as we refer to Table 9.3.

TABLE 9.3 Types of Sampling Designs

Type of Sampling	Brief Description	Advantages	Disadvantages
Probability designs			
A. Simple random	Assign to each population member a unique number; select sample items by use of random numbers.	1. Requires minimum knowledge of population in advance 2. Free of possible classification errors 3. Easy to analyze data and compute errors	1. Does not use knowledge of population that researcher may have 2. Larger errors for same sample size than in stratified sampling
B. Systematic	Use natural ordering or order population; select random starting point between 1 and the nearest integer to the sampling ratio (N/n); select items at interval of nearest integer to sampling ratio.	1. If population is ordered with respect to pertinent property, gives stratification effect, and hence reduces variability compared to A 2. Simplicity of drawing sample; easy to check	1. If sampling interval is related to a periodic ordering of the population, increased variability may be introduced 2. Estimates of error likely to be high where there is stratification effect
C. Multistage random	Use a form of random sampling in each of the sampling stages where there are at least two stages.	1. Sampling lists, identification, and numbering required only for members of sampling units selected in sample 2. If sampling units are geographically defined, cuts down field costs (i.e., travel)	1. Errors likely to be larger than in A or B for same sample size 2. Errors increase as number of sampling units selected decreases
With probability proportionate to size	Select sampling units with probability proportionate to their size.	Reduces variability	Lack of knowledge of size of each sampling unit before selection increases variability
D. Stratified			
1. Proportionate	Select from every sampling unit, at other than last stage, a random sample proportionate to size of sampling unit.	1. Ensures representativeness with respect to property that forms basis of classifying units; therefore, yields less variability than A or C 2. Decreases chance of failing to include members of population because of classification process	1. Requires accurate information on proportion of population in each stratum; otherwise, increases error 2. If stratified lists are not available, may be costly to prepare them; possibility of faulty classification and hence increase in variability

(Continued)

TABLE **9.3** (Continued)

Type of Sampling	Brief Description	Advantages	Disadvantages
		3. Characteristics of each stratum can be estimated, and hence comparisons can be made	
2. Optimum allocation	Same as D1 except sample is proportionate to variability within strata as well as their size.	Less variability for same sample size than D1	Requires knowledge of variability of pertinent characteristic within strata
3. Disproportionate	Same as D1 except that size of sample is not proportionate to size of sampling unit but is dictated by analytical considerations or convenience.	More efficient than D1 for comparison of strata or where different errors are optimum for different strata	Less efficient than D1 for determining population characteristics (i.e., more variability for same sample size)
E. Cluster	Select sampling units by some form of random sampling; ultimate units are groups; select these at random and take a complete count of each.	1. If clusters are geographically defined, yields lowest field costs 2. Requires listing only individuals in selected clusters 3. Characteristics of clusters as well as those of population can be estimated 4. Can be used for subsequent samples because clusters, not individuals, are selected, and substitution of individuals may be permissible	1. Larger errors for comparable size than other probability samples 2. Requires ability to assign each member of population uniquely to a cluster; inability to do so may result in duplication or omission of individuals
F. Stratified cluster	Select clusters at random from every sampling unit.	Reduces variability of plain cluster sampling	1. Disadvantages of stratified sampling added to those of cluster sampling 2. Because cluster properties may change, advantage of stratification may be reduced and make sample unusable for later research
G. Repetitive: multiple or sequential (double)	Select two or more samples of any of the above types, using results from earlier samples to design later ones or determine if they are necessary.	1. Provides estimates of population characteristics that facilitate efficient planning of succeeding sample; therefore, reduces error of final estimate	1. Complicates administration of field work 2. More computation and analysis required than in nonrepetitive sampling

	Description	Advantages	Limitations
		2. In the long run, reduces number of observations required	3. Sequential sampling can be used only where a very small sample can approximate representativeness and where the number of observations can be increased conveniently at any stage of the research
Nonprobability designs			
H. Judgment	Select a subgroup of the population that, on the basis of available information, can be judged to be representative of the total population; take a complete count or subsample of this group.	Reduces cost of preparing sample and field work because ultimate units can be selected so that they are close together	1. Variability and bias of estimates cannot be measured or controlled 2. Requires strong assumptions or considerable knowledge of population and subgroup selected
Quota	Classify population by pertinent properties; determine desired proportion of sample from each class; fix quotas for each observer.	1. Same cost considerations as Judgment (Advantages) 2. Introduces some stratification effect	Introduces bias of observers' classification of subjects and nonrandom selection within classes
I. Convenience	Select units of analysis in any convenient manner specified by the researcher.	Quick and inexpensive	Contains unknown amounts of both systematic and variable error
J. Snowball	Select units with rare characteristics; additional units are referred by initial respondents.	Only highly specific application	Representativeness of rare characteristic may not be apparent in sample selected

Source: Adapted from Russell L. Ackoff, *The Design of Social Research* (Chicago: University of Chicago Press, 1953), p. 124. Used by permission.

Probability Designs

Simple Random Sampling. The single random sampling design is very easy to use and to describe. In a simple random sample (SRS), each element of the specified population has an equal chance of being selected for the final study sample. SRSs are easy to use because of the simple selection procedures, and they require only a listing of the entire population before the sampling process can begin. The procedure for SRS is generally as follows:

1. Obtain a listing of the specified population and the sampling units to be used.

2. Choose a device for randomly selecting the sample from the sampling frame. This device can be the use of a random number table; a random number generator available on most advanced calculators and computers; or a drum of equally weighted, uniquely numbered Ping-Pong balls.

3. The selection procedure itself must not affect probabilities of another element being selected. In other words, the selection of one number must be independent of selecting any other of the numbers in the population.

Example 9.1 illustrates the SRS sampling design process. Example 9.2 then supplements and extends this example by presenting the generalized procedure for sample size selection for SRSs. Space limitations prevent giving detailed examples such as these for each major class of sampling design, but the basic formulae and steps will be given to help you in the sample design selection process. Other texts should be consulted for more in-depth treatments of each design type.[21]

EXAMPLE 9.1

A Descriptive and Computational Example Using the Sampling Process and a Simple Random Sample Design

The Solar Insistence League of Florida (SILOF) wants information from owners of various solar energy devices: the amount they spend on such devices per household and the percentage of people who are satisfied with these devices. They would like to use this information to evaluate energy programs in the state and to formulate new policy directions for the league. Using the sampling process presented in Table 9.1, the research director initiates the process.

STEP 1: Select the population. The population is defined as:

Element	All adopters of solar energy devices
Sampling unit	in households
Extent	in Orlando, Florida,
Time	who purchased a device in calendar year 2005

STEP 2: Select the sampling unit. The sampling unit that is deemed most appropriate is the individual household where solar devices have been adopted.

STEP 3: Select a sampling frame. A listing of 1000 households adopting solar devices in 2005 in Orlando, Florida, is obtained from warranty cards sent in to the Florida Solar Energy Center (FSEC).

EXAMPLE **9.1**

(Continued)

STEP 4: Select a sample design. It is decided that a simple random sample will be selected from the 1000-household listing obtained from FSEC. A simple random sample (SRS) is deemed appropriate because the research objectives of the study require that the researchers must

1. Estimate the average dollars spent within $100 using the interval measurement of dollars spent for energy devices. Here, we are estimating a population mean.

2. Estimate the percentage (proportion) of people satisfied within 3% using the nominal measurement of satisfaction and nonsatisfaction.

3. Ensure that the users of the research can be 95% confident that the population parameter is included in the calculated confidence interval.

An SRS is also deemed appropriate because it is easy to use and administer.

STEP 5: Select the size of the sample. Because SILOF is concerned with measuring the amount spent on solar devices (an interval variable) and with satisfaction (a nominal dichotomous variable: satisfied or not satisfied), two estimates of sample size may be needed to generate the precision and confidence estimates presented in step 4. SILOF management estimates the spending per household to be approximately $6000. Also, management suspects that about 70% of the people are satisfied with their solar energy devices. Given this limited information, the determination of the sample sizes for SRS follows this procedure:

Sample Size Determination Procedure for SRS	SILOF Problem	
	Amount Spent	**Satisfaction**
1. Estimate σ or P depending on the level of measurement.* One needs an estimate of σ (the population's standard deviation) or P (proportion of elements that belong to a class), depending upon the level of measurement.	$\sigma = \$1500$	$P = 0.70$
2. Determine precision, or B. According to the research objectives specified by managers, dollars spent must be within $100; percent satisfied must be estimated within 3 percentage points.	$B = \$100$	$B = 0.03$
3. Determine confidence desired z.** Management wants to be 95% sure that the confidence interval includes the population parameter.	$z_{0.05} = 1.96$	$z_{0.05} = 1.96$
4. Determine sample size using finite population correction factor (FPCF), if appropriate. The relevant equations of SRS size determination are		

(Continued)

EXAMPLE 9.1

(Continued)

Finite population $\qquad n = \dfrac{N\sigma^2}{\dfrac{(N-1)B^2}{z^2} + \sigma^2} \qquad n = \dfrac{NPQ}{\dfrac{(N-1)B^2}{z^2} + PQ}$

Infinite population $\qquad n = \dfrac{z^2\sigma^2}{B^2} \qquad\qquad n = \dfrac{z^2PQ}{B^2}$

where

$\quad n$ = sample size

$\quad N$ = population size

$\quad \sigma^2$ = population variance (or estimate)

$\quad B$ = allowable error (precision)

$\quad z$ = z-score based on researcher's desired level of confidence

$\quad P$ = population proportion (or estimate)

$\quad Q = 1 - P$

Because the population is of known size, the finite equations are used to solve for n.

$$n = \frac{1000(1500)^2}{\dfrac{(1000-1)100^2}{1.96^2} + 1500^2} \qquad n = \frac{1000(0.7)(0.3)}{\dfrac{(1000-1)0.0009}{1.96^2} + (0.7)(0.3)}$$

$$n = \frac{2{,}250{,}000{,}000}{(999)2603 + 2{,}250{,}000} \qquad n = \frac{210}{0.2340 + .021}$$

$$n = \frac{2{,}250{,}000{,}000}{4{,}850{,}397} \qquad n = \frac{210}{0.4440}$$

$$n \cong 464 \qquad\qquad n \cong 473$$

Therefore, a sample size of approximately 480 will ensure that the research objectives are met.

STEP 6: Select a sampling plan. The sample will be selected and the data will be collected in the following manner.

a. All the 1000 solar households will be assigned a number from 1 to 1000. From a table of random numbers, 480 households will be selected (the needed sample size, given the desired precision and confidence).

b. The necessary data will be collected by an online survey of these 480 households.

STEP 7: Select the sample. The investigators get the e-mail addresses for the 480 households from SILOF to obtain the necessary information from the desired sample.

---

*Using Tchebysheff's Theorem, one can estimate the approximate value of a σ in the population by the formula

$$\sigma \cong \frac{\text{range}}{4} = \frac{\$6000}{4} = \$1500$$

Researchers, if they have no a priori inclination for P, often use 0.5. Other methods for estimating σ^2 or P include management judgment or the use of a pilot study. If a pilot study is used, then s^2 (the sample variance) is used to estimate σ. Similarly, p is used to estimate P.

EXAMPLE **9.1**

(Continued)

For a review of Tchebysheff's Theorem, see Irving W. Burr, *Applied Statistical Methods* (New York: Academic Press, 1974), pp. 155–156.

**For a discussion of the use of the *z*-scores and their relevance to the normal curve and sampling, see Jonathan Cryer and Robert Miller, *Statistics for Business: Data Analysis and Modeling* (Belmont, CA: Duxbury Press, 1994), pp. 247–267. The *z*-score is obtained from any statistical table of *z*-scores (see Appendix B), or from almost any statistical software package.

EXAMPLE **9.2**

Sample Size Determination Procedure for SRS for Both Means and Proportions

Where

n = sample size

N = population size

σ = population standard deviation (or estimate)

B = allowable error, precision

P = population proportion (or estimate)

$Q = 1 - P$

z = z-score determined for a specific confidence level

$\sigma_{\bar{x}}$ = standard error of the mean

σ_p = standard error of the proportion

1. Estimate σ or P given the level of measurement.
2. Determine the confidence level desired by choosing the appropriate *z*-score from the normal distribution (Appendix B, Table B.2).
3. Determine the degree of precision, or allowable error for the study, *B*.
4. Determine the following:

   a. Sample size determination for *SRS samples without replacement from an infinite population*

Means	Proportions (where $Q = 1 - P$)
$B = z\,\sigma_{\bar{x}}$	$B = z\sigma_p$
where $\sigma_{\bar{x}} = \dfrac{\sigma}{\sqrt{n}}$	where $\sigma_p = \sqrt{\dfrac{PQ}{n}}$
Then, $B = z\dfrac{\sigma}{\sqrt{n}}$	Then, $B = \dfrac{z\sqrt{PQ}}{\sqrt{n}}$
$\sqrt{n} = \dfrac{z\sigma}{B}$	$\sqrt{n} = \dfrac{z\sqrt{PQ}}{B}$
Then simply solve for	
$n = \dfrac{z^2\sigma^2}{B^2}$	$n = \dfrac{z^2 PQ}{B^2}$

   b. Sample size determination for *SRS without replacement from a finite population*. The initial form of the equations in (a) is the same, except that the finite population correction factor (FPCF) is added to the formulations.

Means	Proportions
$B = z\dfrac{\sigma}{n^2}\sqrt{\dfrac{N-n}{N-1}}$	$B = \dfrac{z\sqrt{PQ}}{n}$
$n = \dfrac{\sigma^2 N}{\dfrac{(N-1)B^2}{z^2} + \sigma^2}$	$n = \dfrac{NPQ}{\dfrac{(N-1)B^2}{z^2} + PQ}$

FIGURE **9.1**

Nomograph for
Determining Size of a
Simple Random Sample
in Estimation Problems
of the Proportion—
Infinite Population

Source: Used with the
permission and through the
courtesy of Audits and
Surveys, Inc.

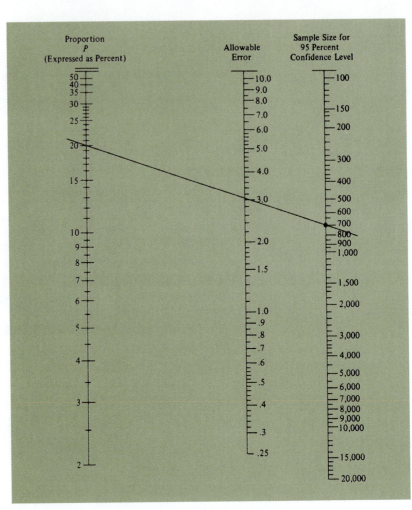

Sometimes it is easier to get an approximation of sample size using a nomograph. For example, to determine the sample size for the proportion of individuals who are satisfied with their jobs, we could use the nomograph in Figure 9.1. Using predetermined researcher criteria of 3 percentage points of tolerable error (precision), a 95% confidence level, and P estimated to be 0.20, we can use a ruler on the nomograph to determine the sample size needed: approximately 700. There are nomographs for all different considerations, using both proportions and means.[22]

Systematic Sampling. Systematic sampling is similar to SRS, except that the units of analysis are chosen in a very predictable fashion. Essentially, one starts with a complete sampling frame specifying all units of analysis, then randomly chooses a constant K, somewhere between zero and the sampling ratio N/n, and systematically takes every Kth element to arrive at a study sample.

For example, suppose a sample of size 200 (n) from a population of 1000 (N) is needed to obtain the necessary precision for the study. Using systematic random sampling, the researcher first finds the sampling ratio (1000/200), which is 5. The researcher randomly selects a number somewhere between 1 and 5 to start the sampling process. Let us suppose the number is 3. The sample chosen would consist of taking the third unit of analysis and every fifth unit after that until the sample of 200 is selected. As you can see, this sampling procedure is merely an extension of SRS. Its ease of use makes it a popular sampling design in business research.

Multistage Random Sampling. Multistage random sampling designs are more complex variations of the SRS and systematic random sample designs. The potential types of multistage designs are too numerous to outline, but the following example should help illustrate the basic process.

Let us suppose we want to study CEOs' opinions of the economic outlook for the next ten years. We are particularly concerned with the reasons why the CEOs with negative outlooks possess them. A two-stage random sample may be very appropriate:

Stage 1: Select, by an SRS, 250 of the 500 CEOs in the *Fortune* 500 companies and send them a paper-and-pencil questionnaire to assess their general economic outlook for the next ten years.

Stage 2: Sort the returned questionnaires into two piles: those CEOs with positive outlooks (let us say 125 are positive) and negative outlooks (let us say 75 are negative). Now we take an SRS of 20 of the 75 CEOs with negative outlooks and we do in-depth personal interviews with these individuals to discover the reasons for their negative views.

There are many potential applications of this form of sampling in business. See Table 9.3 for a summary of the primary advantages and disadvantages of this form of sampling design.

Stratified Sampling. SRS and the two other designs discussed may not be useful or desirable for all research problems. In some instances, the researcher might want to increase the quantity of information obtained for a given cost. Here, stratified random sampling may aid the researcher.

In *stratified random sampling*, the population is separated into nonoverlapping groups of elements, called *strata*, and then a random sample is selected from each stratum.[23] This sample design is often employed when the researchers feel that certain strata differ from others on an important characteristic and want to make sure this difference is taken into account in the study. The ultimate results of this process can be (1) a more representative sample on some characteristic of interest, (2) a reduction in the standard error of the estimates because of the increased homogeneity within strata, and (3) the production of better estimators at a given cost.

To illustrate the stratified sampling process, a general procedure for drawing a stratified sample is presented in Example 9.3. The overall steps in the computational process are outlined and briefly described. An example of determining sample size using means appears in Example 9.4. Not all the subtleties of the allocation problem (assigning sample sizes to strata) and cost concerns can be dealt with here, but the primary equations are presented for your convenience. For more detailed treatments of this subject, refer to some of the excellent discussions on these aspects of sampling.[24]

EXAMPLE 9.3

General Procedure for Stratified Random Sampling

Where

N = population size

n = sample size

L = number of strata

N_i = population size in stratum i

n_i = sample size allocation in stratum i

σ_i = population standard deviation of stratum i

w_i = allocation weight for stratum i

1. Determine the number of strata, L.
2. Estimate, or determine, the number of sampling units, N_i, in each stratum i, as i goes from 1 to L.
3. Determine the total number of sampling units in the population N, where

$$N = \sum_{i=1}^{L} N_i$$

4. Estimate the variance (σ_i^2 for means, or $p_i q_i$ for proportions) of the characteristic measure of interest within each stratum. This estimation, as in SRS, can be done by prior judgment, a pilot study, or some rule of thumb such as Tchebysheff's Theorem.
5. Denote allocation weight, W_i, for strata, where

$$\sum_{i=1}^{L} W_i = 1.00$$

The allocation of weights can be proportionate to a stratum's size, disproportionate for some analytical or convenience reason, or optimum where the weight selected is proportionate to the variability as well as the size of the strata.*

6. Determine the confidence level desired by choosing the appropriate z-score for the normal distribution.
7. Determine the degree of precision, B, or allowable error, for the study.
8. Determine an estimate of the variance of the population estimator using the value B in the following equations:

Mean	Proportion
$Z\sqrt{\sigma_{\bar{x}}^2} = B$	$Z\sqrt{\sigma_p^2} = B$

then

$\sigma_{\bar{x}} = \dfrac{B^2}{z^2} = D$	$\sigma_p = \dfrac{B^2}{z^2} = D$

(Continued)

EXAMPLE **9.3**

(Continued)

9. Calculate n, the sample size, needed to estimate the population estimator within the tolerable limits prespecified by B.

$$n = \frac{\displaystyle\sum_{i=1}^{L} \frac{N_i^2 \sigma_i^2}{W_i}}{N^2 D + \displaystyle\sum_{i=1}^{L} N_i \sigma_i^2} \qquad n = \frac{\displaystyle\sum_{i=1}^{L} \frac{N_i^2 p_i q_i}{W_i}}{N^2 D + \displaystyle\sum_{i=1}^{L} N_i p_i q_i}$$

10. Calculate $n_i, n_2, \ldots, n_i$ based on allocation procedures

*A number of allocation procedures can be used to determine specific stratum sample sizes. For a discussion of these procedures, plus a discussion of cost considerations in sample size selection determination, see Leslie Kish, *Survey Sampling* (New York: Wiley, 1965), pp. 75–142, 254–291; and William G. Cochran, *Sampling Techniques* (New York: Wiley, 1977).

EXAMPLE **9.4**

Estimating Sample Sizes Needed for Stratified Random Sampling (Means)

We would like to estimate the mean number of hours per week heads of household spend reading business publications.

1. $L = 3$
2. $N_1 = 155$ (City A)
   $N_2 = 62$ (City B)
   $N_3 = 93$ (City C)
3. N total = 310 households in all 3 cities
4. Prior survey indicates that variances are approximately

   $\sigma_1^2 = 25 \qquad \sigma_2^2 = 225 \qquad \sigma_3^2 = 100$

5. Equal allocation is assumed: 1/3, 1/3, 1/3
6. 95% confidence is assumed.
7. Allowable error is 2 hours.

Sample sizes required:

$$\sum_{i=1}^{3} \frac{N_i^2 \sigma_i^2}{W_i} = \frac{(155^2)(25)}{(1/3)} + \frac{(62^2)(225)}{(1/3)} + \frac{(93^2)(100)}{(1/3)} = 6{,}991{,}275$$

$$\sum_{i=1}^{3} N_i \sigma_i^2 = 155(25) + 62(225) + 93(100) = 27{,}125$$

8. $N^2 D = (310)^2 (1) = 96{,}100$

9. $n = \dfrac{\displaystyle\sum_{i=1}^{3} \dfrac{N_i^2 \sigma_i^2}{w_i}}{N^2 D + \displaystyle\sum_{i=1}^{3} N_i \sigma_i^2} = \dfrac{6{,}991{,}275}{96{,}100 + 27{,}125}$

$$= 56.7$$

10. Therefore, the researcher should take $n = 57$ total observations. The city samples are calculated as:

   $n_1 = 1/3(57) \quad = 19$
   $n_2 = \qquad\qquad = 19$
   $n_3 = \qquad\qquad = 19$

Cluster Sampling. In *cluster sampling*, the population of interest is grouped into aggregates (clusters) based on physical proximity. Each cluster is therefore supposed to be a miniature representation of the entire population. The study sample is drawn by randomly selecting a subset of clusters, then either using all elements in those clusters (one-stage cluster sampling) or randomly selecting subsequent clusters for study (multistage).

Cluster sampling is one of the more widely used complex probability sampling designs in large-scale survey research. The primary reason for its popularity is that sample efficiency is generally relatively high when compared with the sampling designs described earlier. Procedurally, the one-stage cluster sampling method can be outlined as follows:

> **Step 1.** Divide the population into mutually exclusive and exhaustive subgroups. For example, voting precincts in the city of Chicago could be used to divide the registered voters into exclusive and exhaustive subgroups.
>
> **Step 2.** Select a random sample of clusters and use all elements in these clusters as the study sample. For example, precincts 4, 8, 6, and 9 are randomly selected from 50 voting precincts. All registered voters in these four precincts are used as the study sample.

Alternatively, multistage clustering designs can be used, depending on the needs of the researchers and the specific problem situation.[25]

Stratified Cluster Sampling. Another complex probability sampling design combines the characteristics of stratified and cluster sampling. In essence, stratified cluster sampling uses clusters at random from prespecified strata. In other words, all strata are sampled, but only through the randomly selected clusters chosen.

Repetitive: Multiple or Sequential Sampling. Multiple or sequential sampling designs are in essence a combination of two or more samples drawn from the cluster and stratified cluster designs presented earlier. Multiple sampling designs are typically used in research applications where the research objectives require highly specialized samples with specific characteristics. The efficiency of these designs must be evaluated on a case-by-case basis.

Nonprobability Designs

Generally, three types of nonprobability designs are used in practice. These designs are designated judgment, convenience, and snowball types. A *nonprobability* sample is one in which chance selection procedures are not used. Its basic shortcoming is high variable error, and it lacks the characteristics that would allow for an estimate of this error. Researchers should be acutely aware of these inadequacies when deciding on a sampling design.

Judgment Sampling. *Judgment samples* are those that the researcher or manager chooses as being representative of the population of interest. Judgment samples, although statistically unappealing, are often used because they can be inexpensive and the procedure takes very little time. Many field studies that are exploratory in nature use judgment samples.

A subtype of this sampling method is the *quota sample*, in which certain specified proportions of types of sampling units are included in the study sample. For example, if researchers were to study the differences between high- and low-achieving production managers, they might wish to specify that 50% of the sample ultimately chosen be of each type. The disadvantage of this sampling procedure, as with all nonprobability procedures, is that the errors introduced by the selection procedure are unknown and unmeasurable.

Convenience Sampling. Convenience samples are usually used because they can be chosen quickly and inexpensively. *Convenience samples* are those samples whose units of analysis are chosen in any convenient manner specified by the researcher. Convenience samples are often used in exploratory and descriptive research where time and money are critical constraints.

Snowball Sampling. The last of the nonprobability designs to be mentioned is the *snowball sampling design*. It has a highly specific application, in cases where the units of analysis with a desired characteristic are difficult to find.[26] For example, to study business organizations that have difficulty in hiring competent first-level managers, researchers might want to use this procedure. They would attempt to identify an organization that has had this problem, then ask the general manager of that organization to name other companies that have had similar problems. The design is usually used only when the desired units of analysis are difficult to find.

PRACTICAL CONSIDERATIONS IN SAMPLING

Although it is relatively easy to outline the general process of the sampling decision, it is difficult, if not impossible, to give a framework that will provide the best sample under all conditions. In practice, no sample is without some limitations.[27] To illustrate this fact, Figure 9.2 presents a page from an informational brochure by Survey Sampling, Inc.® (an industry leader in survey research sampling), describing some of the intricacies and pitfalls of obtaining random-digit telephone samples. These difficulties in sampling are then dealt with by specific actions by SSI. Similarly, creative solutions must usually be applied to sampling problems that are often unique. Where does one start?

First, following the sampling process outlined in Table 9.1 is a good starting point for researchers in applied settings. However, other considerations

FIGURE **9.2**

Intricacies for Random-Digit Telephone Sales, from Survey Sampling, Inc.

Source: From Survey Sampling, Inc.®, One Post Road, Fairfield, CT 06430, informational brochure, 1991, pp. 8, 9. Used with permission.

RANDOM DIGIT TELEPHONE SAMPLES

Projectable research demands random digit sampling

Today, nearly 30 percent of American households can't be reached from directory samples, and this group is quite different from the general population. This segment of the population is very young, for example. Samples drawn entirely from directories or from directory seed numbers increase the potential for inaccurate measurement. Simply designed "plus-one" techniques based on directory seed numbers often end up under-representing unlisted households.

SSI has recognized the need for projectable samples and has developed high-tech methodologies that accurately represent both listed and unlisted households. Using element and cluster techniques, we've developed SSI Random Digit Samples that are systematically drawn and geographically precise.

Random digit samples from SSI are well-known for their consistent reliability. They're accurate, timely and projectable ... *and our methodology is unsurpassed in efficiency.* Data is developed from smaller, more accurate samples. You'll experience fewer dialings, faster completion and improved interviewer morale.

In 1977, SSI recognized another key problem inherent in random digit methodology—the unavoidable inclusion of non-working or disconnected phone numbers. Consequently, we developed methods to reduce the associated wasted cost.

Eliminate business numbers from your sample

First, we introduced a process for eliminating business numbers and the wasted calls and callbacks that they cause. Our business telephone number database contains nearly 13,000,000 unique numbers for purging random digit samples.

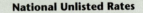

National Unlisted Rates

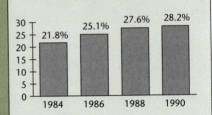

As the national unlisted rate approaches nearly one-third of all U.S. telephone households, the need for representative random digit sampling techniques is becoming more critical than ever.

SSI's Sample Selection Process

1. Maintain an up-to-date database containing virtually every directory of listed U.S. household telephone numbers.

2. Remove duplicate numbers from the database.

3. Identify all working telephone exchanges and working blocks of numbers (the first 2 digits after the exchange).

4. Assign each exchange to a single county.

5. Stratify the sample by county proportionate to estimated households and within county by working blocks of all exchanges.

6. Systematically select the sample for the geographic area specified.

7. Eliminate business telephones and wasted callbacks by passing the sample against the SSI database of nearly 13 million business listings.

Benefits: Smaller, more accurate samples. Fewer dialings. Faster completion. Improved interviewer morale.

FIGURE **9.2**

(Continued)

The Method to Fit Your Needs

Random A . . . provides the most representative random digit sample available. Each exchange will have a probability of selection equal to its share of active working blocks. Numbers are protected against re-use for a period of one year. Business numbers are eliminated.

Random B . . . provides the most efficient random digit sample available. Each exchange will have a probability of selection equal to its share of listed telephone households. Selected telephone numbers are protected against re-use for a period of one year. Business numbers are eliminated, significantly reducing the proportion of unproductive numbers in your sample.

Random C . . . provides samples with low up-front cost to researchers who are most interested in lowering initial out-of-pocket sampling costs. Samples will be accurate and projectable, but will provide a lower level of sample efficiency.

Cluster . . . used by many pollsters and other researchers who need overnight answers to their time-critical surveys. Cluster Samples are designed to reach a representative set of respondents without the need for callbacks or interviewing by replicate.

Screened vs. Standard Random Digit Sampling

	Unscreened Sample	Screened Sample
Sample Size	5000	5000
Busy	81	79
Completed	413	410
Callback/ans machine	168	165
Deaf/language	89	89
No answer	852	837
No respondent	273	272
Non-user	33	33
Over quota	19	19
Refused	686	678
Resp unavailable	9	8
Terminated	61	60
Disqualified	837	827
Business	309	294
Not dialed	20	–
Disconnected	**1150**	**580**
Total	**5000**	**4351**
Saved Dialings		**649**
Overall Efficiency	**70.7%**	**79.9%**
Increase in Efficiency		**+13%**

This sample disposition shows the increase in efficiency found by one of SSI's clients in a side-by-side test conducted in early 1990, based on three dialing attempts per number.

Eliminate over-surveyed respondents from your sample

We further reduce wasted calls through the *SSI Protection System*. The phone numbers in every Random Digit Super Sample we select are removed from further selection for approximately 12 months. This means we don't deliver the same phone number twice in any given year—to you or anyone else. Which means you won't receive stale, over-surveyed respondents in your next SSI sample.

Estimate your sample size with accuracy

We then recognized that variations in contact rates, cooperation rates and incidence rates create difficulty with estimating scheduling and determining sample size. That's why we maintain a series of databases that have been designed to help determine just the right amount of sample for each area in the study. With this information, you'll be able to produce sharper, more reliable cost estimates for your next survey.

Eliminate non-working numbers from your sample

More recently, SSI introduced its *Sample Screening Service*, a stand-alone post-production process that identifies non-working numbers before you dial them.

On average, our *Sample Screening Service* eliminates about half of the non-working numbers without affecting the integrity of your sample. Customers report working phones rates that are comparable to a directory-listed sample. The bottom line on SSI's Screened Random Digit Sample is unrivaled representation and unmatched efficency. It's like having your cake and eating it, too!

must then come into play. What will the nonresponse rate be for the study? How many elements will have to be sampled in the general population to obtain a usable sample? How will the fieldwork be conducted—by telephone, mail, the Internet, or some other means? Will the sample allow the researcher to do the analysis that is planned to answer the research question? These and other questions are often difficult to answer at the start of the sampling endeavor. Experience and/or relying on experts in the field are often the only guides that a researcher has in identifying and answering these difficult questions.

The sampling process is directly influenced by such factors as the objectives of the study, resource constraints, sampling efficiency, and researcher preference. In sum, the researcher faces many organizational and personal considerations in designing and realizing a usable sample. Other opportunities and problems arise because of such trends as low incidence and response rates; increasing internationalization, making for more complex research environments; and online sampling design issues. These three areas are discussed in the following sections.

Incidence and Response Rates

One particularly troublesome area in obtaining usable samples is incidence and response rates.[28] *Incidence* is the percentage of the population that possesses the trait (or traits) to be included in the study. For example, if we were studying electric blanket owners, the incidence rate for the population of adults in the United States would be 50%, because 50% of U.S. residents over 18 own electric blankets.

Response rate, or cooperation rate, is the percentage of respondents contacted who participated in the study. Alice K. Sylvester, vice president and senior associate media research director of J. Walter Thompson of Chicago, has gone so far as to say that nonresponse is the single biggest problem facing the research industry today.[29] For example, if we sent a mail survey to a random sample of 5000 U.S. adults to study their electric blanket needs and received only 4000 surveys back, our response rate would be 80%.

Incidence and response rates have a profound effect on the eventual sample drawn in practice. For example, assume that we have determined through our computations that a simple random sample of 500 blanket owners will be needed to accomplish the objectives of our blanket study. With a 50% incidence rate and an 80% response rate, we can calculate the actual number of questionnaires we would have to mail in order to get 500 back from blanket owners:

$$\text{Actual sample} = \frac{\text{required sample}}{(\text{incidence} \times \text{response})}$$

$$= \frac{500}{(0.50 \times 0.80)}$$

$$= 1250$$

However, this will result in 500 or more responses only about half the time. If one uses basic statistics, the actual sample size can be calculated with a specified level of confidence (assume 95%).

$$\text{Actual sample} = \frac{2X + Z(ZQ + \sqrt{(ZQ)^2 + 4XQ})}{2P}$$

where

X = required sample

$\quad$ = 500

P = incidence × response rate

$\quad$ = $0.50 \times 0.80 = 0.40$

$Q = 1 - P$

$\quad$ = $1 - 0.40 = 0.60$

Z = the z value for the desired confidence level from the standard normal distribution (assume 95% confidence, one-tailed)

$\quad$ = 1.65

Therefore, where

$$ZQ = 0.6 \times 1.65 = 0.99$$

$$\text{Actual sample} = \frac{2(500) + 1.65(0.99 + \sqrt{0.99^2 + 4(500)(0.60)})}{2(0.40)}$$

$$= 1324$$

Therefore, an actual mail-out sample of 1324 could be expected to produce a usable study of 500 or more electric blanket owners 95% of the time.

Another approach that is used by many firms to compensate for low incidence rates is to *screen* respondents for the traits before they are actually chosen to participate in a study. Screening questions are usually very short, their sole purpose being to identify a representative sample with the characteristic of interest. Screens can be delivered by just about any medium. For example, in a study of adult golfers (approximately 21% of adults golf), a random sample of adults was called (1) to screen for golfers, and then (2) if they were golfers, to ask if they would participate in a golfing study. If yes, they were sent a rather extensive written questionnaire on golfing.

Screening as a sampling tool is widely used and usually increases response rate because the respondent has been prenotified. The downside of screening (particularly with very low incidence rate samples, such as 1% or 2%) are the costs and resources involved in obtaining a sample of sufficient size. The other alternative is to use some form of snowball sampling (ask other golfers to identify other golfers) or convenience sampling (all golfers at Cocoa Beach Golf Course on some weekend). The obvious disadvantage with these alternatives is the lack of representativeness of the final study sample.

Internationalization of the Marketplace

As Exhibit 9.1 highlights, there are significant additional concerns that a researcher must address when applying the sampling process in an international setting. For example, even the basic step of defining the elements of a population can get difficult. A household (all those people living under the same roof) in one country or culture may consist of an extended family, whereas in another, it may consist of a nuclear family. Similarly, sampling frames may be inadequate or nonexistent; certain parts of the population

EXHIBIT **9.1**

Some Practical Considerations in Applying the Sampling Process in an International Setting

1. **Select the population.** The specification of the population required to answer the research question is more difficult in international research. For example, language differences make for different interpretations of what constitutes a household, a family, and so on. Furthermore, cultural and political differences make some classifications meaningless, so different population specifications may be necessary to accomplish the research objectives.

2. **Select the sampling units.** Because relevant study populations may differ across countries, the selection of a universal sampling unit is often not possible. As a sampling unit, the U.S. household is not directly comparable to a household in Asia. The goal to strive for is comparable data, regardless of differences in the sampling unit.

3. **Select a sample frame.** International sample frames are either nonexistent, incomplete, or not comparable across countries. Much more judgment often goes into the sample selection process. The researcher must be alert for bias coming from differences in sample frames.

4. **Select a sample design.** As a result of differences in steps 1 through 3, the same sample design may not be appropriate in multinational research. Probability samples may be used appropriately in one country but not another because of inadequate population information. Bias may creep into the study because of differences in sample type.

5. **Select the size of the sample.** The homogeneity of the area to be studied greatly affects sample size determination. How different are England, Spain, Germany, and Greece when it comes to looking at the EC market? Do we need different sample sizes and plans for each country, or is one EC sample enough? Assumptions as to homogeneity will greatly affect the usefulness of the results.

6. **Select a sampling plan.** The specific procedures by which the samples are drawn may differ because of cultural and or other considerations. Telephones may be appropriate in Japan, but totally inappropriate in Sudan. Government data in Sweden may allow easy identification of farmers but be totally inadequate in Thailand.

7. **Select the sample.** The procedures for actually arriving at the final sample may vary substantially from the original plan. Different protocols, varying levels of completeness of data, and differences in willingness to participate in research endeavors will all have an effect on the soundness of the final sample. Critical evaluation of these differences and their potential effect on the sample's value should be undertaken.

may be largely inaccessible to researchers; or, even if accessible, those segments of the population might not respond to research inquiries.

Here, the researcher must be alert to problems in sampling that can adversely affect the quality of the sample and ultimately the data collected from the sample. The question of whether the data is comparable must be continually asked—otherwise, another international research blunder will find its way into the literature.

In response to this internationalization of the marketplace, significant changes are going on in the business research community. Globe-trotting businesses have created a significant need for global research firms with both the ability and capability of conducting research in multiple worldwide markets.[30] Previously, even large research firms were rather parochial in their view of their markets, largely sticking to their home turf. But because businesses and their money are going global, so are the major research companies.

A couple of changes are very apparent. First, research organizations are merging at an unprecedented rate. At the end of 1997, the top 25 marketing, advertising, and public opinion research firms had revenues of $6.6 billion, controlling more than 60% of worldwide spending for these services.[31] Furthermore, if these top 25 firms continue to acquire at the same rate, their combined share of the world market for these research services would be 70% going into 2000. This change is creating truly global research capabilities with a vast amount of highly concentrated knowledge.

In addition, specialized suppliers of sampling services, such as Survey Sampling, Inc. [http://www.worldopinion.com], are becoming more global in their focus. As of late 2003 Random Digit Dialing samples were available from 21 countries. More were to come in the near future. The net result is that it will also be easier for companies and smaller research organizations to obtain specialized services in the very near future.

ONLINE SAMPLING DESIGN ISSUES

The fast-growing acceptance of online research methods is likely to continue to accelerate in the foreseeable future. Quite frankly, the advantages of speed, cost, and ease of use are far too great to justify not developing online research technologies. In this regard, the online conduct of research poses some unique problems insofar as sampling is concerned. As with other more traditional types of mail and telephone samples, one must be constantly aware of unknown bias that can creep into online samples.

The first major issue is that of online access. Currently, about 70% of U.S. households have access to the Internet and this trend is growing.

Internationally, however, Internet penetration is mixed at best and thus creates some inherent problems in getting truly random samples of the generalized population. However, early research suggests that online samples do in fact compare favorably with more traditionally obtained sampels.[32] In other words, the sampling frame is not yet fully specified. On the other hand, for specialized samples and for many research questions reliable and effective sampling strategies are possible.

Online sample respondents can be obtained/recruited from a variety of sources, including customer/client lists, company databases, Web site promotions, on-site intercepts, online newsgroup posting areas, and even offline methods such as mail and telephone.[33] Still another option is to commission one of the many online research panels that are emerging. Major traditional mail panel vendors (such as NFO (http://www.nfow.com/worldgroupfamily.asp) have now helped to convert these more traditional panels into interactive ones that can be sampled quickly and easily. Other companies such as FGI Research (http://www.fgiresearch.com/online.htm), Harris Interactive (http://www.harrisinteractive.com), Decision Analyst (http://www.decisionanalyst.com/online.asp), and MarketTools' Zoomerang (http://www.zoomerang.com/panelpurchase/Index.htm), among others, maintain various types of panels that can be accessed relatively easily for a fee. The reader is encouraged to explore these sites to understand the nature of these panels.

Regardless of how the online sample is obtained, it is critical the researcher evaluates its adequacy using the same criteria used for more traditionally obtained personal, mail, or telephone samples. The critical questions here include who is the sample supposed to represent, does it in fact represent the population of interest, how were the sample members included in the sample (paid, opt-in by pop-up recruitment, panel member, etc.), and are there any obvious biases given the sampling frame and obtained list?

Overall, the Internet holds much promise in the sampling arena. It makes possible access to sampling frames that were often buried in endless paper or were simply unavailable previously. Initial results also suggest that online respondents are more likely to complete surveys, thus effectively turning around the nonresponse problem that has been plaguing researchers for years. Another potential advantage is quick identification of low-incidence samples that can be quickly and easily accessed. These possibilities, accompanied by the appropriate safeguards, should continue to accelerate the acceptance of online samples in the future.

In conclusion, the sampling decision is indeed complex. Researchers are faced with numerous decisions, unknowns, and tradeoffs in their attempt to generate the best possible information for decision making. As always, the manager and researcher should be aware of false promises. For this reason, it is vital for managers to be involved and familiar with the difficulties facing sampling design. The next section deals with managerial considerations in sampling.

MANAGERIAL CONSIDERATIONS

As in research design, managers should heed certain key considerations in sampling if they hope to acquire the information they desire. These considerations involve the nature and appropriateness of sampling in the overall study's design and the actual procedures chosen for the selection of the study sample.

The decision to sample requires important judgments to be made by the researchers. Not only are researchers faced with the question of how to sample, they must contend with the more basic question of whether to sample at all. Resource constraints often dictate crucial decisions in the sampling process. Managers must require that researchers clearly justify the costs of the methods they are using and the crucial decisions about sampling, so as to ensure the study will achieve the stated objectives.

The specifics of sampling can become complex and confusing to the nonstatistical manager. The mathematics of sample designs and their properties are often beyond the grasp of the average manager. Therefore, managers should encourage a peer review system where important and costly research projects' sampling designs are evaluated. This should help ensure that the sampling design and plan chosen are truly appropriate to the study at hand.

Sampling decisions affect the type and quality of inferences that can be made from a study. The importance of this fact cannot be overstated. The selection of probability versus nonprobability design is a critical decision because of the properties inherent in each sampling type. Nonprobability designs have the inherent weakness of the lack of error estimation. If the errors of sampling cannot be estimated, then the degree of confidence the manager can have in a study's results is questionable indeed. Seemingly concrete research results can lead to serious management errors in decision making. The manager must be aware of the limitations of a study's sampling procedures.

Probabilistic sample designs do not ensure correct research results. Although the bulk of this chapter focused on the nature and advantages of probability sample designs, their use does not guarantee infallible research results. Probabilistic samples only increase the confidence that researchers and managers can have in a particular study. In addition to the sampling errors that can be made, a multitude of other errors can invalidate any particular investigation's results. Therefore, the merits of each study must be evaluated on a case-by-case basis.

Online samples should be considered if appropriate to the problem and to the design chosen to solve the research problem. The inherent advantages of using online samples (primarily speed and cost) necessitate that managers take a serious look at using these types of samples. However, these samples should be scrutinized carefully to

EXHIBIT 9.2

Key Managerial
Questions Pertaining to
Sampling

1. Has the population been defined in a manner that is consistent with the research problem?
2. Does the sample frame represent the population completely and accurately?
3. Given the research objectives, is the sampling design chosen appropriate for the investigation?
4. If management has specified particular levels of precision and confidence, is the probability sampling design chosen capable of generating estimators with these properties?
5. If a specific probability design is chosen for use in the study, is it statistically efficient? Does it possess a high degree of sample efficiency?
6. If a nonprobability sampling design is chosen for use in a study, will it give management the type of information required for that problem situation?
7. Have online samples been considered in the research design process? Are they appropriate to answer the research question under study?

ensure their appropriateness to the study at hand. In other words, do not use them simply because of their convenience. Require adequate justification for their use.

These concerns should be kept in mind when evaluating the sampling plan for any study. More detailed questions, such as those presented in Exhibit 9.2, should be asked in the sampling design process. These concerns and questions, if dealt with directly, should increase the probability of obtaining usable research findings.

SUMMARY

This chapter begins with a discussion of the terminology and rationale for sampling. The differences between a census and a sample are discussed. Three basic reasons are given for the use of sampling in business research: resource constraints, accuracy, and the often destructive nature of measurement.

Next, the sampling process, or the means to select a particular study sample, is presented. This seven-step process includes defining the population, selecting the sampling units in the population, selecting a sample frame, choosing a sample design, selecting the sample size needed to accomplish the research objectives, designing a sampling plan, and selecting the study sample.

Next is an introduction to the philosophy of sampling, including a brief review of basic statistical concepts relevant to sampling. This background provides the framework for a discussion of the major types of sample designs.

Sample designs are divided into two major types: probability and nonprobability designs. There follows an overview of the sample design subtypes available to the researcher. Next, a descriptive and computational example illustrates the sampling process using a simple random sample design. The chapter ends with practical and managerial considerations in sampling.

DISCUSSION QUESTIONS

1. Define the following terminology of sampling design:
   a. element, UOA   b. population
   c. sampling units   d. sampling frame
   e. sample   f. parameter
   g. statistic   h. sampling error
   i. statistical efficiency   j. sample efficiency
   k. sampling plan

2. Define the appropriate population for sampling to determine the attitudes of women toward sexual harassment in colleges and universities. Feel free to redefine the project further and to adjust your answer likewise.

3. Discuss the major difference between probability and nonprobability sampling.

4. A local business has asked you to select a sample of 75 potential managerial professionals (PMPs) from your campus. The manager wants these PMPs to evaluate their new, easy-to-use microcomputers for business applications. The store carries five different computers. Describe the sampling process that you would use.

5. Why are sample statistics generated from research efforts important to the overall completion of a research plan? (Remember, sample statistics estimate population parameters.)

6. Why are stratified and cluster sampling methods more efficient than a simple random sampling method?

7. Describe the major nonprobability sampling procedures. When is their use justified?

8. Any researcher who uses a probability sample to conduct research can be assured of fulfilling the research process in an error-free manner. Discuss this statement. Is it true?

9. What are the major reasons for using sampling instead of a complete canvas of the population in business research?

10. The manager of a large department store chain is interested in the average amount of time that cashiers actually are not checking out customers per day. A systematic sample will be drawn from an alphabetical list of cashier records, where $N = 2500$. In a similar survey conducted in the previous year, the sample variance was found to be $s^2 = 2$ hours. Determine the sample size required to estimate μ, the average amount of time the cashiers are not checking out customers per day, with a bound on the error of estimation of $B = 0.5$ hour.

11. A manager is interested in expanding the facilities of the company health and exercise center. A sample survey will be conducted to estimate

the proportion of employees that will use the facility. The employees are divided into two groups, users and nonusers of the facility. Some employees live close by, and some live far away. Four strata are isolated: 1, 2, 3, and 4. Approximately 90% of the present users and 50% of the present nonusers would use the expanded facility. Approximately 70% of the close-by employees and 50% of the far-away employees would use the expanded facility. Existing records indicate that

$$N_1 = 97, N_2 = 43, N_3 = 145, N_4 = 68$$

Find the appropriate sample size and allocation needed to estimate the population proportion, with a bound of 0.05 on the error of the estimation.

12. Assume a survey sampling organization has determined that a minimum usable sample of 600 respondents is needed to complete a study of women's attitudes toward microwave cooking (only women who have used microwaves are desired in the final sample). Given that previous studies have indicated that 40% of the general population of women will not respond to surveys of this type, and that 75% of all women have actually used microwaves, what is the actual sample that will need to be drawn to reach the required sample figure of 600?

13. Go to Survey Sampling's Web site (http://www.worldopinion.com) and investigate the information available to the interested researcher/manager. Summarize this information and be able to contribute to an active class discussion about the information available on this site.

14. Using a research engine of your choice, search the Web for the latest developments in sampling. Include in this search the latest developments in both technological and global sampling processes. Identify and reference all sites from which information was obtained.

NOTES

1. There are several comprehensive sources of information on the sampling process in the literature. The reader is referred to the following texts for further study in the field of sampling: Leslie Kish, *Survey Sampling* (New York: Wiley, 1995); William G. Cochran, *Sampling Techniques* (New York: Wiley, 1977); Richard L. Scheaffer, William Mendenhall, and Lyman Ott, *Elementary Survey Sampling*, 5th ed. (Belmont, CA: Duxbury Press, 1996); and Seymour Sudman, *Applied Sampling* (New York: Academic Press, 1976).

2. Scheaffer, Mendenhall, and Ott, op. cit. note 1.

3. Earl R. Babbie, *The Practice of Social Research*, 9th ed. (Belmont, CA: Wadsworth, 2001), p. 94.

4. Babbie, op. cit. note 3, pp. 194–196.

5. For a discussion of the inadequacies of sampling frames, see Alvin C. Burns and Ronald F. Bush, *Marketing Research* (Upper Saddle River, NJ: Prentice-Hall, 1998), pp. 360–361.

6. Thomas C. Kinnear and James R. Taylor, *Marketing Research*, 3rd ed. (New York: McGraw-Hill, 1987), p. 266.

7. Babbie, op. cit. note 3, p. 189.

8. See, for example, Donald S. Tull and Del I. Hawkins, *Marketing Research*, 6th ed. (New York: Macmillan, 1993), pp. 69–70, for one possible breakdown of these errors.

9. Kish, op. cit. note 1, p. 26.

10. This framework is adapted from Sheldon R. Olson, *Ideas and Data* (Homewood, IL: Dorsey, 1975), Chapter 9.

11. When we begin discussions of sample size determination, we again assume a basic familiarity with elementary statistical concepts. For a review, refer to any good basic statistics text. For example, see Gene V. Glass and Julian Stanley, *Statistical Methods in Education and Psychology* (Englewood Cliffs, NJ: Prentice-Hall, 1970); or William L. Hays, *Statistics* (New York: Holt, Rinehart, and Winston, 1981).

12. Julian L. Simon, *Basic Research Methods in Social Science* (New York: Random House, 1969), Chapter 12.

13. See, for example, Gary T. Henry, *Practical Sampling* (Newbury Park, CA: Sage, 1990).

14. See Scheaffer, Mendenhall, and Ott, op. cit. note 1, and Kish, op. cit. note 1.

15. Adapted from C. M. Backstrom and G. D. Hursh, *Survey Research* (Evanston, IL: Northwestern University Press, 1963), p. 24.

16. For discussions relating to this decision, see Cochran, op. cit. note 1, pp. 73–81; and Thomas T. Semon, "Basic Concepts," in R. Ferber, Ed., *Handbook of Marketing Research* (New York: McGraw-Hill, 1974), pp. 2:217–2:229.

17. For a discussion of the mathematics of these distributions, see Hays, op. cit. note 11; or George Snedecor and William Cochran, *Statistical Methods* (Ames, IA: Iowa State University Press, 1971).

18. Scheaffer, Mendenhall, and Ott, op. cit. note 1, pp. 27–29.

19. Glass and Stanley, op. cit. note 11, p. 242.

20. Glass and Stanley, op. cit. note 11, pp. 244–245.

21. See, for example, Kish, op. cit. note 1.

22. See, for example, Tull and Hawkins, op. cit. note 8, pp. 578–580.

23. Scheaffer, Mendenhall, and Ott, op. cit. note 1, p. 59.

24. See Kish, op. cit. note 1, Chapter 3.

25. Kish, op. cit. note 1, Chapter 5.

26. Paul Green and Donald Tull, *Research for Marketing Decisions*, 4th ed. (Englewood Cliffs, NJ: Prentice-Hall, 1978), pp. 210–211.

27. See Sudman, op. cit. note 1, p. 22.

28. The discussion on incidence and nonresponse is largely drawn from Tull and Hawkins, op. cit. note 8, pp. 586–587.

29. Lynn G. Coleman, "Researchers Say Nonresponse Is Single Biggest Problem," *Marketing News* (7 January 1991), 25:1, p. 32.

30. Cyndee Miller, "Research Firms Go Global to Make Revenue Grow," *Marketing News* (6 January 1997), pp. 1, 22.

31. Jack Honomichl, "Research Growth Knows No Boundaries," *Marketing News* (17 August 1998), p. H2.

32. See Insight Express, "Proven: Online Research's Time has Come," May 2002, downloaded November 11, 2003, http://www.insightexpress.com/proven/wp_register.asp and WebSurveyor, "Comparison of Traditional vs. Online Survey Methods," downloaded September 12, 2003, http://www.websurveyor.com/pdf/webvsmail.pdf

33. Council of American Survey Research Organizations, "New Methodologies for Traditional Techniques," 1998, downloaded October 22, 2003, http://www.casro.org/faq.cfm

SUGGESTED READING

The Nature of Sampling

Babbie, Earl, *The Practice of Social Research*, 9th ed. (Belmont, CA: Wadsworth, 2001).

> *A good research text overall. Chapter 7 is especially good on the logic of sampling, complementing the chapter in this text very well.*

Kish, Leslie, *Survey Sampling* (New York: Wiley, 1995).

> *Without a doubt, Kish is the leader in the field of survey sampling. In this Wiley classic, the author emphasizes sampling use and usefulness rather than mathematical aspects. We would urge all to acquire this book for a personal library.*

Kalton, Graham, and Steven Heeringa, Eds., *Leslie Kish: Selected Papers* (New York: Wiley-Interscience, 2003).

> *This book honors Professor Leslie Kish posthumously by publishing a collection of his papers that deal with the theory and application of survey sampling. This collection provides a valuable insight into the art and practice of survey sampling.*

An Introduction to the Philosophy of Sampling

Hansen, Morris H., William N. Hurwitz (Editor), William G. Madow, and Morris N. Hansen, *Sample Survey Methods and Theory*, vols. 1 and 2 (New York: Wiley, 1993).

A general-purpose work on sampling method and theory. Volume 1 gives a simple, nonmathematical discussion of principles and their practical applications. Volume 2 covers theory and proofs.

Scheaffer, Richard L., W. Mendenhall, and L. Ott, *Elementary Survey Sampling*, 5th ed. (Pacific Grove, CA: Duxbury Press, 1996).

An introductory text written for students with a simple statistics background. It is very well suited for business research because of the wide array of business sampling problems used in the various chapters.

Thompson, Steven K., *Sampling* (New York: Wiley, 2002).

This 367-page book examines many of the sample designs referred to in this text. It is a useful, in-depth reference as far as practical sampling problems are concerned.

Sample Designs and Practical Considerations in Sampling

Henry, Gary T. *Practical Sampling* (Newbury Park, CA: Sage, 1990).

This 139-page book gives a number of examples of practical sample designs. The book is written in an easy-to-read format to help the researcher understand the major sampling decisions.

Lohr, Sharon L., *Sampling: Design and Analysis* (Belmont, CA: Brooks Cole-Duxbury, 1999).

Provides a nice introduction to the field of sampling. Various types of designs are presented and discussed in a contemporary setting.

The reader is encouraged to visit the Web sites of Survey Sampling International (http://www.surveysampling.com) and World Opinion (http://www.worldopinion.com) to check out the latest news and links dealing with practical sampling design issues.

A COMPUTATIONAL EXAMPLE ILLUSTRATING THE PROPERTIES
OF THE CENTRAL LIMIT THEOREM[1]

Suppose Cort Industries, which has divisions worldwide, wants to estimate the number of hours a day its six midlevel managers spend on nonprogrammed decision making. Mr. Imperatus, the research manager, is instructed by his boss, Ms. Martin, that although there is a need for accurate and timely information, he is under a very tight budget constraint. Specifically, he is limited to two long-distance calls on the company phone system.[2] Otherwise, Mr. Imperatus could take the company phone list, call the six executives on the phone, and ask them the simple question concerning their decision-making behavior. This census would probably provide very accurate data.

However, Mr. Imperatus must get as accurate and precise information as possible by talking to only two of the six executives. Is this possible? With the help of the Central Limit Theorem, he can do just that. To begin with, suppose we actually know that the daily hours spent by each of our six executives in nonprogrammed decision making is as follows:

Executive (i)	Daily Hours (x)
Ms. R	1
Mr. S	2
Mr. F	3
Mr. C	3
Ms. D	4
Ms. M	5

From this information we can define μ, where

$$\mu = \sum_{i=1}^{N} x_i/N = \frac{1 + 2 + 3 + 3 + 4 + 5}{6} = 3$$

where

μ = population mean

x_i = daily hours of nonprogrammed decision making for executive i

N = population size

Similarly, we can take all possible combinations of the 15 simple random samples of size 2 from this population as defined by the following combinatorial formula:

$$N^c n = \frac{N!}{(N-n)!n!} = \frac{6!}{4!2!} = 15 \text{ possible samples of size 2}$$

and show that the mean of all of these possible samples is equal to μ, or 3 hours. This is simply done as follows:

Sample of Executives	Sample Total Daily Hours of Nonprogrammed Decisions	Average Daily Hours
Ms. R – Mr. S	$1 + 2 = 3$	1.5
Ms. R – Mr. F	$1 + 3 = 4$	2.0
Ms. R – Mr. C	$1 + 3 = 4$	2.0
Ms. R – Ms. D	$1 + 4 = 5$	2.5
Ms. R – Ms. M	$1 + 5 = 6$	3.0
Mr. S – Mr. F	$2 + 3 = 5$	2.5
Mr. S – Mr. C	$2 + 3 = 5$	2.5
Mr. S – Ms. D	$2 + 4 = 6$	3.0
Mr. S – Ms. M	$2 + 5 = 7$	3.5
Mr. F – Mr. C	$3 + 3 = 6$	3.0
Mr. F – Ms. D	$3 + 4 = 7$	3.5
Mr. F – Ms. M	$3 + 5 = 8$	4.0
Mr. C – Ms. D	$3 + 4 = 7$	3.5
Mr. C – Ms. M	$3 + 5 = 8$	4.0
Ms. D – Ms. M	$4 + 5 = 9$	4.5
		$\dfrac{45.0}{15} = 3.00$

Now, any of the 15 possible simple random samples of size 2 would provide Mr. Imperatus with a point estimate of the mean decision-making hours. However, this estimate does not reflect the precision of the estimate or the confidence that the researcher can have in it. To do this, we must use the notion in the Central Limit Theorem concerning the known properties of the normal curve.

Let us suppose Mr. Imperatus wanted to be 95% confident that he will correctly estimate the true number of decision-making hours with the random sample that he drew, which consisted of Ms. R and Mr. C. Using basic statistics, the process is as follows:

Study Sample ($n = 2$)	Daily Decision Hours (x)	$(x - \bar{X})$	$(x - \bar{X})^2$
Ms. R	1	$(1 - 2)$	1
Mr. C	3	$(3 - 2)$	1

$$\sum_{i=1}^{n} x_i = 4$$

$$\sum_{i=1}^{n} (x_i - \bar{X})^2 = 2$$

where the sample mean is defined as

$$\overline{X} = \frac{\sum\limits_{i=1}^{n} x_i}{n} = \frac{4}{2} = 2$$

and the variance of the sample is

$$s^2 = \frac{\sum\limits_{i=1}^{n} (x_i - \overline{X})^2}{n-1} = \frac{2}{1} = 2$$

and the *finite population correction factor* (FPCF)[3] is

$$\frac{N-n}{N} = \frac{6-2}{6} = \frac{4}{6} = 0.66$$

Therefore, we can define the boundaries of the confidence intervals (with 95% confidence) as follows:

$$z\sqrt{\frac{s^2}{n} \frac{N-n}{n}} = 1.96\sqrt{1(0.66)} = 1.59$$

Then the following defines the relevant confidence interval:

$$\overline{X} - 1.59 \text{ to } \overline{X} + 1.59$$

$$\text{or } (2 - 1.59) \text{ to } (2 + 1.59)$$

$$\text{therefore, } 0.41 \leq \mu \leq 3.59$$

In other words, the results of calculations indicate that the true population mean (daily decision-making hours) will fall between 0.41 and 3.59 approximately 95 out of 100 times (i.e., in replications of the study).

Now suppose we increase the sample size to 3, 4, or 5. What will happen? We will leave the calculations up to you, but it should be intuitively obvious that as n increases relative to N, we will come closer to estimating the true population mean. The important lesson of this example is that probability sampling can provide good and efficient estimators of population parameters while at the same time reducing research time and cash outflow.

NOTES

1. This example presumes some familiarity with the notions found in basic statistics texts.

2. This is obviously a fabricated example, but it does serve to demonstrate the importance and implications of the Central Limit Theorem in sampling design.

3. The FPCF is a correction for SRS when there is a finite population. See Leslie Kish, *Survey Sampling* (New York: Wiley, 1965), pp. 43–45, for an explanation of the role and rationale of this factor in sampling.

10

PDC Using Survey Instruments

Overview

The Nature of Primary Data Collection (PDC)

Active PDC Using Survey Instruments

Offline Methods

Online Methods

A Comparison of Collection Methods

PDC Vendors

Panel Vendors

Managerial Considerations

Summary

Discussion Questions

Notes

Suggested Reading

OVERVIEW

As we saw in Chapter 4, the general rule of thumb in conducting business research is to exhaust all the possibilities of identifying and collecting secondary data that could possibly solve the research decision problem. Frequently, however, secondary data proves to be inadequate for the problem at hand or is simply unavailable. If this is the case, the researcher must generate primary data to use in conjunction with the study formulation and research design.

Primary data, as you will recall, is collected for a specific purpose from original sources. Determining the type of PDC method to use in any investigation is an integral part of the business research process. This chapter focuses on discussing PDC in general, then concentrates on the procedural and managerial aspects of collecting data using survey instruments (commonly called *quantitative research*). The following chapter in the text deals with PDC issues using observation, focus groups, and other forms of qualitative research (*qualitative research*).

Given this background, the chapter introduces the nature of PDC. Major methods of primary data collection are presented. Here, PDC methods are distinguished as to whether they are active (questioning) or passive (observational) methods and whether they are largely quantitative or qualitative type studies. The possible variants of each type of collection method are presented.

Once the distinction is made between PDC types, the remainder of the chapter focuses on the conduct and control of active PDC using survey instruments. The methods of collection are compared, PDC and panel vendors are isolated, and managerial considerations are outlined.

THE NATURE OF PRIMARY DATA COLLECTION (PDC)

Once secondary data sources have been searched and found to be inadequate for the information needs, efforts must turn to the collection of primary data. Primary data, by definition, is collected from original sources for a specific purpose. The data does not currently exist in a compiled form; it is the task of the researcher to collect it in an efficient and useful format for decision making.

For example, suppose top management of Dow Chemical poses the following problem to its research staff: "What are the current attitudes of the European consumer toward Dow Chemical and its products?" At this point, the researchers would most likely begin a secondary data search, both internally and externally, to find any studies that would address this research problem. If the data located is outdated or inadequate, it must be decided whether or not to collect primary data.

The primary data collection decision involves selecting the method(s) of obtaining the information needed. As emphasized throughout the text, this decision is an integral part of the research process. The decision is interrelated with the other steps in the research process, and ultimately with the quality of the information obtained from any study. You need to understand the two chief methods of data collection to make intelligent decisions about the appropriateness of a data collection method in a particular study.

As portrayed in Figure 10.1, the two principal means of PDC are active and passive. The distinction between these two methods has to do with whether or not questions are asked during the data collection process. *Active data collection* involves the querying of respondents by personal or nonpersonal means (for example, an online survey). *Passive data collection* involves the observation of characteristics, by human or nonpersonal means, of the elements under study.

This distinction is not to be confused with the difference between qualitative and quantitative research. For the most part, *qualitative research* consists of studies that cannot be meaningfully quantified. These studies are typically in-depth analyses of one or a few observations; they involve less structured questioning or observation of the respondents. Conversely, *quantitative research* typically uses large samples and involves structured survey questioning of some type that is subsequently numerically and statistically analyzed. Although these distinctions are by no means clear-cut, they are a useful classification for studying primary data collection methods. More will be said about this in the next chapter.

Passive data collection is useful in obtaining data from both people and other types of study elements. It involves the observation of characteristics of individuals, objects, organizations, and other entities of interest. Active collection is broader than passive collection. Active collection methods require respondents to participate actively in the process of obtaining data, whereas passive methods do not. In passive methods, the researchers oversee specific characteristics without querying individuals.

For example, if we collect data on managers' leadership styles by observing the actions of managers in the work environment, then it is passive data collection. Conversely, if we ask managers to complete a paper-and-pencil questionnaire asking them about their leadership style, it is active data collection. When people are the primary response units under study, researchers generally have the choice of utilizing either of the two methods of PDC. When the study units of analysis are other than people, researchers must usually rely on passive collection methods. The only exception is when the characteristics of some entity are inferred from the responses of some designated individuals. For example, if I were to ask the presidents of all the *Fortune* 500 companies to rate their organizations on five dimensions of organizational climate, I would be inferring the characteristics of an entity (their organizations) through the use of active data collection.

The possible variations in each type of collection method are almost limitless. Many studies use both active and passive methods to accomplish

FIGURE 10.1

PDC Decision Framework

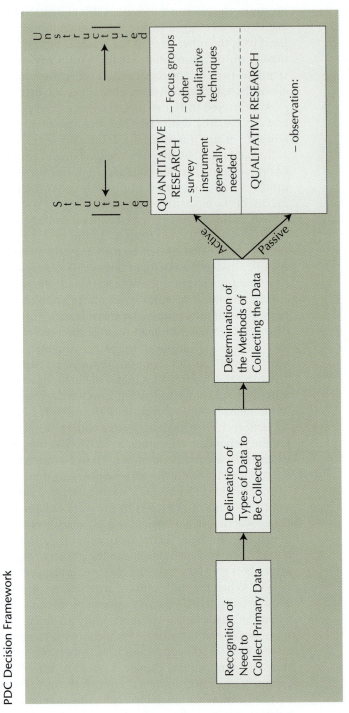

the objectives of a particular study. Regardless, both methods can be subdivided on the basis of three dimensions: the degree of disguise, the degree of structure, and the method of collection.

The *degree of disguise* has to do with whether or not the purpose of the study is known to the respondent. In certain instances, it may be desirable to hide the purpose of the study from respondents because of some expected bias. For example, the whole field of motivation research is based on the fact that people are unable or unwilling to describe their innermost feelings. Therefore, the objectives of the exercises are masked from respondents so that their motivations and interests can be revealed in an unbiased fashion to the researchers.[1] It must be noted that the disguise issue poses a series of problems with which researchers must contend.[2]

The *degree of structure* involves the amount of formalization of the data collection process. Observational studies as well as querying studies can be relatively structured or unstructured. In structured observational studies, researchers are looking for specific acts or characteristics according to a predetermined checklist. In unstructured observation, researchers must actively view a situation, looking for whatever they think are the salient aspects. Structured active methods typically use highly structured questionnaires in the data collection process. Unstructured active methods use no formalized questions or only loosely structured ones to help guide the researchers.

The final dimension of interest has to do with the method of collection. *Method of collection* refers to the means by which the data is obtained from the units of analysis in the study. Human or some other means can be used to obtain data using passive collection methods. An example of human observation is observing the number of cars passing an intersection on certain days to help determine the attractiveness of a potential retail location site. Similarly, a computerized counter on the road could do the same observation. Although not all observations can be done electronically, many lend themselves to computerization.

In active data collection, the methods are slightly different. They are classified according to the means of administering the questions posed by the researchers, these being the more traditional offline and the newer online (Internet-assisted) methods. The more traditional offline methods include personal interview, telephone, and mail, which may or may not be computer-aided. The newer online, Internet-assisted methods are quickly gaining acceptance in the research community. Each of these methods will be explored in more depth later in this chapter.

The researcher is faced with the decision as to which of many possible data collection methods will be best for the research problem under study. This decision is becoming more complex because the research environment is changing rapidly. Such factors as the globalization of markets, increased concerns about privacy and the intrusion of research into people's lives, and changes in technology must be considered in deciding which method to choose.[3]

Unfortunately, there is no definitive plan by which we can select the ideal data collection method. Instead, we must choose the appropriate method based on the data needs of the investigation and practicality/cost considerations.

ACTIVE PDC USING SURVEY INSTRUMENTS

Contemporary business research relies heavily on the use of active primary data collection methods. Because business is largely a social phenomenon, dealing with people, much of the data needed to make decisions has to come from people themselves. Active data collection methods using survey instruments are designed specifically to obtain large amounts of information from human respondents.

One of the first decisions in the collection process is which method will be used to obtain the data needed to answer the problem. The unique characteristics of each method affect the cost and quality of the data obtained. The discussion of the active methods is organized around the two major methods of collection, these being offline and online methods. Online methods are those that rely on the Internet as a method of delivery and data collection. Online methods are treated separately because of the vast potential for this collection method.

Offline Methods

For our purposes, offline methods are the more traditional methods of personal, telephone, and mail data collection. These methods may or may not be computer-aided, as is discussed after the three major delivery offline delivery methods are presented.

1. *Personal interviewing* can be defined as person-to-person discourse, initiated by the interviewer for the purpose of obtaining relevant research information.[4] The interview is actually a personal relationship between the investigator and a set of respondents under study. The interviewer usually possesses a plan of interrogation, often in written form, containing questions that focus on the research problem. The personal interview as a data collection method is equally applicable to ex post facto and experimental research designs.

Like other forms of data collection, the personal interview is an implementation of the measurement process described in Chapter 7. It usually involves the eventual quantification of data for analytical purposes. Therefore, the researcher must be concerned with various types of errors that might potentially affect the results of the investigation. These errors can be broken down into two general classes: nonresponse and response errors.

Nonresponse Errors. A nonresponse error occurs when a respondent is included in the study design but is not actually reached. In personal interviews, it is often a problem to get the respondent to answer your questions or even to contact the individual in the first place. For example, if we were going to personally interview the presidents of *Fortune* 500 companies to obtain data on their managerial styles, we would probably find some of these executives refusing to answer our questions because of busy schedules or issues of confidentiality, or we simply might not be able to locate them during the time set aside for data collection.

We should make every attempt at reducing these nonresponse errors through our study design. Past research suggests some guidelines for doing this.[5] For example, callbacks (recontacting refused interviews or not-at-homes) have been found to be effective in reducing nonresponse rates in research.[6] Similarly, prior introduction of the interviewers to respondents and the absence of formal consent procedures to participate in the study also appear to reduce nonresponse in personal interviewing.[7]

Other researchers suggest that the data collection plan include specific time frames for conducting the interviewing process so as to minimize the not-at-home problem. Specifically, it has been found that someone is most likely to be home during the week between Sunday and Friday from 5:00 to 9:00 P.M. (approximately 65% to 80% of the time) and on Saturdays between 12:00 and 4:00 P.M. (approximately 60% to 65% of the time). Overall, however, Saturday is the best day of the week for finding potential in-home respondents at home (60% of the time).[8]

There are numerous other suggestions for reducing nonresponse in personal interviewing, but it can rarely be completely eliminated.[9] Other writers have suggested techniques for estimating nonresponse errors through a variety of methods.[10] Although these techniques are beyond the scope of this text, you should be aware that such tools do exist in the research literature. The general guideline is that all attempts should be made to reduce nonresponse errors.

Response Errors. These errors are present when a discrepancy exists between reported data and the true value of the variable of interest. Although we can never fully estimate such errors, some knowledge of their sources can help us minimize them. The response errors inherent in personal interviewing appear to fall into four categories: interview variability, question structure and sequence, methods of administration, and respondent errors.[11]

> **Interview variability.** These errors relate to the variability in the interview situation and the characteristics of the interviewers themselves. Several studies have attempted to relate interview errors to characteristics of the interviewer (sex, age, weight, and so on). For the most part, the effects are trivial except for sex and race characteristics.[12] These characteristics do appear to affect responses in some systematic and unknown fashion.[13] Overall, it appears that the

quality of the interrogation depends more on interviewer abilities and the standardization of the interview situation itself, rather than the specific characteristics of the interviewer. Thus, the selection, training, and monitoring of interviewers are essential in reducing interview errors.

Question structure and sequence. Errors due to the format and order of questions in an interview can also bias a study's results. There is confusion about the proper structuring and sequencing of questions, as we saw in Chapter 8. Although there are many guidelines in the art of asking questions, it appears that the critical concern in personal interviewing is the actual communicative abilities of the interviewers themselves.[14]

Method of administration. Personal interviews may elicit responses that are different from those obtained from other forms of active primary data collection. For example, research suggests that personal interviews may actually stimulate answers that appear to be more socially acceptable.[15] It has also been suggested that response errors may actually be inversely related to the degree of structure in the questioning process during the interview, but the evidence is very inconclusive at this time.[16]

Respondent error. Any errors that are due directly to inaccuracies, either intentional or unintentional, on the part of the respondent are called *respondent errors*. Such errors come about from either deliberate distortion of answers or lack of knowledge. This type of response error is the most difficult to assess in the personal interviewing process. The primary method of minimizing the effect of these errors on results is the meticulous development and pretesting of scales and questionnaires, as presented in Chapters 7 and 8.

Planning the Interview. Personal interviewing is a very costly method of primary data collection. However, it can be a very effective method as well because of the high degree of personal interaction between the investigator and the respondent. The primary means of ensuring an efficient procedure is the selection and training of competent interviewers, and the proper monitoring and conduct of the interview situation itself.[17]

Generalizations about what makes a good interviewer and interview are dangerous because of oversimplification, but some guidelines are useful. First, interviewers should be selected on the basis of the intellectual competence of the respondents they will be interviewing. Second, they should have basic communicative abilities that will allow them to properly empathize with these respondents. Third, they should be anxious to undertake the task.

Once the interviewers are selected, training them becomes the next step. Basic guidelines can help in the pursuit of successful personal interviews. Exhibit 10.1 outlines a number of important training factors that should be

EXHIBIT 10.1

Training Factors That
Should Be Covered in
Any Personal Interview
Training Program

Source: Adapted from Thomas
J. Bouchard, Jr., "Field
Research Methods: Inter-
viewing, Questionnaires,
Participant Observation,
Unobtrusive Measures," in
Marvin Dunnette, Ed.,
*Handbook of Industrial and
Organizational Psychology*
(Chicago: Rand McNally,
1976), p. 378; and C. Cannell
and R. L. Kahn, *The Dynamics
of Interviewing: Theory,
Technique, and Cases* (New
York: Wiley, 1957), p. 586.

1. Provide the new interviewers with the principles of measurement and give them an intellectual grasp of the data-collecting function and a basis for evaluating interviewer behavior.

2. Teach techniques of interviewing.

3. Provide opportunity for practice and evaluation by actually conducting interviews under controlled conditions.*

4. Offer the careful evaluation of interviews, especially at the beginning of actual data collection. Such evaluation should include a review of interview protocols.

5. Provide training for coding of interview responses.

6. Involve the interviewer in the construction of the interview schedule if possible.

_____
*See E. Blair, "Using Practice Interviews to Predict Behavior," *Public Opinion Quarterly* (1980), *44*, p. 257.

covered in any formal program. In addition, it has often been found useful to prepare a manual on interviewing for each research project.

Monitoring the personal interviewing process involves the assessment of interviewer bias and cheating in actual data collection. *Cheating* usually involves the interviewer making up all or some part of the interview. Monitoring allows the researcher to assess the systematic biases created during personal interviewing. The assessment of these biases is difficult, but techniques such as spot checks, telephone follow-ups, and postcard follow-ups have been found useful in minimizing interviewer cheating and other forms of response bias.[18]

The final area of concern in the planning of personal interviews relates to the actual conduct of the interview. A generalized model depicting the conditions for successful interviewing is presented in Figure 10.2.[19] Essentially, the model suggests that four conditions must exist before a successful interview can happen: access, trust/goodwill, expertise, and motivation.

FIGURE 10.2

Interactive Conditions
Required for Successful
Interviewing

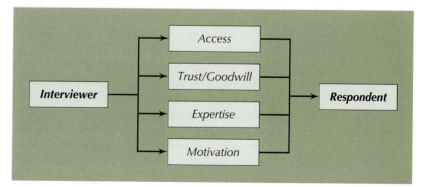

EXHIBIT 10.2

1. Maximize privacy of data collection setting.
2. Know your respondent's name.
3. Maintain neutrality.
4. Maintain confidentiality.
5. Listen.
6. Don't ask for elaboration of sensitive issues.
7. Inform respondents how and why they were chosen.
8. Identify yourself and your organization, if applicable.

The first condition, *access*, is the ability of the respondent to retrieve and convey the information asked by the interviewer. Usually, this is a matter of interview and question structure. Specifically, if the questions are framed in such a way that they are not offensive and are easily understood by the respondent, open access should be present.

The second condition, *trust/goodwill*, requires the interviewer to establish a rapport with the respondent, a feeling of trust and goodwill between the two interview participants. This condition is usually met if the interview situation is comfortable to the respondent and if ego-threatening questions are avoided.

The third condition is *expertise* of the interviewer. In order to establish legitimacy and credibility, the interviewer must be viewed as an expert. However, respondents also need to feel that they can contribute something in the interview process. In general, as the status or expertise of the respondent increases, the interviewer's qualifications and status must also increase.

The final condition, *motivation*, is the willingness of an interviewee to provide the information requested by the interviewer. Motivation is linked to the previous three conditions. A number of tactics designed to help ensure respondent motivation[20] are presented in Exhibit 10.2.

2. *Telephone interviewing* can be used for either structured or unstructured and disguised or undisguised investigations. It is perhaps best suited for structured, undisguised data collection. Unlike personal interviews, telephone interviews are seldom used for unstructured, disguised interrogation because they lack the visual cues and the intimacy of personal contact.

The advantages of telephone data collection include cost-effectiveness and speed of data collection. The consensus appears to be that the telephone interview is at least equal to other active methods of primary data collection in obtaining accurate and valid data.[21] In telephone interviewing, researchers can instantly and constantly monitor the interview process. Some researchers claim that this method can result in more valid data than the personal interview because the telephone limits the effects of visual cues (i.e., image, age, appearance) on a respondent's answers.[22]

One difficulty is that not all individuals have telephones (or, at least, listed numbers). Internationally, the problem is worse: In many developing countries, few people have telephones, and those systems are often unreliable. In such cases, telephone interviewing is often not possible. For example, it is almost impossible to conduct phone surveys in Mexico. Telephone penetration in Mexico City is around 55%, and in many other Mexican cities it is as low as 35%.

Recent technological trends have compounded problems in telephone interviewing. The explosive growth of Internet usage, cellular phones, paging equipment, modems, and fax machines have all dramatically increased the demand for telephone numbers around the world.[23] In fact, in the United States alone, 91 new area codes were implemented in the four-year period between 1995 and 1999. In the United Kingdom, eight million customers got new telephone numbers during the Easter weekend in 1999. This situation drives down the working phone rate (meaning more calls to get to your sample) and tends to decrease response rates.

Other issues, such as increased call screening, security and safety issues (post 9/11), privacy concerns, and increased regulatory involvement in unfettered access to potential respondents, are all helping to change the conduct of research. For example, while the Do Not Call lists maintained in the United States do not prohibit calls for research and polling purposes, it has further sensitized respondents and researchers alike to the "intrusive" methods of data collection used in the past. As a result, the traditional data collection paradigms are currently under scrutiny and are in a state of transition.

These situations result in nonresponse errors. In practice, attempts have been made to address the resulting bias created by these methodological shortcomings.[24] One example is the use of random-digit dialing to locate unlisted numbers. This procedure is now well respected in the research community, and there is an important body of literature on this aspect of telephone interviewing.[25]

One illustration of how a major commercial survey sampling firm (Survey Sampling, Inc.®) uses random-digit dialing (RDD) was presented in Figure 9.2. This illustration also dealt with the nuances and reasons for RDD in the context of practical sampling. Figure 10.3 extends this topic by presenting "A Survey Researcher's View of the United States," outlining such critical information as unlisted rates, contact rates, and cooperation rates throughout the United States.

The conduct of the telephone interview is similar in many ways to the personal interview; many of the factors and tactics in personal interviewing are equally applicable here. Additional considerations for telephone interviewing are presented in Exhibit 10.3.

3. *Mail Interviewing* is interrogation using a written questionnaire delivered by mail. This method of data collection is widely used by business researchers to obtain data on various subjects. Generally speaking, the mail interview gives the researcher flexibility at a low cost. Additional advantages such as respondent anonymity, confidentiality, and leisureliness of response made this an attractive data collection mechanism in the past.

FIGURE **10.3**

A Survey Researcher's View of the United States

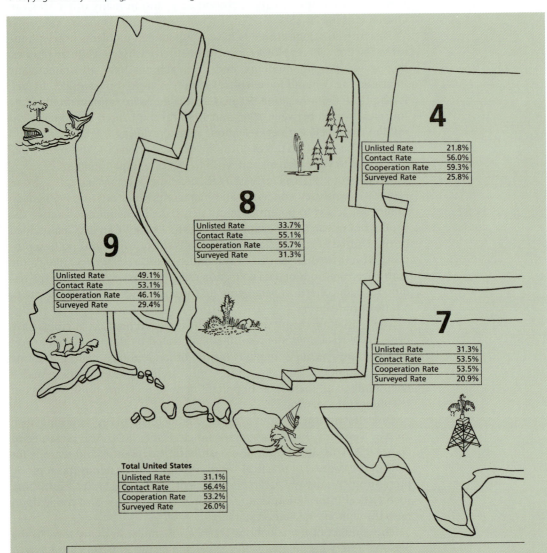

4

Unlisted Rate	21.8%
Contact Rate	56.0%
Cooperation Rate	59.3%
Surveyed Rate	25.8%

8

Unlisted Rate	33.7%
Contact Rate	55.1%
Cooperation Rate	55.7%
Surveyed Rate	31.3%

9

Unlisted Rate	49.1%
Contact Rate	53.1%
Cooperation Rate	46.1%
Surveyed Rate	29.4%

7

Unlisted Rate	31.3%
Contact Rate	53.5%
Cooperation Rate	53.5%
Surveyed Rate	20.9%

Total United States

Unlisted Rate	31.1%
Contact Rate	56.4%
Cooperation Rate	53.2%
Surveyed Rate	26.0%

Unlisted Rates

The number of homes with unlisted phones is steadily increasing. Our latest research indicates that nearly 1/3 of all U.S. households have unlisted numbers and in some markets it's as high as 60%. In addition, demographics tend to differ somewhat between listed and unlisted households. All of which can impact survey representation. Survey Sampling's Random Digit methodology offers you the most cost efficient solution to this growing problem. Our Random Digit Super Samples always allow you to contact both listed and unlisted numbers—and yield 65% or better working phone numbers.

Contact Rates

Reaching households in some markets can be tricky—particularly in the major metropolitan areas, where people tend to be out more often. We know, because we randomly dialed 600,000 households over a 14-month period and made note of working phone rate and the proportion of calls that resulted in contact with an English-speaking interviewee. Now you can take advantage of this "reachability factor" when you order your next sample. Just ask for our *Contact and Cooperation Rate Adjustment* when you call.

FIGURE **10.3**

(Continued)

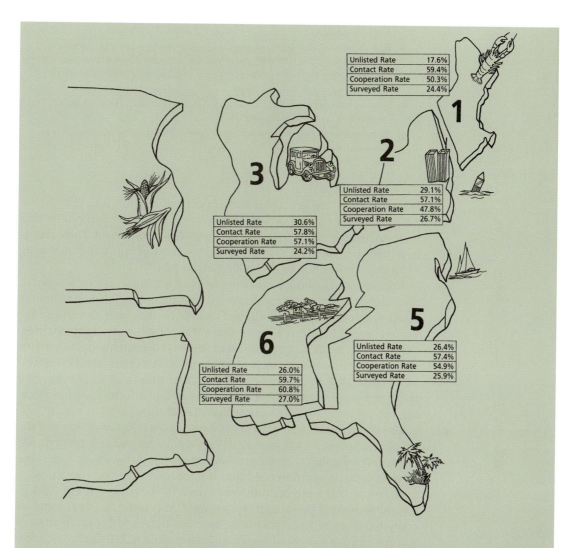

Unlisted Rate	17.6%
Contact Rate	59.4%
Cooperation Rate	50.3%
Surveyed Rate	24.4%

Unlisted Rate	29.1%
Contact Rate	57.1%
Cooperation Rate	47.8%
Surveyed Rate	26.7%

Unlisted Rate	30.6%
Contact Rate	57.8%
Cooperation Rate	57.1%
Surveyed Rate	24.2%

Unlisted Rate	26.4%
Contact Rate	57.4%
Cooperation Rate	54.9%
Surveyed Rate	25.9%

Unlisted Rate	26.0%
Contact Rate	59.7%
Cooperation Rate	60.8%
Surveyed Rate	27.0%

Cooperation Rates

Some test markets are more cooperative that others. And refusals to cooperate appear to be on the rise—making it difficult to estimate sample size.

That's why Survey Sampling developed the *Contact and Cooperation Rate Adjustment.* Based on more than a year of nationwide research, over 600,000 dials and 65,000 interviews, it lets you know in advance how your test markets are likely to cooperate. So, if you want to stop paying for more sample than you need—and avoid the hassle of running out of sample before you reach quota—just ask for our *Contact and Cooperation Rate Adjustment* the next time you order.

Survey Frequency

Frequently surveyed markets can add to a researcher's difficulties. Some respondents refuse to cooperate. Others join in too eagerly and the value of their responses becomes questionable.

Survey Sampling's *Protected Sample* is the answer. The phone numbers in our Random Digit Super Samples are protected for up to 12 months. Which means we don't deliver the same number twice in any year—to you or anyone else. Overall, we protect over 30 million records a year—and guarantee no reduction in statistical efficiency. So you won't get stale, over-surveyed respondents in your next sample. Be sure to ask for our *Protected Sample* when you order.

EXHIBIT 10.3

Tactics for Successful
Telephone Interviewing

1. Use a simply structured interrogation design. Carefully repeat responses to questions if necessary.
2. Train the interviewer to introduce the respondent briefly to the purpose of the study. Ask very short questions first (preferably dichotomous or multichotomous).
3. Train the interviewer to concentrate on getting through the first few questions in an interview. After this, he or she will probably develop a rapport with the respondent that will enable interviewer to continue and complete the interview.
4. Have all interviewers keep track of refusal phone numbers. Have the better interviewers recall these numbers.
5. Train, supervise, and monitor your interviewers for increased accuracy of measurement.

However, there are problems. Response rates can run less than 5% in random samples of the general population. Data collection can be very slow, particularly where international samples are concerned. These factors have had the effects of raising the effective cost per unit and increasing researcher suspicion as to the validity of the data.[26]

Response and nonresponse errors for mail interviews were discussed in the chapters dealing with measurement, instrument design, and sampling. Again, such errors can be minimized, mainly by thorough planning of the various aspects of research design that are applicable to mail interviewing. Other specific suggestions deal with the use of inducements and the adjustment of bias through the use of techniques such as time trend extrapolation. Table 10.1 outlines some of the chief means of increasing mail response rates using inducements.[27] Table 10.2 lists techniques for checking and adjusting for nonresponse bias in mail surveys.[28]

These three traditional methods of data collection can all involve the aid of computers. As was noted earlier, the Internet-assisted methods will be treated separately. *Computerized interviewing* techniques use some type of electronic or computerized question process. The techniques range from completing a self-administered interview contained on a diskette to computer-assisted personal interviewing (CAPI) systems. Other forms of computer-assisted interviewing include computerized fax-fax/mail-back surveys, computer-assisted telephone interviewing (CATI), and completely automated telephone surveys (CATS). All of these are variations on the computer-based delivery system.[29]

CATI and CAPI are semiautomated software programs designed to supplement the interview process. CATI will automatically dial the respondent's telephone, then a human interviewer reads questions off the computer screen and keys in responses. Open-ended responses are usually taped for future reference. CAPI systems usually require remote machines and the interviewer usually helps the respondent in person to key in the appropriate response.

TABLE 10.1 Summary of Findings on Response Rate Inducements

Inducement	Influence	Comments
1. Prenotification Techniques	Consistent increase in response	Precontact by telephone most effective, but expensive; no difference between postcards and letters
2. Concurrent Techniques		
a. Personalization	Consistent increase in response	
b. Monetary incentives	Consistent increase in response	
c. Return postage	Consistent increase in response	Stamps preferred to business reply
d. Source sponsorhip	Increase, but situation-specific	Professors and public may be more responsive to university, whereas members of a professional society may respond more to the society's sponsorship
e. Appeal in cover letter	Increase, but situation-specific	University researchers may get better results with science or social utility appeal, whereas business firms might best appeal to respondent's ego
f. Address location	No influence on returns	
g. Specification of deadline	No influence on returns	May accelerate the rate of returns, but will not reduce nonresponses
h. Form type, color, length, and precoding	No influence on returns	
i. Factor interaction	No influence on returns	
j. Outgoing postage	Inconclusive	Data suggests higher class (first-class metered or stamped) better, but not conclusive
3. Follow-up Techniques	Consistent increase in response	Higher-class postage (certified and special delivery) most effective with, at some point, enclosure of replacement questionnaire

Source: W. Jack Duncan, "Mail Questionnaires in Survey Research: A Review of Response Inducement Techniques," *Journal of Management* (1979), *5,* p. 52. Reprinted by permission of the publisher and updated with David Jobber and Daragh O'Reilly, "Industrial Mail Surveys: A Methodological Update," *Industrial Marketing Management* (1998), *27,* pp. 95–107.

For example, Anheuser-Busch used Sawtooth Technologies' (http://www.sawtooth.com) computer-assisted multimedia interviewing product Sensus Q&A to design a 45- to 50-minute survey to conduct an image study for its beer products' group. The study was conducted with 3500 respondents in over 38 locations with only one respondent refusing to do the study because the computer seemed too challenging.[30] Carrier Corporation reported a similar experience using the interviewing product for an advertising study.

To illustrate Sensus Q&A's capabilities, Figure 10.4 shows two sample screen shots of the Sensus Q&A interviewing product. As one can see, both audio and visual capabilities exist in the survey experience. One can almost hear the tire sounds at different speeds and see the tire up close in the zoom shot. Question responses are also easy, with data analysis not far behind the collection process.

TABLE 10.2 Techniques for Adjusting for Nonresponse Bias in Mail Surveys

Technique	Description
1. Comparison method	Compare results of a given survey with known values for the population. Estimation of direction of bias is then done on the basis of these comparisons.
2. Subjective estimates	Subjectively estimate the nonresponse bias by comparing respondents to nonrespondents on the characteristics of interest. Adjustments for bias are then made using these estimates.
3. Extrapolation methods	These methods are based on the assumption that those individuals who respond less readily are more like nonrespondents. Three methods—successive waves, time trend, and method of concurrent waves—estimate nonresponse bias by extrapolating the responses of earlier respondents.

Source: Adapted from J. Scott Armstrong and Terry S. Overton, "Estimating Nonresponse Bias in Mail Surveys," *Journal of Marketing Research* (August 1977), *14*, pp. 396–397.

CATS uses an interactive voice response technology to conduct interviews without human interviewers. The recorded voice of the interviewer asks the questions, and the respondent answers using a touch-tone phone. The system obviously lacks the richness of CATI and CAPI systems. In the last decade, essentially all major research organizations have been forced to adopt some type of computer-assisted interviewing or risk becoming obsolete in the industry.

Online Methods

Internet-based survey methods and focus groups (which will be discussed in the next chapter) are currently the rage. The explosive growth of the Internet throughout the world has enabled the development of a multitude of online delivery options, particularly in the highly penetrated Internet markets of the United States and Western Europe. Online surveys have the advantages of being easy to administer, interactive, timely, capable of including multimedia, and low in cost. In addition, online methods appear to be becoming increasingly representative (in highly penetrated markets) of the general population and hold the promise of increasing response rates over the more traditional offline methods because of privacy, security, and legal issues that have created a much more difficult data collection environment.[31]

The evidence is coming from the successful use of online technologies by a wide variety of organizations. For instance:

- Florida Power & Light Co. in Juno Beach, Florida, conducted an Internet-based survey that produced valuable, replicable results in the case of a human resource problem in the company.

FIGURE 10.4

Sensus Multimedia Screen Captures

Reprinted by permission of Sawtooth Technologies, Inc.

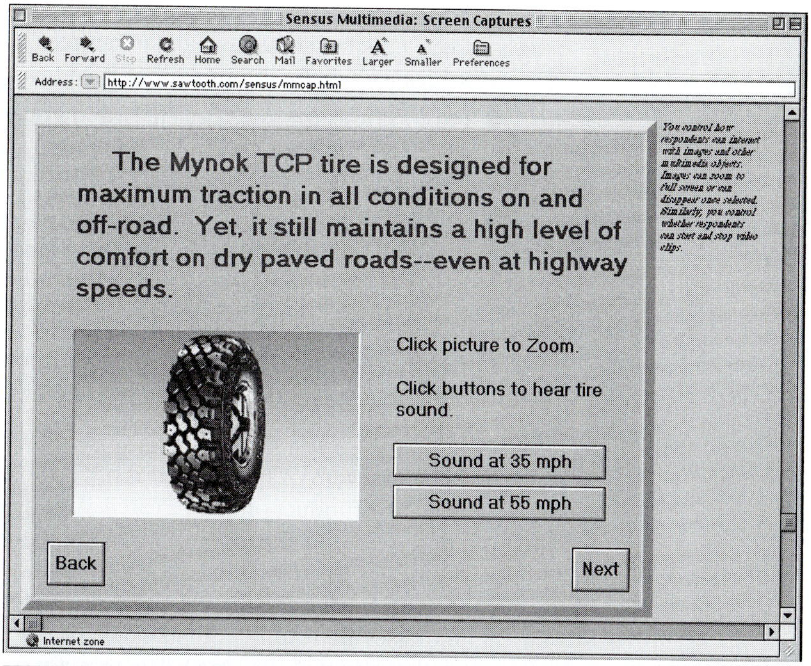

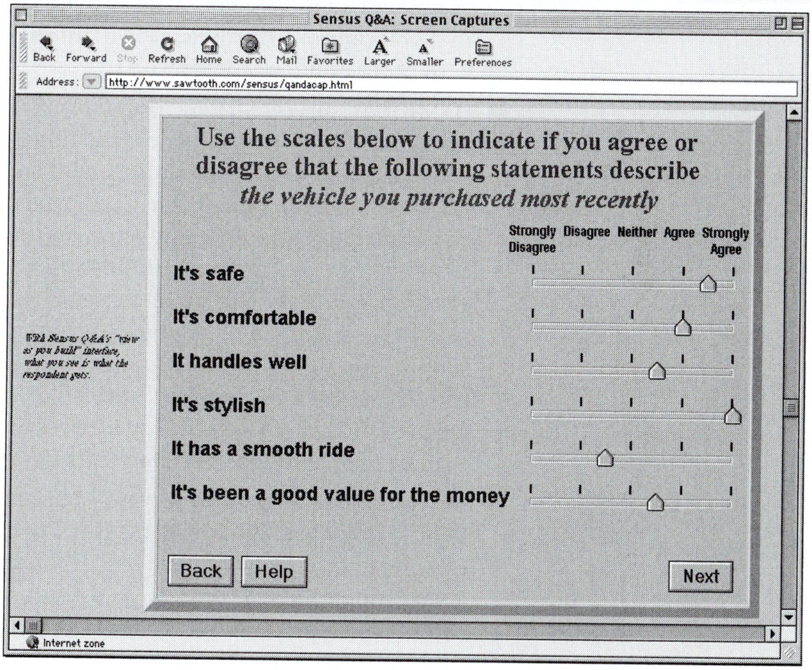

- When the Schick division of Pfizer Inc. needed customer satisfaction information for its newly launched Xtreme 3 triple blade disposable razor, it turned to InsightExpress to help design and collect online data from a pool of subjects who had sampled the product. With 48 hours, Schick had all the information it needed on the particular problem.[32]

- Pitney Bowes has used Perseus' Internet-based software solutions to conduct a variety of online research. It claims the organization has saved both time and resources.[33]

- PeopleSoft utilized Zoomerang's zPro to design "self-service" online surveys for both internal and external research problems. The experience was both rewarding and productive and is described in the Zoomerang case study presented in Figure 10.5.

Currently, the vast majority of online research is being conducted using three basic delivery methods.[34] Each method has strengths and weaknesses, but all three are viable alternatives. These methods are:

- E-mail surveys—This method is often the easiest and simplest with which to conduct online research. All one needs is a set of e-mail addresses and a means to analyze the returned data. As a result, this method is frequently used with internal studies (i.e., employee satisfaction, training evaluation, etc.) and external studies lists of e-mail addresses are readily available (i.e., product registration cards, resale customer e-mail lists, etc.). The primary downside to this method is that you cannot incorporate the more advanced features available with other online methods.

- HTML form surveys—These forms are very flexible but require considerably more programming and setup time. These forms can recruit respondents to a given Web address and then have them self-administer the survey. Subjects can be recruited from on-site intercepts, pop-up ads, e-mail invitations, or any other way you can direct potential respondents to the survey instrument. The advantage of this type of survey over the simpler e-mail survey is that you can build skip logic into the questionnaire, randomize answers, and include graphics and sounds, among other, more sophisticated features. However, as noted earlier, "self-service" suppliers (such as Zoomerang and WebSurveyor) are providing easy-to-use design services to allow researchers and managers alike to access primary data quickly and cost effectively.

- Downloadable interactive surveys—This method requires the respondent to download an executable file. The advantage here is that you can create extremely interactive files for use with many of the available Windows-based controls. As a result, very sophisticated surveys can be developed. The problems in using this method include the substantial time and programming needed to develop this type of survey and the reluctance of respondents to download an executable file because it may contain a crippling virus.

FIGURE **10.5**

zoomerang

PeopleSoft clicks with Zoomerang

Zoomerang zPro is delivering fast, responsive surveying for internal and external audiences at PeopleSoft.

At PeopleSoft Inc., the worldwide leader in enterprise software technology based in Pleasanton, California, their Marketing team is learning how Zoomerang zPro can help them better understand the needs of their sales force, customers and business partners.

Scott Danish, Director of Marketing, PeopleSoft Supply Chain Management, started using Zoomerang zPro recently. "I received a survey, and I thought the interface was attractive, so I looked up their information on their web site," he says. "When I saw the templates, and how easy it was to use, I subscribed right away."

Zoomerang zPro "was a really attractive interface... It created the professional image I wanted to project."
—Scott Danish, Director of Marketing, PeopleSoft Supply Chain Management

Online surveys become easier

Prior to using Zoomerang, surveys at PeopleSoft required the use of an internal party and PeopleSoft's regular HTML. "It was a comprehensive process," says Danish, because every project depended on the availability of other people in Marketing to help build the survey.

Their existing system meant that online surveys "could take several days to build, and you had to set up a separate process to access the results," he says. So when he saw what Zoomerang offered, he was immediately interested.

Background

Previously, when PeopleSoft wanted to develop online surveys for their customers, business partners, or internal staff, they needed to work through their Marketing group. This process required up to two weeks lead time and did not allow immediate access to the results.

Results

Within thirty minutes of trying Zoomerang, Scott Danish was able to understand the range of options available to him. With its intuitive interface, he was able

First-time use is simple and intuitive

Danish's initial Zoomerang experience sold him on the tool. "They had a really robust FAQ section, which answered my questions," he says. The first time he used the tool, it took him about 20 to 30 minutes to click around and understand how to create the kind of survey he wanted. While he thought it was intuitive and easy-to-use at first, "it was even faster the second time," he says.

For first-time users, Zoomerang is able to "bring you up the learning curve really fast," Danish says.

Its range of question options allowed him the versatility he was looking for. "I was able right away to set up the rating system for the questions," he says, creating for example, a 1 to 10 ranking, where the surveyor determines the value of the scale.

East of use and good response reinforce frequency

To date, Danish has created two surveys with Zoomerang. One survey was for an external audience. "It went to about 500

(Continued)

FIGURE 10.5

(Continued)

to create a brand-compatible survey that was rapidly deployed and that delivered the research results he needed, when he needed them.

high-level procurement professionals," he says and about 80 recipients responded, or about 16 percent. "I was really impressed with its rapid deployment," he says. "The data gathered from this survey will provide us with a general understanding of what they find valuable in a two-day conference."

The other survey was internal, to PeopleSoft's own sales force on account profiling. The information gathered from this survey will provide the Marketing team with an understanding of the in-depth account information on key prospects and customers that the sales force finds most useful.

"Going into a sales call with a deep level of account information can make or break a deal," Danish says. "Digging beyond just contacts and competitive solution information can uncover facts that guarantee them an edge. We're interested in utilizing a specialized team of telemarketers to conduct profiling at a much deeper level. So we posed several issues (business pains, company roles, customer sales cycle, software buying history, competition and more) and asked the reps to rank them as 'must have in detail,' 'nice to have some detail' and 'not necessary.' From there, we can take the top 'must have' responses and run a specialized telemarketing effort."

According to Scott Danish, features that make Zoomerang ideal for surveys, include its:
- *Attractive interface.*
- *Range of surveying options.*
- *Rapid deployment.*

Going forward; Danish plans to use Zoomerang, to survey prospects, PeopleSoft's existing customer base, its business partners, and its sales force.

Companies appreciate professional-looking interface

One of the benefits of Zoomerang that Danish discovered immediately was the tool's appearance. "It was a really attractive interface," he says. "I liked the look and feel of it. It created the professional image I wanted to project."

From first-timer to regular user

Within half-an-hour, Danish moved from novice to skilled Zoomerang zPro user. By exploring the tool's intuitive interface, he was able to understand the range of options available to him and quickly create the survey he needed, when he needed it. Now online surveys are an ongoing part of his research plans.

Zoomerang is a product of MarketTools, Inc. For more information please contact us at:

Zoomerang by MarketTools, Inc.
One Belvedere Place
Mill Valley, CA 94941
(415) 462-2200
www.zoomerang.com

When using online research data collection and sampling methods, the astute research manager must be concerned with the usual error issues. The sampling frame issue must be foremost in the researcher's mind. Here, the uneven penetration of Internet access throughout the world poses significant questions about the true representativeness of the sample collected online. Although one may be able to match the population under study by key characteristics such as age, gender, etc., one must be continually aware of the bias created, by the fact that everyone does not have access to the Internet. But as other countries approach 70% to 80% penetration levels of the more developed nations, this problem becomes less of an issue.

Figure 10.6 presents. "A Survey Researcher's View of the U.S. in the Age of the Internet," which contrasts with the more traditional perspective presented in Figure 10.3 illustrating telephone data collection methods. The figure shows that the online survey panel is beginning to match the broader profiles of the U.S. population. As is discussed later in the chapter, online panels are becoming established worldwide, thus helping to resolve the sampling frame problem.

Similarly, given the increased difficulties of reaching respondents with the more traditional offline data collection methods, online methods of data collection are generally producing better response rates than offline methods. Given that personal and telephone interviews are getting more difficult to conduct because of security, privacy, and access issues and that mail interviews are often viewed as junk mail, permission-based online access is creating very positive response rates. One study suggests that generic online surveys are producing response rates at twice the level of telephone surveys and 17 times the level of mail, while at the same time substantially reducing both the cost per survey and the turnaround time to top-line results—a win, win, win anyway it is analyzed.[35]

Online research also has the potential to reduce response errors. Given the permission-based nature of online research, respondents have the ability to answer the survey when and where they want to, if they want to. This freedom of action and the nonintrusive nature of the medium may prove to be the silver bullet for many of the problems of the past.

Finally, other developments, such as virtual reality simulators, are beginning to make their way into the field.[36] These simulators can offer a diverse array of research applications, allowing companies and individuals to explore behavior in a controlled and private setting. Advances in computer graphics and three-dimensional modeling enable researchers to create different environments, from shopping centers to board offices. The respondent can enter virtual reality and react to stimuli presented by the researcher. For example, one could go shopping in a virtual store, walking down aisles, looking at shelves, buying products, and putting them in their shopping basket. The potential for virtual reality in business research is almost limitless.

In conclusion, given the proliferation of the Internet, cable and satellite networks, and powerful personal computers, more research interviewing

FIGURE 10.6

A Survey Researcher's View of the U.S. in the Age of the Internet

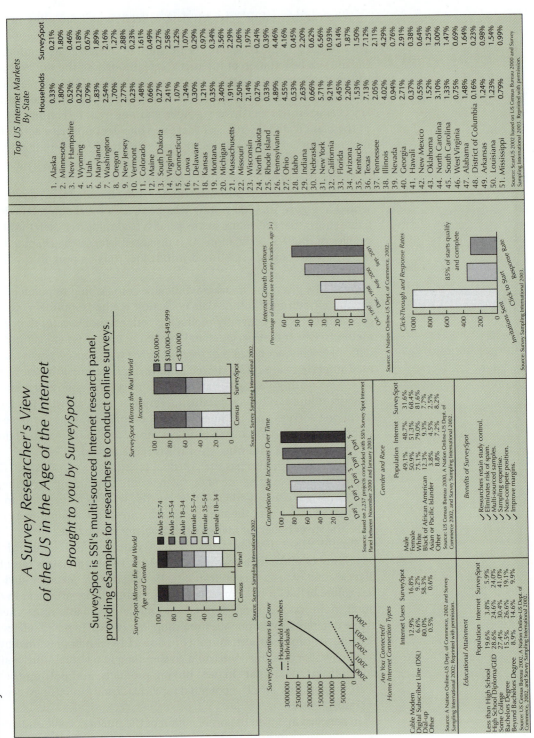

A Survey Researcher's View of the US in the Age of the Internet

Brought to you by SurveySpot

SurveySpot is SSI's multi-sourced Internet research panel, providing eSamples for researchers to conduct online surveys.

TABLE 10.3 A Comparison of Data Collection Methods

	Method			
	Offline			Online
Dimension	**Personal Interview**	**Telephone Interview**	**Mail Interview**	**Internet Interview**
Respondent identification	Excellent	Good	Fair	Good
Flexibility	Excellent	Good	Fair	Good
Anonymity of respondent	Poor	Fair	Excellent	Fair
Accuracy on sensitive data	Fair	Fair	Good	Good
Control of inter-viewer effects	Poor	Fair	Excellent	Excellent
Rigidity of schedul-ing requirements	Poor	Fair	Excellent	Excellent
Time required	Fair	Good	Fair	Excellent
Probable response rate	Good	Fair	Fair to Poor	Good/Excellent
Cost	Poor	Good	Good	Excellent

will certainly be done by online methods in the future. This trend will also be fueled by the inhibiting influence of concerns about personal issues such as security and privacy and the overall intrusive feeling that the traditional offline methods of data collection have elicited, which have made these methods less cost-efficient.

A Comparison of Collection Methods

No one data collection method is best for all situations. Researchers must evaluate the data needs of each problem, in conjunction with the time and cost constraints of the project, to arrive at the best solution. The collection method is chosen in light of each technique's strengths and weaknesses.

Table 10.3 presents a general evaluation of data collection methods on nine dimensions. As you can see from this summary, the various collection methods have different strengths and weaknesses. Researchers often combine methods so that the strengths of each can be exploited.

Figure 10.7 outlines some frequently used strategies in the collection of data using multiple methods. Multiple collection methods are used primarily for two reasons: (1) to increase overall response rate and (2) to prescreen potential respondents. For example, strategy 3 in Figure 10.7 is often used to improve response rates in mail surveys. Specifically, telephone follow-ups are used to urge the potential respondent to fill out the questionnaire and return it, or to ask the questions directly while the individual is on the telephone.

FIGURE 10.7

Frequently Used
Strategies in the
Collection of Data

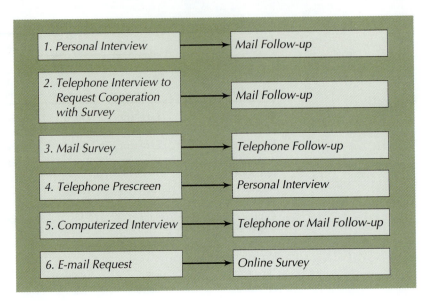

Similarly, strategies 2, 4, and 5 in Figure 10.7 are often used to prescreen potential respondents before inclusion in the final study sample. For example, to study the degree of consumer satisfaction among current owners of electronic video games, we would first have to identify current owners. One way to accomplish this is to follow strategy 2: A telephone survey would prescreen owners versus nonowners of these games. Once the owners are identified, they are sent a longer, more complex mail survey to assess their degree of satisfaction. Strategy 6 is commonly used in online surveys to invite the respondent to participate in the study.

The researcher must creatively design the data collection method(s) to suit the particular needs of the study situation. The benefits and costs of each method should be thoroughly evaluated in light of the study's objectives. As we have emphasized throughout this text, the goal of the research process is to provide information to the manager for decision-making purposes. If the data collection procedures are poorly planned or executed, this goal is not likely to be achieved.

PDC VENDORS

Once primary data needs are recognized within the organization, it must be decided who will be responsible for the design and management of the research process to solve the problem. This "make-or-buy" decision has to do with which individuals and organizations will be involved in the various stages of the research project. The decision alternatives of this problem

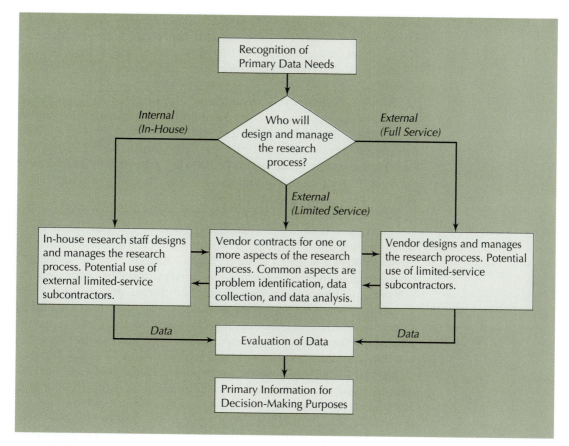

FIGURE **10.8**

Decision Alternatives
in the Make-or-Buy
Decision

are illustrated in Figure 10.8. There are basically two alternatives, with a myriad of possible variations.

The first alternative is that the research process for a particular problem will be designed and managed in-house; the internal research staff will direct and execute the project. The project can be completed entirely internally, or parts of it may be subcontracted to vendors or research suppliers that provide specific research services commercially.[37] Such subcontractors often bid on one or more of the steps in the research process (problem identification, research design, data collection, data analysis).

To illustrate, let us examine a possible research scenario. Suppose General Foods is considering developing a new brand of ready-to-eat natural cereal but is concerned about consumers' attitude toward the product. The research staff undertakes to identify these attitudes. They formally develop their research problem and are given the go-ahead by management to initiate a study. Most of the project is kept in-house for confidentiality reasons, but other organizations are thought to be better suited to actually collect the data. The staff considers some research suppliers that could collect the data desired: NFO World Group (WFOG), Market Research Corporation of America (MRCA), and Market Facts, Inc. It is decided that NFO is the most appropriate for the particular study as designed by the research staff. NFO collects the data and returns it to General Food's staff for analysis.

The second alternative is to contract the entire research project to an external full-service vendor. The vendor is then wholly responsible for the design and management of the research process. However, vendors may subcontract specific portions of the project to limited-service vendors, as the in-house research staff would do. The full-service vendor is responsible for administering the subcontractors, whereas the in-house research staff might not be.

For example, suppose General Foods management feels that an outside vendor could just as easily design and conduct the complete ready-to-eat cereal investigation. The usual procedure would be to issue a request for proposals (RFP) to research vendors. The RFP states the general problem to be investigated and the specific data required by the buyer of the research. The potential research suppliers then submit formal proposals like those described earlier in the text. These proposals would be evaluated according to their appropriateness, and one supplier would normally be awarded the project. It is then the supplier's responsibility to ensure that the data is collected as prescribed in the original proposal.

Table 10.4 provides a sampling of vendor organizations. Full-service vendors are those that provide complete research services (everything from problem development to consulting on the implementation of the research) across a range of customers (health care to banks). However, some of these full-service houses specialize in an industry or area to gain a competitive advantage. Full-service vendors often contract out specialized services, but most would prefer to provide the total solution.

On the other hand, many vendors specialize by research function (focus groups, survey research, human resource management, sampling, and so on) or by customer type (consumer versus business-to-business, health care, computer industry, hotels, and so on). The diversity of vendors can be a boon but also a problem, in that it makes it difficult to choose a vendor.

The decision as to which alternative to use depends on a number of factors. Many small and medium-size businesses have little choice but to contract out the research because no formal research staff exists within their organizations. If research is to be done, outside suppliers must do it.

However, since the late 1990s, a new form of research company has emerged—self-service research organizations. These firms (for example, Zoomerang, http://www.zoomerang.com, and WebSurveyor, http://www.websurveyor.com) provide online design services that enable customers to quickly and easily design studies to address their needs. One example of such an experience—PeopleSoft with Zoomerang—was illustrated in Figure 10.5.

TABLE 10.4 Some Primary Data Collection Vendors

Organization	Description
Interviewing Services of America 16005 Sherman Way, Suite #209 Van Nuys, CA 91406-4024 Phone: 818-989-1044 Fax: 818-782-1309 http://www.isacorp.com	One of the largest established WATS telephone interviewing centers in the United States. In its 20+ years of service to the research community, ISA has expanded to include not only domestic consumer and executive interviewing, but international interviewing. Full cultural, conceptual, and idiomatic translation services.
MRD Co., Ltd. U.S.A. Liason Office No. 375, 6331 Fairmount Ave. El Cerrito, CA 94530 (San Francisco) Phone: 510-525-9675 Fax: 510-525-9671 http://www.mrd.co.jp	Full-service research, Japan and international, qualitative and quantitative, consumer and business, customer satisfaction, focus group facilities with test kitchens, TV program and celebrity popularity (Japan only), mail panel, telephone interviewing facilities, in-house tabulation, full or partial service.
TNS NIPO—The Dutch Institute for Public Opinion and Market Research P.O. Box AE Amsterdam, The Netherlands Phone: + 31 20 522 54 44 Fax: + 31 20 522 53 33 http://www.nipo.com	The largest full-service agency in the Netherlands. Quantitative, qualitative, ad hoc, and telephone (CATI) interviewing. Specializing in electronic data gathering. Telepanel: 1000 households with home computers; CAPIBUS interviewers collect data by portable PC (CAPI) 2000 households weekly; Business Monitor: 20,000 establishments yearly.
P. Robert and Partners Ltd. International Marketing Research Ave. de Lonay 19 1110 Morges, Switzerland Phone: + 4 21 802 84 85 Fax: + 4 21 802 84 85	A technology-driven, full-service agency with extensive experience in integrated qualitative and quantitative multicountry research. A leader in Pan-European telephone interviewing. Over 600 native interviewers fluent in 25 languages conduct worldwide research.

Source: Abstracted from *The 1998 American Marketing Association International Member & Marketing Services Guide.* Copyright © 1997 American Marketing Association. Reprinted by permission. (http://www.ama.org)

Internationally, the choice is more complicated because of the logistical problems inherent in conducting multinational research.[38] In the United States, Western Europe, and Japan, the research infrastructure is fairly complete, so it is not difficult to identify and manage research vendors on a particular project. On the other hand, if multinational research is being conducted in China, Russia, or the developing countries, the infrastructure is incomplete or nonexistent. In these cases, the conduct of research is a major management problem. Who does the research? How do we make sure suppliers in different countries produce comparable results? Who is going to oversee the entire project? All of these factors must be considered in the decision process.

In organizations where research staffs do exist, four factors affect the choice: confidentiality, expertise, objectivity, and resources.

Confidentiality. Often, businesses want to test new products and other ideas that need to remain confidential for competitive reasons. Such research should be done in-house, to minimize the risk of information leaks.

Expertise. Research suppliers specialize by industry, by functional area (finance, marketing management, and so on), and by geographic region. If the buying organization lacks expertise in a particular subject, it may be advantageous to follow expert opinion.

Objectivity. Objectivity can be an important reason to buy research services. Outside suppliers can frequently bring unbiased perspective to the research situation. If management suspects that pet projects or preconceived ideas exist in a research endeavor, it may be wiser to go outside the organization to obtain a more unbiased view.

Resources. Resource constraints are a major factor in the make-or-buy decision. Because research suppliers are specialists, they can sometimes produce (or at least collect) primary data more quickly and cheaply than can an in-house research staff. The prudent manager should be alert to such shortcuts in resource-tight research budgets.

Ultimately, however, the make-or-buy decision rests on which alternative is expected to generate the most useful and cost-effective information for decision-making purposes.

PANEL VENDORS

Among the important primary data collection vendors are those that supply the services of a *panel:* a group of individuals, households, or other entities who are measured repeatedly over time for research purposes.[39] With a *true panel,* the same variables are measured each period; with an *omnibus panel,* different variables are measured each collection period. Panel data is extremely useful in many types of research endeavors because this technique can provide continual monitoring of the same respondents. This allows longitudinal and tracking studies to be conducted.

The data can be obtained from panel members by any of the four data collection methods outlined in the active PDC section earlier in this chapter. One vehicle for data collection using panels is the *diary:* a self-administered mail questionnaire on which the respondent records the relevant behavior or actions as they happen. Such organizations as the Market Research Corporation of America (MRCA), National Family Opinion (NFO), and Market Facts use variations of the diary concept to track household consumption of various goods and services.

TABLE **10.5** Some Panel Vendors

Vendor	Panel Characteristics
A. C. Nielsen Stamford, CT (203) 961-3330 http://www.acnielsen.com	Nielsen operates ScanTrack, a weekly data service on packaged-goods sales, market share, and retail prices from 4800 super-markets, and the ScanTrack National Electronic Household Panel, a data service based on 119,000 households in 15 countries that collects packaged-goods purchase data via in-home scanner. In other countries throughout the world, Neilsen has ScanTrack, Electronic Household, and other panel services available.
Decision Analyst, Inc. 604 Avenue H East Arlington, TX 76011-3100 (800) 262-5974 http://www.decisionanalyst.com	DA maintains a major consumer panel and numerous specialty panels. The American Consumer Opinion Panel has over 3.5 million consumers in the United States, Canada, Europe, Latin America, and Asia (representing over 150 countries).
Information Resources, Inc. Chicago, IL (312) 726-1221 http://www.infres.com	IRI provides international data services covering over 20 countries, with data delivered in consistent formats and supported by an international service team. IRI offers consumer behavior tracking through its proprietary consumer panel in the United States and through an alliance with Europanel®, the leading European panel operator.
NFO Worldwide, Inc. Greenwich, CT (203) 629-8888 http://www.nfow.com	NFO provides a 1.3 million household consumer panel, categorized by demographics and product ownership, which makes it possible to survey target samples of consumers. In addition, NFO maintains specialized panels such as the Mover Panel, the Baby Panel, and a Chronic Ailment panel.

Source: Abstracted from Jack Honomichl, "Honomichl Global Top 25," *Marketing News* (17 August 1998), *32: 17*, pp. H1–H11.

Increasingly, computerized forms of panel data collection are being used. Included here are electronic devices to measure television and radio audiences and UPC (Universal Product Code) scanning technologies, both in the home and in business establishments. A. C. Nielsen, Arbitron, and Information Resources all use some combination of these technologies. Table 10.5 presents four panel vendors and a brief description of their panels.

Although panel vendors are discussed here in the context of primary data collection, these organizations generally combine some common database with unique information collected specifically for the client. For example, Arbitron uses PRIZM®, a model of U.S. consumer behavior based on census demographics, and a statistical technique known as *clustering,* to help marketers identify groups of consumers who have the highest potential for their particular products. Similarly, NFO uses an ongoing database, categorized by demographics and product ownership, to help researchers pinpoint samples with low incidence rates.

Panel vendors, like other research suppliers, offer a wide variety of services. If used judiciously with existing in-house research capabilities, these organizations can provide useful information to the decision maker. We now turn to managerial considerations in the collection and use of primary data.

MANAGERIAL CONSIDERATIONS

Managers should be concerned with the collection of primary data, given the large resource commitments usually required to collect such data. Specifically, the manager needs to be involved with the following major considerations.

Develop a clear understanding of the need for the data and how it will be used before PDC expenditures are authorized. Far too often, primary data is collected that looks good on paper but that has very little relevance for decision-making purposes. Therefore, before any primary data collection is authorized, there should be a thorough justification of the procedures and expenditures, so that the data collected will be in a format that will be useful for decision making. To this end, managers must develop an adequate understanding of what research can and cannot do. This is particularly true in the case of international research. Significant limiting factors such as religious concerns, lack of a reliable mail system, and inadequate telephone systems sometimes preclude the use of a specific data collection method. Managers should ask for reasons why the selected PDC method is appropriate given the nature of the research environment. Exhibit 10.4 outlines some of the factors that should be considered in this analysis.

Develop control procedures to ensure that the PDC process goes as planned. PDC is a process that must be planned, executed, and controlled. The manager should expect periodic progress reports and even conduct independent data checks if the situation so dictates.

If using external research suppliers, check references and develop open lines of communication to facilitate the proper conduct of the portion(s) of the study contracted. When contracting for research services, the manager should be concerned with the background of the supplier. There are literally thousands of research vendors available to the average firm. The references and research capabilities of these organizations should be checked before a decision is made to contract with a particular vendor. Once a contract is let to a vendor, managers should open the lines of communication with the supplier to facilitate the proper conduct of the study. Written and oral communication should be initiated by the manager showing the course of the investigation.[40]

EXHIBIT **10.4**

International Considerations in the Selection of a Primary Data Collection Method

Source: Adapted from Susan P. Douglas and C. Samuel Craig, "Marketing Research in the International Environment," in Ingo Walter and Tracy Murray, Eds., *Handbook of International Management* (New York: Wiley, 1988), pp. 5.28–5.32. Reprinted by permission of John Wiley & Sons, Inc.

Active vs. passive.	The determination of whether to use active or passive methods is largely driven by the objectives of the research and the nature of the investigation. Regardless of what methods are used, the manager should be aware of cultural bias that may result during the collection of primary data. Ethnocentric interpretations of multinational data often lead to incorrect, or at least misleading, conclusions. Generally, it is desirable to have multiple interpreters from different cultural and linguistic backgrounds to minimize any cultural bias in the data collection process.
Personal interviews.	In many developing countries, personal interviewing is the primary means of data collection. This is largely due to illiteracy rates, cultural unwillingness to respond to other types of inquiry, and concerns about the invasion of privacy. Personal interviewers must be adequately trained to deal with the nuances of the culture in which the research is going to be conducted.
Mail interviews.	Mail surveys can often be successfully used in developed nations, where literacy rates are high and market information is generally available. In less developed countries, high illiteracy, poor market information (no available listing of addresses), and inefficient mail systems can make mail surveys impractical. For example, in several Latin American countries, it is estimated that a substantial portion of domestic mail is never delivered.
Telephone interviews.	The practicality of conducting telephone interviews varies substantially from country to country. For example, in Egypt and Morocco, there are less than 2 telephones per 100 inhabitants, and in Chile and Brazil, there are 6 and 8 telephones per 100 inhabitants, respectively. On the other hand, telephone interviewing has been used successfully in Europe to track consumer acceptance and satisfaction with automobiles.
Online interviews.	The level of computer technology varies widely from country to country and from culture to culture. How this affects the use of online interviews and computer-assisted interviews should be thoroughly explored before choosing this type of interview procedure.

Evaluate primary data before it is used for decision-making purposes. Just as with secondary data, primary data must be evaluated before it is used in decision making. In essence, the criteria are the same. The data must be (1) timely, (2) relevant, and (3) accurate.

The astute manager knows the difficulties of verifying the quality and value of primary data collected through the research process. It is important for the manager to keep abreast of the research procedures used in any

1. Has the need for primary data collection been established before the research endeavor begins?
2. What major method(s) of data collection should be used to obtain the data needed to solve the problem?
3. Should the research project (or portions thereof) be contracted out to primary data collection vendors? Why or why not?
4. What control procedures have been instituted to ensure that the primary data is collected according to the research plan?
5. Is the primary data collected being evaluated before it is used for decision-making purposes?

study. The key managerial questions are summarized in Exhibit 10.5 as a reminder to managers about primary data collection during the research process.

SUMMARY

Primary data is collected for a specific purpose from original sources. Primary data is usually collected only when it has been found that secondary data is inadequate to answer the research problem. The judicious collection of this type of data can provide important information for decision-making purposes. The nature and procedural aspects of primary data collection are reviewed in this chapter.

There are two main methods of collecting primary data: passive and active. Passive methods involve the observation of characteristics, by human or nonpersonal means, of the elements under study. Active methods involve the questioning of respondents by personal or nonpersonal means. Each method can be subdivided on three important dimensions: (1) the degree of disguise inherent in the method, (2) the degree of structure, and (3) the method of administration.

A distinction is made between quantitative and qualitative research. Quantitative research typically uses large samples, involves structured survey questioning of some type, and is subsequently numerically and statistically analyzed. This type of active PDC is the focus of the present chapter. Qualitative research is the primary focus of the next chapter.

The discussion moves to the make-or-buy decision. The focus is on describing the alternatives important in this decision and the factors that influence the ultimate selection of a specific alternative. Various PDC vendors are described. Particular attention is given to panel vendors, because of the increasing importance of this data collection method in research. The chapter ends with the managerial considerations in primary data collection.

DISCUSSION QUESTIONS

1. Primary data is generated for a specific research purpose—it is data needed to solve a business research problem. Elaborate on the different types of primary data we could collect for business research purposes. Can you develop a classification scheme that relates type to purpose or use (demographic data, socioeconomic data, and so on)?

2. Compare and contrast active and passive primary data collection.

3. What is the major reason that business research depends heavily on active primary data collection? What problems does this reason pose for a business researcher?

4. The versatility of the active communication method is readily apparent. Also, the accuracy of the passive observational method is obvious when it can be employed. Discuss the rationale behind these statements.

5. The personal interviewing technique of collecting primary data is a very complex technique because it involves a sensitive interactive relationship between an interviewer and a respondent. What kinds of problems can arise from this interaction that could hinder your data collection efforts?

6. Compare and contrast response rate and response accuracy of the personal interview, the telephone interview, and the mail survey.

7. You must determine the effect of secretaries' interaction with one another on their productivity during the average workday. Can you devise an active and passive approach to this data collection problem?

8. Nonresponse bias has a very detrimental effect on research efforts. Describe it, and discuss ways to prevent and measure its effects on active primary data collection.

9. Discuss the components of a make-or-buy research decision in the firm.

10. Discuss some of the major considerations in the decision of what primary data collection procedure should be used in international research. Identify significant factors that can influence the decision and describe how they impact the data collection process.

11. Using http://www.amazon.com or http://www.barnesnoble.com, go in under the subject search and identify some contemporary books that deal with business survey research. Pick one of these books and provide a brief synopsis of the book. Relate the contents to the present chapter.

12. Using any sources you wish, identify some software suppliers for computer-aided data collection and provide an overview of the offered software's capabilities. (Use the Web to effectively answer this question.)

NOTES

1. George H. Smith, *Motivation Research in Advertising and Marketing* (New York: McGraw-Hill, 1971); reproduction of 1954.

2. For example, ethical problems arise because individuals as principal study units have the right not to participate in a research project or be observed if they do not want to. This right falls under the respondent's basic rights (1) to know and (2) to privacy. These subjects will be treated in more detail in Chapter 16.

3. See, for example, Linda Piekarski, Gwen Kaplan, and Jessica Prestegaard, "Telephony and Telephone Sampling: The Dynamics of Change," May 20, 1999, downloaded from http://www.worldopinion.com/news?cm=item&id=3966, November 11, 2003; Peter Tuckel, "The Vanishing Respondent in Telephone Surveys," May 17, 2001, downloaded from http://www.worldopinion.com/reference.taf?f=refi&id=1477, November 11, 2001; and Insight Express, "Proven: Online Research's Time has Come," May 2002, downloaded from http://www.insightexpress.com/proven/wp_register.asp., November 16, 2003.

4. This definition is adapted from Charles Cannell and Robert Kahn, in Gardener Lindzey and Elliot Aronson, Eds., *The Handbook of Social Psychology,* 2nd ed. (Reading, MA: Addison-Wesley, 1968), p. 562.

5. See, for example, W. W. Daniel, "Non-Response in Sociological Surveys: A Review of Some Methods for Handling the Problem," *Sociological Methods and Research* (February 1975), *3,* pp. 291–307; Joseph A. Bellizzi and Robert E. Hite, "Face-to-Face Advance Contact and Monetary Incentives on Mail Survey Return Rates, Response Differences, and Survey Costs," *Journal of Business Research* (February 1986), p. 99; and M. Chris Paxton, "Increasing Survey Response Rates," *Hotel and Restaurant Administration Quarterly* (August 1995), *36:* 4, pp. 66–73.

6. William C. Dunkelberg and George S. Day, "Nonresponse Bias and Callbacks in Sample Surveys," *Journal of Marketing Research* (May 1973), *10,* pp. 160–168; and David Jobber and Daragh O'Reilly, "Industrial Mail Surveys: A Methodological Update," *Industrial Marketing Management* (1998), *27,* pp. 95–107.

7. See Duane Alwin, "Making Errors in Surveys," *Sociological Methods and Research* (November 1977), *6,* p. 136; and Cannell and Kahn, op. cit. note 4, p. 578.

8. B. Weeks et al., "Optimal Times to Contact Sample Households," *Public Opinion Quarterly,* (1980), *44,* pp. 101–114.

9. See, for example, Charles S. Mayer, "Data Collection Methods: Personal Interviews," in Robert Ferber, Ed., *Handbook of Marketing Research* (New York: McGraw-Hill, 1974), pp. 2–82, 2-89; and Lee Andrews, "Interviewers: Recruiting, Selecting, Training and Supervising," in Robert Ferber, Ed., *Handbook of Marketing Research* (New York: McGraw-Hill, 1974), pp. 2-124–2-132.

10. For a review of nonresponse estimation procedures that can be applied to personal interviewing, see D. Hawkins, "Estimation of Non-Response Bias," *Sociological Methods and Research* (May 1975), *3,* pp. 461–488; and J. S. Armstrong and T. S. Overton, "Estimating Non-Response Bias in Mail Surveys," *Journal of Marketing Research* (August 1977), pp. 396–402.

11. This framework was adapted from S. Sudman and N. Bradburn, *Response Effects in Surveys* (Chicago: Aldine, 1974).

12. Ibid., pp. 109–117.

13. See S. Hatchett and H. Schuman, "White Respondents and Race of Interviewer Effects," *Public Opinion Quarterly* (Winter 1975–76), *39,* pp. 523–528; and H. Schuman and R. M. Converse, "The Effects of Black and White Interviewers in White Respondents in 1968," *Public Opinion Quarterly* (Spring 1971), *35,* pp. 44–68.

14. For example, see S. L. Payne, *The Art of Asking Questions* (Princeton, NJ: Princeton University Press, 1951; paper 1980).

15. S. Wiseman, "Methodology Bias in Public Opinion Surveys," *Public Opinion Quarterly* (Spring 1972), *36,* pp. 105–108.

16. Sudman and Bradburn, op. cit. note 11, p. 36.

17. See Andrews, op. cit. note 9, pp. 2-124–2-132, for a good discussion of the specifics of these aspects of personal interviewing.

18. See P. B. Case, "How to Catch Interviewer Errors," *Journal of Advertising Research* (1971), *11,* pp. 39–43.

19. This framework is adapted from Thomas J. Bouchard, Jr., "Field Research Methods: Interviewing, Questionnaires, Participant Observation, Unobtrusive Measures," in Marvin Dunnette, Ed., *Handbook of Industrial and Organizational Psychology* (Chicago: Rand McNally, 1976), pp. 367–370; and Cannell and Kahn, op. cit. note 4, pp. 535–540.

20. See Cannell and Kahn, op. cit. note 4, pp. 538–539.

21. See Theresa Rogers, "Interviews by Telephone and in Person: Quality of Responses and Field Responses and Field Performance," *Public Opinion Quarterly* (Spring 1976), *40,* pp. 51–65; John J. Brock, "Phone Interviews Equal In-person Methodology," *Marketing News* (7 January 1986), p. 29; and Richard Lysaker, "Data Collection Methods in the U.S.," *Journal of the Market Research Society* (October 1989), *31,* pp. 477–488.

22. Robert Morton and David Duncan, "A New Look at Telephone Interviewing Methodology," *Pacific Sociological Review* (July 1978), *21,* pp. 259–273; and Vivian J. Biscomb, "Detailed Planning Makes Telepolling Effective," *Marketing News* (3 January 1986), p. 5.

23. See "Working Block Density Declines," *The Frame* (September 1996; Survey Sampling, Inc., Fairfield, CT); and "Lockheed Martin Takes Over North American Numbering Plan," *The Frame* (February 1998; Survey Sampling, Inc., Fairfield, CT); and visit http://www.worldopinion.com.

24. Michael J. O'Neil, "Estimating the Nonresponse Bias Due to Refusals in Telephone Surveys," *Public Opinion Quarterly* (1979), *43,* pp. 218–232.

25. See, for example, William Lyons and Robert Durant, "Interviewer Costs Associated with the Use of Random Digit Dialing in Large Area Samples," *Journal of Marketing* (Summer 1980), *44,* pp. 65–69; K. Michael Cummings, "Random Digit Dialing: A Sampling Technique for Telephone Surveys," *Public Opinion Quarterly* (1979), *43,* pp. 232–244; Joseph Walisberg, "Sampling Methods for Random Digit Dialing," *Journal of the American Statistical Association* (March 1978), *73,* pp. 40–46; and Robert Groves, "An Empirical Comparison of Two Telephone Sampling Designs," *Journal of Marketing Research* (November 1978), *15,* pp. 622–632.

26. See Lysaker, op. cit. note 21, p. 486; and Charlotte Steeh, "Trends in Nonresponse Rates 1952–1979," *Public Opinion Quarterly* (Spring 1981), *45,* pp. 40–57, for discussions of these issues.

27. See W. Jack Duncan, "Mail Questionnaires in Survey Research: A Review of Response Inducement Techniques," *Journal of Management* (1979), *5,* pp. 39–55; M. Chris Paxson, "Follow-Up Mail Surveys," *Industrial Marketing Management* (August 1992), *21,* pp. 195–201; A. Diamantopopoulos, Bodo B. Schlegelmilch, and Lori Webb, "Factors Affecting Industrial Mail Response Rates," *Industrial Marketing Management* (November 1991), *20,* pp. 327–339; and Jobber and O'Reilly, op. cit. note 6.

28. See Robert Ferber, "The Problem of Bias in Mail Returns: A Solution," *Public Opinion Quarterly* (Winter 1948–49), pp. 668–676.

29. See Peter DePaulo and Rick Weitzer, "Interactive Phone Technology Delivers Survey Data Quickly," *Marketing News* (6 June 1994), *28,* pp. H33–H34; and "1998 Marketing Technology Directory: Software/Internet Services," *Marketing News* (13 April 1998).

30. "Multimedia Interviewing Makes Advertising Research A Breeze," and "Anheuser-Busch Services Up Survey with a Twist," [http://www.sawtooth.com] (7 October 1998).

31. See Council for American Survey Research Organizations, "New Methodologies for Traditional Techniques," copyright 1998, downloaded from http://www.casro.org/faq.cfm, October, 22, 2003; WebSurveyor, Comparison of Traditional vs. Online Survey Methods," copyright 2003, downloaded from http://www.websurveyor.com/pdf/webvsmail.pdf, November 18, 2003, and Insight Express, op. cit. note 3.

32. Insight Express, "Pfizer: Schick," downloaded from http://www.insightexpress.com/success/schick.asp, November 18, 2003.

33. Perseus, "Customer Testimonials," downloaded from http://www.perseusdevelopment.com/corporate/testimonials.html, November 18, 2003.

34. Council of American Survey Research Organization, op. cit. note 31.

35. WebSurveyor, op. cit. note 31.

36. "The Virtual Store: A New Tool for Consumer Research," *Retailing Review* (Summer 1994), pp. RR1–RR2.

37. For directories listing these vendors, see *1999 Marketing Yellow Pages* (Chicago: American Marketing Association, 1999); and Paul Wasserman and Janet McLeon, Eds., *Consultants and Consulting Organizations Directory* (Detroit: Gale Research), updated periodically.

38. See Thomas T. Semon, "Red Tape is Chief Problem in Multinational Research," *Marketing News* (14 February 1994), *28,* p. 7; Cyndee Miller, "China Emerges as Latest Battle Ground for Marketing Researchers," *Marketing News* (14 February 1994), *28,* pp. 1–2; and Naghi Namakforoosh, "Data Collection Methods Hold Key to Research in Mexico," *Marketing News* (29 August 1994), *28,* p. 28.

39. For a discussion of panels, see Seymour Sudman and Robert Ferber, *Consumer Panels* (Chicago: American Marketing Association, 1979).

40. See Robert Peterson and Roger Kerin, "The Effective Use of Marketing Research Consultants," *Industrial Marketing Management* (February 1980), pp. 69–73; and James Rothman, "Acceptance Checks for Ensuring Quality in Research," *Journal of the Market Research Society* (July 1980), pp. 192–204, for discussion on the use and evaluation of research suppliers.

SUGGESTED READING

The Nature of Primary Data Collection (PDC)

Dillman, Don A., *Mail and Internet Surveys*, 2nd ed. (New York: John Wiley, 2000).

This book presents the Tailored Design Method for the design and application of self-administered surveys in a changing environment. A good read for those concerned with the issues of mail and Internet-based surveys.

Rea, Louis, M., Richard A. Parker, Alan Shrader, Eds., *Designing and Conducting Survey Research: A Comprehensive Guide*, 2nd ed. (New York: Jossey-Bass Public Administration Series, 1997).

Overviews the important aspects of designing and conducting survey research. Also discusses how focus groups can be used with survey research.

Active PDC Using Survey Instruments

Alreck, Pamela, L., and Robert B. Settle, *The Survey Research Handbook,* 2nd ed. (Burr Ridge, IL: Irwin Publishing, 1997).

In down-to-earth language, the handbook provides methods and guidelines for conducting practical surveys.

Fowler, Floyd J., *Survey Research Methods,* 3rd ed. (Thousand Oaks, CA: Sage, 2002).

A widely accepted basic survey research methods text for anyone who collects, analyzes, or examines survey data.

PDC Vendors and Panel Vendors

For the most up to date listings of vendors, go to: http:// www.greenbook.org—New York AMA Communication Services *Green Book, Vol. 1* http://www. marketingpower.com/live/directory-display.php— American Marketing Association's *The M Guide— Marketing Services Directory.*

11

PDC Using Observation, In-Depth Interviews, and Other Qualitative Techniques

Overview

Nature and Uses
of Qualitative Research

Passive PDC Using Observation

Active PDC Using Qualitative
Research Techniques

Individual In-Depth Interviews

Focus Groups

Other Qualitative Research
Techniques

Qualitative Research Vendors

Managerial Considerations

Summary

Discussion Questions

Notes

Suggested Reading

Whereas the last chapter focused on quantitative research using survey instruments, the present chapter deals with primary data collection using qualitative research methods. As noted earlier, *qualitative research* deals with matters that cannot be meaningful quantified. These types of studies generally focus on understanding the behavior or phenomenon of interest and are generally characterized by small nonprojectable samples and "messy" data from an analytical standpoint.

Because of the nature of these studies, they are often viewed as *soft* research, implying to many that the research has little value in the business setting. Quite to the contrary, however, qualitative research is an integral part of the researcher's arsenal of tools to answer management's questions.

The chapter begins with a discussion of the nature and uses of qualitative research. The differences between qualitative and quantitative research are further highlighted, as are the complementary natures of the two data collection methodologies.

Both passive and active methods of data collection are used in this type of research. Passive PDC using observation and active PDC using focus groups and other qualitative research techniques are possible. These two approaches are frequently used in conjunction with the formation and/or interpretation of the results of quantitative studies discussed earlier.

Once the primary methods of qualitative research are introduced, the discussion switches to the services of qualitative research vendors and developments in the field. Here, technology is playing a major role in changing how qualitative research studies are being conducted throughout the world.

The chapter ends with a discussion of the implications of these changes and the major managerial considerations in the conduct of qualitative research.

NATURE AND USES OF QUALITATIVE RESEARCH

There are significant differences between qualitative and quantitative research. Each type of research and data collection mechanism has its place in the business research environment. Table 11.1 highlights the nature and potential applications of each. The major differences have to do with the substance and approach to the problem at hand. Generally, purely quantitative studies are used to measure specific characteristics through structured data collection procedures, often with large samples (more than 100 observations), so that the results can be projected to the entire population

TABLE 11.1 Comparison of Quantitative and Qualitative Research

Area	Quantitative	Qualitative
Objective	Quantification of characteristics or behavior	In-depth understanding of characteristics or behavior
Approach	Structured	Largely unstructured
Sample size	Large	Small (fewer than 12)
Representativeness to population	Yes, if random	No
Interviewer skill required	Moderate to low	High
Length of interview	Relatively short (generally less than 30 minutes)	Longer (more than 30 minutes)

(assuming random samples). Interviewer skill is usually moderate to low (mail samples and online interviewing require little, if any, interviewer intervention), and the interview is relatively short (30 minutes). As noted in Chapter 6, surveys are common in such investigations and generally involve higher levels of understanding on the part of the researcher.

Conversely, qualitative studies are usually in-depth investigations of an unstructured nature, using a very limited sample (8 to 12 individuals are usually sought in focus groups). The analyses make no attempt to be representative of the entire population; they are frequently meant to be impressionistic rather than definitively analytical. The goal is to understand the nature of the phenomenon in a holistic sense, rather than to numerically measure and analyze a predetermined set of variables. This requires a high degree of skill on the interviewer's part and usually involves long interview sessions.

Qualitative research should not be viewed as a competitor but rather as a complement to quantitative research.[1] However, the distinction between the two is always not clear. Some qualitative approaches can be quantified and statistically analyzed, whereas open-ended questions on quantitative survey instruments often produce soft data that must be interpreted judiciously. This situation is further confused because (passive) observational and (active) querying data collection can (and often should) be used side by side in the same study. Generally though, passive PDC using observation techniques and active PDC using focus groups or some other small-sample data collection technique is considered a qualitative study.

Although the nature of qualitative research makes it ideal in exploratory stages of the research process because of the generally low degree of understanding of the problem situation, it has also been used successfully in all stages of the problem understanding process. For example, when Atlantis Submarines International conducted a quantitative product concept test of a new underwater-themed attraction for market potential assessment, focus groups were used subsequent to the study to clarify issues identified from the survey instrument. Similarly, Psychological Motivations, Inc., specializes in using two forms of qualitative research techniques to help such companies as Monsanto, FMC Corporation, and Shell Oil & Chemical better understand the real reasons people behave or think as they do (see Exhibit 11.1).

Furthermore, the preeminence and use of qualitative research techniques varies by culture and researcher preference. An article in the *Harvard Business Review* notes that the Japanese rely on data collected from qualitative research endeavors much more than their American counterparts.[2] However, the acceptance and use of these techniques throughout the world has grown recently and predictions for increased use abound.[3] Reasons for this growth appear to center around the versatility of the techniques themselves and changes in technology. Qualitative research has been found to be useful in studying the development of new products and advertisements; to understand the nuances of consumer service in health care settings; to identify problem areas within profit and nonprofit organizations; and to investigate behaviors that are considered to be sensitive and socially undesirable, such as compulsive shopping and shoplifting.[4] Essentially, the application of this type of research is almost limitless.

PASSIVE PDC USING OBSERVATION

We are all continually observing what goes on around us. It is perhaps one of the most pervasive activities in our daily lives and forms the basis for many of our beliefs and actions. However, the type of observation we are concerned with differs from everyday observation. In passive data collection, the observation meets the following criteria:[5]

1. It serves a research purpose.
2. It is systematically collected and recorded.
3. It is subjected to checks and controls on validity and reliability.

Passive methods of data collection using qualitative research, although not the predominant method of collecting data in contemporary business research, can serve a wide variety of research purposes.[6] They may be used in an exploratory fashion to gain insights that may be tested later with active data collection methods. They may be used in conjunction with active techniques to develop a complete picture of some situation. Or they may even be used as the primary data collection device to describe a situation or test specific hypotheses formulated by the researchers. For example, in developing its Fairfield Inns, Marriott sent researchers into the field to observe the strengths and weaknesses of its competitors' lodging designs and then used the information to create the physical environment of its offerings.[7] The precise use of passive data collection, and thus observation, depends on the nature of the research undertaking and the individual preferences of the researchers.

Passive collection methods are often the only means by which certain data can be collected. In the case of inanimate objects or entities, where

questions cannot be asked of the study element, this type of data collection is the only method available. Similarly, even human beings are sometimes not willing or able to report certain activities to the researchers.

For example, people are often not willing to divulge specific information about their lifestyles or even their shopping habits to researchers. Disguised observation can then be used to collect data on an individual's shopping habits. (However, there are important ethical considerations, discussed in Chapter 16.)

Collection methods generally vary on three dimensions: the degree of disguise, the degree of structure, and the method of collection. Table 11.2 gives examples of each method in the context of a study that is designed to uncover the reactions of a respondent to a commercial for a new brand of beer. The examples in this table give a basic idea as to the differences of each approach.

Passive data collection methods have long been a primary means for collecting useful information, but they are not sufficient to generate all the types of data that are necessary in business research.[8] There are two main reasons for this. First, passive methods require that the characteristic, behavior, and so on be observed when it happens. This is not always possible in the business research setting. For example, many expenditure patterns of households have been found to be related to demographic and other life-history variables of the household. Passive collection methods are not well suited to the assessment of such variables.

The second reason passive methods often need to be supplemented by other data collection methods is that the data collected is an *interpretation* of what the observer sees. Different observers may give different meanings for the same phenomenon. This raises a question as to the validity of the judgments of the observers.

Given the limitations of passive collection methods, they are often used in conjunction with active, or communicative, types of data collection. The next section explores the nature of active qualitative data collection methods in business research.

ACTIVE PDC USING QUALITATIVE RESEARCH TECHNIQUES

The two main qualitative research techniques are individual in-depth interviews and group in-depth interviews, also known as focus groups.[9] By all accounts, in-depth interviews, particularly in the form of focus groups, are the most popular data collection methods in this type of research endeavor.

Each approach generally uses an unstructured, undisguised format and can produce extremely valuable data if conducted correctly.[10] The *in-depth interview* is an intensive, unstructured personal interrogation in which the interviewer attempts to get the respondent to talk freely about the subject

TABLE **11.2** Examples of Passive Primary Data Collection Materials

Variations in Methods	Human		Computerized	
	Structured	Unstructured	Structured	Unstructured
Disguised	Unknown to the movie patron, the observer watches the viewer and records the following: a. attentiveness to commercial b. whether or not the entire commercial was viewed	Unknown to the movie patron, the individual sitting beside her is actually an observer who is attempting to identify reactions to the commercial.	Unknown to the movie patron, the seat is wired to record movements and galvanic skin response (GSR) during the showing of the advertisement.	Unknown to the movie patron, a videotape is being made of the individual's reaction to the beer advertisement being shown at the start of the movie.
Undisguised	The observer, who describes the purpose of the study to respondents, watches the viewer and records the following: a. attentiveness to commercial b. whether or not the entire commercial was viewed	The observer, who describes the purpose of the study to respondents, views all aspects of a viewer's actions to identify reactions to the commercial.	The observer, who describes the purpose of the study to respondents, attaches a pupillometer and GSR unit to the individual viewer to monitor a. pupil reactions b. skin response	The observer, who describes the purpose of the study to respondents, videotapes individual viewers' reactions to the advertisement. The tapes are later viewed for salient reactions.

of interest. Usually, the interviewer begins the process by asking a broad question about the area of interest. When the respondent answers, the interviewer probes with additional questions designed to elicit more detailed responses.

For example, assume we are interested in what executives really think about the economic outlook. An in-depth interview might start with a broad question such as, "Do you personally feel that the present economic condition of the country will affect the outlook for the future?" Then the interviewer attempts to probe, to get the executive to express more specific opinions about the economic outlook. In-depth interviews are particularly suited for exploratory research.

A related personal interview technique is the *focus group*, a variation of the in-depth interview. It usually uses a similar questioning format, but more than one respondent participates in the interview. If we extend our example just given, a focus group would involve questioning a small group of executives instead of just one. Currently widely used in business research, focus groups are becoming more popular both nationally and internationally because this method has proved to be extremely enlightening in the early stages of research.[11] In fact, focus groups are used by every *Forbes* 500 company, including AT&T, General Motors, Xerox, and many other multinational firms.[12]

Individual In-Depth Interviews

As the name suggests, an individual in-depth interview is used for intensive probing into one person's feelings about some subject. An interview may last an hour or even longer. Usually, the interviewer does not have a set of structured or prespecified questions. Rather, the interviewer is given considerable flexibility to probe and stimulate responses from the subject. Frequently, the interview is either videotaped or audiotaped so that the researcher can review the details later. The researcher must take extreme care not to lead the interviewee during the questioning process. If this happens, the data and conclusions drawn will be biased.

Several questioning strategies have been proposed to help the interviewer obtain the best possible information.[13] These strategies are laddering, hidden issue questioning, and symbolic questioning. *Laddering* is a strategy designed to find the underlying attributes that distinguish things, people, or organizations. For example, an interviewee could be asked, "What differentiates Ford Motor Company from General Motors and Chrysler?" The interviewer would then identify salient attributes as the interviewee "ladders" from GM to Chrysler. These attributes would subsequently be probed with further questioning to identify the significance and meaning of the attributes to the image of Ford Motor Company.

Hidden issue questioning is designed to uncover an individual's feelings about sensitive issues such as employee theft, job security, or shoplifting. The questioning process would attempt to uncover common themes across various interviewees.

Finally, *symbolic questioning* can be fruitfully used to identify underlying meanings. For example, a respondent could be asked to name the opposite of yogurt. If the response is candy, the interviewer probes for the attributes of candy. Common responses might be sweet, unhealthy, fattening, and bad for you. Conversely, then, yogurt would have the symbolic meanings of not sweet, healthy, nonfattening, and good for you. Thus, contextual meaning is given to the food class "yogurt."

Although individual in-depth interviews can be very enlightening, there are significant drawbacks. The cost and time associated with one-on-one interviews can be prohibitive from a resource standpoint. In addition, the demands on the interviewer are tremendous. It does not take long for fatigue and information overload to cloud the interpretation ability of one interviewer. For this reason, focus groups are perhaps the most popular of all qualitative research methods.

Focus Groups

Much has been written about focus groups in recent years.[14] These writings cover everything from the philosophical underpinnings of the technique to how to make focus groups more productive.

A focus group usually consists of 8 to 12 interviewees, drawn from a sampling of the population under study.[15] Groups smaller than 8 tend to be too small, with little diversity and insufficient group interaction. Groups larger than 12 can be difficult to control because of sheer size, and often group interaction is not achieved.

The sample is drawn with the objectives of the study in mind. Group composition is critical because of the small number of respondents in the study, so the selection and recruitment of interviewees is all-important for the successful completion of the study.

A typical focus group lasts from 1 1/2 to 2 1/2 hours. It is becoming common to use specialized focus group facilities, with one-way mirrors for client observation, that are neutral and nonthreatening for the interviewee. Generally, the interviews are videotaped for playback and review by moderators and clients alike. Generally, the research organizations also provide field recruitment and selection services. An example of such a service, provided by Consumer Surveys of Arlington Heights, Illinois, is presented in Exhibit 11.2.

In recent years, technological changes have allowed innovations in the conduct and delivery of focus groups and in-depth interviews alike. Perhaps the major change here is in the conduct of online focus groups over the Internet. Online focus groups have been successfully used to test new products, evaluate the performance of existing ones, brainstorm hot ideas, and test traditional market research to see if any questions need to be reworded or expanded.[16]

Using online focus groups, participants can interact with each other and facilitators, while at the same time pull up video and/or graphics to

EXHIBIT **11.2**

Focus Group Services Provided by Consumer Surveys Company

Source: Advertisement from Consumer Surveys, Northpoint Shopping Center, 304 Rand Road, Suite 220, Arlington Heights, IL 60004-3147. Used with permission.

MEMO TO: All Marketing Research Professionals, Moderators and Clients

FROM: CONSUMER SURVEYS COMPANY

MESSAGE: You Deserve To Be Pampered!

You get off a cramped airplane, drive through an unfamiliar city to arrive, barely on time, at a research facility where you spend time in a cramped, too warm / too cold viewing room, craning your neck to see around a cameraman who is busily videotaping your group.

Next time, pamper yourself at our new focus facility at the Northpoint Shopping Center. Relax in our spacious client lounge. Monitor the group in the lounge, or in our tiered, airy, observation room along with your 15 agency and corporate traveling companions. You'll feel refreshed by our separate air / heating system. View your group through an insulated one-way window stretching from the desktop writing ledge to the ceiling while a cameraman records the session from a remote video system located at the rear of the room, thereby providing you an unobstructed view.

While you are comfortably watching your moderator capture every nuance, you can periodically jump up and utter those immortal words . . . "That's right! That's exactly what I've been telling you!"

Come join us in a research environment designed to provide both comfort and quality research. After all, you are worth it.

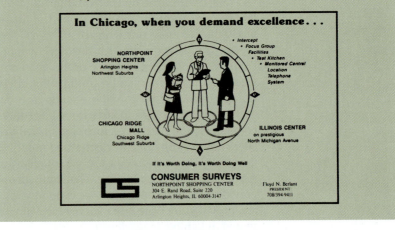

clarify their understanding of the key issues being studied. Variants of these cyberspace focus groups include Internet-based and teleconference focus groups at remote locations beamed to the client's place of business, the development of chat rooms and bulletin boards to gauge opinions on specific topics and subjects, and multiple global focus groups interacting on an idea or set of ideas.[17] One of the major reasons for the growing popularity of these online focus groups is the ability to obtain real-time information from a geographically dispersed group at a very low cost.

Research organizations such as Decision Analyst, Inc. (http://www.decisionanalyst.com) and Harris Interactive (http://www.harrisinteractive.com) provide a full array of online qualitative research services. An investigation of these Web sites will help one understand the qualitative research services currently available. In addition, if an organization has the capability of conducting qualitative research, it can use standard Web-based services such as Microsoft's live meeting (http://www.my.placeware.com) to collect self-service qualitative research data online.

The moderator is essential to the successful execution of a focus group interview. A good moderator is usually highly experienced in group dynamics and personal interaction. A moderator should neither lead nor lose control of the respondents in the focus group; he or she stays in charge of what is happening in the group while maintaining an objective view so that personal bias does not affect communication among the participants.

The moderator provides the structure for the focus group participants. More structure means less spontaneity from the participants, but their comments are more directed to the researcher's goals. Less structure means more opportunity for respondents to speak, but what they say is not necessarily of particular interest to the study's objectives. What is the ideal structure? It depends on the objectives agreed upon between the moderator and the client before the focus group is conducted.

All in all, focus groups have great potential to be used and abused in contemporary business research. This data collection method is extremely flexible, can produce insights other collection methodologies cannot, and tends to be more readily accepted by managers than many quantitative approaches because managers are frequently more involved in the focus group research.

On the other hand, serious abuses have occurred with focus groups. Inadequate involvement of management in the research process and the lack of generalizability can spell disaster for the unwary. For example, Bic Corporation apparently spent thousands of dollars in focus groups and name-generation research to name a new line of pens aimed at students.[18] The name that they arrived at was Spaz, and the product was to be launched in the fall of 1994. A concerned mother of a 14-year-old boy contacted Bic and said she was upset over the use of the name Spaz because it was a derogatory slang term often aimed at people with disabilities. After all the research, the company failed to pick up on the negative connotations of the name. Bic has subsequently dropped the Spaz line.

Similarly, Gerry Linda, a principal in the Kurtzman/Slavin/Linda ad agency in Northfield, Illinois, recalled a client who had him run a focus group.[19] The focus group was extremely successful in that case, and the manager was sold on the accuracy of the method. Not long after, when another problem arose, the client wanted to do another focus group because of the success of the previous one. However, a focus group was not appropriate for that particular problem. Just because it worked in one research setting does not necessarily mean it will work again.

Other Qualitative Research Techniques

Although focus groups and in-depth interviews are the most widely used qualitative research techniques in business, there are other techniques that have very specialized applications and are used when the problem situation so dictates.[20] Some of these techniques include protocol analysis and projective techniques.

Protocol techniques place a person (or persons) in a decision-making situation and then have the interviewer ask the subject to verbalize the important considerations in the decision-making process. The idea of these types of techniques is to tap into the underlying aspects important to the person(s) in making a purchase decision.

Projective techniques involve placing study subjects into simulated situations and then asking them to describe or explain what is happening in the situation. One type of projective technique is word association. *Word association* techniques suggest a word to respondents, then ask them to provide the words that are most appropriate given the stimulus word. For example, if I gave the stimulus BMW, the respondent may associate the words *German, dependable, fast*, and *stylish* with the automobile brand. Similarly, *sentence completion, role playing*, and *picture tests* are all versions of projective techniques that have been used in business research.

Again, these are not widely used techniques, but they do have very specific applications and can be helpful in understanding a particular problem situation. The reader is encouraged to do further reading on these techniques if the need arises.

QUALITATIVE RESEARCH VENDORS

As noted, the use and growth of qualitative research throughout the world is growing. The demand for this type of research by clients has fueled the development of a host of specialized qualitative research suppliers and has reinvigorated the attention paid to this type of research by the more traditional full-service research suppliers. The services provided by these suppliers are quite impressive indeed.

For example, Figure 11.1 presents an ad that outlines M.O.R.-PACE, Inc., focus group services and facilities. Their services include state-of-the-art interviewing facilities, with spacious viewing rooms, floor and ceiling observation mirrors, wet bar, refrigerator, and private office. Other services and facilities include video conferencing with quality video, multiple focus group locations, and moderating and recruiting support.

Table 11.3 presents a sampling of other qualitative research vendors and a brief description of their services. As shown in the table, the variety and sophistication of the services offered vary by supplier. Generally, however, the trends one sees in supplier offerings include more reliance

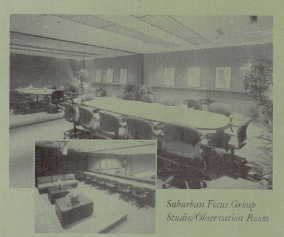

FIGURE 11.1

Ad Depicting M.O.R.-PACE Focus Group Services and Facilities.
Reprinted by permission of M.O.R.-PACE, Inc., Farmington Hills, MI 48334.

TABLE 11.3 A Sampling of Qualitative Research Vendors

Vendor	Description
Decision Analyst, Inc. 604 Avenue H East Arlington, TX 76011-3100 Phone: (817) 640-6166 Fax: (817) 640-6567 http://www.decisionanalyst.com	DA conducts both face-to-face and online qualitative research worldwide. Through the use of Internet panels, online qualitative research can be conducted across time zones in multiple countries and in multiple languages.
Lieberman Research Worldwide 1900 Avenue of the Stars Los Angeles, CA 90067 Phone: (310) 553-0550 Fax: (310) 553-4607 http://www.lrwonline.com	Lieberman is an industry leader in qualitative research. It conducts more than 700 focus groups a year and a substantial volume of in-depth interviews, anthropological in-home studies, and other types of qualitative research.
Harris Interactive 135 Corporate Woods Rochester, NY 14623-1457 Phone: (585) 272-8400 http://www.harrisinteractive.com	Harris Interactive provides both personal and online focus group technology for clients throughout the world. Utilizes chat room and bulletin-board-style qualitative research using its professionally trained moderators or yours.
Focus Central 162 Fifth Avenue New York, NY, 10010 Tel: (212) 989-2760 Fax: (212) 647-7659 http://www.focuscentral.com	Focus Central developed *Professional Qualitative Support Service (Pro-Qual),* which provides field and project management services. It specializes in coordinating qualitative research projects, including reserving facilities around the country, designing screeners, recruiting respondents, managing invoicing and collections, and other related services.

on technology-delivered services (more Internet, teleconferencing, and interlinking sites), more globalization (more international sites with increased usage of multilingual moderators and translators), and an overall increase in the quality and sophistication of the services offered to their research clients.

MANAGERIAL CONSIDERATIONS

When undertaking qualitative research, managers must be aware of some very specific concerns in order to maximize the information from the study. As noted throughout this text, the manager must become active in the specification and evaluation process if usable decision-making information is to be obtained from the investigation. The following six major concerns should help the manager maximize the information gained from any qualitative investigation.

Clearly state the research objectives of the study. As with any investigation, clearly defined research objectives are a must. This is doubly so with qualitative research because of its ill-defined and unstructured nature.

Without a clear understanding of what is to be achieved from the investigation, it is highly unlikely management will ever be satisfied with the output of the research process.

Ensure the qualitative research technique used is the most appropriate given the objectives of the study. As noted in this chapter, a variety of techniques are available to researchers. Each technique has its advantages and disadvantages and can be structured in any number of ways. Often researchers can fall into the trap of conducting a particular type of study because it "worked in the past." Managers should always require sound justification for the technique being proposed in light of the research objectives.

If quantitative research is also being undertaken, ensure the two types of studies are complementary. Qualitative research is frequently undertaken in conjunction with quantitative investigations. This combination of research methodologies can be very powerful if the roles of both types of studies are clearly delineated and the two types are made to be complementary. For example, focus groups can be used to develop measurement instruments for further refinement in quantitative studies. Similarly, in-depth interviews can be used after an extensive survey to dig into the why of the results obtained from an extensive survey. The upshot is: Ensure both managers and researchers alike understand exactly what it is they are getting from the proposed investigation.

Investigate the investigators. High-quality quantitative research cannot be conducted by just anyone. There is a significant learning curve involved in conducting and subsequently interpreting the results of *soft data* from qualitative studies. Check the references, experience, and capabilities of the moderators, interviewers, and facilitators that are responsible for conducting the investigation. Without competent and capable study personnel, the quality of the information will be "iffy" at best.

If international or multicultural data collection is needed, be sure that the aspects of culture, language, and customs are taken into consideration in the design and conduct of the study. Culture, language, religion, and customs are extremely important in the conduct of qualitative research because these aspects have much more opportunity to directly influence the conduct and interpretation of a study. In many cultures, nuances in pitch and inflection, body language, and cultural mores play an important role in fully interpreting the meaning of actions and/or responses.

For example, in several Asian countries, unlike Latin America and the Mediterranean, it is acceptable to put men and women on the same panel, depending upon the woman's role as decision influencer.[21] In Malaysia, women have more influence than men do when it comes to appliances, electronics, and financial services. The man's role is to accommodate the woman's decision and pay for the product.

EXHIBIT **11.3**

1. What are the objectives of the qualitative study being undertaken?
2. Is the qualitative research technique the most appropriate one given the objectives of the study?
3. If quantitative research is also being undertaken, what role will the quantitative research play in maximizing the results of the survey research?
4. Are the quantitative research suppliers experienced and knowledgeable in the conduct and interpretation of the results of the study?
5. If international and/or multicultural data collection is needed, have the aspects of culture, language, and customs been taken into consideration in the design and conduct of the study?
6. Has adequate planning and development of contingencies gone into the design of the qualitative research before any data collection takes place?

Not only is the makeup of the study panel affected by culture, but the size of panel is influenced too. For example, in Asia it is very difficult to conduct one-on-one interviews. This situation is largely a result of the cultural propensity for privacy and the simultaneous reluctance to go anywhere "strange" alone. Here, dual-depth interviews would be appropriate because it uses two respondents and provides a culturally and emotionally safe environment that is socially acceptable.

Ensure that adequate planning has gone into the design of the qualitative research undertaking before any data collection takes place. The effort put forth for advanced planning always pays in terms of the overall quality of the completed study.[22] Here, have the objectives between the client and researcher been thoroughly discussed and agreed upon? Are the facilities, recruitment procedures, and processes for data collection in place and assigned to the appropriate personnel? Has the content and flow of the actual data collection process been scrutinized with the appropriate contingencies made for "surprises" in the conduct of the study? These and other operational concerns must be dealt with in planning stages of the investigation in order to maximize the worth of the investigation.

These managerial considerations are summarized in Exhibit 11.3 in a series of six questions. These questions will help ensure the results obtained from qualitative studies will produce significant information for decision-making purposes.

SUMMARY

There are significant differences between qualitative and quantitative research methods. Generally, quantitative studies are used to measure specific characteristics using structured data collection procedures, often with large samples, so that the results can be projected to the entire population.

Conversely, qualitative studies are usually in-depth investigations of an unstructured nature, using a very limited sample, with no attempt made to be representative of the population. The methods should be viewed as complementary rather than competitive in nature. Each approach has its unique contribution to make to the field of business research.

Qualitative research can use both passive and active forms of data collection. Passive forms are observational in nature. Often observation is the only means to collect the data to solve a particular research problem. Observational methods can vary by the degree of disguise, the degree of structure, and the method of collection. To be scientific, observation must serve a research purpose, be systematically collected and recorded, and be subjected to checks and controls on validity and reliability. Other forms of data collection frequently supplement observation in order to solve a particular research problem.

Active forms of PDC in qualitative research include individual in-depth interviews, focus groups, and other forms of data collection techniques. Two other forms of qualitative research techniques include protocol techniques and projective techniques. Focus groups and in-depth interviews are perhaps the most widely used PDC devices using qualitative research methods. In-depth interviews are intensive, unstructured personal interrogation in which the interviewer attempts to get the respondent to talk freely about the subject of interest. Focus groups are a variation of the in-depth interview, but there is more than one participant in the interview. Technological innovations and the globalization of business have changed both the delivery and consumption of the information derived from these research techniques.

The discussion then turned to the nature of the services and facilities provided by qualitative research vendors. A sampling of vendors and their services are presented. The chapter ends with the key managerial questions in the conduct of qualitative research.

DISCUSSION QUESTIONS

1. Discuss the differences between quantitative and qualitative research. Is one type of research inherently better than the other? Why or why not?

2. What is the difference between passive and active data collection as it relates to qualitative research? Generally, when are passive methods used and when are active methods used?

3. What is an individual in-depth interview? What are its primary strengths and weaknesses?

4. What is a focus group? What advantages do focus groups have over in-depth interviews?

5. List and discuss three major questioning strategies used in individual in-depth interviews.

6. How is technology affecting the conduct and worth of focus groups? Discuss some of the implications of these changes for both managers and researchers alike.

7. What problems/opportunities exist in the field of qualitative research for researchers in a global marketplace? Provide examples of these problems/opportunities.

8. What are some of the concerns managers should have in the conduct of any qualitative research? Be specific.

9. Using the Internet, conduct a search for qualitative research vendors (like those in Table 11.3). Identify a vendor and collect information on both its capabilities and facilities. Be prepared to discuss this vendor in class.

10. Conduct a search on the Web to identify global trends that will affect the conduct of qualitative research in the future. Enumerate these trends and discuss their importance to both managers and researchers alike.

NOTES

1. Ellen Day, "Qualitative Research Course Emphasizes Understanding Merits and Limitations," *Marketing News* (4 August 1997), p. 9; Howard Furmansky, "Debunking the Myths About Focus Groups," *Marketing News* (23 June 1997), p. 22; and Sang Kim, "Qualitative, Quantitative Methods Combine for the Best Online Research: Marketers Must Answer 'What' and 'Why' for Best Research Results," *Dartnell's Selling*, August 2003, downloaded from http://www.invoke.com/documents/Selling.pdf, November 24, 2003.

2. Johny K. Johansson and Ikujiro Nonaka, "Market Research the Japanese Way," *Harvard Business Review* (May–June 1987), pp. 16ff.

3. See Barnett A. Greenberg, Jac L. Goldstucker, and Danny Bellenger, "What Techniques Are Used by Marketing Researchers in Business?" *Journal of Marketing* (April 1997), pp. 62–68; A. Parasuraman, "A Study of Techniques Used and Clients Served by Marketing Research Firms," *European Research* (October 1982), pp. 177–185; and Thomas L. Greenbaum, "Focus Group Spurt Predicted for the '90s," *Marketing News* (8 January 1990), *24*:1, pp. 21ff.

4. For example, see David W. Stewart and Prem N. Shamdasani, *Focus Groups: Theory and Practice*, vol. 20 (Newbury Park, CA: Sage, 1990), pp. 122–139;

and Holly Edmunds, *Focus Group Research Handbook* (New York: McGraw-Hill, 2000).

5. From Claire Selltiz, Lawrence S. Wrightsman, and Stuart W. Cook, *Research Methods in Social Relations* (New York: Holt, Rinehart, and Winston, 1976), p. 252.

6. For good discussions of the nature and use of observational studies, see John W. Creswell, *Research Design: Qualitative, Quantitative, and Mixed Methods Approaches*, 2nd ed. (Thousand Oaks, CA: Sage, 2003) and Lee Sechrest, Richard D. Schwartz, Eugene J. Webb, and Donald T. Campbell, *Unobtrusive Measures*, rev. ed. (Thousand Oaks, CA: Sage, 2000).

7. Stephen J. Grove and Raymond P. Fisk, "Observational Data Collection Methods for Services Marketing: An Overview," *Journal of the Academy of Marketing Science* (Summer 1992), *20*:3, p. 218.

8. For a discussion of some of these problems, see Jum C. Nunnally, *Introduction to Psychological Measurement* (New York: McGraw-Hill, 1970), pp. 375–380; and John P. Dean, Robert L. Fichhorn, and Lois R. Dean, "Observation and Interviewing," in John T. Doby, Ed., *An Introduction to Social Research*, 2nd ed. (New York: Appleton-Century-Crofts, 1981), pp. 274–281.

9. Different authors use slightly different classification schemes for this discussion. For a discussion that parallels this approach, see Danny N. Bellenger, Kenneth L. Bernhardt, and Jac L. Goldstucker, *Qualitative Research in Marketing*, monograph series no. 3 (Chicago: American Marketing Association, 1976). For an interesting discussion and alternative viewpoint, see Yvonna S. Lincoln and Egon G. Guba, *Naturalistic Inquiry* (Beverly Hills, CA: Sage, 1985).

10. For a more complete discussion, explanation, and critique of the in-depth interview and focus group, see William Wells, "Group Interviewing," in Robert Ferber, Ed., *Handbook of Marketing Research* (New York: McGraw-Hill, 1974), pp. 2-133-2-146; Bobby J. Calder, "Focus Groups and the Nature of Qualitative Marketing Research," *Journal of Marketing* (August 1977), *14*, pp. 353–364; and Thomas L. Greenbaum, *The Handbook for Focus Group Research*, 2nd ed. (Thousand Oaks, CA: Sage, 1998).

11. See David Aaker, V. Kumar, and George Day, *Marketing Research*, 6th ed. (New York: John Wiley, 1998), and Greenbaum ibid.

12. See Jeffrey A. Tractenberg, "Listening the Old-Fashioned Way," *Forbes* (5 October 1987), p. 202.

13. For example, see J. T. Durgee, "Depth-Interview Techniques for Creative Advertising," *Journal of Advertising Research* (December 1985), pp. 29–37; and Donald S. Tull and Del I. Hawkins, *Marketing Research: Measurement & Method*, 6th ed. (New York: Macmillan, 1993), pp. 442–445.

14. See David L. Morgan, *Focus Groups as Qualitative Research*, 2nd ed. (Thousand Oaks, CA: Sage, 1997), and Greenbaum, op. cit note 4.

15. Much of this section is drawn from Stewart and Shamdasani, op. cit. note 4, pp. 1–14.

16. Alexia Parks, "On-Line Focus Groups Reshape Market Research Industry," *Marketing News* (12 May 1997), p. 28.

17. See Thomas L. Greenbaum, "Focus Group by Video Next Trend of the 90s," *Marketing News* (26 July 1996), p. 4; and Cyndee Miller, "Anybody Ever Hear of Global Focus Groups?" *Marketing News* (27 May 1991), p. 14.

18. See "Penning the Wrong Word" and "Bic Drops 'Spaz' After Mother of Disabled Boy Is Offended," *Marketing News* (23 May 1994), *28*, pp. 4, 7.

19. See Howard Schlossberg, "Good Design Critical to Focus Group Success," *Marketing News* (27 May 1991), *25*:11, pp. 1ff.

20. For other discussions about the use of these and other techniques, see Alvin C. Burns and Ronald F. Bush, *Marketing Research*, 4th ed. (Upper Saddle River, NJ: Prentice-Hall, 2003), pp. 229–236; and Tull and Hawkins, op. cit. note 13, pp. 452–460.

21. Hal Daume, "Making Qualitative Research Work in the Pacific Rim," *Marketing News* (12 May 1997), p. 13.

22. Thomas Greenbaum, "10 Tips for Running Successful Focus Groups," *Marketing News* (14 September 1998), p. 13.

SUGGESTED READING

Nature and Uses of Qualitative Research

Denzin, Norman K., and Yvonna S. Lincoln, Eds., *Handbook of Qualitative Research*, 2nd ed. (Thousand Oaks, CA: Sage, 2000).

This collection of readings reviews the major strategies and paradigms in qualitative research. The issues of data collection, interpretation, and the future of qualitative research are addressed.

Lee, Raymond M., *Unobtrusive Methods in Social Research*, (Philadelphia, PA: Open University Press, 2000).

Provides an interesting look at passive PDC and its potential implications in the research environment. It also looks at the Internet as a potentially rich source for unobtrusive data collection.

Passive PDC Using Observation

Creswell, John W., *Research Design: Qualitative, Quantitative, and Mixed Methods Approaches*, 2nd ed. (Thousand Oaks, CA: Sage, 2003).

This book presents an interesting discussion of how mixed methods of qualitative and quantitative research can be useful in the research process. A short, but straightforward approach to a difficult subject.

Sechrest, Lee, Richard D. Schwartz, Eugene J. Webb, Donald T. Campbell, *Unobtrusive Measures*, rev. ed. (Thousand Oaks, CA: Sage, 2000).

An interesting continuation of a classic introduction to unobtrusive measures. Provides a creative and thought provoking look at alternative ways to capture information.

Active PDC Using Qualitative Research Techniques

Bader, Gloria E., and Catherine A. Rossi, *Focus Groups: A Step-By-Step Guide*, 3rd ed. (The Bader Group, 2002).

A short, easy-to-read guide for conducting focus groups. Not an in-depth treatment of focus group management, but a short and sweet treatment of the nuances of focus group management.

Edmunds, Holly, *Focus Group Research Handbook*, (NY: McGraw-Hill, 2000).

A useful text for those that have to order and specify the nature of qualitative research. Particularly useful is the chapter that deals with conducting focus groups with international and unique samples, such as children or teenagers.

Krueger, Richard A., and Mary Anne Casey, *Focus Groups: A Practical Guide for Applied Research*, 3rd ed. (Thousand Oaks, CA: Sage, 2000).

Provides a useful guide to conducting focus groups. Its engaging style makes it easy to read, yet provides practical tips on conducting focus groups.

Qualitative Research Vendors

For the most up-to-date listing of vendors, go to the following: http://www.greenbook.org—New York AMA Communication Services *Green Book, Vol. 2.* http://www.marketingpower.com/live/directory-display.php—American Marketing Association's *The M Guide—Marking Services Directory.*

12

Planning for Data Analysis

Overview

Planning Issues

Selecting the Appropriate Analytical Software or Online Research Supplier

Analytical Software

Online Research Suppliers

The Preanalytical Process

Step 1: Data Editing

Step 2: Variable Development

Step 3: Data Coding

Step 4: Error Check

Step 5: Data Structure Generation

Step 6: Preanalytical Computer Check

Step 7: Tabulation

Basic Analytical Framework for Business Research

Managerial Considerations

Summary

Discussion Questions

Notes

Suggested Reading

Data analysis is a carefully planned step in the business research process. This step should stem from a consideration of the purpose of the analysis, which is to provide information to solve the problem under study. There are seemingly endless ways to analyze data, but all analysis must begin with certain preparatory steps. These will be discussed in this chapter.

We begin the chapter by discussing basic issues in planning for data analysis. The researcher must develop a data analysis plan to ensure that the raw data is converted into usable information. A generic four-stage plan is outlined for discussion purposes. The first stage deals with the identification and selection of analysis software for the actual summarization and reporting of the data. Some data analysis packages are briefly reviewed for informational purposes.

As was noted in Chapter 8, on instrument design, online Internet-based and stand-alone research-oriented software is becoming increasingly integrated throughout the research process. These developments are helping aid the researcher to design, collect, and analyze data more quickly than ever before. We review breakthroughs in this area from the perspective of data analysis planning.

The discussion then switches to the second stage of the plan, the conduct of the preanalytical process. This process essentially purifies the data. The seven steps in this process, which include editing, coding, and developing data for final analysis, are described and explained.

Finally, a basic analytical framework for business research is outlined. This framework is organized around the primary purpose of the analysis. The framework sets the stage for the next four chapters, which deal with specific data analysis methods and techniques. The chapter ends with managerial considerations in planning for data analysis.

PLANNING ISSUES

As noted earlier, to obtain the desired information, the researcher must be thinking of a data analysis strategy in the development of the research proposal and the selection of the research design. If inadequate thought is given to data analysis procedures early in the research process, a lot of useless data may result; no one may know how to analyze it to answer the research question. Thus, it is important to develop a plan that outlines how the researcher is going to prepare the data to accomplish the research objectives.

The strategy for analyzing the data should come directly from the research problem and objectives. Researchers must ask what they are trying to achieve and what analytic method will get them to that goal.

Many statistical methods and assumptions are available; the researcher must continually assess what approaches are best to answer the research question. The plan must be flexible, to allow for unanticipated problems in the research design.

For example, in a recent consulting project, the response rates were not sufficient to conduct the analyses that were planned in the study. Essentially, there were two options: Go back and collect more data, or adjust the analysis method to accommodate the current data and achieve the research objective. After discussion with the client, the second option was chosen.

To facilitate planning, it is very useful to create a plan that systematically addresses the major concerns in data analysis. These concerns include selecting the software or online research supplier, preparing for final analysis, deciding which analyses to actually run, and deciding how to present the results of the analyses to achieve the research purpose. Exhibit 12.1 presents a summary of these concerns in the form of an outline of a data analysis plan.

EXHIBIT 12.1

General Outline for a
Data Analysis Plan

1. **Select appropriate analytical software or online research supplier.** Most all data analysis in today's environment uses some type of software to help in data analysis and reporting. If you choose to outsource your research requirements to an online supplier, then frequently the software decision is made for you. The software requirements generally include a statistics component, a graphical component, and a word processing/desktop publishing component. With the explosion of sophisticated, integrated PC software packages in recent years, this decision often comes down to an all-purpose program.

   However, as usage of the Internet continues to grow, one should expect to see more online suppliers like Zoomerang (http://www.zoomerang.com) and SurveySolutions (http://www.perseusdevelopment.com) presented earlier in Exhibit 8.2 who are providing integrated design, collection, analysis, and reporting services for their clients. Regardless, the choice of software or outsource supplier should be driven by the research purpose.

2. **Conduct preanalytical steps.** The preanalytical steps are elaborated in Table 12.2. These seven steps are designed to purify and refine the data for subsequent final data analysis. A carefully orchestrated preanalytical process helps reduce analytical errors and leads to a better understanding of the nature of the study data. Here, again, many software and online service providers are beginning to provide "smart" software that automates most, if not all, of the preanalytical steps.

3. **Identify specific statistical procedures.** This step isolates possible analytical strategies to run on the variables in the study. The basic framework at the end of this chapter, and the four analytical chapters that follow, help in identifying the appropriate statistical procedures to use. Again, research objectives dictate what procedures should be conducted. However, the process of research often leads to new insights and new analyses that may not have been planned in the initial stages of the project's design.

4. **Identify summary tables, graphs, and outlines to organize results.** The plan should include a general framework for presenting the results. This section should include an outline of the findings and a preliminary listing of tables and graphs that might be useful to organize the results. Flexibility is the key.

Each section of Exhibit 12.1 deals with issues that can affect the quality of the results of the study. Steps 1 and 2 are presented in more detail in the next sections of this chapter. This chapter also begins to cover step 3 by introducing a basic analytical framework for business research. Chapters 13 through 15 explore more specific statistical techniques. Step 4 is covered in more detail in Part 6 of this text, which presents the basics of research reporting.

SELECTING THE APPROPRIATE ANALYTICAL SOFTWARE OR ONLINE RESEARCH SUPPLIER

The selection of the appropriate analytical software or online research supplier is often dictated by availability and familiarity. This text will not recommend one analytical package or supplier over another; there are many good ones, and they all seem to be improving in capabilities and ease of use.[1] Researchers must ultimately choose and evaluate one on the basis of their needs and the research problem at hand.

Analytical Software

Unique research problems often necessitate custom software—that is, software designed for that specific purpose. This text cannot do justice to such specialized packages, nor can it cover extremely comprehensive packages such as SPSS®, SAS®, and Excel® in any depth.[2] The online documentation and references for these and other packages fill entire books. Most of these packages, however, are accompanied by excellent tutorials (especially the student versions of software such as SPSS, which actually coaches the student in the use and interpretation of the statistics in the package). The presentation here only lists sources and vendors that produce off-the-shelf software for analytical purposes and then introduces a few major software packages. For the analytical software packages SPSS, SAS, and Excel, we simply illustrate their general capabilities and features. Other customized and integrated statistical/graphics packages are available and are often equally appropriate for the researcher's needs.

The trend in software development is to totally integrate statistical, graphical, and written/voice presentations into multimedia formats to take advantage of the growing capabilities of microcomputers. A partial listing of microcomputer statistical analysis vendors and their products is presented in Table 12.1. These computer packages have widespread acceptance in the marketplace. The reader is encouraged to consult some of the sources presented here to explore the nature and capabilities of these packages.

TABLE 12.1 Selected Micro Statistical Analysis Packages and Vendors

Vendor	Package	Description
Creative Research Systems 411 B Street, Ste. 2 Petaluma, CA 94952 (707) 765-1001 http://www.surveysystem.com	The Survey System	A software solution that includes interviewing techniques, data entry, analysis, and presentation capabilities
Minitab, Inc. 1829 Pine Hall Road State College, PA 16801 (814) 238-3280 http://www.mintab.com	Mintab	Provides general-purpose systems for solving statistical problems
NCSS 329 N. 1000 E. Kaysville, UT 84037 (801) 546-0445 http://www.ncss.com	NCSS	Performs a wide range of statistical procedures for research purposes
Manugistics, Inc. 9715 Key West Ave. Rockville, MD 20850 (800) 592-0050 http://www.statgraphics.com	STATGRAPHICS	Easy-to-use program with more than 250 parametric and nonparametric procedures
StatPac Inc. 4425 Thomas Ave. S. Minneapolis, MN 55410 (612) 925-0159 http://www.statpac.com	STATPAC	Integrated, user-friendly, and comprehensive survey analysis system

Voice recognition and "point and shoot" analytic systems are in development. More and more systems are menu-driven and have packages that will increase the access to sophisticated data analysis systems. Traditional spreadsheet programs such as Lotus® and Excel are consistently increasing their data analysis capabilities.[3] In addition, because of improved microcomputer technology, almost any data analysis is now feasible from a networked computer.

Thus, serious researchers must stay current on the capability of the software packages available in the marketplace. The changes in software capabilities today are phenomenal; a few integrated packages are discussed in the following section.

SPSS (Statistical Package for the Social Sciences). SPSS has been one of the most widely used statistical analysis packages for the serious business researcher. SPSS offers a full line of data analysis products ranging from data collection through analysis and modeling to report presentation. Because it is modular, one can have a fully integrated survey analysis system or use individual modules to design a unique software solution for specific applications. The latest Windows version of SPSS is very

FIGURE 12.1

Sample Windows and Menus in SPSS for Windows

Source: SPSS® for Windows® developed by SPSS Inc., 444 N. Michigan Ave., Chicago, IL 60611-3962. All rights reserved.

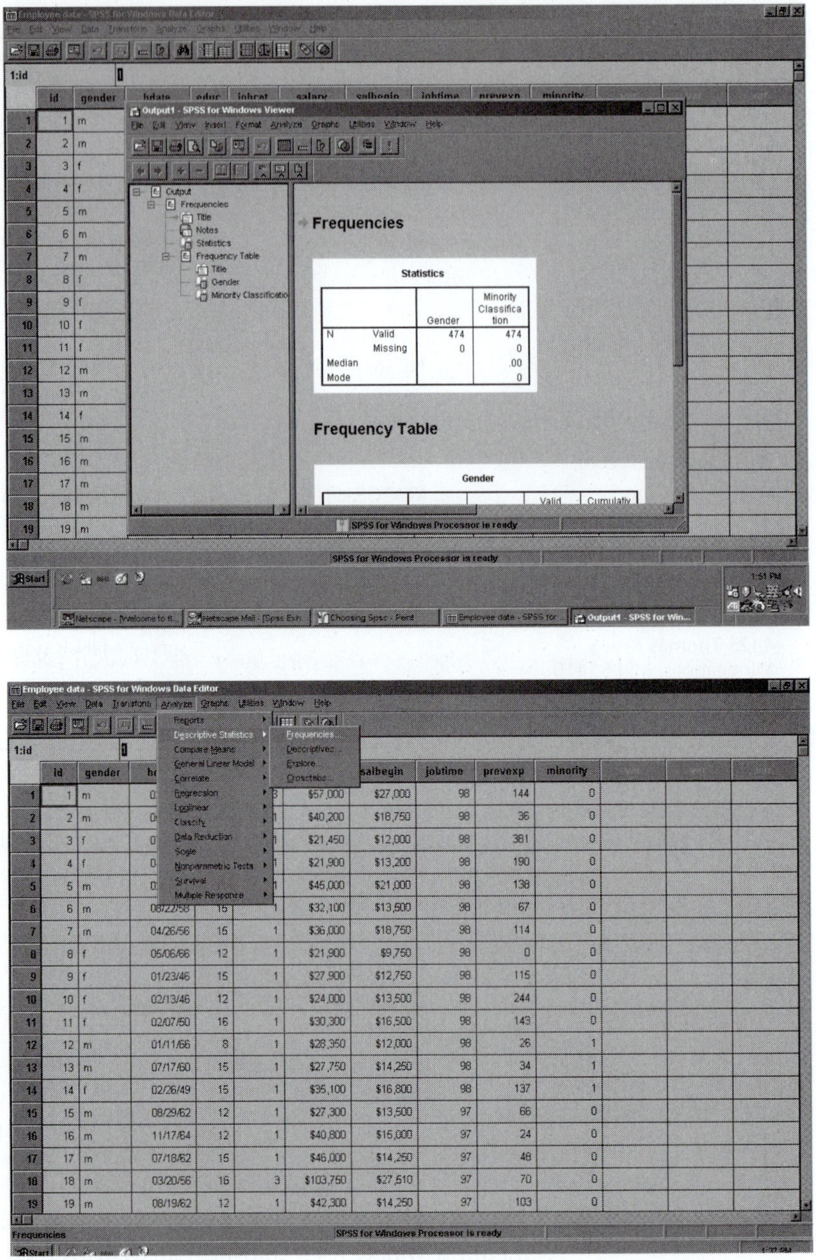

user-friendly, with easy-to-use but sophisticated procedures.[4] The windows and menu-driven commands illustrated in Figure 12.1 provide spreadsheet-like data input, "point and shoot" menu commands, and integrated graphics capabilities. However, as noted previously, these developments are not unique to SPSS.

In addition, integrated add-on modules perform all the statistical procedures mentioned in this text and most other research and statistics texts. The programs have extensive tutorials, usually with statistics and/or results coaches to help the analyst correctly choose and interpret the statistical procedure needed for analysis. A student version and graduate pack are low-cost alternatives for learners who want to try full-functioned versions of the base program modules. The tutorials on these discs are self-explanatory and allow the student to get up and running in a short period of time.

Some of the more exciting developments with the SPSS package are in the supplementary modules that one can obtain with the base statistical system. A few of these modules are stand-alone systems, whereas others need to be integrated into the base statistics and data management module to be functional. For example, SPSS Data Entry Builder™ is a tool for survey design and data collection that offers a drag-and-drop interface for developing both printed and online data entry forms.[5] This program allows users to create rules to ensure that only valid data are entered. In addition, it helps prepare users for data analysis because when the questions and responses that will appear on the data collection instrument are specified, users must also define the variables that will be used to run analyses. Other modules of note include TextSmart™, which uses a set of algorithms to quantify open-ended survey responses, and Missing Values Analysis™, which examines the missing value patterns in data and helps users estimate these values. In essence, these programs automate the preanalytic process, thus saving the researcher valuable time and resources.

The interested reader is encouraged to explore these programs. A myriad of reference manuals and supporting documentation illustrates how to use these powerful tools. A good starting point would be a visit to SPSS's Web site (http://www.spss.com). A sampling of manuals and appropriate contact points is presented in the Suggested Reading section at the end of this chapter.

SAS (Statistical Analysis System). The SAS system is another widely used software package for data analysis and report writing.[6] It includes modules that can be integrated for accessing, entering, and managing data; graphics generation; and forecasting capabilities. It also allows complex statistical analyses such as conjoint analyses, logistic regressions, and other multivariate procedures to be conducted.

The SAS system is one of the more complete and useful statistical packages available today. The serious business researcher should at least become familiar with the capabilities of this package by reading one of the introductory guides. Again, a visit to the Web site of SAS (http://www.sas.com) would be extremely informative. Appropriate contacts and partial referencing of documentation are presented in the Suggested Reading section.

Excel[7]. Spreadsheets are one of the most popular and indispensable forms of business software and are fast becoming essential tools in the research community. Their popularity is largely due to the expanding number of statistical and analytical tools that allow the user to import data, mathematical formulas, text, and graphics into a single report or workbook. Excel by Microsoft is the leader in this trend. Although Excel's statistical capabilities in no way compare to the breadth and complexity of comprehensive analytical packages like SPSS and SAS, Excel, with its various statistical add-ins, provides a good basic analysis package that would suffice for many data analysis situations.

The standard Excel program comes with 81 statistical and 59 mathematical functions. The add-in Analysis ToolPak, which is bundled with Excel, contains another 12 or so functions, including correlation, descriptive statistics, multiple linear regression, and others. A StatPlus add-in provides another 20 or so tools to increase the analytical capabilities of Excel. This total package would be sufficient to cover a basic statistics course, but in no way would fulfill the needs of a researcher in need of advanced multivariate techniques.

As the screen capture in Figure 12.2 shows, to access the basic statistical functions, users must click the Tools command, then bring the cursor down

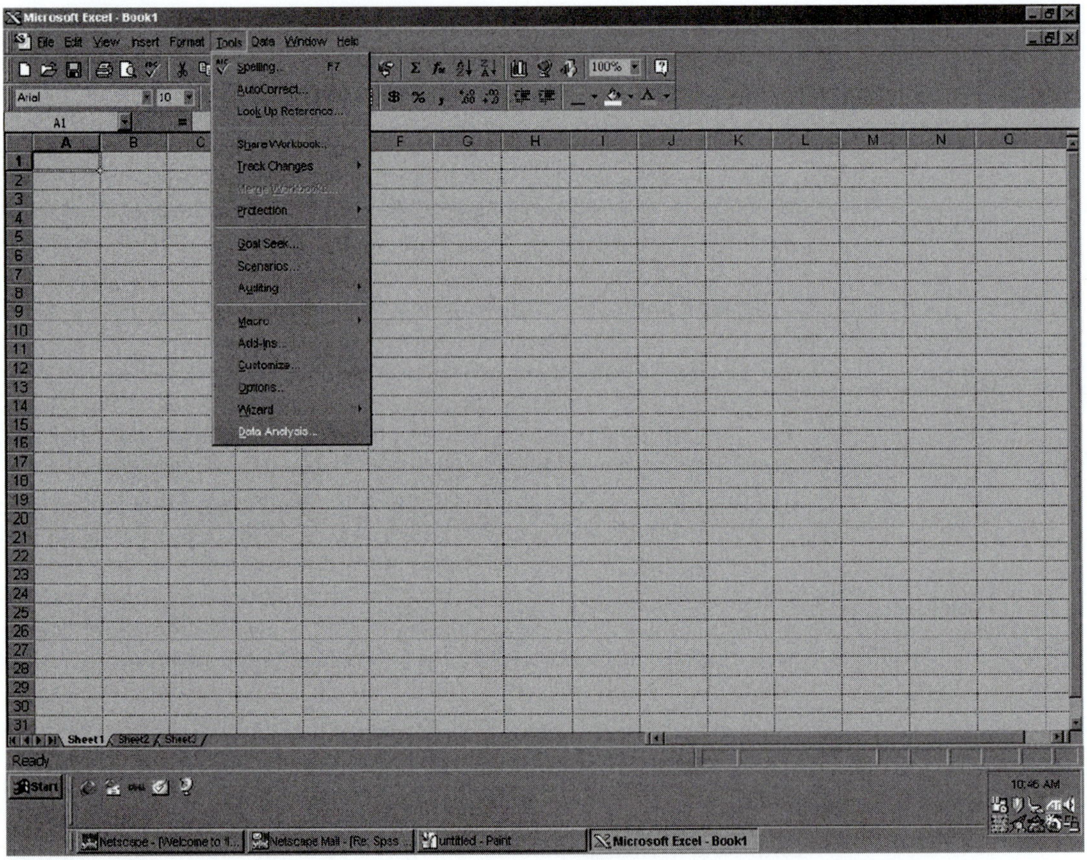

FIGURE 12.2

Screen Capture Illustrating Where to Find the Primary Statistical Functions in Excel. Reprinted with permission of Microsoft Corporation

to Data Analysis to access the specific analytical function. Once the appropriate analysis is found in the pull-down menu in data analysis, users double-click on it to begin the procedure. Users are advised to get into Excel and begin to familiarize themselves with the capabilities of its ToolPak.

Online Research Suppliers

As noted throughout this text, an increasing number of companies are beginning to provide online research support. This growth in online researching and online service providers will continue into the future, as increased technological capabilities and highly specialized and knowledgeable service providers make outsourcing of specific research tasks highly desirable.

Research suppliers vary widely in the range of services they provide. For example, Perseus Development (http://www.perseusdevelopment.com) offers a hosted online survey solution called Perseus Enterprise Hosting where customers enjoy the combined benefits of SurveySolutions Enterprise technology, a state-of-the-art managed data center, and the guidance of dedicated research professionals to ensure the success of the research project. Such organizations as Pitney Bowes, MSN, Pfizer, and General Electric have all found Enterprise solution useful in their research efforts. As Exhibit 12.2 illustrates, the company is able to provide specialized services and benefits for companies that may not have the expertise or time to conduct the full range of research activities for themselves.

Furthermore, *outsourcing*, or the strategic use of outside resources to perform activities traditionally handled by internal staff and resources, can be of significant benefit to business organizations when dealing with the research function. According to the Outsourcing Institute (http://www.outsourcing.com), the top five reasons companies outsourced were (1) the company could reduce and control operating costs, (2) outsourcing improved company focus, (3) the company gained needed expertise, (4) resources were freed for other needs, and (5) the resources were not available internally. Given these benefits, researchers and managers alike need to seriously examine the value of outsourcing when planning the data analysis process specifically and the research process generally. A word of caution, though: Researchers must take an active management role if a task is outsourced to ensure the research is of the highest quality and is addressing the problems at hand.

THE PREANALYTICAL PROCESS

Many books and articles have been written about the preanalytical process.[8] Basically, this process incorporates personal and electronic means to ensure that the data collected is purified before analysis, thus reducing inaccuracies and fallacious elements in the data. The steps in this process

EXHIBIT **12.2**

Perseus Development Sample Product Offerings & Service

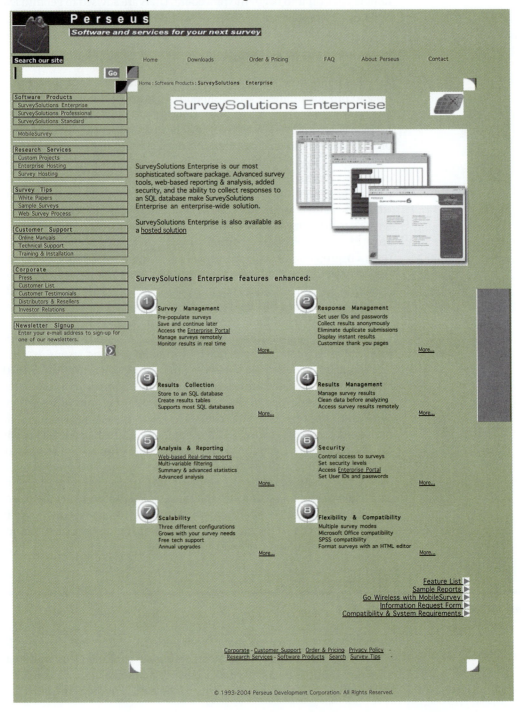

TABLE 12.2 Preanalytical Steps, Description, and Related Issues

Steps	Description	Issues
1. Data editing	The process of giving clarity, readability, consistency, and completeness to a set of collected data.	a. Will the collected data create conceptual and/or technical problems in the analysis? b. Will the collected data logically justify the interpretation of the results? c. Is the data clear, consistent, and complete for coding?
2. Variable development	The specification of all variables of interest subsumed in the data.	a. Are all the variables of interest subsumed in the data?
3. Data coding	The translation of the data into codes (usually numerical) for the purpose of transfer to a data storage medium and subsequent analysis.	a. Are the coding categories consistent with the design and purpose of the study? b. Are the data transfer, storage, and retrieval methods consistent with the analytical hardware and software?
4. Error check	The coded, collected data is checked before transfer to data storage.	a. Have the previous steps been completed without serious errors?
5. Data structure generation	The data structure is finalized to include all data needed for analysis and then transferred to a storage medium.	a. Is the data stored in a manner consistent with its ultimate use? b. Is there missing data that was not previously accounted for? c. How will missing data be handled in the analysis? d. Have all data transformations been planned and/or executed?
6. Preanalytical computer checks	The finalized stored data structure is prepared for computer analysis and subjected to preanalytical computer check for consistency and completeness.	a. Has a computer-processed error check been completed? b. Have the errors inherent in the analytical software been considered?
7. Tabulation	This lists the number of individuals who have answered a question in a particular way. It can also be used to create descriptive statistics for variables and/or crosstabulations.	a. Are there any outliers in the data? b. What are the breakpoints for converting interval data to categorical data? c. Generally, is there a good feeling about the data?

Used with permission.

are presented in Table 12.2. Because computer programs now conduct many of the preanalytic steps for the researcher, these steps to clean and purify data before it is submitted for analysis may be undertaken by either the researcher or the computer. Keep this in mind as we begin the process.

Step 1: Data Editing

A carefully thought-out research plan requires pretesting of the data collection procedures, especially the data collection instrument. This creates a small set of data that can be examined to ensure that any conceptual or technical problems in the subsequent analysis can be eliminated.

Some important aspects of the pretest edit include (1) substitution of questions or scales that reflect better measurements of the important concepts needed in the study, (2) removal of foreseeable inconsistencies in measuring a concept across respondents, and (3) a recheck of the variables to ensure that the scales, categorizations, and so on can be properly analyzed to obtain interpretable results. If oversights are found beforehand, the editing of the collected data and the subsequent analysis are made more efficient.

After the data is collected—no matter whether it is secondary, primary, or mixed—it must immediately be edited by a qualified person. Personal interviews require more immediate editing than survey data because of comprehension and memory problems that may arise from the personal interview. In any event, the data edit is a check for clarity, readability, consistency, and completeness of a set of collected data before the data is transferred to a data storage medium. Editing the data for clarity and readability makes it easily understandable, so the transfer to storage can be more efficient. Generally, the editor will look for ambiguities in classification variables on collection instruments. Also, illegible handwriting on collection instruments is clarified. In the case of personal interviews, a quick call to the interviewer can solve an editing problem. However, editing the survey instrument in cases of misclassifications and illegible answers is a judgment call for the editor. The editor can only interpret the data as it stands.

The consistency edit can detect obviously incorrect answers to questions. For example, the respondent indicates that he or she takes out 30 books a month from the library, yet reads very seldom. This is an obviously inconsistent response and should be clarified. Consistency checks can also be made by a computer; this topic will be discussed in a later section.

Completeness is also checked at this stage. How much of the data is missing from a questionnaire or an interview? Missing data could have resulted from a respondent's refusal to answer certain questions, or a respondent may have inadvertently forgotten to answer a question. The completeness edit requires a check for incomplete or missing data. Even with the utmost planning and caution, researchers seldom get 100% complete data in the collection process. Incomplete questionnaire sections or questions do not make the rest of the respondent's replies useless. The completeness edit allows the editor to make judgments as to what to do about missing information. Although all missing information is suspect (it could be due to design or data collection errors), it is nevertheless useful. Therefore, except for extraordinary omission, all information should be coded for analysis.

Step 2: Variable Development

Often the business researcher is interested in constructing indices, developing composite variables, or creating variable transformations from the basic data. Such variables are generated for specific analyses related to the objectives in the study design. The basic data must be examined to ensure that all the desired variables to be used in the analysis are included or can be constructed. If not, the major premise behind the study could be invalidated. For example, if your aim is to develop an index of occupational happiness from the summation of the scores of three simple scales, and you have inadvertently left out one of the scales on the questionnaire, your entire study might be negated unless a surrogate variable can be determined. Another means of variable development can evolve from mathematical transformations or planned analytic or statistical procedures performed on the basic data set.[9] This discussion is beyond the scope of this text.

Step 3: Data Coding

Data coding involves the translation of the collected data into codes (usually numerical) for the purpose of transferring the data to a data storage medium for subsequent computer analysis. Coding simply transforms data from one form (marks on a survey, punches on a touch-tone phone, and so on) into another (numbers to be processed by the computer). For example, a yes-or-no answer can be numerically coded 1 = yes, 2 = no. From a practical standpoint, the translation of the variables into codes facilitates computer analysis. Computers can recognize an alphanumeric language but can more efficiently manipulate the numeric codes.[10]

Most collection instruments are planned with the coding of variables in mind. These are sometimes referred to as *precoded* questionnaires. From the questionnaire, a codebook is constructed. Some microcomputer analysis programs can actually generate a codebook for you if you input the format, variables, and coding devices. A sample codebook for the questionnaire that produced the data in Table 12.4 is displayed in Table 12.3. For example, the variable *age*, which is variable #2, will start in column 4 in the data structure, takes up two columns total, and will be numeric in nature (in this case, interval). Because the code structure in this example was thoroughly planned with the research problem in mind, there was little trouble in translating it to a storage medium. Most coding problems occur when the questions are open-ended or the categorization of answers is not preplanned.

Sometimes the use of open-ended questions makes it necessary to create categories not originally anticipated. The three types of coding situations are presented in Exhibit 12.3. The structured question has categories that have been built into the answer. As illustrated, these answers are easily coded. For example, a family with an income of $40,000 to $49,999 is represented by the numeric code 3. Because five respondents answered this question by checking this income category, there are five codes of 3 for that particular question. Similarly, in the preplanned open-ended question, coding is just as easy. Five respondents have income levels of $40,000 to $49,999. Therefore, using the coding rule, the income variable is coded as a 3 for these families.

TABLE 12.3 Codebook for Questionnaire that Produced Data in Table 12.4

Codebook Listing—EXP.BK	
Variable #1 Name/Label	"ID" Respondent Number
Starting column	1
Number of characters	3
Variable type	N
End of variable symbol	@
Variable #2 Name/Label	"AGE" Current Age
Starting column	4
Number of characters	2
Variable type	N
End of variable symbol	@
Variable #3 Name/Label	"WORKEXP" Work Experience
Starting column	6
Number of characters	1
Variable type	N
Value Label	1=YES
Value Label	2=NO
End of variable symbol	@
Variable #4 Name/Label	"FULTIM" Working Full-Time
Starting column	7
Number of characters	1
Variable type	N
Value Label	1=YES
Value Label	2=NO
End of variable symbol	@
Variable #5 Name/Label	"SAT" Satisfaction with Job
Starting column	8
Number of characters	2
Variable type	N
End of variable symbol	@

Source: Courtesy of Walonick Associates, Inc., *STATPAC Computer Programs for the PC,* 3814 Lyndale Ave., Minneapolis, MN 55409, 1988.

Open-ended questions, with no planned coding rules, present two problems for the researcher. First, determining the number of mutually exclusive and exhaustive classes is not an easy task. Second, "data dredging" should only be used in exploratory or descriptive research with no strict hypotheses. Even the use of open-ended questions with preplanned coding can present problems for the researcher. If there are several coders, the consistency of the coded data is reduced. If a significant number of anomalous responses are found in the questions, another category might have to be added. If at all possible, open-ended questions should be avoided in quantitative research.

TABLE 12.4 Data Structure

Respondent Number	Variable Name (SCALE)	AGE (1–∞)	WORKEXP (Y or N)	FULTIM (Y or N)	SAT (5–25)
1		26	1	2	12
2		22	2	2	15
3		30	1	1	15
4		30	1	2	20
5		19	2	1	20
6		22	2	1	14
Row 7		22	1	1	22
8		20	2	2	24
9		18	1	1	15
10		55	1	1	5

Column

Codes

Age: 1–∞

Work experience: Y = 1, N = 2

Satisfaction with current job: 5 (low) to 25 (high) satisfaction

Finally, after the data is coded and readied for transfer to a storage medium, a codebook is developed, containing the general instructions on how each item or variable in the data was coded. The codebook also contains the method of transfer and the position of each item or variable in the storage medium.

Step 4: Error Check

The error check involves two tasks. The first task is to check that all the previous preanalytical steps were performed correctly. Second, the coded data must be rechecked visually to detect any possible clerical errors. This is the final human check before the data is transferred to a storage medium.

Step 5: Data Structure Generation

The information age allows flexibility in the analysis of data. However, computer analysis depends on the data structure—the way the respondent's information is positioned in a storage medium. As noted earlier, significant advances in data technology transfer have made direct electronic transfers commonplace in today's research environment. However, data is often still entered manually using spreadsheetlike formats.

EXHIBIT **12.3**

Coding Problems

STRUCTURED QUESTION (PREPLANNED CODING)

How much was your total family income last year (both spouses considered, excluding children)? Check the appropriate box.

	Number of Responses	Codes
Under $30,000	1	1
$30,000–$39,999	3	2
$40,000–$49,999	⑤	③
$50,000 or more	1	4

OPEN-ENDED QUESTION (PREPLANNED CODING)

How much was your total family income last year (both spouses considered, excluding children)?

Responses	Codes	Responses	Codes
$19,876	1	$42,693	③
$32,500	2	$48,000	③
$33,500	2	$49,000	③
$38,000	2	$46,500	③
$41,500	③	$54,000	4

Coding Rule

1 = under $30,000
2 = $30,000–$39,999
3 = $40,000–$49,999
4 = $50,000 or more

OPEN-ENDED QUESTION (NO PLANNED CODING)

How much was your total family income last year (both spouses considered, excluding children)?

Responses	Codes ???
$19,876	
$32,500	
$33,500	
$38,000	
$41,500	
$42,693	
$48,000	
$49,000	
$46,500	
$54,000	

Generally, the creation of a data structure simply involves positioning numbers on a storage medium. A record is created for each respondent. A record is a line of data that contains positioned numbers that represent the codes (classifications) or actual values (for interval or ratio data) of answers to particular questions. Normally, there are at least as many records as respondents. Once the finalized data structure is stored in the medium, various transformations are possible, depending on the research problem and the capabilities of the analytical software. In fact, new data structures can be created that include transformations. Data structures are referred to as *files* after they have been stored in a medium. It is not uncommon to have several files for a research project, as well as a backup file or files. For purposes of analysis, three points regarding data structure should be considered: (1) The stored data must be consistent with the software used for analysis. (2) The data must contain codes for missing values consistent with the software used for analysis. (3) All coding in the data structure must conform to the specifications in the codebook.

For example, a survey instrument usually contains some unanswered questions, but the researcher does not normally discard the entire set of respondent's answers. Usually, these unanswered questions are coded as missing values. Computer software packages handle these missing values differently. With their unique coding, they are properly handled by the internal workings of the program in the computer analysis.

Step 6: Preanalytical Computer Check

The preanalytical computer check subjects the stored data to its final screening for completeness and consistency. Most often, this check can be done using the analytical software chosen for the analysis. Usually, it involves generating a series of frequency tabulations on the entire set of questions to check for inconsistencies and missing data. Errors detected can be corrected either by the coder or, in the case of actual missing data, by the computer software. Another preanalytical computer check involves the software itself. This check is used to establish the precision and reliability of the software. It is not uncommon to find differences in the outputs of certain procedures. In general, the reliability of software is established by its longevity and use in the marketplace. However, if you are not sure of the quality of your software, seek expert advice. The proper interpretation of your results depends on it.

Step 7: Tabulation

The tabulation of data is not usually thought of as an analytical procedure in scientific research because it does not express the relationships in the data.[11] However, as discussed in Chapter 1, much business research is directed at clarifying problems and discovering relationships. Although this does not fit the social-scientific model of research, it is nevertheless important to management. Tabulation provides a frequency count of an item or a numerical approximation of the distribution of the item. The former is usually referred to as *frequency analysis,* the latter as *condescriptive analysis.* Tabulation is a business analytical tool in itself, and it is essential to understanding the complex analytical techniques that will be discussed later. Also, tabulation helps to discover outliers—quirks in the data that might change an analysis and ultimately the interpretation of the results. Finally, tabulation can provide the researcher with a means of developing categories when transforming interval variables into nominal classifications.

Whereas step 6 is designed to find errors and make corrections in the database and software selection, step 7 begins the actual understanding of the data itself. Preliminary summaries and descriptive analyses of the variables are used to help the researcher understand the characteristics of the data collected for further analysis. Tabulation is a very important last step before more sophisticated statistical procedures are applied to the data at hand.

An example can facilitate the discussion of tabulation. Suppose the final data structure looks like the one in Table 12.4. Ten respondents answered four questions: their age (interval scale), whether they have previous paid work experience (nominal scale), whether they are currently employed fulltime (nominal scale), and satisfaction with their current job (summated scale—assumed interval scale). The variables are labeled AGE, WORKEXP, FULTIM, and SAT. The rows represent the respondents' answers to all four questions. The columns represent the entire set of respondents' answers to a particular question. A frequency count of all the variables is not beneficial because it is likely to produce as many categories as points on the measurement scale for an interval variable such as age. However, it is sometimes useful to perform a frequency count on an interval variable if you suspect there is an outlier (an answer from a respondent that is totally irregular).

Most often, frequency counts are computed only on categorical or rank data. Descriptive statistics (condescriptive) are computed for all other data. All of the pertinent frequencies and condescriptive statistics are shown in Exhibit 12.4, a review of the more common basic statistics. The use of this information will prove more important when we discuss the basic elements of analysis.

Generally, the following can be determined from an examination of Exhibit 12.4: (1) The answers of respondent 10 might be suspect. This particular respondent's answers might have to be eliminated from the analysis, or the data might also be treated in a special way. (2) The median and mean are meaningless measures of central tendency for nominal data. (3) The range and frequencies are of special use when deciding on the number of categories. (4) Finally, the standard deviation explains the spread of the distribution of continuous measures. Both the mean and standard deviation of the age and satisfaction data are affected by the outlier.

EXHIBIT 12.4

Tabulation of Data in
Table 12.4

Variable: AGE	AGE	FREQUENCY	CUM/FREQ %	%
Scale: Interval	18	1	10	10
Descriptive Statistics:	19	1	20	10
Mean = 26.4	20	1	30	10
Median = 22	22	3	60	30
Mode = 22	26	1	70	10
Range = 18–55	30	2	90	20
S = 10.895	<u>55</u>	<u>1</u>	100	<u>10</u>
Assessment:				
Respondent 10,				
AGE = 55 may				
prove to be an				
outlier		Total 10		100%

Variable: WORKEXP	WORKEXP	FREQUENCY	CUM/FREQ %	%
Scale: Nominal	Yes	6	60	60
Descriptive statistics:				
Mode = Yes	No	<u>4</u>	100	<u>40</u>
Assessment: No				
problems		Total 10		100%

Variable: FULTIM	FULTIM	FREQUENCY	CUM/FREQ %	%
Scale: Nominal	Yes	6	60	60
Descriptive statistics:				
Mode = Yes	No	<u>4</u>	100	<u>40</u>
Assessment: No				
problems		Total 10		100%

Variable: SAT	SAT	FREQUENCY	CUM/FREQ %	%
Scale: Interval	5	1	10	10
Descriptive statistics:	12	1	20	10
Mean = 16.2	14	1	30	10
Median = 15	15	3	60	30
Mode = 15	20	2	80	20
Range = 7–24	22	1	90	10
S = 5.534	24	<u>1</u>	100	<u>10</u>
Assessment:				
Respondent 10,				
SAT = 5 may				
prove to be an				
outlier		Total 10		100%

(Continued)

EXHIBIT **12.4**

(Continued)

COMMON FORMULAE

Measures of Central Tendency

Ungrouped mean: $\bar{X} = \sum_{i=1}^{n} \frac{X_i}{n}$

where X_i is a measure of an interval variable X for subject i and n is the total number of numerical values in the set.

Grouped mean: $\bar{X} = \sum_{i=1}^{h} \frac{f_i X_i}{n}$

where X_i is a measure of an interval variable X for subject i, h is the number of categories, f_i is the frequency of X_i, and n is the total number of numerical values in the set.

Median: The midpoint of an array of values arranged from lowest to highest value. In the case of an even number of values in the array, it is the average of the two middle values.

Mode: The most frequently appearing value in an array.

Measures of Dispersion

Range: The highest minus the lowest value in an array. Sometimes it is simply reported as the lowest and highest boundary values in an array.

Variance

$$S^2 = \sum_{i=1}^{n} \frac{(X_i - \bar{X})^2}{n}$$ [For an unbiased estimator of σ^2, divide by $(n - 1)$ instead of n.]

where X_i is a measure of an interval variable X for subject i, $\bar{X}$ is the mean of this variable, and n is the total number of subjects.

Standard Deviation

$$S = \sqrt{S^2}$$

BASIC ANALYTICAL FRAMEWORK FOR BUSINESS RESEARCH

As discussed in Chapters 1 and 6, the researcher has specific goals when planning the design of a study. The primary goals are exploratory, descriptive/predictive, and causal. After the research problem is specified and the research goals set, the research process leads to a solution of the problem. Some of the primary goals of business researchers are not obtained in a strict

FIGURE 12.3

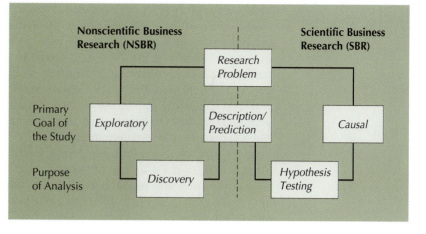

scientific environment. Therefore, for the purpose of developing an analytical framework, two areas of business research are defined. Figure 12.3 depicts this framework.

The first area includes research studies that are systematic but not strictly scientific. The research that takes place in this area is referred to as *nonscientific business research* or simply NSBR. The goal of this research can be exploratory and/or descriptive/predictive. Generally, the analytical emphasis is on uncovering phenomena of interest to management. These phenomena may be used to clarify and/or specify future research directions for the problem under study. Most often, the problem is systematically studied by posing research questions that may be answered by information provided by some analytical method. The answers to the research questions are not dependent on strict statistical tests of hypotheses because the outcomes might not be generalized. However, the analytical methods are often the same ones used in the second area of business research. This notion will become clearer later in the chapter.

Research in NSBR can take on many forms, but the intent—discovery—is always the same. For example, a firm might want to get an idea of the characteristics of certain markets before developing a full-blown segmentation study. Or the same firm might want to explore the general decision-making styles of its executives before it considers studying the effects of incentives on decision making in the firm. Or an organization might simply want to define all of the personnel positions (as to purpose, requirements, and so on) in the firm. Last is the case study. The case study approach is used to examine one or more units of analysis in an intensive fashion (such as all organizational behavior in one or two firms) or the behavior of one or more individuals in a certain setting.

In the second area of business research, researchers are guided by a systematic and scientific analytical plan. The research that takes place in this area is referred to as *scientific business research* or simply SBR. The goal of the research can be either descriptive/predictive or causal. Generally, the analytical emphasis is on obtaining results that can be generalized. To accomplish this, data obtained is subjected to the theoretical and philosophical rules of probability. Most often, the problem is systematically and scientifically studied by posing a research hypothesis, developing this hypothesis into statistically testable hypotheses, and using statistical analytical methods to ensure that the results are generalizable. For example, a survey is taken to determine whether consumers intend to buy your product. The results are then generalized to the population after rigorous statistical testing. Or it may be believed that the relationship between worker cohesion and productivity does not apply to your business. You would like to test this hypothesis to guide your managers in future personnel selection.

The element in the analytical framework that differentiates the two areas of business research is the purpose of the analysis. Additionally, the analytical method chosen is somewhat determined by the statement of purpose. Keep in mind that even though the method used may be the same in the two areas of research, the results of the analysis are used differently. In *discovery,* one fishes, snoops, and hunts. In *hypothesis testing,* strict ideas about the phenomena under study are tested. More will be said about these two analytical approaches in the next chapter.

MANAGERIAL CONSIDERATIONS

An astute manager should take an active role in how the data is prepared for analysis, as well as the way it is to be analyzed. These considerations are fundamental to the materials presented in this chapter and the subsequent chapters on advanced analysis for the following reasons:

Without good data analysis planning, the results of the study may be of little value to the manager. Regardless of how good the measurement and data collection processes are, errors of incompleteness and inconsistency in the data could ruin the analysis and ultimately corrupt the results of a study.

The way the data is analyzed actually sets limits on how the results of the study can be (a) generalized and (b) used to make decisions. As discussed earlier, some business research does not require generalization. However, if generalization of results is desired, the analytical process must be well thought out and implemented.

Once the data is stored in a data storage medium and the analysis run, the GIGO rule applies. The old adage—garbage in, garbage out—applies to more than data. It also means that the proper analysis must be used for the level of measurement in the data. If the data is improperly analyzed, it can never become useful decision information.

How can managers get involved in the data analysis planning stage of the research process? They can implement guidelines that the doers (the researchers) must follow before the actual analysis is conducted. These guidelines are essential to a good analysis.

Require a clearly defined preanalytical plan. Such a plan will ensure completeness, clarity, and consistency in the collected data before it is analyzed. Also, it will increase the likelihood that the data will be converted to useful decision information.

Require that a data analysis plan be drawn up before the actual analysis is performed. This will ensure that the analysis will be in line with the research problem, and the information provided in the analysis will be useful in solving the problem. Additionally, the plan will provide a last-minute check on the adequacy of the analysis, before computer runs are made.

Require tabulations and/or scatter diagrams to be run on all variables whenever possible. This will ensure the last-minute accuracy of the data before the analysis. Additionally, valuable information, useful for the interpretation of the results, can be obtained from this basic analysis.

Require the researcher to explain the basic tabulations or scatter diagrams of the variables and how they might affect the analysis. This makes the researcher think about the analysis again. It helps ensure that the analysis will provide the results that are needed.

Require the researcher to justify all results and conclusions from the analysis. This forces management and the researcher to communicate at the same level. It ensures that the analysis yields the information that management needs to solve the problem.

This framework can also be stated in the form of key managerial questions pertaining to the management process, presented in Exhibit 12.5.

1. Is the purpose of the analysis discovery or testing a hypothesis?
2. Is the collected data what is needed to answer the research problem?
3. Is the collected data clear, complete, and consistent?
4. Is the collected data in the form needed for analysis?
5. How will missing data, "no opinion," and "don't know" be handled in the analysis?
6. What are the analytical hardware and software needs?
7. Is the analytical technique the correct one for the level of measurement of the data?
8. Will the information generated in the analysis accomplish the objectives of the study and answer the research question(s)?

SUMMARY

Planning for data analysis usually involves both personal and computerized means of ensuring that the data collected is ready for subsequent analysis. The chapter begins by outlining primary planning issues in this important function. Researchers should develop a data analysis plan, a systematic approach in preparing and analyzing data to provide appropriate information for decision-making purposes. This process entails (1) selecting the appropriate analytical software or online research supplier, (2) conducting preanalytical steps to establish clarity, completeness, and consistency in the data, (3) identifying the statistical procedures for subsequent data analysis, and (4) establishing a general framework to organize the results for presentation purposes.

At the beginning of the chapter, software packages were surveyed to provide insight into the first stage of the planning process. The discussion then moved to a seven-step preanalytical process for data purification. Then a basic analytical framework for business research was presented to guide the reader in selecting statistical procedures for data analysis.

Business research has two possible purposes. One purpose, discovery, requires systematic but nonscientific analytical methods. The other purpose, hypothesis testing, requires both systematic and scientific analytical methods. What differentiates the two systems of analysis is generalizability. Scientific analytical methods, coupled with scientific implementation of the entire research process, are the only means of generalizing results of a study.

The chapter ends with managerial concerns in analysis. Three major reasons for management's concern are outlined, and a five-point guideline for a manager's involvement in the analytical process is proposed. A list of key questions pertaining to the data analysis process provides managers with guidelines for implementing this important planning function in the research process.

DISCUSSION QUESTIONS

1. What is a data analysis plan? What are the major components of this plan? Describe their importance to the conduct of good business research.

2. Many statistical packages are available for analytical purposes. Choose a major software package and do an analysis on its major capabilities and limitations. Be sure to check appropriate outside sources in any analysis you undertake.

3. If your firm employs a large data collection agency to collect its data, how can your firm be assured that correct data analysis steps are taken?

Can you be sure of what occurred at each step? If you do follow these steps, should the results be incorporated into the final report?

4. Why should one always insist on pretest edits? What are the potential costs and losses of failing to do this? What are the gains?

5. How does one ascertain the presence of inaccuracies or ambiguities in a pretest instrument? Can there be doubt as to their presence? How does one resolve such doubts if they occur?

6. What types of errors can result from poorly planned data encoding? If data is improperly coded, what information losses occur? Can all of these be properly traced? Can they all be properly fixed? What costs are associated with such problems? Will you ever really know these problems if you employ data collection firms to do the work and you don't audit a copy of the original questionnaire yourself?

7. What are the uses of data tabulation? Can these be extended to include checks of coding and question wording accuracy? Can this step discover inaccuracies? Can this step solve business problems and, if so, what are some examples?

8. What is the difference between scientific and nonscientific business research? Can nonscientific research have hypotheses? If nonscientific research can have hypotheses, can they be resolved without using science?

9. Using Table 12.1, pick a statistical analysis package, then go to its Web site and write a synopsis on the capabilities of the package. Be prepared to present and discuss your findings in class.

10. Go to http://www.outsourcing.com, the Web site of the Outsourcing Institute. Using this site, summarize and evaluate the information provided by the Institute. Be prepared to discuss your findings in class.

NOTES

1. For a listing of some of these packages, see "1998 Marketing Technology Directory: Software/Internet Services," *Marketing News* (13 April 1998), 28.

2. These packages are registered trademarks: SPSS for SPSS, Inc.; SAS for SAS Institute, Inc.; and Excel for Microsoft Corporation.

3. Lotus 1-2-3 is a registered trademark of the Lotus Development Corporation. For a good book explaining newer data analysis capabilities with Excel, see Michael R. Middleton, *Data Analysis Using Excel* (Belmont, CA: Duxbury Press, 2000).

4. See the latest versions of *SPSS for Windows Base System User's Guide* and related manuals for the plug-in modules available from SPSS Inc., 444 North Michigan Ave., Chicago, IL 60611 (http://www.spss.com).

5. See *SPSS Data Entry Builder™ 1.0: User's Guide* (Chicago, IL: SPSS Inc., 1998).

6. See SAS Institute, *SAS Learning Edition 1.0*, cd-rom version (Cary NC: SAS Institute, 2002), and Geoff Der and Brian S. Everitt, *Handbook of Statistical Analyses Using SAS*, 2nd ed. (Boca Raton, FL: Chapman & Hall, CRC Press, 2001).

7. This section is largely drawn from Kenneth N. Berk and Patrick Carey, *Data Analysis with Microsoft Excel* (Pacific Grove, CA: Duxbury Press, 2000).

8. See John Sonquist and William C. Dunkelberg, *Survey and Opinion Research* (Englewood Cliffs, NJ: Prentice-Hall, 1977), and Neil J. Salkind and Delbert C. Miller, *Handbook of Research Design and Social Measurement* (Thousand Oaks, CA: Sage, 2002).

9. See, for example, Earl Babbie, *The Practice of Social Research*, 9th ed. (Belmont, CA: Wadsworth, 2001), Chapter 6; and Paul E. Green, Donald S. Tull, and Gerald Albaum, *Research for Marketing Decisions*, 5th ed. (Englewood Cliffs, NJ: Prentice-Hall, 1988), Chapters 8 and 10.

10. There are no hard-and-fast rules for what numbers to assign to categories of a variable, including missing values, "don't know," or "does not apply." These are usually left up to the researcher or determined by the analytical software used.

11. See Fred N. Kerlinger and Howard B. Lee, *Foundations of Behavioral Research*, 4th ed. (Fort Worth, TX: Harcourt Publishers, 2002).

SUGGESTED READING

Selecting Appropriate Software

As noted earlier, there are many good analytic packages that are fully referenced and documented. Contact information on the packages covered in this section:

SPSS
SPSS Inc.
444 North Michigan Ave.
Chicago, IL 60611
Tel: (312) 329-3500
     (800) 543-6609
FAX: (312) 329-3668
http://www.spss.com

SAS
SAS Institute Inc.
SAS Campus Drive
Cary, NC 27513
Tel: (919) 677-8000
FAX: (919) 677-8123
http://www.sas.com

Good introductory texts for these and other packages:

Der, Geoff, and Brian S. Everitt, *Handbook of Statistical Analyses Using SAS,* 2nd ed. (Boca Raton, FL: Chapman & Hall, CRC Press, 2001).

Green, Samuel B., and Neil J. Salkind, *Using SPSS for the Windows and Macintosh: Analyzing and Understanding Data,* 3rd ed. (Upper Saddle River, NJ: Prentice-Hall, 2002).

Hamilton, Lawrence C., *Statistics with STATA,* 4th ed. (Pacific Grove, CA: Duxbury Press, 2002).

Ryan, Barbara F., and Brian L. Joiner, *Minitab Handbook,* 4th ed. (Belmont, CA: Duxbury Press, 2001).

SAS Institute, *SAS Learning Edition 1.0,* CD-ROM version (Cary, NC: SAS Institute, 2002).

The Preanalytical Process

Salkind, Neil J., and Delbert C. Miller, *Handbook of Research Design and Social Measurement* (Thousand Oaks, CA: Sage, 2002).

A useful handbook to assess earlier stages in the analytical process. Describes the determination of the types of research that are often undertaken.

Sonquist, John A., and William C. Dunkelberg, *Survey and Opinion Research* (Englewood Cliffs, NJ: Prentice-Hall, 1977).

Chapters 1–6 offer an excellent overview of the preanalytical processes, consistent with our presentation but in more depth.

Basic Analytical Framework for Business Research

Albright, Christian, Wayne Winston, and Christopher Zappe, *Managerial Statistics* (Pacific Grove, CA: Brooks/Cole, 2000).

Provides a practical and easy-to-read introduction to managerial statistics. Focuses on the frameworks and techniques to help managers make better decisions.

Keller, Gerald, and Brian Warrack, *Statistics for Management and Economics,* 6th ed. (Pacific Grove, CA: Brooks/Cole, 2003).

Chapters 1 and 2 provide a nice introduction to the preanalytical steps in the business research process.

13

Basic Analytical Methods

Overview

Classification of Analytical Methods by Purpose

Exploratory Data Analysis

Basic Methods of Assessing Association

Crosstabulation

Contingency Correlation

Spearman Rank Correlation

Pearson's *r*

Basic Methods of Assessing Differences

Chi-Square (χ^2) Test

Z-Test for Differences in Proportions

t-Test for Differences in Means

Summary

Discussion Questions

Notes

Suggested Reading

Appendix: Additional Basic Analytical Techniques

OVERVIEW

ata becomes meaningful only after analysis has provided a set of descriptions, relationships, and differences that are of use in decision making. Thus, all data analysis is driven by the purpose of the investigation and the needs of the decision maker. The researcher must choose the appropriate analytical methods and approaches for the given research purpose. This is the focus of this chapter.

The chapter begins with the classification of analytical methods in terms of the purpose of the investigation. This classification system is based on the analytical framework for business research presented in Chapter 12. Research conducted for discovery purposes is distinguished from research conducted for hypothesis testing. Research conducted for discovery purposes uses descriptive statistics, and research conducted for hypothesis testing uses inferential statistics. The primary purpose of descriptive statistics is to describe and characterize the sample under study. On the other hand, inferential statistics is designed to draw conclusions about a population from sample information. The same statistical techniques can be used in either type of statistics, but they differ in the purpose and goal of the analytical process and the way in which the techniques are applied.

Once the classification scheme is fully developed, exploratory data analysis is briefly presented and explored. Next, basic methods of assessing associations and differences are presented. The methods of association discussed are crosstabulation, the contingency correlation, the Spearman rank correlation, and Pearson's r. The methods of assessing differences covered are the chi-square (χ^2) test, the Z-test, and the t-test. Presentation of these basic analytical methods sets the stage for the more complex statistical methods in the following two chapters.

CLASSIFICATION OF ANALYTICAL METHODS BY PURPOSE

The purpose of any analytical method is to convert data into information needed to make decisions. Scientific (SBR) and nonscientific business research (NSBR) differ in the level of generalization of outcomes. In SBR, the goal is to make generalizations to the population of interest. In NSBR, the results of the study cannot be applied to some larger population. The results merely describe the sample, and no inferences can be made to a larger population.

For example, management might want to describe the buying characteristics of customers in certain markets. A study is designed to profile all customers in these selected markets, summary statistics are calculated, and the customers are profiled. In this case, there is no intent to generalize this profile to all of the firm's customers—nor would it be appropriate to do so.

TABLE **13.1** Classification of Analytical Methods by Purpose

	Purpose	
	Discovery	**Hypothesis Testing**
Definition	Uncovering important phenomena that may describe or be related to a situation in some way.	Purposefully and statistically testing phenomena that are hypothesized to describe a situation or be related in some way.
Level of satisfaction	Descriptive; no generalizations to the population.	Inferential; generalizations to the population.
Examples	1. Description of age, sex, and attitudinal characteristics of a group of nurses (summary statistics).	Inferring that the age, sex, and attitudinal characteristics are represented in the population (statistical inference).
	2. Is there a possible relationship between age and attitude in the sample? (cross-breaks, Pearson's r)	Is there a statistical relationship between age and attitude in the sample? Can we generalize the relationship to the population? (contingency chi-square, or statistical measures of association)
	3. Are there any differences in age and attitude across gender in the sample? (tests for differences such as the t-test and χ^2—no interpretation of significance)	Are there any differences in age and attitude across gender in the sample? Can we generalize these differences to the population? (tests for differences such as the t-test and χ^2—interpretation of significance)

However, if the stated goal is to show that these characteristics differ among markets, then the researcher must set stricter guidelines for the analysis. In this instance, a test of hypotheses can be used to generalize the outcomes to real market conditions—inferential generalizations to the population. In Table 13.1, this distinction is listed as the level of statistics.

This distinction will become clearer in the following example. Suppose a group of 40 nurses is asked three questions: (1) How old are you? (2) Are you male or female? (3) Do you like or dislike your present job? The goal of the study is to get a general view of the age and sex composition of the nurses in the hospital and a general idea of their job satisfaction. The results are shown in Table 13.2.

TABLE **13.2** Results of Nursing Study: Discovery

Question	1		2		3	
Variable	Age		Sex		Job Satisfaction	
Category	≤30 yrs	>30 yrs	Male	Female	Like	Dislike
Number	15	25	10	30	10	30
Percentage of total	38%	62%	25%	75%	25%	75%

Suppose the goal of the study is exploratory and descriptive, and the purpose of the analysis is discovery. The analytical method used is tabulation, and the summary statistic in each instance is a percentage. Age and job satisfaction could have been measured on an interval scale, and the summary statistics for these two variables would have been averages. In any event, from the analysis we have *discovered* the compositional nature of the group of nurses. Sixty-two percent are over 30 years old; 38% are 30 years old or less. Seventy-five percent are female, and 25% are male. It is also discovered that 75% of the nurses dislike their job, whereas 25% like it. Has the goal of the study been accomplished? Certainly, it was discovered that the nursing staff might be older, more female, and more dissatisfied. But can the information be useful? In this form, the information cannot be used to generalize to all the nurses in the hospital. Why? Because it was not subjected to rigorous study in SBR. However, from the discovered information, a research hypothesis might be proposed for each of the findings. Then it could be analyzed using *inferential statistics* and a *statistical test of hypothesis.*

A research hypothesis states the expectations of the researcher. It is a tentative declarative statement about a phenomenon. Most often, it is an educated guess, based on a theory or a model, or derived from the exploratory phase of research. To be useful, a research hypothesis must be supported by a statistical test of hypothesis and a logical defense based on the collected data.

Usually, the statement of the research hypothesis follows directly from a well-thought-out and properly specified problem definition. A research hypothesis should be declarative and operational. It should reflect a guess at a solution that is based on some knowledge, previous research, or identified needs of the population under study. However, most important, a research hypothesis must be testable.

Inferential statistics forms the basis of generalizing results from a probability sample to its corresponding population. Sample statistics are used to infer population parameters. Additionally, inferential statistics enables a researcher to perform the required statistical test of hypothesis in SBR.

The *statistical test of hypothesis* (STOH) follows from the research hypothesis, but it is the specific form that is actually tested. It is sometimes thought of as the empirical hypothesis because it is stated in terms of statistics or numbers. It is frequently stated in null form (no relationship or difference)—thus, the term *null hypothesis,* which is encountered in statistics texts. The null hypothesis is a convention postulating a structure that must be torn down. This convention allows the researcher to test the hypothesis that the relationship or difference is due to chance or sampling error. Then the assumption is subjected to a strict probability test. If the researcher rejects the null hypothesis, there *might* be an actual relationship or difference that can be generalized. Similarly, if the researcher cannot reject a null hypothesis, there *might not* be an actual relationship or difference that can be generalized.

EXHIBIT **13.1**

Steps in Scientific
Hypothesis Testing in
Business Research

1. Formulate a research hypothesis to solve a specific problem.
2. Develop some of the decision consequences, in operational terms, from the possible results of testing the research hypothesis.
3. Collect the data of interest.
4. Determine if the data generally supports the research hypothesis.
5. If so, and the hypothesis is to be tested statistically, state the null and alternative hypotheses.
6. Select a test procedure that is appropriate for the collected data and will yield evidence regarding the truth of the null hypothesis.
7. Conduct the test on the data.
8. Make a decision about the veracity of the null hypothesis—the significance criterion.
9. Statistical significance may enable you to substantiate your research hypothesis and implement your decision.

The discussion of hypothesis testing continues later in this chapter. For now, remember this from the previous example: Exploratory research indicates that a problem may exist with nurses' job satisfaction (see Table 13.2). Why? Maybe the morale is low or the work environment is not up to par. For whatever reason, management is interested in determining whether job satisfaction is low. Management consensus in the hospital is that about 65% of the nurses should like their jobs and that only 35% should dislike their jobs. They agree on testing the situation in the SBR using a research hypothesis and subsequent STOH. The research hypothesis is simply stated: *Nurse job satisfaction might be low in the hospital.*

To test this hypothesis, data are collected by asking a random sample of 50 nurses if they like their jobs. (A random sample is used because of cost and efficiency.) The proper summary statistics are calculated. The statistical null hypothesis is formed. Then the statistical hypothesis is tested using an analytical method. This analytical method is chosen so that evidence can be obtained to determine the truth of the null hypothesis. The general process for testing hypotheses in the SBR is summarized in Exhibit 13.1. The actual conduct of a statistical test of hypotheses follows another set of steps. These can be generalized as follows:

1. Calculate the needed statistics from the collected data.
2. State the null and alternative hypotheses.
   a. Consider the direction of the test.
   b. Use parameters in the statements.
3. Determine the appropriate statistical test and set the level of alpha (the level of significance for the STOH).
4. Conduct the test and determine whether the null hypothesis should be rejected.

In our SBR example, the results from the random sample of 50 nurses are as follows: 30% of the nurses reply that they like their jobs, and 70% reply that they dislike their jobs. Remember, management believes that 65% of the nurses should like their jobs. Therefore, management consensus is that only 35% should dislike their jobs. This 35% figure is the population proportion of dissatisfied nurses (according to management). Therefore, the data generally support the scientific research hypothesis that nurse satisfaction in the hospital is low. Can more support be added to the findings? Yes, if the data is subjected to a statistical test of hypothesis. These statistical tests of hypothesis are in the realm of inferential statistics and rely on the philosophy of sampling distributions discussed in Chapter 9. The steps for the testing of the satisfaction data are as follows:

1. Calculate p:

    $p = f/n$

    where f is the number of sample units that possess the characteristic in question

    $$p = \frac{35 \; (disliked \; job)}{50 \; (total \; nurses)} = 0.7$$

    This is the sample statistic p and is an estimate of π, the population proportion.[2]

    $$\hat{\sigma}_p \; (\text{estimate of standard error of } p) = \sqrt{\frac{p(1-p)}{n}}$$

    $$= \sqrt{\frac{0.7(0.3)}{50}} = \sqrt{0.0042}$$

    $$= 0.0648$$

2. State the null and alternative hypotheses.

    H_0: $\pi = 0.35$ (no difference), where π is the population proportion

    H_1: $\pi \neq 0.35$ (difference)

3. The appropriate test is a hypothesis test involving one proportion. Alpha is set at 0.01. Approximate z critical values relative to $\alpha = 0.01$ for this test are −2.58 and 2.58. (See Appendix B.)

    $$z = \frac{p - \pi}{\sqrt{\dfrac{p(1-p)}{n}}}$$

4. Conduct the test and determine if p is significantly different from 0.35 at the $\alpha = 0.01$ level.

    $$z = \frac{0.7 - 0.35}{\sqrt{0.7(0.30)/50}} = \frac{0.35}{0.0648} \cong 5.40$$

If z calculated is greater than positive z critical or less than negative z critical, reject H_0 and accept H_1. So we conclude that there may be a significant difference between what management believes and what actually is. Can management conclude that 70% of all nurses in the hospital actually dislike their jobs? Not really; it is never known whether the null or the alternative hypothesis is true. In some instances, the researcher may have been misled by sample information. This situation is sometimes called the *researcher's dilemma:* It involves the risk of making a decision error in rejecting or accepting the null hypothesis.

For a business decision maker, making a decision mistake, by relying on a weak test of a decision proposition, could mean substantial losses for the company. Similarly, a wrong decision could also mean opportunity loss. The decision maker is generally more aware of the dollar losses than the opportunity losses simply because dollar losses are more visible. This is why most researchers and decision makers go all out to avoid a Type I error—the kind committed when the researcher rejects H_0 and it is actually true. The probability of making this error is alpha (α). A Type II error is committed when the researcher accepts H_0 when it is actually false. A researcher can manipulate the probability of making a Type I error by decreasing the value of alpha (α).

For example, in the nursing problem, reduce alpha from $\alpha = 0.01$ to $\alpha = 0.001$. This in turn increases the value of beta (β), the probability of making a Type II error. Beta can be reduced by increasing the sample size. Because the business researcher can determine the size of α and the sample size, some control over the STOH decision error can be established. Because costs of sampling are a constraint in business research, alpha is usually manipulated rather than the sample size being increased.

In the nursing example, a very small alpha was used, so the risk of making a Type I error is very small. This consideration is most important to management. Given this information, can it now be concluded that 70% of all the nurses in the hospital dislike their jobs? Yes, but with extreme caution. Working with probability can often create illusions. In this illustration, what the researchers have going for them is the fact that the data generously support the research hypothesis. This dispels much of the risk of illusion. In summary, a research hypothesis must be substantiated by the results of *both* the sample data analysis and the statistical test of hypothesis. Now management can do something to improve the nurses' morale and change their job satisfaction.

This fairly elementary example has served to highlight the differences in the use of basic analytical methods in descriptive and inferential statistics. It should be obvious by now that analytical techniques can be used for either purpose of business research—discovery or hypothesis testing. The major difference is whether the results are to be generalized to a population, and if so, how.

Although many schemes are used to categorize statistical techniques, the most widely accepted is the parametric–nonparametric scheme. This classification is important because it deals directly with the underlying assumptions of the data itself. *Parametric statistics* is based on rather rigid assumptions related to the normality of the variables under study. On the other hand, *nonparametric statistics* is used with categorical data; it does not require an underlying normal distribution of the data. Nonparametric methods are sometimes called *distribution* or *sample-free* statistics.

Nonparametric statistics are extremely important in business research because they allow analysis of nominal-level and ordinal-level data. However, if the underlying data meets the assumptions of parametric techniques, it is desirable to use the more powerful methods of analysis because they generally have greater power in detecting real similarities and differences in the variables. Because this classification method is widely accepted, the parametric–nonparametric classification is the one used in later sections of the text.

The scale of measurement in which the data has been collected is very important to the way it can be analyzed. This is why the tenets of the research process stress the interrelationship of steps. Every step is planned to provide the most powerful yet efficient analysis. For example, one of the goals of the measurement process is to create variables that can be analyzed or manipulated in the study design. Therefore, this must be planned in conjunction with the analysis step of the research process. Also, the choice of descriptive and inferential statistics is dictated by the type of measurement or data collected in the study. (Refer to Chapter 7, Table 7.1.)

As one can see, a number of significant decisions go into the selection of an analytical and/or modeling procedure to answer research questions. Previously, the only real guides were largely the experience of the researchers themselves and often fragmented literature on the subject. However, as was noted in Chapter 12, many statistical packages now provide "statistical coaches" to help the user/researcher pick modeling tools to achieve the objectives of the research.

Figure 13.1 presents screen captures from SPSS's Statistics Coach showing the selection of analytical tools available in answering the question, What do you want to do? As you will note from the questions in the first screen of the Statistics Coach, basic exploratory or descriptive choices are possible, as are choices relating to the differences and relationships between variables and/or groups. Other, more sophisticated choices are available (groups of similar cases and groups of similar variables) if the research purpose so dictates.

The remainder of this chapter and the following two chapters deal with basic statistical modeling techniques in a systematic fashion. First, exploratory data analysis will be briefly examined as a means to summarize, describe, present, and gain insight into data for further analysis. Next, the two approaches of analyzing basic associations and differences will be reviewed by examining several basic statistical methods used in each approach. Subsequent chapters will deal with more sophisticated modeling techniques currently used in business research.

However, before we examine the basic statistical methods for assessing associations and differences, let us take a look at some of the more visual ways to look at the data of interest—namely, exploratory data analysis (EDA).

FIGURE 13.1

Screen Captures
Illustrating the Use of
SPSS, Inc., Statistics
Coach

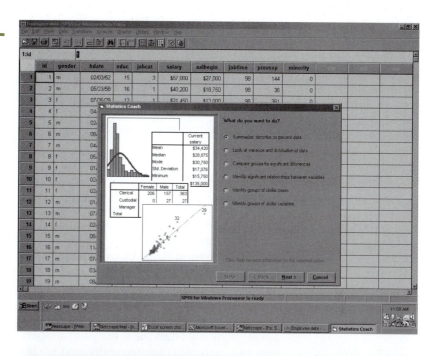

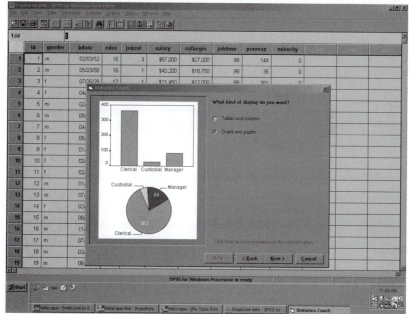

EXPLORATORY DATA ANALYSIS

Exploratory data analysis (EDA) is both an analytical approach to gain insight into the nature of the data and the mechanism to bring meaning to the data at hand.[3] EDA is largely data-driven, preferring graphical and visual displays to summary statistics. Visual displays are seen to bring meaning where summary statistics often hide underlying patterns and significant relationships of interest to the manager and researcher alike. Essentially, EDA is the forerunner to many of the more sophisticated data-driven mining techniques.

Its purpose is to explore the nature and characteristics of the data at hand in the most simplistic manner possible.

Although most of the techniques useful to EDA are far from new to the field of research, their value is widely recognized. First, simple *frequency tables, bar charts,* and *histograms* have been found to be extremely valuable in visualizing trends and distributions in the data that give insights to further analysis. As one can see in Figure 13.1, charts and graphs are an integral part of data analysis. In fact, Excel, SAS, STATA, and most other analytical packages all come with some type of exploratory data analysis tools.

One of the primary goals of visualization is to ensure that prior assumptions are correct, and if they are not correct, to decide how the new assumptions will affect the subsequent analysis chosen. For example, *boxplots* and *scatterplots* are frequently used to check the underlying distributions and skewness of data before more sophisticated analyses are undertaken. Without these basic visual checks, researchers can make significant mistakes in the application and interpretation of more sophisticated analytical modeling tools available to the researcher.[4]

BASIC METHODS OF ASSESSING ASSOCIATION

How are things related? The media bombard the public with relationships: between smoking and lung cancer, between eating foods high in saturated fat and heart problems. Are these statements useful and truthful? Finding associations helps us understand the relationship better, and knowing the strength of the relationship can lead us to a whole new realm of scientific research.[5] Also, a knowledge of the basics of association is fundamental to the understanding of more complex analytical procedures. Some of the commonly used methods of association are outlined in Table 13.3. This chapter covers four of these methods: crosstabulation, the contingency correlation, Spearman rank correlation, and Pearson's r. The reader is encouraged to investigate other methods of association as the need arises.[6] Remember that any of these methods can be applied for either of the two analytical purposes discussed earlier.

TABLE 13.3 Appropriate Method of Association Given Combinations of Measurement

Measurement of Variable X_1	Measurement of Variable X_2		
	Nominal (Dichotomous)	Ordinal	Interval or Ratio
Nominal (dichotomous)	Crosstabulation Phi coefficient Contingency correlation*	Rank–biserial correlation	Point–biserial correlation
Ordinal	Rank–biserial correlation	Spearman rank Kendall tau	Convert X_2 variable to ranks, and use Spearman or Kendall tau.
Interval or ratio	Point–biserial correlation	Convert X_1 variable to ranks, and use Spearman or Kendall tau.	Pearson product-moment correlation

*Special case when there are more than two categories of each variable.

Adapted from *Statistical Methods in Education and Psychology,* pp. 155–191. Copyright © 1970 by Prentice-Hall, Inc. Reprinted by permission.

Crosstabulation

The easiest way to look for association is to use crosstabulation, frequently called *cross-break.* The cross-break can be used with any level of measurement, because it uses categorical data that can be easily formed from other than nominal data. Crosstabulation of data requires minimum quantitative knowledge; all that is needed is the ability to calculate percentages. Essentially, the calculation of percentages in the cross-break table follows from the earlier discussion of one-way tabulation of frequencies. The analysis of cross-breaks simply involves two-way frequency tabulation.

For example, suppose a researcher draws a sample of 100 workers to study whether a relationship exists between previous work experience (WORKEXP—a yes-or-no response) and whether the respondents are presently working full-time (FULTIM—a yes-or-no response). These variables are both nominally scaled and contain two categories of measurement. A crosstabulation is simply formed with this data by crossing all levels of the variable WORKEXP with all levels of the variable FULTIM. The resultant table is shown in Table 13.4. It is a two-way cross-break (crosstabulation) because there are two variables in the analysis. This crosstabulation expresses the simplest possible relationship.

The left margin (the rows) of the crosstabulation matrix is where the independent variable is positioned; the top margin (the columns) is where the dependent variable is positioned. The table is interpreted by examining the cell frequencies and the marginal frequencies. For example, row 1 of the table refers to the answer yes to the question concerning full-time job status. In total, 55 respondents answered yes to this question (a row marginal

TABLE 13.4 Full-Time Job and Previous Work Experience: A 2 x 2 Crosstabulation

Previous Work Experience (WORKEXP)	Currently Hold Full-Time Job (FULTIM)			Totals
	Yes		**No**	
Yes	33		22	55
		60%		40%	
	A²	(11)¹	B²	(12)¹	
No	17		28	45
		38%		62%	
	C²	(21)¹	D²	(22)¹	
Totals	50		50		100

¹These numbers are cell identifications—that is, (11) is row 1, column 1; (22) is row 2, column 2; and so on.
²These letters refer to calculations in Exhibit 13.2.

frequency total). Also, each cell in the row represents additional information. In cell 11, for example, 33 of those who answered yes to the full-time job status question also had work experience. Furthermore, 22 of the respondents with previous work experience do not presently hold a full-time job, and so on, until we exhaust all possible interpretations of cell and marginal frequencies.

It is easier to see the relationship between the variables and assess its strength if the frequencies are converted to percentages. The percentages are calculated along the rows of the matrix, or from the independent variable to the dependent variable. In some studies, it might not be possible to designate an independent and a dependent variable, but the row rule still applies. There is a statistical rule for doing this, which will become more apparent later in the chapter. For example, the frequencies in cell 11 would be converted to a percentage by dividing: $33/55 = 0.6$ or 60%, and so on. Most often, if the relatively higher percentages in a 2×2 table are on the diagonal, there is a strong relationship between the two variables. In Table 13.4, the data (percentages) seem to exhibit an inverse relationship between previous work experience and currently holding a full-time job. If the variables have been designated as independent and dependent variables, the relationship can be further interpreted as positive or inverse.

The interpretation of cross-breaks gets more complicated when the number of categories for each variable increases and if an additional variable is added to the analysis to refine the interpretation. Although important, these topics are beyond the scope of this text.[7]

Contingency Correlation

Although the percentages in Table 13.4 indicate that a relationship exists between the two variables, they do not provide a summary indicator of the strength of relationship. One such summary value for assessing the strength of relationship in a 2×2 crosstabulation table is called the *phi coefficient*. If the frequencies are arranged in a table of cell and marginal sums, phi can be calculated directly from the data. Exhibit 13.2 illustrates the phi (ϕ) calculation from the data in Table 13.4. Also illustrated is the general form for calculating strength of association for any cross-break table. This general form is called the contingency correlation coefficient (C), and it indexes the strength of association between categorical variables. Both measures of association, phi and C, have the general interpretation of showing stronger relationships as they approach 1. Strictly speaking, the range of phi is -1 to 1,

EXHIBIT **13.2**

Calculation of the Phi and Contingency Correlation Coefficients

PHI (ϕ)

Step 1   Put frequencies in tabular form

		Variable Y				FULTIM		
		B	A	B + A		22	33	55
Variable X		D	C	D + C	WORKEXP	28	17	45
Totals		B + D	A + C	N		50	50	100

Step 2   Calculate using formula

$$\phi \text{ (phi)} = \frac{BC - AD}{\sqrt{(A + C)(B + D)(B + A)(D + C)}}$$

$$= \frac{22(17) - 33(28)}{\sqrt{50(50)(55)(45)}} = \frac{-550}{2487} = -0.22$$

CONTINGENCY CORRELATION (C)

Step 1   Calculate expected cell frequencies

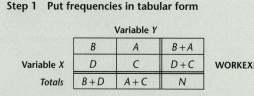

$$B_e = \frac{(B_o + A_o)(B_o + D_o)}{N} \qquad A_e = \frac{(B_o + A_o)(A_o + C_o)}{N}$$

$$D_e = \frac{(D_o + C_o)(B_o + D_o)}{N} \qquad C_e = \frac{(D_o + C_o)(A_o + C_o)}{N}$$

and so on, where o subscript means observed frequency and e subscript means expected frequency.

Step 2   Calculate contingency coefficient from formula

$$C = \sqrt{\frac{\chi^2}{\chi^2 + N}}$$

(Continued)

EXHIBIT 13.2

(Continued)

where

$$\chi^2 = \sum_{\text{over cells}} \left[\frac{(o-e)^2}{e} \right]$$

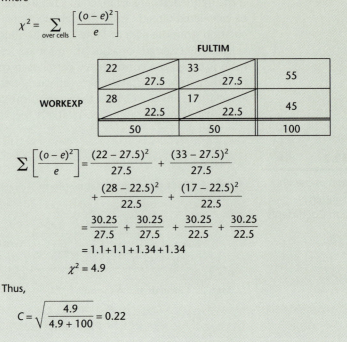

FULTIM

WORKEXP	22 / 27.5	33 / 27.5	55
	28 / 22.5	17 / 22.5	45
	50	50	100

$$\sum \left[\frac{(o-e)^2}{e} \right] = \frac{(22-27.5)^2}{27.5} + \frac{(33-27.5)^2}{27.5}$$

$$+ \frac{(28-22.5)^2}{22.5} + \frac{(17-22.5)^2}{22.5}$$

$$= \frac{30.25}{27.5} + \frac{30.25}{27.5} + \frac{30.25}{22.5} + \frac{30.25}{22.5}$$

$$= 1.1 + 1.1 + 1.34 + 1.34$$

$$\chi^2 = 4.9$$

Thus,

$$C = \sqrt{\frac{4.9}{4.9 + 100}} = 0.22$$

whereas the range of C is 0 to approaching 1. However, they must be interpreted with caution, because they have mathematical and statistical quirks.[8] For example, C does not actually ever reach 1, has no sign, and is not a strict analog of the Pearson r. Nevertheless, they are both useful indicators of relationships using categorical (nominal) measures.

Both the calculations in Exhibit 13.2 indicate a moderate relationship between the variables. The calculation of C relies on the χ^2 test of independence, which is explained in a later section in the chapter.

Spearman Rank Correlation

Sometimes it may be necessary to establish the strength of relationship between two ordinal variables. If the researcher cannot assume that these variables have interval properties (such as with a rating scale), then the appropriate form of assessing strength of association is the Spearman rank coefficient or Kendall tau.[9] To calculate these coefficients, the measurements must be ranked for each variable and difference scores calculated (ties must be handled in a special manner). These difference scores form the basis for computing the coefficients. As measures of association, these coefficients are excellent, producing about the same coefficient as the Pearson r on the same data.[10] For example, the formula for computing Spearman rank (rho) is as follows:

$$\text{Rho} = 1 - \frac{6\Sigma d^2}{N(N^2 - 1)}$$

where

  N = number of ranks

  d = algebraic difference for each rank in the two distributions of ranks

Biserial correlation and point–biserial coefficients have several disadvantages. Most often, if strength of association must be assessed, it is better to transform the data to a lower scale of measurement and use C or ϕ. As discussed in the next section, measures of association are subject to limits of interpretation. Without exception, the previous methods are subject to the same limitations as the Pearson r, the most familiar measure of association. They should be used with caution. Second, the contingency coefficient (C) is affected by sample size (as are all statistical tests); it must be corrected before any interpretations can be made. The correction for continuity will be discussed in the section on χ^2 in this chapter. Finally, use cross-breaks only when it is actually necessary to do so. Otherwise, use the more powerful mode of analysis on the data (for example, do not calculate C for the determination of association between two interval variables).

Pearson's *r*

When the data collected possesses the properties of interval measurement, the appropriate indicator of association between the two variables is the Pearson product-moment (PM) correlation coefficient. The PM coefficient possesses important features that make it most useful in research. Not only can it be used to establish the strength of relationship, but it can also help the researcher determine the direction (positive or negative) of the relationship and the percentage of variation explained by the relationship. The importance of the first two properties will be made apparent. The third property is more useful in advanced analysis, thoroughly discussed in Chapters 14 and 15.

The summary statistic for the strength of relationship is called Pearson's r or PM. Its limits are (–1, 1) for determining strength and direction of relationship. The first step in calculating r is to plot the data. (Most often, a computer can do this for you.) The understanding of the visual relationship is most important in the interpretation of r. A positive relationship is shown in Figure 13.2(a) and a negative relationship in Figure 13.2(b). Because PM correlation is a linear relationship, positive and negative relationships are established by linear mathematical methods. A curvilinear relationship, as shown in Figure 13.2(c), requires the use of a different measure of association.

The plot of the data is necessary for two reasons: first, to determine whether there are outliers in the data that will interfere with the analysis; and second, to reveal any violation of the assumption of linearity in the data. Obviously, these factors both affect the interpretation of the coefficient.

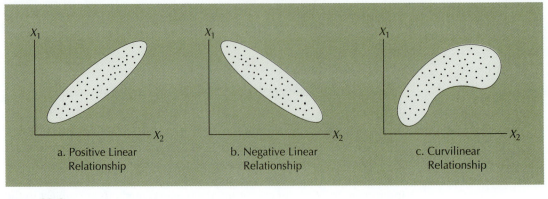

| a. Positive Linear Relationship | b. Negative Linear Relationship | c. Curvilinear Relationship |

FIGURE 13.2

Graphical Plots of Data

The PM coefficient is calculated mathematically by using linear algebra. It becomes a statistical value when we try to generalize it to the population. It must then be interpreted with caution, because the result could be spurious (occurring by chance alone). A statistical test of significance for r is affected by the sample size. Therefore, the significance of r is not very useful in generalizing r; only common sense can assist you in the generalization of the strength and direction of a relationship.

The data in Exhibit 13.3 is used to calculate r. In this instance, the strength of relationship between age and satisfaction, two intervally scaled variables, is assessed. The plot of the data in Exhibit 13.3 seems to indicate an outlier in the data that could change the interpretation. This is item number 10 (55, 20). If this pair is left in the analysis, it will produce a positive relationship when the rest of the data shows a more negative relationship. To illustrate this, the calculation of r is shown in Exhibit 13.3 with and without the outlier. The calculation indicates that the PM without the outlier is -0.216, whereas with the outlier it is 0.106 positive—a substantial shift in interpretation.

Another use of the PM correlation coefficient is in prediction, when an independent and dependent variable are specified. This procedure is called *simple linear regression*. Also, by squaring the simple correlation coefficient r, it is possible to assess the amount of variance or change in the dependent variable reflected by changes in the independent variable. This is called r^2, or the coefficient of determination. The range of this value is from 0 to 1. In the example, if r is squared $[(-0.216)^2 = 0.05]$, this shows that about 5% of the variance is explained in the system—not very much. These topics are important to the study of regression models of analysis, which are discussed in Chapter 14. Most often in business research, except in unusual circumstances, the data does not indicate very strong relationships. An r^2 value of 0.05 could be significant at a 0.01 level and be of little value to a researcher. The common-sense analysis of the data might indicate otherwise.

Sometimes the PM coefficient cannot be used because of the curvilinear nature of the data (see Figure 13.2(c)). In these instances, the more general form of correlation should be used. This coefficient is called *eta*. The calculation of eta is beyond the scope of this text, but a good statistics book will provide the necessary logic and formulae. The use of eta will be discussed in Chapter 14, when we develop analysis of variance techniques.[11]

EXHIBIT **13.3**

Product-Moment
Correlation—Pearson's *r*

DATA

Number	Age (X)	Satisfaction (Y)	XY	X^2	Y^2
1	26	12	312.00	676.00	144.00
2	22	15	330.00	484.00	225.00
3	30	15	450.00	900.00	225.00
4	30	20	600.00	900.00	400.00
5	19	20	380.00	361.00	400.00
6	22	14	308.00	484.00	196.00
7	22	22	484.00	484.00	484.00
8	20	24	480.00	400.00	576.00
9	18	15	270.00	324.00	225.00
10	55	20	Possible outlier		

$\Sigma X = 209$ $\Sigma Y = 157$ $\Sigma XY = 3614$ $\Sigma X^2 = 5013$ $\Sigma Y^2 = 2875$ (without outlier)

$\Sigma X = 55$ $\Sigma Y = 20$ $\Sigma XY = 1100$ $\Sigma X^2 = 3025$ $\Sigma Y^2 = 400$ (outlier)

$\Sigma X = 264$ $\Sigma Y = 177$ $\Sigma XY = 4714$ $\Sigma X^2 = 8038$ $\Sigma Y^2 = 3275$ (with outlier)

PLOT OF DATA

Correlation of Age with Satisfaction
Correlation Example

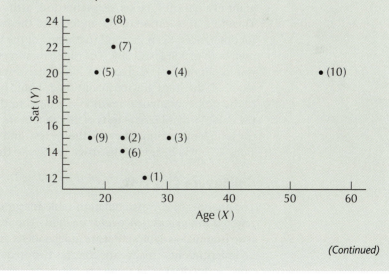

(Continued)

EXHIBIT **13.3**

(Continued)

CALCULATION

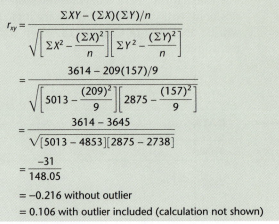

$$r_{xy} = \frac{\Sigma XY - (\Sigma X)(\Sigma Y)/n}{\sqrt{\left[\Sigma X^2 - \frac{(\Sigma X)^2}{n}\right]\left[\Sigma Y^2 - \frac{(\Sigma Y)^2}{n}\right]}}$$

$$= \frac{3614 - 209(157)/9}{\sqrt{\left[5013 - \frac{(209)^2}{9}\right]\left[2875 - \frac{(157)^2}{9}\right]}}$$

$$= \frac{3614 - 3645}{\sqrt{[5013 - 4853][2875 - 2738]}}$$

$$= \frac{-31}{148.05}$$

$= -0.216$ without outlier

$= 0.106$ with outlier included (calculation not shown)

BASIC METHODS OF ASSESSING DIFFERENCES

Business researchers are often interested in finding differences in data. Earlier in this chapter, an example was presented using the one-sample case, where the researcher tests the difference between a sample statistic and a population parameter. This was the one-sample test involving a population proportion. Although important, the one-sample test does not give the researcher the tools needed to detect group differences.

The detection of group differences is very useful. Managers can obtain enormous amounts of useful information from this type of analysis (program evaluation, market segmentation, attitude differences, and so on). Therefore, it is important to understand the fundamental univariate tests for differences. Keep in mind that many of the techniques discussed can be used both in NSBR and SBR. The differentiating factor is generalization, as it was in our discussion of association. When a statistical test of hypothesis is required, the steps previously outlined should be followed. Also, the choice of the proper analytical method is related to the level of measurement of the data. Two univariate tests of differences are discussed with examples in the following section. They are the χ^2 test, and the *Z*- and *t*-tests. Others in Table 13.5 will be briefly discussed relative to their importance.

Chi-Square (χ^2) Test

The analysis of a crosstabulation table for group differences is sometimes referred to as *contingency table analysis*. The test is used for nominal data (two nominal variables) that are independent. All events in the table should be independent. This means that no two frequencies can be based on the same individual. The same assumptions are true when the relationship between two nominal variables is assessed (as shown in the previous section).

TABLE 13.5 Univariate Tests for Differences—One or Two Samples

Measurement Level	One-Sample Case	Two Samples Independent Samples	Two Samples Dependent Samples	Statistical Classification
Nominal	Hypothesis test involving a sample proportion χ^2 goodness-of-fit test	Hypothesis testing involving two sample proportions Contingency table analysis	McNemar test	Nonparametric
Ordinal	Kolmogorov–Smirnov	Mann–Whitney Median test Kruskal–Wallis	Wilcoxon signed-rank test	Nonparametric
Interval and ratio	μ: Hypothesis test involving a sample statistic (t-test)	t-test of differences	t-test ($\bar{d}$)	Parametric
	σ^2: χ^2 test	F-test (independent)	F-test (dependent)	Parametric

The χ^2 statistic is calculated as in Exhibit 13.4, but it is used in a test of hypothesis for group differences. Usually, the expected frequencies in the analysis should be greater than 5. Otherwise, a hypothesis test for two proportions might prove more useful. This is not a problem when N is large, but if N is small, the χ^2 statistic needs a correction factor.[12]

EXHIBIT 13.4

Chi-Square Test for Differences

Tardiness	Sex Male	Sex Female	Totals
Tardy	12	28	40
Nontardy	30	30	60
Totals	42	58	100

H_0: tardiness is independent of sex of employee
H_1: tardiness is dependent on sex of employee

where

$\alpha = 0.05$
d.f. $= (R - 1)(C - 1) = 1$ d.f.
tabled value of χ^2 with 1 degree of freedom is 3.84 (See Appendix B.)
$R =$ number of rows
$C =$ number of columns

Decision rule: If χ^2 calculated ≤ 3.84, accept H_0.
If χ^2 calculated > 3.84, reject H_0.

$$\chi^2 = \sum_{\substack{\text{across} \\ \text{cells}}} \frac{(o - e)^2}{e} = 3.94$$

Therefore, accept H_1. The events of the case may be dependent. Sex and tardiness may be dependent. Females might be tardy more frequently than males.

The null hypothesis in a contingency test is not based on a probability model but on the rules of probability for independent events. Therefore, it is also called a test of independence. Suppose a researcher would like to examine if there are any differences between males and females in a company with regard to tardiness. Management needs this information to implement new work rules. Records are checked for 100 individuals and sex (male or female) and tardiness (tardy or nontardy employee) are determined. The data is crosstabulated, and a χ^2 test is performed. The calculations are shown in Exhibit 13.4.

Notice that a new concept has been introduced: degrees of freedom (d.f.). This statistical concept is inherent in all tests of group differences using χ^2, t-, and F-distributions. Generally, degrees of freedom are used to provide unbiased estimators of parameters used in statistical analysis. *Unbiased* in this case means the *best* estimator. We can explain the use of degrees of freedom with the simple example of calculating the sample variance as an estimator of the population variance. The formula for S^2, as you remember, has a denominator of $(n-1)$ when used as an estimator. The reason that 1 is subtracted from n is because the mean $(\overline{X})$ is one of the values in the array (sample) that is used to calculate S^2. For example,

$$\frac{\sum_{i=1}^{n} (X_i - \overline{X})^2}{n-1}$$

Therefore, we must eliminate it from the calculation of S^2 by subtracting it out of the sample size. S^2 then becomes the unbiased estimator of σ^2. The use of unbiased estimators is the norm for all statistical analysis.[13]

From the analysis, it can be generally concluded that the variables are dependent. Because the calculated χ^2 is greater than the tabled value, the null hypothesis is rejected. Therefore, females might be tardy more frequently, because 28 out of 58 sampled were designated as tardy, whereas only 12 out of 42 males were designated as being tardy.

Z-Test for Differences in Proportions

Many tests of univariate differences apply the theory of difference distributions. These distributions are analogous to sampling distributions (Chapter 9) but involve two populations. Because these distributions approximate the shape of the normal distribution, the researcher can use the power of this distribution in the analysis. (See the appendix at the end of this chapter.) All group tests follow the general format of the difference scores of means (or proportions) divided by the standard error of the difference of means (or proportions).

The z-test of proportions is one of this class of statistical tests. It is used when the data is dichotomous and the samples are assumed to be independent. It is a two-tail test of hypothesis. The formula for calculating the z-statistic is

$$z = \frac{p_1 - p_2}{\sqrt{\bar{p}(1 - \bar{p})\left(\frac{1}{n_1} + \frac{1}{n_2}\right)}}$$

where

$$\bar{p} = \frac{n_1 p_1 + n_2 p_2}{n_1 + n_2}$$

and p_1 and p_2 are the proportions being tested.

The test uses the z table in any basic statistics text to determine the tabled z-value. The null and alternative hypothesis are stated in nondirectional terms. The decision rule requires that the calculated z exceed the tabled z to reject the null hypothesis of no differences between proportions.

For example, in a survey of 500 people from city A, 95 are found to earn less than \$10,000 per year. In city B, a survey of 300 people contains 60 who are earning under \$10,000 a year. Using this data, a researcher would like to determine if the proportion of respondents earning less than \$10,000 is the same in both cities.

We conduct the test, H_0: $\pi_1 = \pi_2$ against H_1: $\pi_1 \neq \pi_2$, where π_1 is the true proportion of individuals earning under \$10,000 in city A and π_2 is the true proportion of individuals earning under \$10,000 in city B. The p_1, p_2 proportions are sample estimates of these population parameters. Therefore,

$$p_1 = \frac{95}{500} = 0.19 \qquad \text{and} \qquad p_2 = \frac{60}{300} = 0.20$$

In addition,

$$\bar{p} = \frac{500(0.19) + 300(0.2)}{500 + 300} = \frac{95 + 60}{800} = 0.194$$

Then, using the formula for z,

$$z = \frac{p_1 - p_2}{\sqrt{\bar{p}(1 - \bar{p})\left(\frac{1}{n_1} + \frac{1}{n_2}\right)}} = \frac{0.19 - 0.20}{\sqrt{0.194(0.806)\left(\frac{1}{500} + \frac{1}{300}\right)}}$$

$$= \frac{-0.01}{0.0289} = -0.346$$

If the level of significance is set at $\alpha = 0.05$, the tabled z from the area under the normal curve (see Appendix B) is $z = \pm 1.96$. Because it is a nondirectional (two-tail) test, the null hypothesis H_0: $p_1 = p_2$ is rejected if the calculated z does not fall in the boundary set by ± 1.96. The calculated $z = -0.346$. Therefore, we must accept the null hypothesis and accept the claim that the proportion of individuals earning under \$10,000 is probably the same for both cities.

t-Test for Differences in Means

The *t*-test for differences is another test that uses difference distributions. It is a test of mean differences used on interval-scale measures. The major assumption of the test is that the samples are independently drawn from normal populations with equal population variances. If the independent sample assumption is violated, the major consequence is correlated sample means, which could seriously increase the probability of making a Type II error. If the researcher purposely uses correlated samples, a different t-statistic is calculated. Instances of both conditions are presented in Exhibit 13.5, where a researcher wants to test the statistical hypothesis that there are no differences in job satisfaction between workers with previous experience and those with no previous experience. Assume that the sample sizes of the groups of workers—previous experience and no previous experience—are $n_1 = 10$ and $n_2 = 10$, respectively. The satisfaction scores are shown for each group. The null hypothesis is that there is no difference in mean satisfaction between experience groups. The alternative hypothesis is that such

EXHIBIT **13.5**

The *t*-test

INDEPENDENT SAMPLES

1. $H_0: \mu_1 = \mu_2$

$H_1: \mu_1 \neq \mu_2$

$$\bar{X}_1 = \sum_{i=1}^{n_1} \frac{X_{i_1}}{n_1} \quad \text{and} \quad \bar{X}_2 = \sum_{i=1}^{n_2} \frac{X_{i_2}}{n_2}$$

where

X_{i_1} = measure of a characteristic of ith unit of analysis for group 1

X_{i_2} = measure of a characteristic of ith unit of analysis for group 2

2. H_0 is tested against H_1 by means of the following test statistic:

$$t = \frac{\bar{X}_1 - \bar{X}_2}{\sqrt{\dfrac{(n_1 - 1)S_1^2 + (n_2 - 1)S_2^2}{n_1 + n_2 - 2}\left(\dfrac{1}{n_1} + \dfrac{1}{n_2}\right)}}$$

where

$\bar{X}_1$ and $\bar{X}_2$ are sample means for groups 1 and 2

S_1^2 and S_2^2 are unbiased estimates of the common population variance, σ^2

$$S_1^2 = \sum_{i=1}^{n_1} \frac{(X_{i_1} - \bar{X}_1)^2}{n_1 - 1}$$

$$S_2^2 = \sum_{i=1}^{n_2} \frac{(X_{i_2} - \bar{X}_2)^2}{n_2 - 1}$$

n_1 and n_2 are the size of samples 1 and 2

3. Degrees of freedom for the test are

$n_1 + n_2 - 2$

4. *Decision rule.* If t calculated is greater than t tabled with $n_1 + n_2 - 2$ d.f. and a selected alpha level, then reject H_0 and accept H_1.

EXHIBIT **13.5**

(Continued)

5. *Data*

Group 1: Previously Experienced Workers	Group 2: Not Previously Experienced Workers
X_{i_1}, satisfaction scores $\begin{cases} 20 \\ 24 \\ 24 \\ 24 \\ 23 \\ 23 \\ 22 \\ 20 \\ 19 \\ 16 \end{cases}$	$\begin{cases} 14 \\ 24 \\ 22 \\ 19 \\ 18 \\ 18 \\ 18 \\ 16 \\ 15 \\ 14 \end{cases}$ X_{i_2}, satisfaction scores

$$\Sigma X_{i_1} = 215 \qquad \Sigma X_{i_2} = 178$$

$$\bar{X}_1 = 21.5 \qquad \bar{X}_2 = 17.80$$

$$S_1^2 = 7.167 \qquad S_2^2 = 10.844$$

$$n_1 = 10 \qquad n_2 = 10$$

$$t = \frac{21.5 - 17.8}{\sqrt{\frac{9(7.167) + 9(10.844)}{10 + 10 - 2}\left(\frac{1}{10} + \frac{1}{10}\right)}} = 2.76$$

Because 2.76 is greater than ±2.101, reject the null hypothesis.

DEPENDENT SAMPLES

1. $H_0: \mu_1 - \mu_2 = 0$

$H_1: \mu_1 - \mu_2 \neq 0$

2. a. Degrees of freedom = (number of pairs −1).

  b. Set α level.

  c. Determine tabled t-statistic (for example, if $\alpha = 0.05$, $n = 10$, then $t = \pm 2.262$).

3. Calculate

$$t = \frac{\bar{d}}{S_d \sqrt{n}}$$

where

  $\bar{d}$ = the average difference between pairs of scores

  n = number of pairs

$$S_d = \sqrt{\sum_{i=1}^{n} \frac{(d_i - \bar{d})^2}{n - 1}}$$

where d_i is the difference between matched pairs of scores.

4. If t calculated > t tabled, reject H_0, accept H_1.

differences exist. The researcher chooses an alpha value of $\alpha = 0.05$. The tabled t-statistic (Appendix B, Table B.3) for an $\alpha = 0.05$ for a two-tail test with 18 d.f. $(10 + 10 - 2)$ is ± 2.101. Because 2.76 (see Exhibit 13.5) exceeds ± 2.101, the researcher rejects the null hypothesis and concludes that there is a statistically significant difference in job satisfaction between experienced and nonexperienced workers and that the previously experienced workers are probably more satisfied.

A t-test of significance is a very robust test. In statistics, *robustness* means the efficiency of the test even when the assumptions are violated. Only two factors are of concern when using a t-test. First, the means are not correlated (if so, use an alternative test, which is shown in the second part of Exhibit 13.5). Second, the population variances are not equal (then use the Behrens–Fisher test).[14] This second factor can be ignored if the sample sizes, n_1 and n_2, are approximately equal. However, if an inappropriate independent t-test is performed with dependent groups, significant differences will often be masked as nonsignificant.

The appropriate t-test for dependent samples or paired data is calculated by the formula in Exhibit 13.5. The test might be used where the same subjects are measured at two time intervals, or store sales of a particular product with two differentiating features are examined. In the first case, the test assesses differences in the means for the two different time periods. In the latter case, the test assesses the difference in sales based on the differentiating features of the products. In each instance, the data is paired on a unit of analysis, so it is assumed to be dependent.

Other tests of differences mentioned in Table 13.5 are the Mann–Whitney U test, the median test, the McNemar test, and the Wilcoxon matched-pairs signed-ranks test. The Mann–Whitney U test examines differences in independently sampled ordinal data. The U statistic is interpreted from tables of U values. The median test is a series of tests to detect differences in groups based on the median of each group. When groups are related and the level of measurement is nominal or ordinal, the appropriate tests of hypothesis are the McNemar test for nominal data and the Wilcoxon test for ordinal data.[15]

SUMMARY

The subject of this chapter was the nature of basic analytical methods in business research. First, the analytical methods were classified by the purpose of the research being undertaken. The first purpose is discovery, for which descriptive statistics are appropriate. For the second purpose, hypothesis testing, inferential statistics must be used so that the sample information can be generalized to some relevant population. Particular statistical techniques can be used in either descriptive or inferential statistics, depending on the purpose of the study. The main steps in scientific hypothesis testing were presented, with an example provided to demonstrate the process.

Parametric statistics are based on rather rigid assumptions related to the normality of the variables under study. Nonparametric statistics are also called *distribution* or *sample-free* statistics because of their relaxed assumptions. This distinction is important because the level of measurement of the variable(s) under study is critical in the selection of an appropriate statistical method for analysis purposes.

The discussion then turned to some of the advancements made in the selection and use of analytical methods. Here, the concept of software-aided "statistics coaches" was introduced, and an example using SPSS software was given. This led into a discussion of exploratory data analysis techniques.

There followed a discussion of the basic methods of assessing association of variables and differences between groups on a specific variable. The basic methods of assessing association reviewed include crosstabulation, contingency correlation, Spearman rank correlation, and Pearson's *r*. The methods of assessing differences include the chi-square (χ^2) test, the Z-test for differences in proportions, and the *t*-test for differences in means. This chapter sets the framework for coverage of more advanced analytical methods in Chapters 14 and 15.

DISCUSSION QUESTIONS

1. What is inferential statistics? Someone once said, "The value of the inference depends on the data generated as well as the sampling." Do you agree? Why?

2. Have you done exploratory research? Were you systematic? If you had to do it again, what would you do differently? (For example, choosing a college to attend.)

3. What are Type I and Type II errors? Can these errors both decrease if you were to increase sample size? Can an increase in sample size decrease one while having no effect on the other? How?

4. Is there a relationship between the level (for example, nominal, ratio) of measurement and the choice of parametric versus nonparametric statistical tools? What impact does the resulting empirical level of data have on the choice of a statistical technique to test a hypothesis?

5. Is it possible that a researcher who intends to gather interval data winds up with ordinal data instead? What does it do to his or her plans for analysis if he or she winds up with mixed types?

6. When choosing a statistical test for a contingency table, such as those given in Table 13.5, what are the basic rules of thumb one should

follow? Suppose an analyst presented you (the decision maker) with a contingency table and accompanied it with a test you are not familiar with. What would you do?

7. How does sample size affect the usefulness of the Z-test and the t-test used to test the differences between groups?

8. What rules of thumb could you suggest to a manager who just received a long technical report based on survey data and who was overwhelmed by the statistical figures in the report?

9. A vice president of marketing of AT&T once asked, "What's so magic about using an α level of 0.05? Why should I believe you just because you say it is statistically significant at the 0.05 level?" Can you help him out?

10. The data in the table that follows was collected by a large car manufacturer concerning a new prototype car called the CIELO. Thirty carefully selected respondents were shown the car and fully briefed about its capabilities. Here is the data, which includes age (intervally scaled), sex (nominal), social status (interval scale ranging from 10, low status, to 30, high status), attitude toward CIELO (interval scale ranging from 6, negative attitude, to 30, positive attitude) and intention to purchase CIELO (nominal, yes or no).

   a. Is there any difference in male and female attitudes toward the CIELO?

   b. What is the relationship between intention to purchase and gender?

   c. Is there anything unusual about this data?

Respondent #	Age	Sex	Social Status	Attitude Score	Intention to Purchase
1	20	M	15.6	16	Y
2	19	F	17.5	15	Y
3	34	M	28.2	14	N
4	23	M	24.6	6	N
5	42	M	16.5	28	Y
6	55	F	12.2	24	N
7	24	F	12.5	19	Y
8	26	M	29.6	14	N
9	35	F	27.6	23	N
10	41	M	23.2	16	N
11	43	F	21.2	26	Y
12	51	M	20.2	28	N
13	56	M	19.4	14	N
14	62	F	18.6	12	Y

(Continued)

Respondent #	Age	Sex	Social Status	Attitude Score	Intention to Purchase
15	43	F	10.2	11	Y
16	51	F	14.3	10	N
17	28	M	12.6	22	Y
18	19	M	14.8	24	N
19	24	M	29.6	19	N
20	26	F	26.5	18	Y
21	28	F	23.2	20	N
22	35	M	17.9	23	N
23	38	F	19.9	23	N
24	23	M	20.1	25	Y
25	42	F	18.6	17	N
26	41	F	18.6	17	N
27	30	F	24.6	16	Y
28	29	M	26.6	13	N
29	19	M	14.5	22	Y
30	26	F	12.9	21	N

11. What is EDA? Go to the Web page or documentation of your choice and identify the techniques that would be helpful in EDA. Be prepared to discuss your findings in class.

NOTES

1. Most statistical tests of hypothesis in business research are nondirectional. Because directional tests are rarely used, we will discuss only nondirectional tests. The reader should be aware that some statistical procedures are designed to test nondirectional hypotheses using one-tail tests for decision making (for example, χ^2 tests). These will be covered in this and later chapters.

2. The finite correction factor was not used here because it provided no useful information. See Leslie Kish, *Survey Sampling* (New York: Wiley, 1965), pp. 43–45.

3. For discussions of exploratory data analysis, see John W. Tukey, *Exploratory Data Analysis* (Reading, MA: Addison-Wesley, 1977); and David C. Hoaglin and John W. Tukey, Eds., *Understanding Robust and Exploratory Data Analysis* (New York: John Wiley, 1983).

4. See Paul F. Velleman and David C. Hoaglin, *Applications, Basics, and Computing of Exploratory Data Analysis* (Boston: Duxbury Press, 1981).

5. Some researchers and scientists believe that there is no true scientific method to establish cause and effect. However, others believe that associative data can truly yield this cause-effect information.

6. See, for example, Jean Dickinson Gibbons, *Nonparametric Statistical Inference*, 2nd ed. (New York: Marcell Dekker, 1985); and Gilbert A. Churchill, Jr., *Basic Marketing Research*, 2nd ed. (Fort Worth, TX: Dryden Press, 1992), pp. 732–739.

7. See Jonathan D. Cryer and Robert B. Miller, *Statistics for Business* (Belmont, CA: Duxbury Press, 1994), pp. 128–131.

8. See William Lee Hays, *Statistics,* 5th ed. (Fort Worth, TX: Holt, Rinehart & Winston, 1997).

9. See W. J. Conover, *Practical Nonparametric Statistics* (New York: Wiley, 1980).

10. Jum C. Nunnally, *Psychometric Theory*, 2nd ed. (New York: McGraw-Hill, 1978), pp. 134–135.

11. See W. L. Hays and J. J. Kennedy; "The Eta Coefficient in Complex ANOVA Designs," *Educational and Psychological Measurement* (1970), 30, pp. 885–889.

12. $\chi^2 = \Sigma((|o - e| - {}^1/_2)^2/e)$ See Chris A. Theodore, *Managerial Statistics* (Boston: Kent Publishing, 1982), pp. 234–239.

13. See Fred N. Kerlinger and Howard B. Lee, *Foundations of Behavioral Research*, 4th ed. (Fort Worth, TX: Harcourt Publishers, 2002) for a brief discussion of degrees of freedom.

14. Ronald A. Fisher, *Statistical Methods and Scientific Inference* (New York: Hafner, 1959), pp. 93–97.

15. Examples and calculation of these statistics can be found in any good statistics book that stresses nonparametric statistics. The classic book is S. Siegel, *Nonparametric Statistics* (New York: McGraw-Hill, 1956), and its updated version, S. Siegel and N. J. Castellan, *Nonparametric Statistics for the Behavioral Sciences,* 2nd ed. (New York: McGraw-Hill, 1988).

SUGGESTED READING

Classification of Analytical Methods by Purpose

Albright, S. Christian, Wayne L. Winston, and Christopher Zappe, *Data Analysis and Decision Making with Microsoft Excel*, 2nd ed. (Pacific Grove, CA: Duxbury Press, 2003).

> *The text features an example-based spreadsheet approach to data analysis and modeling. Multiple purposes are evident in the combined analytical style.*

Babbie, Earl, *The Practice of Social Research*, 9th ed. (Belmont, CA: Wadsworth, 2001).

> *Chapter 17 on social statistics is a great complement to this chapter. It outlines and discusses descriptive and inferential statistics in a very understandable fashion.*

Exploratory Data Analysis

Hoaglin, David C., and John W. Tukey, Eds., *Understanding Robust and Exploratory Data Analysis* (New York: John Wiley, 1983).

> *Discusses the attitudes and philosophy underlying exploratory data analysis methods and examines the connections among exploratory techniques, conventional techniques, and classical statistical theory.*

Hermann, Hans Bock, and E. Diday, Eds., *Analysis of Symbolic Data: Exploratory Methods for Extracting Statistical Information from Complex Data (Studies in Classification, Data Analysis, and Knowledge Organization)* (New York: Springer-Verlag, 2000).

> *Examines recent methods for analyzing symbolic data and generalizes classical methods of exploratory, statistical, and graphical data analysis.*

Associations and Differences in Data Analysis

Most elementary statistics books cover the assessment of associations and differences, but the author particularly likes the following books:

Hays, William L., *Statistics*, 5th ed. (Fort Worth, TX: Holt, Rinehart & Winston, 1997).

Keller, Gerald, and Brian Warrack, *Statistics for Management and Economics,* 6th ed. (Pacific Grove, CA: Brooks/Cole, 2003).

Siegel, S., and N. J. Castellan, *Nonparametric Statistics for the Behavioral Sciences*, 2nd ed. (New York: McGraw-Hill, 1988).

ADDITIONAL BASIC ANALYTICAL TECHNIQUES

The Normal Distribution

The normal probability distribution is a bell-shaped distribution, discussed in most statistics books. One use of the normal distribution in research is significance testing. The baseline of the curve is labeled in standard deviation units (or Z-units). The curve for this distribution is shown in Figure 13A.1.

An additional interpretation of the baseline for the curve is the measurement axis for the characteristic of interest. In univariate descriptive statistics, this is a characteristic of interest with interval properties, or X_i. For example, if X_i referred to a normally distributed attitude characteristic with a mean of 25 and a standard deviation of 5, the X_i axis would look like that in Figure 13A.1. The standard deviation scores for an attitude of 25, 20, and 30 are 0, −1, and +1, respectively. One-half of a standard deviation above the mean, in raw scores (X_i), is $25 + 0.5(5) = 27.5$. In standard deviation scores (z), it is $0 + 0.5 = 0.5$.

In practical usage, the standard deviation scores are called standard scores (z) and range from −3 through 0 to about +3. If z-scores are to be used, the total area under the curve is set to equal 1.00, and the curve is said to be in standard form. Because the probability area is 1, percentages of area under the curve can be approximated, as depicted in the figure. The area between −1 and +1 represents 68% of the scores on the characteristic, or 20 to 30, and so on.

With reference to the Central Limit Theorem (Chapter 9), the normal distribution provides very interesting probability properties for statistical

FIGURE **13A.1**

Normal Probability
Distribution

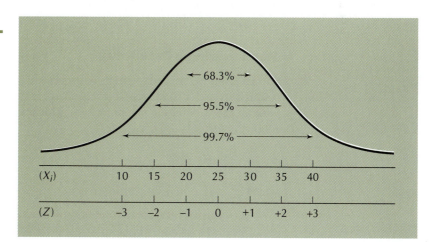

significance testing. In fact, the z-, t-, F-, and χ^2 distributions are all somewhat related and follow the same basic ideas just discussed. The only exception is their construction and use (correlation uses a bivariate normal distribution from two populations).

Calculation of the Sample Mean $\bar{X}$ and Sample Variance S^2

Although there are various methods for calculating the mean and variance from sample data, the following example shows the calculations for ungrouped data using the basic standard formulae.

Respondent Number (i)	Hourly Wage ($\$$) X_i	$(X_i - \bar{X})$	$(X_i - \bar{X})^2$
1	1.90	−0.60	0.3600
2	2.40	0.10	−0.0100
3	1.75	−0.75	0.5625
4	4.05	1.55	2.4025
5	2.65	0.15	0.0225
6	2.90	0.40	0.1600
7	1.85	−0.65	0.4225
	$\sum\limits_{i=1}^{7} = 17.50$	$\sum = 0$	$\sum\limits_{i=1}^{7}(X_i - \bar{X})^2 = 3.9400$

$$\bar{X} = \sum_{i=1}^{n} \frac{X_i}{n} = \sum_{i=1}^{7} \frac{X_i}{7} = \frac{17.50}{7} = \$2.50$$

$$S^2 = \sum_{i=1}^{n} \frac{(X_i - \bar{X})^2}{n-1} = \sum_{i=1}^{7} \frac{(X_i - \bar{X})^2}{6} = \frac{3.940}{6} = 0.66$$

Standardizing a Data Array

For a variety of reasons, which will become obvious in later chapters, researchers prefer to analyze a *standardized* data structure: a data structure with all the variables in standard form. This is easily accomplished by the following formula:

$$z = \frac{X_i - \bar{X}}{S}$$

where

X_i = the ith measurement of an individual, object, and so on

$\bar{X}$ = the mean of all the measures

S = the standard deviation

For example, if the mean of five measures of a certain characteristic is 15 and the standard deviation is 5, the z-scores are computed as follows:

X_i	$(X_i - \bar{X})/S$	z
20	$(20 - 15)/5$	1.0
15	$(15 - 15)/5$	0
10	$(10 - 15)/5$	-1.0
20	$(20 - 15)/5$	1.0
25	$(25 - 15)/5$	2.0

Covariation (Covariance)

When research questions or hypotheses are put in the following form, it is an easy way of asking if the variables covary (that is, vary together in a systematic fashion). For example, do managers with MBAs from private institutions perform better than those with MBAs from public institutions? Or do managers with high IQs have high business achievement and managers with lower IQs have low business achievement?

Researchers often talk about covariance—the variance that two or more variables share. Actually, covariance is used in the calculation of simple correlation. The Pearson product-moment correlation can be obtained by dividing the covariance of two variables by the product of their standard deviations. This is the standardized relationship, which goes from −1 through 0 to +1. It is expressed as

$$r_{xy} = \frac{S_{xy}}{S_x S_y} = \frac{\Sigma XY - (\Sigma X)(\Sigma Y)/n}{\sqrt{\left[\Sigma X^2 - \frac{(\Sigma X)^2}{n} \right]\left[\Sigma Y^2 - \frac{(\Sigma Y)^2}{n} \right]}}$$

where the covariance between variable x and variable y is expressed for convenience as S_{xy}, $S_x S_y$ as the product of the standard deviations of variables x and y, and r_{xy} as the correlation between variables x and y.

$$S_{xy} = \frac{\Sigma (X - \bar{X})(Y - \bar{Y})}{n - 1}$$

Actually, covariance is not a good measure of the strength of relationship because the larger the variance of X and Y, the larger the amount of covariance in the system. Standardization of the covariance is obtained in a manner like that just described. The result is the Pearson coefficient, which is easily interpreted.

14

Analysis of Variance and Regression Techniques

Overview

The Nature of Variance Decomposition

Linear Models

One-Way Analysis of Variance (ANOVA)

Two-Way Analysis of Variance (ANOVA)

Linear Regression

Analysis of Covariance

Nonparametric ANOVA

Summary

Discussion Questions

Notes

Suggested Reading

Appendix: The Use of Dummy and Effect Coding to Examine Group Differences Using Multiple Linear Regression

The research problems that are encountered in a business environment often require more complex analysis than previously discussed. The simple analytics in Chapter 13 provide the building blocks for such advanced analysis. This chapter deals with some widely applied analytical procedures using multiple variables.

Multiple-variable analysis involves a single dependent variable and one or several independent variables. Multivariate analysis, which is the basis of Chapter 15, involves two or more dependent variables and one or several independent variables. This distinction is not observed by many authors, but it provides a framework for the discussion of advanced analytical techniques. Actually, all of the topics subsumed under parametric multiple-variable analysis fit a general analytical approach called the *general linear model,* which forms the basis of discussion in the next two chapters.

Any intelligent discussion of advanced linear analysis of data must begin with partitioning of variance.[1] After this concept is explained, a general linear model for data analysis is presented. Examples are then given of the most widely used linear analysis techniques: one-way analysis of variance, two-way analysis of variance, linear regression, and analysis of covariance. It is not uncommon for linear regression analysis to be used for any linear analysis. This approach is discussed in the appendix of this chapter.

Sometimes a researcher is less restrictive about the assumptions of an analysis of variance. Most often, this less restrictive attitude is a function of the level of measurement in the data. Nonparametric analysis of variance techniques can be used in these instances. These useful techniques, including the Kruskal–Wallis test, are discussed in the final section.

THE NATURE OF VARIANCE DECOMPOSITION

Relations are the crux of most business research. In the previous chapter, the direction (positive or negative) and the magnitude (covariance, strength of relation, positive or negative) of simple relations were discussed. In this chapter, the focus is on multiple-variable relationships. Multiple-variable relationships tend to be more technically complex than those previously discussed. To simplify this discussion, we begin in this section by presenting the technical concept of variance partitioning.

The term *variance* refers to the spread of a distribution of scores—basically, the extent to which the scores differ from each other. This is the most common understanding of variance, but it comes in a number of forms, all with different characteristics and purposes. To design and understand research, one must be exposed to all the notions of variance. A great deal of analysis in research consists of comparing variances. Whether for discovery or testing hypotheses, the comparison of variances is essential.

Generally, the decomposition of variance consists of breaking down the total variance in the design into two components. The decomposition must satisfy the following relationship in a variety of different forms:

Total variance = explained variance + error variance

A researcher tries to explain as much of the total variance as possible. To accomplish this, the error variance (that variance which is unexplained) must be controlled, either by the researcher's control in the design or by a statistical method.[2] Theoretically, a good design maximizes explained variance to show similarities or differences in the variables. Some examples of variance decomposition can help illustrate the concept.

If a researcher is analyzing the differences in group means to find systematic differences due to the experimentation (manipulation of an independent variable) or nonexperimentation (survey data), between-group variance is important. In this instance the total variance equation is

Total variance = between-group variance + error variance

The researcher finds differences between groups if between-group variance is greater than error variance. Error variance is analogous to within-group variance in this case; it reflects things like sampling variance and other nuisance variation.

Researchers are also interested in finding relationships. In the relationship between two variables, the important element of variance is the common-factor variance or shared variance. Two variables, X_1 and X_2, may have a seemingly strong relationship but share little variance. If $r_{x_1x_2} = 0.30$ (the correlation of X_1 with X_2), then the square of the correlation coefficient ($r^2_{x_1x_2}$) indicates the amount of shared variance. Here, $r^2_{x_1x_2}$ equals 0.09. This indicates that only 9% of the variance in the system is shared. Two or more variables can share variance; this concept will be used later in the chapter. Sometimes, the relationship between two variables is expressed in a functional form: $Y = f(X)$. Here the researcher is interested in determining the amount of variance accounted for in Y due to the independent variable X. In both instances, the total variance equation is

Total variance = shared variance + error variance

or

Total variance of Y = variance accounted for in Y by X
$\qquad\qquad$ + variance not accounted for in Y by X

When multiple independent variables are used in the analysis of a relationship, the independent variables are assumed to share very little variance with each other, but together they account for much of the variance in the dependent variable.

All these examples tend to emphasize two points. First, keeping unaccountable variance small is essential to all research. Second, determining the components of variance enables the researcher to estimate what variance is accounted for (that is, Total variance – error variance = explained variance). Therefore, the decomposition of variance enables the researcher to interpret statistically significant differences and relationships with more meaning.

In behavioral research, as well as business research, most relationships are discovered and tested using linear models.[3] These are algebraic models that fit patterns in the data; they are useful in modeling relationships. This is not to say that the data itself is linear, because this may not always be the case. However, the models assume linearity as a convention, for ease of analysis. The general linear model involves the algebraic representation of relationships among observable characteristics of a unit of analysis. Its general form is

$$y_i = \sum_{j=i}^{p} a_j x_{ji} + \varepsilon_i$$
$$= a_1 x_{1i} + a_2 x_{2i} + \cdots + a_p x_{pi} + \varepsilon_i$$

In this form, y_i is an individual measure of a criterion variable (dependent variable) y. The values of x_j may be observations of a categorical variable x or an ordinal, interval, or ratio measure of variable x, or combinations of both types.

The x_{ji} is the observation of variable x for individual i in group j. The process of fitting a linear model to data requires the determination of a set of coefficients a_j that, when multiplied by the appropriate x_{ji}, will reproduce y_i as closely as possible for the set of data. Up to now, the discussion has been mathematical. The values of ε_i in the mathematical model are the residuals in the estimation of y_i from $\Sigma a_j x_{ji}$. If y is a random variable and x_{ji} represents a fixed, known value of x_j, then ε will also be a random variable. With this condition, the model becomes a statistical linear model. In this form, different linear hypotheses can be tested using the model.

When the x_j are entirely categorical, the linear model is usually referred to as *analysis of variance*. In this instance, the fit of the model is expressed in terms of whether there are significant differences among the means of the jth populations represented by samples. When the x_j are at least interval measures, then the model is referred to as *regression*. The fit of the model is now expressed as how much of the variation in y can be attributed to one or more of the x_j. If some of the x_j are categorical and at least interval, then the model is called *analysis of covariance*.

Satisfying different assumptions relative to the model justifies the use of the model in an analytic situation.[4] Remember that the use of the model is not restricted to artificially formed groups (as is true in experimentation). The general model can also be used with naturally occurring groups. That is why it is so important to the business researcher.

Having introduced the partitioning of variance and the general linear model, we now discuss a variety of analytical techniques previously mentioned.

TABLE **14.1** Tests for Group Differences—One Dependent Variable

Measurement Level	Samples	
	Dependent Samples	Independent Samples
Nominal	Cochran-Q	X_2 for samples
Ordinal	Friedman—two-way ANOVA	Kruskal–Wallis—one-way ANOVA
Interval/ratio	Linear models	Linear models

We begin with analysis of variance and linear regression. Then, analysis of covariance will be illustrated. All of these are very important to field research. As before, the choice of the linear model as a method of testing group differences is related to the level of measurement in the data. (See Table 14.1.) In all cases, the researcher is concerned with the measurement level of the dependent variable.

LINEAR MODELS

One-Way Analysis of Variance (ANOVA)

A researcher often wants to know whether two or more groups differ on a specific dependent variable. The groups can either be naturally occurring or created by the researcher for purposes of the study; the latter is known as the *experimental* approach.

Furthermore, these groups or classification variables, often called *factors* in ANOVA, can be categorized as *fixed factors* or *random factors*. Fixed factors contain all levels of the classification of interest; random factors contain only a random sample of all possible levels of a factor being studied. An ANOVA design that contains only fixed factors is called a *fixed effects design,* and one that contains only random factors is called a *random effects design.* Occasionally, fixed and random effects may be used in the same design, and this is called a *mixed effects design.* The type of design is important because the computations for significance testing are different. The reader is encouraged to consult an appropriate statistical text (see Suggested Reading at the end of the chapter) if contemplating the use of a random or mixed effects design.

To illustrate a fixed effects model, suppose a survey is conducted concerning aerospace company presidents' attitudes toward government economic policy. Furthermore, suppose that there are 16 firms in the United States: four small, four medium, four large, and four very large. The researcher is interested in determining whether attitudes are different given the size of the corporation. The 16 executives are questioned and the results tabulated. The higher the attitude score, the more favorable the attitude toward economic policy. Assume the population means of the four company size groups are $\mu_1 = 50$, small; $\mu_2 = 55$, medium; $\mu_3 = 60$, large; $\mu_4 = 65$, very large. The population variance for each group is $\sigma_j^2 = 10$. Because the means are population means, we can conclude, without a hypothesis test, that the means are different. Therefore, we can conclude that the four groups differ in attitude.

With equal numbers in each group, a grand mean (μ) can be calculated by taking the average of the group means, as follows:

$$\mu = \frac{\sum_{j=1}^{p} \mu_j}{p} = \frac{\mu_1 + \mu_2 + \mu_3 + \mu_4}{p} = \frac{230}{4} = 57.5$$

where

p = number of groups

μ_j = individual group mean

The grand mean describes the effects of being a member of a particular size company. An effect is defined as the difference between a group mean and a grand mean. Furthermore, each group effect, α_j, is represented by $\mu_j - \mu$. For example, the effect for the smallest company, α_1, is obtained by subtracting μ from μ_1. That is,

$$\alpha_1 = \mu_1 - \mu = 50 - 57.5 = -7.5$$

Each effect can be interpreted as either adding to or reducing an attitude score, depending on the sign of the effect and the particular group. For example, the value of -7.5 is interpreted as: Being in a small company reduces an executive's attitude score by 7.5 units.

A mathematical model can be used to describe the score of each executive who participates in the study. If each executive's score were equal to the grand mean, then the mathematical model would be $Y_{ij} - \mu$, where Y_{ij} represents the ith executive's score in the j group. This is not the case, however, because the group means are not equal. Further specification of each executive's score is possible if the effect term for group membership is included in the model as $Y_{ij} = \mu + \alpha_j$. However, this convention allows each executive's score in a group to be the same. This is a very unlikely outcome because of individual differences among the executives. Because of these individual differences, an error term must be added to the model. The error term, e_{ij}, is found by subtracting the $Y_{ij} = \mu + \alpha_j$ from the actual attitude score of each executive. Thus, the specified model is

$$Y_{ij} = \mu + \alpha_j + e_{ij}$$

The model is linear and additive. Because this discussion involves population parameters, there is no need to test statistical hypotheses with the model. However, researchers almost always use samples, not populations, in studies. Also inherent in all studies are components of e_{ij} (error). The sample statistics and the model assumptions enable the researcher to test hypotheses about the parameters in the model. The assumptions in the model for hypothesis testing are as follows:

1. The scores are sampled at random.
2. The scores are sampled from normal populations.
3. The populations have equal variances, σ^2.
4. The different samples are independent.

The researcher estimates the components of the model using the sample data by the method of least squares.

In contrast to the previous population example, suppose that a sample survey is taken concerning corporate presidents' attitudes toward governmental economic policy. The researcher is interested in determining whether the attitudes are different given the size of the corporation. Four corporate executives are sampled from four different-size companies in similar types of business. The higher the attitude score, the more favorable the executive's attitude toward the policy. The data is shown in Exhibit 14.1.

1. Data

		Size of Company (j)		
	Smallest			Largest
	1	2	3	4
Executive (i)	Observations			
1	37	36	43	76
2	22	45	75	66
3	22	47	66	43
4	25	23	46	62
$n =$	4	4	4	4

2. Computations

Total $= \sum\limits_{i=1}^{n} Y_{ij} =$	106	151	230	247
Group mean $= \overline{Y}_{.j} =$	26.50	37.75	57.50	61.75

$$\text{Group mean} = \overline{Y}_{.j} = \frac{\sum\limits_{i=1}^{n} Y_{ij}}{n}$$

where

Y_{ij} = score of the individual i
$(i = 1, 2, \ldots, n)$ in group j
$(j = 1, 2, \ldots, p)$

(Continued)

EXHIBIT **14.1**

(Continued)

i = individual score or observation

j = group

p = total number of groups

n = number of scores within each group

N = total number of scores in the design

$$\text{Grand mean} = \bar{Y}_G = \sum_{j=1}^{p} \sum_{i=1}^{n} \frac{Y_{ij}}{N} = \frac{\text{sum of all the scores}}{N} = \frac{734}{16} = 45.87$$

$$\text{Between-groups } SS = n \sum_{j=1}^{p} (\bar{Y}_{.j} - \bar{Y}_G)^2$$

$$\text{Within-groups } SS = \sum_{j=1}^{p} \sum_{i=1}^{n} (\bar{Y}_{ij} - \bar{Y}_{.j})^2$$

$$\text{Total } SS = \sum_{j=1}^{p} \sum_{i=1}^{n} (\bar{Y}_{ij} - \bar{Y}_G)^2$$

3. Computational algorithm

   a. Grand total = $\displaystyle\sum_{j=1}^{p} \sum_{i=1}^{n} Y_{ij} = 106 + 151 + 230 + 247$

   $= 734$ (sum of all totals for groups)

   b. Sum of the squared observations n

   $$= \sum_{j=1}^{p} \sum_{i=1}^{n} Y_{ij}^2 = (37)^2 + (22)^2 + (22)^2 + (25)^2 + (36)^2 + (45)^2 + (47)^2$$

   $$+ (23)^2 + (43)^2 + (75)^2 + (66)^2 + (46)^2 + (76)^2 + (66)^2$$

   $$+ (43)^2 + (62)^2 = 38,792$$

   (Square each individual score for each group and add them together.)

   c. Sum of the squared group totals divided by n

   $$= \frac{1}{n} \sum_{j=1}^{p} \left(\sum_{i=1}^{n} Y_{ij} \right)^2 = \frac{1}{4}[(106)^2 + (151)^2 + (230)^2 + (247)^2]$$

   $$= \frac{147,946}{4} = 36,986.5$$

   (Square each group total, add them together, and then divide by group size.)

4. Grand total squared and divided by total sample size = correction term (*CT*)

$$CT = \frac{1}{N} \left(\sum_{j=1}^{p} \sum_{i=1}^{n} Y_{ij} \right)^2 = \frac{(734)^2}{16} = 33,672.25$$

EXHIBIT 14.1

(Continued)

5. SS total $= \sum\limits_{j=1}^{p} \sum\limits_{i=1}^{n} Y_{ij}^2 - CT$ (sum of squared observations − correction term)

$$= 38{,}792 - 33{,}672.25 = 5119.75$$

6. SS between $= \dfrac{1}{n} \sum\limits_{j=1}^{p} \left(\sum\limits_{i=1}^{n} Y_{ij} \right)^2 - CT$ (sum of the squared group totals divided by n minus the correction term) $= 36{,}986.5 - 33{,}672.25 = 3314.25$

7. SS within $= SS$ total $- SS$ between

$$= (SS_T) - (SS_B)$$
$$= 5119.75 - 3314.25$$
$$= 1805.50$$

ANOVA Table

Source	SS	df	MS	F
Between groups	3314.25	$(p-1) = 3$	$\dfrac{SS_B}{p-1} = 1104.75$	
Within groups	1805.50	$p(n-1) = 12$	$\dfrac{SS_W}{p(n-1)} = 150.46$	$\dfrac{MS_B}{MS_W} = 7.34$
Total	5119.75	$N - 1 = 15$		

TEST OF HOMOGENEITY OF VARIANCES

Attitude Score

Levene Statistic	d.f. 1	d.f. 2	Sig.
1.245	3	12	0.337

ANOVA

Attitude Score

	Sum of Squares	d.f.	Mean Square	F	Sig.
Between groups	3314.250	3	1104.750	7.343	0.005
Within groups	1805.500	12	150.458		
Total	5119.750	15			

(Continued)

EXHIBIT 14.1

(Continued)

POST HOC TESTS

Multiple Comparisons

Dependent Variable: Attitude Score
Scheffé

(I) Company Size	(J) Company Size	Mean Difference (I − J)	Std. Error	Sig.	99% Confidence Interval	
					Lower Bound	Upper Bound
Small	Medium	−11.25	8.673	0.651	−47.90	25.40
	Large	−31.00	8.673	0.029	−67.65	5.65
	Very Large	−35.25	8.673	0.013	−71.90	1.40
Medium	Small	11.25	8.673	0.651	−25.40	47.90
	Large	−19.75	8.673	0.214	−56.40	16.90
	Very Large	−24.00	8.673	0.105	−60.65	12.65
Large	Small	31.00	8.673	0.029	−5.65	67.65
	Medium	19.75	8.673	0.214	−16.90	56.40
	Very Large	−4.25	8.673	0.970	−40.90	32.40
Very Large	Small	35.25	8.673	0.013	−1.40	71.90
	Medium	24.00	8.673	0.105	−12.65	60.65
	Large	4.25	8.673	0.970	−32.40	40.90

The null hypothesis for the test is that the population means are equal. The alternative hypothesis is that they are not equal.

$H_0 = \mu_1 = \mu_2 = \mu_3 = \mu_4$

H_1: Not all μ's are equal.

The proper way to perform this test is to employ a procedure known as *analysis of variance* (ANOVA). The linear model for the one-way analysis of variance is

$Y_{ij} = \mu + \alpha_j + e_{ij}$

where Y_{ij} represents the ith attitude score of an executive in the jth group.

The estimates of the model parameters based on the collected data can replace the model parameters. The result is

$Y_{ij} = \overline{Y}_G + (\overline{Y}_{.j} - \overline{Y}_G) + (Y_{ij} - \overline{Y}_{.j})$

where

Y_{ij} = attitude score

$\overline{Y}_G$ = grand mean

$(\overline{Y}_{.j} - \overline{Y}_G)$ = effect of membership in jth group

$(Y_{ij} - \overline{Y}_{.j})$ = individual error

For example, any score in the table can be reproduced by solving this equation. Executive 3 in group 1 has an attitude score of 22, which was calculated as

$$22 = 45.87 + (26.50 - 45.87) + (22 - 26.50)$$
$$= 45.87 - 19.37 - 4.50$$
$$= 22$$

Recall that total variance can be partitioned into two groups—between-group and within-group variation—where sums of squares is the convention used to represent the partitioned variation. By calculating the sums of squares and the appropriate degrees of freedom for each component of variance, the null hypothesis can be tested by using the F-distribution. These formulae are presented in part 2 of Exhibit 14.1. Because these calculations are tedious, direct computational algorithms are presented in part 3 of the example. Of course, any statistical software that can do ANOVA will calculate these statistics for you. To illustrate, the output of the SPSS ANOVA program is presented at the end of the exhibit.

It is very common to see the sums of squares presented in the form of an ANOVA table. The table gives all the pertinent data to test the null hypothesis by calculating the proper test statistic. The actual steps for conducting the test are as follows:

1. State the null and alternative hypotheses.

   H_0: $\mu_1 = \mu_2 = \mu_3 = \mu_4$
   H_1: Not all μ's are equal.

2. Collect the data and calculate appropriate statistics.

   $SS_B = 3314.25$ d.f. = $(p - 1) = 3$
   $SS_W = 1805.50$ d.f. = $p(n - 1) = 12$

   where d.f. represents the degrees of freedom for the test.

3. Choose level of significance, $\alpha = 0.01$.

4. Determine tabled F-value for (3, 12) d.f., $\alpha = 0.01$, (critical $F)_{0.99}F_{3,12} = 5.95$. (See Appendix B, Table B.3, upper 1% points.) It is worth noting that this step is unnecessary with most statistical software because an exact significance level is calculated for you. In our example, the significance (α) level is 0.005 for an observed F of 7.343.

5. Calculate test statistic (observed F).

$$F = \frac{SS_B/(p-1)}{SS_W/(p(n-1))} = \frac{MS_B}{MS_W} = \frac{1104.75}{150.46} = 7.34$$

6. If the observed F is greater than the critical F (Appendix B, Table B.3), the null hypothesis is rejected. Because $7.34 > 5.95$, we conclude that the population means are different. Executives in different-size companies have differing attitudes toward economic policy. We *cannot conclude* from this test where the differences occur among the four groups included in the study. This is one of three additional topics that will be covered in the next section.

What happens if we violate the assumptions of the test? Actually, the one-way ANOVA is very robust in terms of violation of its basic assumptions.[5] However, one should be constantly on the lookout for gross violations of the assumptions of equal variance, normality, and independence of error terms. For example, if you are unsure of the assumption of the equality of variances, use Bartlett–Box's or Levene's test for homogeneity of variance. Using the data in Exhibit 14.1, we can test the assumption via the Bartlett–Box procedure. The null hypothesis is:

H_0: $\sigma_1^2 = \sigma_2^2 = \sigma_3^2 = \sigma_4^2$

As is indicated in the Levene test in the SPSS output, the test is not significant ($\alpha > 0.05$), so we can conclude that the variances are equal.

It is also worth noting that the analysis of variance procedure tests only the overall effects in a model. It provides no information as to which means or combination of means are different when the null hypothesis is rejected. If researchers want to determine which means are different in a set of means, they can use procedures known as multiple comparisons. Multiple comparison procedures such as Scheffé's test, Tukey's honestly significant difference, and Duncan's multiple range test are generally used to test for differences between individual means.[6] If these comparisons are planned in advance, they are conducted in lieu of the overall F-test. If they are not planned, they take the form of discovery—a search for interesting findings to evaluate, with no hypothesis in mind.

The complete description of these procedures is beyond the scope of this text, but the SPSS output in Exhibit 14.1 illustrates the results of a planned comparison using Scheffé's test. Essentially, the table provides the difference scores between the two group means and the probability that the differences could be explained by chance alone. As one can see, the only pairwise statistical differences ($\alpha < 0.05$) are between small firms and very large firms ($p < 0.013$) and medium firms and small firms ($p < 0.029$).

As noted, the explanation of total variance accounted for in the linear model is an important facet of its use. Additionally, the explanation of variance ensures that the researcher will not speculate that significant results are real. A significant F-ratio in a completely randomized design indicates that there is association between X and Y. However, if the sample size is large enough, the differences indicated by statistical significance may be trivial. The strength of the relationship must be determined. An F-ratio can only provide information concerning the probability that the σ_i^2 are equal or not equal to zero. It cannot provide information on the strength of a relationship. In the one-way ANOVA, omega squared, ω^2, can provide information on the strength of a relationship. It can be estimated from the partitioned total variance as follows:

$$\omega^2 = \frac{SS_B - (p-1)MS_W}{SS_T + MS_W}$$

where p = number of groups, ω^2 can be interpreted as the proportion of variance in the dependent variable accounted for by the independent variable (the grouping variable or the treatment effect).

In Exhibit 14.1, ω^2 can be estimated as

$$\omega^2 = \frac{3314.25 - 3(150.46)}{5119.75 + 150.46}$$
$$= \frac{2862.87}{5270.21}$$
$$= 0.5432$$

Almost 55% of the variance in executive attitudes toward economic policies can be accounted for by the size of the company.

Two-Way Analysis of Variance (ANOVA)

In the previous example of a one-way ANOVA, the dependent variable is maintained at the same level across the different groups. This is necessary in order to conclude that the observed differences among the group means can be attributed solely to the independent variable. A problem with this procedure is the generalizability of the results. The particular patterns of the results may be the result of another relevant variable held constant in the study (or the analysis).

The factorial analysis of variance provides one solution to the problem by allowing the effect of an independent variable to be averaged over levels of another relevant variable(s). One important form of factorial ANOVA is the two-way ANOVA. The classification variables in this form of ANOVA are called *factors*, and the categories within factors are referred to as *levels* of a factor. In a two-way ANOVA, there are two classification variables and one dependent variable.

The factorial design contains a very important analytical convention: an *interaction effect*. Two variables interact when the effect of one variable changes at different levels of the second variable. The estimation of the interaction effect is important to a researcher. Sometimes the researcher is not concerned with the effects of any one independent variable on a dependent variable. Therefore, the estimation of main effects (differences across groups) is useless. The two-way analysis of variance enables the researcher to evaluate the combined effects of two independent variables on a dependent variable in one analysis. The importance of interactions in the interpretation of results is succinctly noted by Mason and Bramble:

When there are significant main effects and interactions present in an experiment, the interpretation of the results is not as simple. . . . That is, an interaction might actually be the cause of the significant main effect. In such a case, it would not be appropriate to say that there were main effects, even though the existence of main effects might be supported statistically. . . . Probably one of the most common errors in research using analysis of variance with two or more factors is failure to interpret significant main effects with respect to significant interactions.[7]

Thus the interaction effect is the focus of the two-way analysis.[8]

Suppose the researcher is interested in determining whether the combined effects of the size of a firm and the age of aerospace executives influence attitudes toward government economic policy. A sample survey is conducted, and the data is classified for a 2×3 analysis of variance. It is a 2×3 ANOVA because there are two levels of factor A (age) and three levels of factor B (size of company). This produces six cells in the analysis (rows $\times$ columns). There are three attitude scores in each cell. This data array is illustrated in part 1 of Exhibit 14.2. The cell means from the table are

$a_1 b_1 = 44$
$a_1 b_2 = 46$
$a_1 b_3 = 45$
$a_2 b_1 = 35$
$a_2 b_2 = 45$
$a_2 b_3 = 59$

		B		
		b_1	b_2	b_3
A	a_1	44	46	45
	a_2	35	45	59

The model for the two-way ANOVA is similar to the one-way ANOVA model, with the exception of an added effect term for each factor (the main effects) and an effect term for the combined effects of both factors (the interaction effect). The linear model is

$$i = 1, 2, \ldots, r$$
$$j = 1, \ldots, c$$
$$k = 1, \ldots, n$$
$$Y_{ijk} = \mu + \alpha_i + \beta_j + \delta_{ij} + e_{ijk}$$

where

Y_{ijk} = the kth attitude score for level i of factor A and level j of factor B

μ = grand mean

Main effects α_i = effect of the ith age group

β_j = the effect of the jth size group

Interaction δ_{ij} = combined effect for ijth cell

Error e_{ijk} = difference between μ and Y_{ijk}

1. Data

Factor A (Age Group)		Factor B (Company Size)			
		b_1 (Small)	b_2 (Medium)	b_3 (Large)	Row Means
a_1 (25–39 years)		Cell 42	Cell 50	Cell 49	$\bar{Y}_{1..} = 45.0$
		11 46	12 46	13 52	
		44	42	34	
	Cell means	$\Sigma = 132$	$\Sigma = 138$	$\Sigma = 135$	
		$\bar{Y}_{11.} = 44$	$\bar{Y}_{12.} = 46$	$\bar{Y}_{13.} = 45$	
a_2 (40–65 years)		Cell 36	Cell 50	Cell 62	$\bar{Y}_{2.} = 46.33$
		21 38	22 36	23 60	
		31	49	55	
	Cell means	$\Sigma = 105$	$\Sigma = 135$	$\Sigma = 177$	
		$\bar{Y}_{21.} = 35$	$\bar{Y}_{22.} = 45$	$\bar{Y}_{23.} = 59$	
	Column means	$\bar{Y}_{.1.} = 39.5$	$\bar{Y}_{.2.} = 45.5$	$\bar{Y}_{.3.} = 52.0$	

2. Formulae

Y_{ijk} = individual observation

$\bar{Y}_{ij}$ = mean of each cell

$\bar{Y}_{i..}$ = mean of each row

$\bar{Y}_{.j.}$ = mean of each column

$\bar{Y}_G$ = grand mean

$\bar{Y}_G = \sum_{i=1}^{n} \frac{Y_{ijk}}{n}$

$\bar{Y}_{i..} = \sum_{i=1}^{c} \frac{\bar{Y}_{i..}}{c}$

i = levels of factor A, $i = 1, 2, \ldots, r$

j = levels of factor B, $j = 1, 2, \ldots, c$

r = number of rows

c = number of columns

k = individual, $k = 1, \ldots, n$

n = cell size

$SS_A = cn \sum_{i=1}^{r} (\bar{Y}_{i..} - \bar{Y}_G)^2$

$\bar{Y}_{.j.} = \sum_{i=1}^{r} \frac{\bar{Y}_{.j.}}{r}$

$SS_B = rn \sum_{j=1}^{c} (\bar{Y}_{.j.} - \bar{Y}_G)^2$

$\bar{Y}_{ij.} = \sum_{k=1}^{n} \frac{Y_{ij.}}{n}$

$SS_{A \times B} = n \sum_{i=1}^{r} \sum_{j=1}^{c} (\bar{Y}_{ij.} - \bar{Y}_{i..} - \bar{Y}_{.j.} + \bar{Y}_G)^2 = [SS_T - (SS_A + SS_B + SS_W)]$

$SS_W = \sum_{i=1}^{r} \sum_{j=1}^{c} \sum_{k=1}^{n} (\bar{Y}_{ijk} - \bar{Y}_{ij.})^2$

$SS_T = \sum_{i=1}^{r} \sum_{j=1}^{c} \sum_{k=1}^{n} (\bar{Y}_{ijk} - \bar{Y}_G)^2$

(Continued)

EXHIBIT **14.2**

(Continued)

3. Computational algorithm—Change formulae to reflect additional cells.

Step 1   Add up scores in each cell to obtain cell totals.

Cell 11 = 132 (that is, row 1, column 1)

Cell 12 = 138

Cell 13 = 135

Cell 21 = 105

Cell 22 = 135

Cell 23 = 177

Step 2   Square each score in the entire table and add these values together.

$(42)^2 + (46)^2 + \cdots + (55)^2 = 38{,}824$

Step 3   Add up all cell totals.

$132 + 138 + 135 + 105 + 135 + 177 = 822$

Step 4   Square total of all cells and divide by *N*, where *N* is the total number of scores in the table. This result is the correction term (*CT*).

$$\frac{(822)^2}{18} = 37{,}538$$

Step 5

$SS_R = 38{,}824$ (result in step 2) $- 37{,}538$ (result in step 4) $= \underline{\underline{1286}}$

Step 6   To compute effect of factor *B*, add column sums together.

$132 + 105 = 237$

$138 + 135 = 273$

$135 + 177 = 312$

Square each value and divide by total number of observations in a column. Then add them together.

$$\frac{(237)^2}{6} + \frac{(273)^2}{6} + \frac{(312)^2}{6} = 38{,}007$$

Subtract *CT* from this total.

$38{,}007 - CT = \underline{\underline{SS_B}}$

$38{,}007 - 37{,}538 = \underline{\underline{469}}$

Step 7   To compute effects of factor *A*, add row sums together.

$132 + 138 + 135 = 405$

$105 + 135 + 177 = 417$

Square each value and divide by total number of observations in a row. Then add them together.

EXHIBIT 14.2

(Continued)

$$\frac{(405)^2}{9} + \frac{(417)^2}{9} = 37{,}546$$

Subtract CT from this total. $\qquad 37{,}546 - 37{,}538 = \underline{8} = SS_A$

Step 8 To compute the interactive effects, square each cell sum and divide by cell size. Then add them together.

$$\frac{(132)^2}{3} + \frac{(138)^2}{3} + \frac{(135)^2}{3} + \frac{(105)^2}{3} + \frac{(135)^2}{3} + \frac{(177)^2}{3} = 38{,}424$$

Then subtract (CT), SS_B, SS_A.

$$38{,}424 - 37{,}538 - 469 - 8 = \underline{409}$$

$$SS_{A \times B} = \underline{409}$$

Step 9 To compute SS_W, subtract SS_A, SS_B, $SS_{A \times B}$ from SS_T.

$$SS_T - SS_B - SS_A - SS_{SSA \times B} = \underline{SS_W}$$
$$1286 - 469 - 8 - 409 = \underline{400}$$

ANOVA Table

Source	d.f.	SS	MS	F
A	$r - 1 = 1$	8	$\dfrac{SS_A}{r-1} = 8$	$\dfrac{MS_A}{MS_W} = 0.24$
B	$c - 1 = 2$	469	$\dfrac{SS_B}{c-1} = 234.5$	$\dfrac{MS_B}{MS_W} = 7.04$
$A \times B$	$(r-1)(c-1) = 2$	409	$\dfrac{SS_{A \times B}}{(r-1)(c-1)} = 204.5$	$\dfrac{MS_{AB}}{MS_W} = 6.13$
Within	$rc(n-1) = 12$	400	$\dfrac{SS_W}{rc(n-1)} = 33.33$	
Total	$rcn - 1 = 17$	1286		

BETWEEN-SUBJECTS FACTORS

		Value Label	N
Age Group of Executive	1	25 to 39 Years Old	9
	2	40 to 65 Years Old	9
Company Size	1	Small	6
	2	Medium	6
	3	Large	6

(Continued)

EXHIBIT **14.2**

(Continued)

TESTS OF BETWEEN-SUBJECTS EFFECTS

Dependent Variable: Attitudes Toward Economic Policy

Source	Type III Sum of Squares	d.f.	Mean Square	F	Sig.
Corrected Model	886.000[a]	5	177.200	5.316	0.008
Intercept	37538.000	1	37538.000	1126.140	0.000
AGE	8.000	1	8.000	.240	0.633
CO_SIZE	469.000	2	234.500	7.035	0.010
AGE*CO_SIZE	409.000	2	204.500	6.135	0.015
Error	400.000	12	33.333		
Total	38824.000	18			
Corrected Total	1286.000	17			

[a]R-squared = 0.689 (adjusted R-squared = 0.559).

The estimates of the effects are similar to the one-way ANOVA but are more complex. However, three sources of variability can be isolated that might influence the design mean's variation from the grand mean. These sources of variation correspond to the effects identified in the model: factor A and factor B (main effects) and the interaction effect, $A \times B$.

Essentially, the two-way ANOVA involves partitioning total variance into four components: factor A, factor B, interaction, and error. This partitioning provides information for testing the hypotheses of concern in the model. The assumptions in the model are basically the same as in the one-way ANOVA model. The hypothesis for each source of effects is as follows:

Factor A H_0: $\mu_{1..} = \mu_{2..}$ for all i
            H_1: $\mu_{1..} \neq \mu_{2..}$ for some i

Factor B H_0: $\mu_{.1.} = \mu_{.2.} = \mu_{.3.}$ for all j
            H_1: $\mu_{.1.} \neq \mu_{.2.} \neq \mu_{.3.}$ for some j

$A \times B$ H_0: $\delta_{ij} = 0$ for all i and j
            H_1: $\delta_{ij} \neq 0$ for some i and j

The procedure for testing the hypotheses begins with an examination of the interaction term. The rationale for this will become obvious when the meaning of the interaction is further explained. If $F_{A \times B} = MS_{AB}/MS_W$ is greater than the tabled F-value with $F_{1-\alpha}$; $(r-1)(c-1)$, $rc(n-1)$ d.f., we reject the null hypothesis and accept H_0, that there is a significant interaction. The main effects are tested similarly, where $F_A = MS_A/MS_W$. If F_A is greater than the tabled F-value with $F_{1-\alpha}$; $(r-1)$, $rc(n-1)$ d.f., we reject the null hypothesis and accept H_1, that there are significant A main effects. Finally, if $F_B = MS_B/MS_W$ is greater than the tabled F-value with $F_{1-\alpha}$; $(c - 1)$, $rc(n - 1)$ d.f., then we reject the null hypothesis and accept H_1, that there are significant B main effects.

The necessary formulas to perform the tests are shown in Exhibit 14.2, along with a computational algorithm to partition the sums of squares. An ANOVA table contains the appropriate F-values obtained by dividing the appropriate mean square value by the within-group mean square. From the ANOVA table, the following F-values are calculated (Appendix B, Table B.3):

Source	Calculated F	Tabled F
$A \times B$	6.13	$F_{.95;2,12} = 3.89$
A	0.24	$F_{.95;1,12} = 4.75$
B	7.04	$F_{.95;2,12} = 3.89$

If the researcher sets the alpha level for the test at 0.05, both the interaction and the B main effects are statistically significant (we can reject the null hypothesis in both cases). Whenever an interaction is statistically significant, the interpretation of tests of main effects must be qualified. To accomplish this, the researcher graphs the row cell means with respect to the B factor (treatment levels).[9] This graph is shown in Figure 14.1.

Generally, four types of graphs can result from graphing the cell means in two-way ANOVAs, as shown in Figure 14.2. Of most concern to the researcher are graphs 2, 3, and 4. When there are a significant effect and a main effect present in a study, the researcher must examine the interaction carefully before interpreting the main effect. Often, there is little interest in comparisons among means for main effects. The reason for this becomes

FIGURE **14.1**

Illustration of the
Interaction Present in
Exhibit 14.2

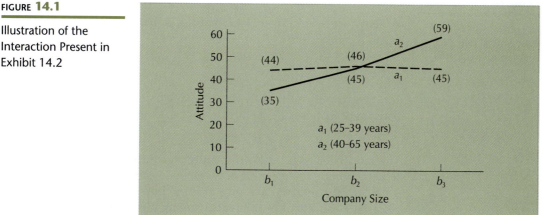

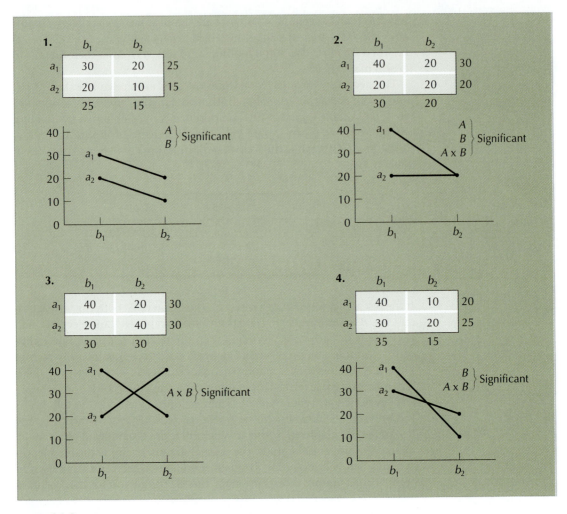

FIGURE 14.2

Graphs of Cell Means for General Two-Way ANOVA Results

evident if we refer to Figure 14.2, graph 3. In the figure, comparisons of $\overline{A}_1$ versus $\overline{A}_2$ and $\overline{B}_1$ versus $\overline{B}_2$ are equal to zero:

$$\overline{A}_1 = \frac{40 + 20}{2} = 30 \qquad \overline{A}_2 = \frac{20 + 40}{2} = 30$$

However, comparisons of $\overline{A}_1$ versus $\overline{A}_2$ at either level of B, as well as comparisons of $\overline{B}_1$ versus $\overline{B}_2$ at either level of A, are not equal to zero:

$$\overline{A}_1 \text{ at } B_1 = 40 \qquad \overline{A}_2 \text{ at } B_1 = 20$$

The researcher might draw the faulty conclusion that there are no differences among the means if only the main effect means were evaluated. Thus, if MS_A and MS_B are insignificant but $MS_{A\times B}$ is significant, the researcher can look for differences among means at specific levels of B (treatment).

CHAPTER **14** Analysis of Variance and Regression Techniques

Similarly, if we examine graphs 2 and 4, the interaction effect is significant, and one or all of the main effects are significant. However, on closer observation of graph 2, for example, although A main effects are significant, the comparison of $\bar{A}_1$ versus $\bar{A}_2$ is appreciably different from zero at level B_1, but not at B_2. Therefore, the A main effects may not truly be present. They are a function of interaction.

When examining a two-way ANOVA, what should you do? (1) The presence of interaction always calls for an examination of the main effects and the data itself. (2) The presence of interaction may be important in and of itself if the purpose of the analysis is to test the significance of the interaction.

The purpose of the study in Exhibit 14.2 is to test the combined effects of company size and age of executive on attitude. The interaction is significant at $\alpha = 0.05$. Also, the B effect is significant. Because we are not concerned with the B effects, we need only interpret the interaction. As can be seen from Figure 14.1, the simple main effects of B are not the same at different levels of A. Simply, this is interpreted as follows:

> In smaller companies, younger executives have more favorable attitudes toward economic policy, whereas in larger companies, the older executives have a more favorable attitude.

As with the one-way ANOVA, the two-way ANOVA can also incorporate multiple comparisons. One such method uses the Tukey T procedure.[10] Also, the ω^2 statistic can be computed for the A, B, and $A \times B$ components. Roughly, this measure can be described as:

$$\hat{\omega}^2 = \frac{\text{accounted-for variation due to factor } A, B, \text{ or } A \times B}{\text{total variation in the criterion (dependent) variable}}$$

Additionally, in both ANOVA models discussed previously, the various levels of factors have been fixed. Two other models can be used: the random effects and the mixed effects models. For example, if we use a random effects one-way ANOVA, we are randomly selecting levels from a population. We are concerned not with differences among levels but with an estimate of variability due to the factor(s).[11]

Finally, all of the designs discussed in the appendix in Chapter 6 can be analyzed with some form of ANOVA model. The complexity of the analysis makes them beyond the scope of our discussion of ANOVA. However, the general linear model and the partitioning of variance are part and parcel of the analysis.[12]

Linear Regression

Simple Linear Regression. In business research situations, it is often useful to conduct prediction or estimation studies. Predictive studies involve the use of current variables to identify a variable of interest at a future time. Estimation, on the other hand, involves the identification of a current attribute from other current traits or attributes. For example, a researcher might want to *predict* the satisfaction of employees based on incremental salary increases. However, if the researcher uses the incremental salary increases of first-year secretaries to gauge their satisfaction during the first year, the purpose of the analysis is *estimation*. Actually, the analytical procedures are the same for each purpose; the terms *prediction* and *estimation* are usually used interchangeably.

The analytical procedure that is widely used to predict or estimate is linear regression. Linear regression is a method of analyzing the change (variability) of a dependent variable by using information available on one or more independent variables, given certain linear assumptions. We seek to answer the question: What are the expected changes in the dependent variable as a result of changes (observed or induced) in the independent variables? For example, what are the expected changes in employee satisfaction as a result of observed changes in their incremental salary increases?

Linear regression uses the same linear modeling approach and variance partitioning as previously discussed. In the simple linear regression model (SLR), we use pairs of observations (X, Y, where X is an independent variable and Y is the dependent variable). In multiple linear regression (MLR), we use sets of observations (X_1, X_2, . . . , X_p, Y, where X_1 to X_p are multiple independent variables and Y is the dependent variable). The model for SLR is

$$Y_i = \alpha + BX + \varepsilon_i \text{ for parameters}$$

where $\alpha + BX$ is fixed, part of the individual score Y_i of an individual at a given level of X, and ε_i is a random component, unique to individual *i*. Alpha (α) (the intercept) is the mean of the population when $X = 0$. B is the regression coefficient in the population or the slope of the regression line. B expresses the change in dependent Y due to changes in independent X. Also, if we standardize the data, B is identical to the correlation of X and Y. The purpose of SLR is to estimate the coefficients of α and B and ε using sample statistics to obtain a statistically correct, useful managerial equation. To accomplish this, we must estimate B with b, α with a, and ε with e. For illustrative purposes, no subscripts are used with the sample model. It is expressed as follows:

$$Y = a + bX + e$$

The estimation of *e* is important for testing the model, but from the manager's standpoint, the managerial regression equation without the *e* term is more practical. It is expressed as

$$Y' = a + bX$$

where Y' is the predicted Y (dependent variable) based on a specific managerially selected X (independent variable).

In this case, *a* (the intercept) and *b* (the slope) are calculated as follows:

$$b = \frac{\Sigma xy}{\Sigma x^2} \qquad a = \overline{Y} - b\overline{X}$$

where

$$\overline{Y} = \frac{\sum\limits_{i=1}^{n} Y_i}{n} \qquad \text{and} \qquad \overline{X} = \frac{\sum\limits_{i=1}^{n} X_i}{n}$$

The assumptions of the model are normality, linearity, independence, and homoscedasticity. A review of these assumptions can be found in advanced texts on regression. They are important considerations for using the model, but their exclusion from this text does not detract from our explanation of linear regression.

Suppose management is interested in predicting employee satisfaction from the amount of commitment to the job. Assume that ten employees are selected at random and questioned concerning the variables of interest. Also assume that management believes commitment to be the only variable that affects satisfaction. The resulting data is shown in Exhibit 14.3. Calculations for estimates of a and b are shown in part 2 of the example, where

$$b = \frac{\Sigma xy}{\Sigma x^2} = \frac{0.924}{1.316} = 0.70213$$

$$a = \overline{Y} - b\overline{X}$$
$$= 3.02 - 0.70213\,(3.18)$$
$$= 0.78723$$

EXHIBIT 14.3

Simple Linear Regression

1. Data

Satisfaction	Commitment			
Y	X	XY	Y²	X²
3.3	3.7	12.21	10.89	13.69
2.8	3.1	8.68	7.84	9.61
3.1	3.0	9.30	9.61	9.00
3.7	3.7	13.69	13.69	13.69
2.6	2.9	7.54	6.76	8.41
3.2	3.5	11.20	10.24	12.25
2.7	2.5	6.75	7.29	6.25
3.0	3.3	9.90	9.00	10.89
2.9	2.9	8.41	8.41	8.41
2.9	3.2	9.28	8.41	10.24
Totals 30.2	31.8	96.96	92.14	102.44
Means 3.02	3.18			

(Continued)

EXHIBIT 14.3

(Continued)

Y'	$Y' - \bar{Y}$	$(Y' - \bar{Y})^2$	$Y - Y'$	$(Y - Y')^2$
3.3851	0.3651	0.1369	−0.0851	0.0081
2.97	−0.02	0.0025	−0.17	0.0289
2.89	−0.13	0.0169	0.21	0.0441
3.39	0.37	0.1369	0.31	0.0961
2.83	−0.19	0.0361	−0.23	0.0529
3.25	0.23	0.0529	−0.05	0.0025
2.55	−0.47	0.2209	0.15	0.0225
3.11	0.09	0.0081	−0.11	0.0121
2.83	−0.19	0.0361	0.07	0.0049
3.05	0.03	0.0009	−0.15	0.0225
Totals 30.2		0.6482		0.2946

2. Some preliminary formulae

Sample variance

$$S_x^2 = \frac{\sum_{i=1}^{n}(X_i - \bar{X})^2}{n-1} = \frac{\sum x^2}{n-1} = \frac{\text{sum of squared deviations from mean}}{\text{sample size minus one}}$$

where

$$\sum x^2 = \sum X^2 - \frac{(\sum X)^2}{n} \quad \boxed{\sum x^2 = \text{Sums of squares}}$$

Sample covariance

$$S_{xy} = \frac{\sum(X - \bar{X})(Y - \bar{Y})}{n-1} = \frac{\sum xy}{n-1} = \frac{\substack{\text{sum of cross-products deviations} \\ \text{of pairs of } X \text{ and } Y \text{ scores} \\ \text{from their respective means}}}{\text{sample size minus 1}}$$

where

$$\sum xy = \sum XY - \frac{\sum X(\sum Y)}{n} \quad \boxed{\sum xy = \text{Sum of cross-products}}$$

Computations

$$\sum x^2 = 102.44 - \frac{(31.8)^2}{10} = 1.316$$

$$\sum y^2 = 92.14 - \frac{(30.2)^2}{10} = 0.936$$

Variance of X

$$S_x^2 = \frac{1.316}{9} = 0.1462$$

Variance of Y

$$S_y^2 = \frac{0.936}{9} = 0.104$$

EXHIBIT 14.3

(Continued)

Covariance S_{xy}

$$= \frac{96.96 - \dfrac{30.2(31.8)}{10}}{9} = \frac{\Sigma xy}{n-1} = \frac{0.924}{9} = 0.1026$$

b model component: $b = \dfrac{\Sigma xy}{\Sigma x^2}$

a model component: $a = \bar{Y} - b\bar{X}$

where: $\bar{Y} = \dfrac{\Sigma Y}{n}$ and $\bar{X} = \dfrac{\Sigma X}{n}$

Thus: $b = \dfrac{0.924}{1.316} = 0.70213$ $a = 3.02 - 0.70213(3.18)$
$$= 0.78723$$

3. $Y' = 0.78723 + 0.70213X$

4.

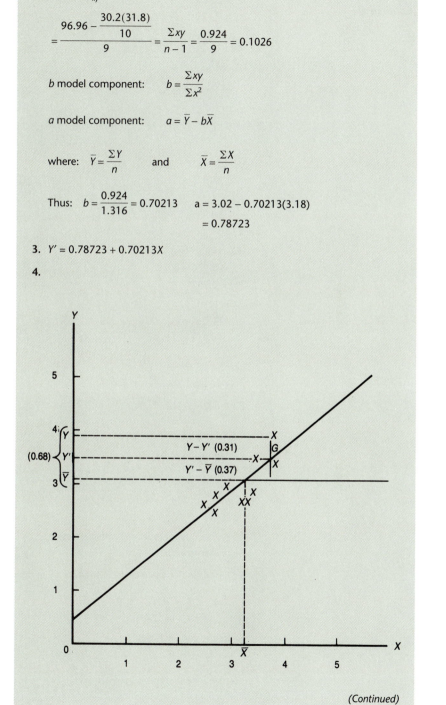

(Continued)

EXHIBIT **14.3**

(Continued)

5. Partitioning sums of squares

$$Y = \overline{Y} + (Y' - \overline{Y}) + (Y - Y')$$

$$\Sigma y^2 = \Sigma(Y' - \overline{Y})^2 + \Sigma(Y - Y')^2$$

$$\Sigma y^2 = SS_{reg} + SS_{res}$$

$$1 = \frac{SS_{reg}}{\Sigma y^2} + \frac{SS_{res}}{\Sigma y^2}$$

where

$$SS_{reg} = \frac{(\Sigma xy)^2}{\Sigma x^2} = b\,\Sigma xy = b^2 \Sigma x^2$$

$$= \frac{(0.924)^2}{1.316} = 0.70213(0.924) = (0.70213)^2(1.316)$$

$$= 0.649 = 0.649 = 0.649$$

and

$$SS_{res} = \Sigma y^2 - SS_{reg}$$

$$= 0.936 - 0.649$$

$$= 0.287$$

Thus,

$$r_{xy}^2 = \frac{(\Sigma xy)^2}{\Sigma x^2 \Sigma y^2} = \frac{(0.924)^2}{1.316(0.936)} = \frac{0.8537}{1.2317} = 0.693$$

$$r_{xy} = \sqrt{r_{xy}^2} = (+0.833)$$

$$1 - r_{xy}^2 = error$$

$$1 - 0.693 = 0.307$$

6. Tests of significance

   a. Regression of Y on X

$$F = \frac{SS_{reg}/d.f._1}{SS_{res}/d.f._2} = \frac{SS_{reg}/p}{SS_{res}/(n - p - 1)}$$

where

   p = number of independent variables
   n = sample size

   b. Proportion of variance accounted for by regression

$$F = \frac{r^2 \Sigma y^2 / p}{(1 - r^2)\Sigma y^2/(n - p - 1)}$$

   c. Regression coefficient

$$t = \frac{b}{S_b}$$

EXHIBIT 14.3

(Continued)

where

$$S_b = \frac{Sy \cdot x}{\sqrt{\Sigma x^2}}$$

where

$$Sy \cdot x = \sqrt{\frac{SS_{res}}{(n - p - 1)}}$$

d. Confidence interval of b

$$b \pm t\,(\alpha/2, \text{d.f.})\,S_b$$

7. Calculation of tests of significance
Regression of Y on X

$$F = \frac{0.648/1}{0.288/(10 - 1 - 1)} = 18.0$$

Proportion of variance accounted for by regression

$$F = \frac{0.693(0.936)/1}{(1 - 0.693)(0.936)/(10 - 1 - 1)} = \frac{0.6486}{0.0359} = 18.07$$

Regression Coefficient

$$t = \frac{0.70213}{\dfrac{\sqrt{0.288/(10 - 1 - 1)}}{\sqrt{1.316}}} = \frac{0.70213}{\dfrac{0.1897}{1.147}} = \frac{0.70213}{0.1653} = 4.25$$

MODEL SUMMARY

Model	R	R-squared	Adjusted R-squared	Std. Error of the Estimate
1	0.833[a]	0.693	0.655	0.1895

[a]Predictors: (Constant), Job Commitment.

ANOVA[a]

Model		Sum of Squares	d.f.	Mean Square	F	Sig.
1	Regression	0.649	1	0.649	18.069	0.003[b]
	Residual	0.287	8	3.590E-02		
	Total	0.936	9			

[a]Dependent Variable: Employee Satisfaction.
[b]Predictors: (Constant), Job Commitment.

(Continued)

EXHIBIT **14.3**

(Continued)

COEFFICIENTS^a

Model		Unstandardized Coefficients		Standardized Coefficients		
		B	Std. Error	Beta	t	Sig.
1	(Constant)	0.787	0.529		1.489	0.175
	Job Commitment	0.702	0.165	0.833	4.251	0.003

^aDependent Variable: Employee Satisfaction.

The regression equation is shown in part 3 of the example:

$$Y' = a + bX$$
$$Y' = 0.78723 + 0.70213X$$

This equation does not include e, which is the error that results from using the regression equation. Simply, $e = Y - Y'$, where Y is the observed score and Y' is the predicted score for each pair of observations. $\Sigma(Y - Y')^2$, the sum of the squared residuals, is what is minimized to obtain estimates of a and b using the mathematical least-squares procedure. This value will be useful later in the discussion.

If managers wanted to predict the satisfaction of a future employee with the equation, they would simply need a measure of their employee's potential commitment. For example, if a new employee was very committed after a few weeks on the job, and the measure of that commitment was 3.3, the manager could use the equation to predict satisfaction.

$$Y' = 0.78723 + 0.70213 \, (X)$$
$$Y' = 0.78723 + 0.70213 \, (3.3)$$
$$Y' \cong 3.10$$

Satisfaction in the case of the new employee would be 3.10. Because we know the X and Y scores of each individual, the accuracy of predicting Y' from X can be estimated. By algebraically transforming $Y' = a + bX$ by substituting $(\overline{Y} - bX)$ for a, we obtain $Y' = \overline{Y} + bX$. Further manipulation divides Y into three components: $\overline{Y}$ the mean of Y; $(Y' - \overline{Y})$, the deviation of the predicted Y from the mean of Y; and finally $(Y - Y')$, actual Y from predicted Y', called error. From the graph in Exhibit 14.3, part 4, we can examine these components more closely. The graph shows the plot of the ten paired (X, Y) observations and the regression equation $Y' = 0.78723 + 0.70213(X)$. In the fourth set of scores, the value of Y can be produced by drawing a perpendicular line representing the mean of $\overline{X}$ and $\overline{Y}$. These values fall somewhere on the regression line. Another perpendicular line is drawn from Y'

to the regression line. It intersects at point G. The fourth value of Y is 3.7. It can be reproduced by the following equation:

$$Y = \bar{Y} + (Y' - \bar{Y}) + (Y - Y')$$

$$3.7 = 3.02 + (3.39 - 3.02) + (3.7 - 3.39)$$

$$3.7 = 3.02 + 0.37 - 0.31$$

The terms in the equation are very important to the further evaluation of the model: $(Y' - \bar{Y})$, the deviation due to regression; and $(Y - Y')$, the deviation due to residual (error). In the example, the individual's Y score deviates 0.68 from the mean of Y. It is the sum of all such deviations that is important to the evaluation of the regression equation. This sum is Σy^2. By partitioning this into two components—residual sum of squares $(Y - Y')^2$ and deviation due to regression sum of squares $(Y - \bar{Y})^2$—we obtain important information. Notice that the residuals get larger as the plots (X, Y) are farther from the regression line. Therefore, if the scatter is large and the residuals are great, the regression equation becomes less useful. Thus,

$$\Sigma y^2 = SS_{regression} + SS_{residual}$$

Four important factors must be considered to evaluate the SLR model:

1. Explained variation (r^2)
2. Statistical significance testing of regression Y on X
3. Statistical significance testing of proportion of variance accounted for by regression (r^2)
4. Testing the regression coefficient, b

The squared product-moment correlation coefficient between X and Y indicates the proportion of the sum of squares of $Y(\Sigma y^2)$ that is due to regression. This is an important calculation in research. It is analogous to the ω^2 (omega squared) estimate used in ANOVA. The formula for this calculation using the sum of squares is

$$r_{xy}^2 = \frac{(\Sigma xy)^2}{\Sigma x^2 \Sigma y^2} = \text{explained variance}$$

$$1 - r_{xy}^2 = \text{error} = \text{unexplained variance}$$

In Exhibit 14.3, part 5, the necessary formulae are displayed for the calculation of r_{xy}^2. Using the data in the example, r_{xy}^2 and error are found to be

$$r_{xy}^2 = \frac{(0.924)^2}{1.2317} = 0.693$$

$$1 - 0.693 = 0.307$$

Sixty-nine percent of the total variance accounted for in Y is attributable to changes in X. We have about 30% unaccounted error variance. The r_{xy}^2 is loosely interpreted as the "fit" of the model. Actually, total "fitting" requires statistical hypothesis testing of the components. As r_{xy}^2 (with bounds 0 to 1) becomes larger, it is assumed that Y can be predicted more accurately from the data. In business research (with the exception of forecasting), this 70% of explained variance is meaningfully large.

Three important statistical tests are used in SLR. These are used in conjunction with r_{xy}^2. For example, it was demonstrated in a previous chapter that the probability of rejecting a null hypothesis increases as the sample size increases. Thus, even with a small r_{xy}^2, there could be significant parameter estimation. The question is whether the regression equation can be useful. Obviously, the answer is no. This would be trivial research.

The first hypothesis test involves testing the regression of Y on X. (See Exhibit 14.3, part 6.) The test involves determining if the effect of commitment on satisfaction is significant. The F-test is again used. At a level of significance of 0.05, the F critical is

$$F_{0.95, p, (n-p-1)} = F_{0.95, 1, 8} = 5.32 \qquad \text{(tabled value in Appendix B, Table B.3, upper 5\% points)}$$

If F calculated is greater than F critical, we reject the null hypothesis and accept the statistical significance of Y on X at the 0.05 level of significance. Calculated F is

$$F = \frac{0.648/1}{0.288/(8)} = \frac{0.649}{0.036} = 18.0$$

Because 18.0 is greater than 5.32, one would conclude that commitment does affect satisfaction ($\alpha = 0.05$).

The next test involves the significance of r_{xy}^2. Again, be cautious in interpreting the statistical significance of r_{xy}^2. The larger the sample size, the greater the probability of finding significance in a meaningless r_{xy}^2. The F-test is again used with the same rationale. At alpha (α) = 0.01, the critical F is

$$F_{0.99, 1, (n-p-1)} = F_{0.99, 1, 8} = 11.26$$

The calculated F is 18.06. Because 18.06 is greater than 11.26, one concludes that the r_{xy}^2 is significant at alpha ($\alpha = 0.01$).

The final test involves the significance of b, the regression coefficient. Like all statistics, the regression coefficient has a standard error associated with it. This calculation is important to the test. For simplicity, we have incorporated this calculation in the formula. The t-test is used for testing the b in the model. The degrees of freedom are ($n - p - 1$), or ($10 - 1 - 1$) = 8 d.f. If t calculated is greater than the critical t value with 8 d.f., then one can conclude that b is significant. If $\alpha = 0.01$, then the critical t is

$$t_{0.99, 8} = 3.3555 \qquad \text{(from Appendix B, Table B.2)}$$

The calculated t, 4.245 is greater than the tabled $t_{0.99, 8} = 3.355$; therefore, one can conclude that the estimated b is significantly different from the population parameter B. A confidence interval can also be constructed for b; it is interpreted in the same way as the confidence interval for any sample statistic.

In summary, the prediction equation in Exhibit 14.3 seems to be adequate for the intended use. Three particular points still need to be mentioned.

First, three factors affect the precision of the regression equation: sample size, scatter of points about the regression line reflected in $\Sigma(Y - Y')^2$, and the range of values selected for X, indicated by Σx^2. Second, the assumptions of SLR are important. Examining both a plot of the observations themselves and a plot of the residuals can help the researcher determine if the assumptions have been violated.[14]

Third, because the b value is estimated from sample information, it should be validated by a procedure called *cross-validation*. This is especially desirable when using linear regression for prediction. It is done either by using two samples or by splitting the original sample in two halves. In either case, the regression equation is calculated from the two parts (samples), and the b values are compared by subtracting the r_{xy}^2 of the first calculation from the r_{xy}^2 of the second. If this value is low, one sample can be used, or the two halves are put together and the b values are estimated from the whole. This cross-validation process is extremely important in multiple regression.

Multiple Linear Regression. Often, a researcher selects more than one independent variable to be included in a study, assuming that more than one variable affects change in another variable. With the addition of variables, the model becomes a multiple regression model. Its generalized form is

$$Y' = a + b_1 X_1 + b_2 X_2 + \cdots + b_k X_k + e$$

where $k = (1, 2, \ldots, p)$ and p = number of independent variables. The objectives of MLR are essentially the same as SLR. The sample data is used to estimate a, b, and the error in the model. Second, we want to know the proportion of variance accounted for in the model. Third, we wish to test the statistical hypotheses concerning the sample estimates. The major difference between the MLR model and the SLR one is the former's determination of the relative contribution of each X in the model in explaining Y. When the actual computations are completed, we refer to values of beta (β) instead of values of b. Beta values (β) are the standardized regression coefficients obtained by standardizing the data before the calculations, or standardizing the b values after the analyses. The model then becomes (disregarding the error term, for simplicity):

$$Y' = \beta_0 + \beta_1 X_1 + \beta_2 X_2 + \cdots + \beta_k X_k$$

The assumptions of the model are essentially the same as those of SLR, with the emphasis on (1) estimating the linearity of the data and (2) the effects of intercorrelation in the independent variables and the residuals. However, for inferential purposes, full disclosure of the assumptions inherent in MLR is necessary to avoid serious errors due to the violation of critical assumptions.[15] The following problem illustrates the analysis with two independent variables.

A large department store serving a large metropolitan area hires many part-time sales personnel. These clerks are standbys who supplement the regular crew. Various departments call the night before to get additional sales help as needs are anticipated. Many factors affect the daily sales of the various departments. The personnel manager tried to isolate a few of these variables so as to minimize the cost of hiring salespeople while maximizing sales opportunities. The following data was selected by the personnel manager as guidelines for hiring policies. The data appears in Exhibit 14.4. The variables that the manager isolated as influencing the daily sales are X_1 (weather) and X_2 (daily advertising dollars). Basically, the partitioning of variance is used to obtain the estimates of the components of the regression equation. The unstandardized regression equation is shown in part 2a of Exhibit 14.4:

$$Y' = 25.262 - 9.429(X_1) + 1.642(X_2)$$

EXHIBIT 14.4

Multiple Regression

1. Data

Observation	Y Average Daily Sales (000 Dollars)	X_1 Average Daily Measurement of Rain, Snow, or Sleet (in inches)	X_2 Daily Advertising Expenditures (00 Dollars)
1	28	0.05	2
2	20	1.02	1
3	32	0.01	4
4	29	0.02	3
5	36	0.01	5
6	33	0.04	7
7	26	1.11	9
8	34	0.03	6
9	32	0.51	8
10	40	0.02	10
11	44	0.07	11
12	46	0.42	12

2. Regression equation
   a. $Y' = 25.262 - 9.429\,(X_1) + 1.642\,(X_2)$
   b.

Independent Variable	Estimated Coefficient	Standard Error	t- Statistic
Constant	25.2623	1.82217	13.8638
X_1	−9.42927	2.09978	−4.49060
X_2	1.64183	0.236032	6.95594

EXHIBIT **14.4**

(Continued)

R-squared = 0.8822
F-statistic (2,9 d.f.) = 33.7134
Durbin–Watson statistic (adjusted for 0 gaps) = 0.9356
Number of observations = 12
Sum of squared residuals = 71.6764
Standard error of the regression = 2.82206

3. ANOVA table

Source	SS	d.f.	MS
Regression	536.9904	2	268.4952
Residual	71.6763	9	7.9640
Total	608.6667	11	

The F-ratio is 33.7135.
Degrees of freedom are 2 and 9.

4. Table of residuals

X_1	X_2	Obs Y	Calc Y′	Residual
0.0500	2.0000	28.0000	28.0745	–0.0745
1.0200	1.0000	20.0000	17.2863	2.7137
0.0100	4.0000	32.0000	31.7354	0.2646
0.0200	3.0000	29.0000	29.9993	–0.9993
0.0100	5.0000	36.0000	33.3772	2.6228
0.0400	7.0000	33.0000	36.3780	–3.3780
1.1100	9.0000	26.0000	29.5723	–3.5723
0.0300	6.0000	34.0000	34.8304	–0.8304
0.5100	8.0000	32.0000	33.5881	–1.5881
0.0200	10.0000	40.0000	41.4921	–1.4921
0.0700	11.0000	44.0000	42.6624	1.3376
0.4200	12.0000	46.0000	41.0040	4.9960

Now we must determine the fit of the model by estimating R^2 and statistically testing all the estimates. The squared multiple correlation coefficient is analogous to r_{xy}^2, the term used in SLR. This value tells us how much of the variance accounted for in Y is attributable to the independent variables. In our example, R^2 is 0.8822. This indicates that a meaningful 88% of the variance in the model is explained; 88% of the variance in sales can be attributed to weather and advertising. The next step in evaluating the equation is to evaluate the set of independent variables as a whole in contributing to the variance in Y. Generally this is done by interpreting an *F*-statistic from the summary ANOVA table (the partitioned variance). This *F*-statistic is shown in part 3 of the example. The appropriate critical *F*-statistic is found in the *F*-table (Table B.3) in Appendix B. At a chosen level of significance (in this case, $\alpha = 0.05$), we test the overall set of independent variables. The critical *F*-value at 2 and 9 d.f. is $F_{0.95,2,9} = 4.26$. The calculated *F*-value is 33.713. Because the calculated *F* is greater than the critical $F_{0.95,2,9}$, one can conclude that the set of independent variables as a whole contributes to the variance in *Y*. In our example, it seems that the variables *weather* and *advertising* are both contributing to the variance in *Y*.

The next step in the evaluation is to test R^2 for significance. The test statistic is

$$F = \frac{R^2/p}{(1 - R^2)/(N - p - 1)} = \frac{SS_{reg}/d.f._{reg}}{SS_{res}/d.f._{res}}$$

In our example, R^2 is significant at a 0.05 level, so we proceed to the test of the statistical significance of the *b* values in the regression equation. Each *b* is tested using a *t*-test, as with SLR. The calculated *t*-statistics are shown in part 2b of Exhibit 14.4. When each *b* is tested for statistical significance, the effects of the other independent variables are controlled. It is possible to obtain a statistically significant R^2 and insignificant *b* values. This might signal that the independent variables are highly intercorrelated, and an analysis of residuals should be conducted. The *b* values in the example are tested as follows:

$$X_1 : b = -9.42, \quad t = -4.49$$
$$X_2 : b = 1.64, \quad t = 6.95594$$

A two-tail *t*-test is used with $(N - p - 1)$ d.f. If $\alpha = 0.05$, the critical *t*-value with 9 d.f. is ± 2.11. Because both *t*-values are greater than ± 2.26, one concludes the individual effects of weather and advertising are statistically significant at a 0.05 level.

The remaining step in the evaluation of the regression equation is to estimate the contribution of each variable in the study. If all the *b* values are significant, the easiest way to interpret the contribution of each variable is to standardize *b*. Most computer software programs with multiple regression routines do this for you. The standardized *b* (beta) values can then be used to interpret the relative contribution of each variable. As simple as this may sound, the interpretation of the relative importance of independent (predictor) variables is a very controversial topic. It should suffice to say that one never interprets the relative importance of the variables from the unstandardized *b* values.

Two additional considerations can prove invaluable to the researcher. If the independent variables are intercorrelated (multicollinearity), it must be decided which variables to leave in the analysis. Multicollinearity makes for highly unstable *b* estimates. Second, if the residuals are correlated (serial correlation), which is a violation of an assumption of the model, undue problems in the analysis can result. Therefore, it is wise to examine the residual plots and conduct a Durbin–Watson test for serial correlation.[16]

A summary of our evaluation of Exhibit 14.4 is in order. Eighty-eight percent of the variance in Y is explained by the independent variables. All tests are statistically significant. Therefore, the fit of the model is statistically significant. Can the regression equation be used to predict the next day's sales? The answer is yes, if two facts have been substantiated: (1) Are there any outliers in the data? (2) Does the model make useful or meaningful sense? We have discussed the issue of outliers earlier in the text. Meaningfulness was also discussed in relation to ANOVA models. It is interpreted the same way in regression. For example, using the data in Exhibit 14.4, we can establish the relationship with weather as the dependent variable and sales and advertising as the independent variables. The R^2 from this new data is 0.6915, and everything is statistically significant at $\alpha = 0.05$. Is the model meaningful and useful? Of course not. Therefore, usefulness and meaningfulness are important managerial considerations.

Two final considerations in multiple regression analysis are the selection of independent variables and the use of partialing in the regression analysis. In business research, many independent variables are initially selected for analysis. How can one determine which variables or sets of variables are best to use? The key is to find the minimum number of variables needed to account for almost the same amount of variance as is accounted for by the entire set. Various procedures are available to accomplish this; each has its advantages and disadvantages. The researcher should be aware of these before using any one of them in an analysis.[17]

One such analysis is called *stepwise multiple regression*. The stepwise procedure inserts variables into the analysis until the regression equation is satisfactory. The purpose is to explain the most variance (*R-square*) with the most important variables. An example of stepwise forward regression (using the data in Exhibit 14.4, processed by SPSS) is given in Exhibit 14.5. Essentially, what this stepwise forward regression does is to enter independent variables as long as individual F-statistics exceed 1.0. The F-statistics for both variables—advertising expenditures, variable x_3, the most important ($F = 20.16$); and measurement of the rain, and so on, variable x_2 ($F = 16.2$)—are substantially larger than 1 and thus significantly contribute to variance in average daily sales ($R^2 = 0.8822$). Naturally, there is not much difference between this analysis and the regular regression in Exhibit 14.4, but with a larger number of variables, the benefits are enormous.

Sometimes a researcher uses one or more categorical variables in a multiple regression. These are called *dummy variables*; they usually present few problems to the experienced user. However, the analyst should be aware that dummy variables are very difficult if not impossible to interpret if they have more than four categories.[18] (See the appendix at the end of this chapter for more on this issue.)

Finally, partialing consists of using partial and semipartial correlations in the analysis for statistical control. Sometimes, when a researcher is studying a relationship among variables, there are confounding effects that hamper the actual mathematical determination of the relationship. These effects can be controlled in an analysis by the method of partialing.[19]

EXHIBIT 14.5

Example of Stepwise Multiple Regression (Using Data from Exhibit 14.4)

Source: SPSS® for Windows® developed by SPSS, Inc., 444 North Michigan Avenue, Chicago, IL 60611. All rights reserved.

MODEL SUMMARY

Model	R	R-squared	Adjusted R-squared	Std. Error of the Estimate
1	0.786[a]	0.618	0.580	$4.82
2	0.939[b]	0.882	0.856	$2.82

MODEL SUMMARY

Model	Change Statistics				
	R-squared Change	F-Change	d.f. 1	d.f. 2	Sig. F Change
1	0.618	16.205	1	10	0.002
2	0.264	20.166	1	9	0.002

ANOVA[c]

Model		Sum of Squares	d.f.	Mean Square	F	Sig.
1	Regression	376.392	1	376.392	16.205	0.002[a]
	Residual	232.275	10	23.228		
	Total	608.667	11			
2	Regression	536.990	2	268.495	33.713	0.000[b]
	Residual	71.676	9	7.964		
	Total	608.667	11			

[a]Predictors: (Constant), Daily Ad Spending (00 $).
[b]Predictors: (Constant), Daily Ad Spending (00 $), Avg. Daily Precipitation (inches).
[c]Dependent Variable: Average Daily Sales (000 Dollars).

EXHIBIT **14.5**

(Continued)

COEFFICIENTS[a]

Model	Unstandardized Coefficients		Standardized Coefficients		
	B	Std. Error	Beta	t	Sig.
1 (Constant)	22.788	2.966		7.683	0.000
Daily Ad Spending (00 $)	1.622	.403	0.786	4.025	0.002
2 (Constant)	25.262	1.822		13.864	0.000
Daily Ad Spending (00 $)	1.642	.236	0.796	6.956	0.000
Avg. Daily Precipitation (inches)	−9.429	2.100	−0.514	−4.491	0.002

[a]Dependent Variable: Average Daily Sales (000 Dollars).

Analysis of Covariance

Analysis of covariance (ANCOVA) is often used (1) to remove extraneous variance from the dependent variable so as to increase measurement precision, or (2) to more fully understand the differences in responses due to the unique characteristics of the respondents in the study.[20] The researcher thus exercises control in the design by introducing a factor or factors as covariates. Essentially, the effects of these covariates are partialed out from the dependent variable; then the dependent variable is tested for differences across the factors of interest in the study.

An effective covariate is one that is highly correlated with the dependent variable but not with the independent variable. If this is the case, the covariate accounts for a large portion of extraneous variance in the dependent variable, and measurement precision thus increases.

The mathematics of ANCOVA are long and complicated. However, ANCOVA is a linear procedure that, when used with strict adherence to assumptions, can be easily implemented using existing computer software. Therefore, we will try to explain its function in the research setting by example. In a study, a sample of corporations was classified into three control configurations based on distribution of stock ownership. The groups were called Direct Family Influence, Management Control, and Indirect Family Influence. One of the purposes of the study was to assess the effects of managerial power in terms of the control configuration groupings on managerial tenure and longevity. The researcher felt that corporate performance of each group of companies might bias the results of the analysis. Therefore, the researchers used corporate performance as a covariate in an analysis of covariance. Thus, they were essentially equating characteristics of the groups.[21]

NONPARAMETRIC ANOVA

When a researcher is interested in analyzing the differences between a dependent variable that does not contain interval properties and an independent variable of a categorical nature, nonrestrictive nonparametric analysis is used. As shown in Table 14.1, various methods are available to the researcher. One that is particularly appealing and widely used is the Kruskal–Wallis ANOVA, which is a generalization of the Wilcoxon rank-sum test. It is analogous to the one-way ANOVA but is less restrictive in assumptions. As with all nonparametric tests, it can be used with any level of measured data.

In Exhibit 14.6, in a sample survey of 12 corporations, the corporations are classified into stages in the corporate development life cycle (SCLC). They are also classified into three groups based on a measure of organization cohesiveness. Corporate life cycle can be any one of eight stages from

EXAMPLE 14.6

Kruskal–Wallis ANOVA

Source: SPSS® for Windows® developed by SPSS, Inc., 444 North Michigan Avenue, Chicago, IL 60611. All rights reserved.

1. Data Cohesiveness

Low A_1		Moderate A_2		High A_3	
SCLC	Rank	SCLC	Rank	SCLC	Rank
8	12	6	9.0	5	6.5
7	11	3	3	4	4.5
6	9.0	2	1.5	6	9.0
5	6.5	2	1.5	4	4.5
	$R_1 = 38.5$		$R_2 = 15$		$R_3 = 24.5$
	$R_1^2 = 1482.25$		$R_2^2 = 225$		$R_3^2 = 600.25$
	$n_1 = 4$		$n_2 = 4$		$n_3 = 4$
	$t_1 = 2$		$t_2 = 3$		$t_3 = 4$

2. Test statistic

$$11 = \left[\frac{12}{n(n)+1} \left(\frac{\Sigma R_j^2}{n_j} \right) - 3(n+1) \right]$$

where

R_j = sum of the ranks in the jth sample group
n_j = number of observations in the jth sample group

EXHIBIT 14.6

(Continued)

n = total number of observations in the k sample groups—that is,

$$n = n_1 + n_2 + n_3 + \cdots + n_k$$

3. Correction for ties

$$H_c = \frac{H}{1 - \dfrac{\Sigma(t_j^1 - t_j)}{n^1 - n}}$$

4. Calculations

$$H = \frac{12}{156}\left(\frac{1482.24}{4} + \frac{225}{4} + \frac{600.25}{4}\right) - 3(13)$$

$$\cong 0.0769(370.56 + 56.25 + 150.06) - 39$$

$$\cong (0.0769)(576.87) - 39$$

$$\cong 5.36$$

$$H_c = \frac{5.36}{1 - [(2^3 - 2) + (3^3 - 3) + (4^3 - 4)]/(12^3 - 12)}$$

$$= \frac{5.36}{1 - 0.05244}$$

$$= 5.65$$

RANKS

	Organizational	N	Mean Rank
SCLC	Low	4	9.63
	Moderate	4	3.75
	High	4	6.13
	Total	12	

TEST STATISTICS[a,b]

	SCLC
Chi-Square	5.510
d.f.	2
Asymp. Sig.	0.064

[a]Kruskal–Wallis Test.
[b]Grouping Variable: Organizational Cohesiveness.

1 (starting out) to 8 (reorganization). Cohesiveness is classified as low (A_1), moderate (A_2), and high (A_3).

The null hypothesis is that the three samples were selected from identical populations. Thus,

H_0: Stage in the corporate life cycle is not different between cohesiveness groups

H_1: Stage in the corporate life cycle is different between cohesiveness groups

An alpha level is set. In this case, alpha equals 0.05. The Kruskal–Wallis statistic is distributed χ^2 with (groups – 1) = degrees of freedom. Therefore, if H is less than 5.99, accept H_0. If $H \geq 5.99$, reject H_0. The Kruskal–Wallis H is calculated from the data in Exhibit 14.6, part 4, and $H = 5.36$. H is less than 5.99, so we must correct the test statistic H_0 for ties in the data. If H was significant, this need not be done. Here $H_c = 5.65$. Note the slight differences in calculations between the SPSS output and the hand calculations. Obviously, the degree of precision is better with the software. Therefore, one concludes that stage in the corporate life cycle is the same across cohesiveness classifications.

SUMMARY

The chapter begins with a discussion of the importance of variance and variance decomposition in advanced analysis. Several total variance decompositions are examined with reference to explained and unexplained variance. Generally, the more variance you can explain in a design, the more evidence there is that the research is meaningful.

Next, the general linear model for analysis is discussed. All analysis using the model requires the estimation of components of the model to reproduce individual scores. Various compositions of the linear model are then illustrated. One-way ANOVA, two-way ANOVA, linear regression, and analysis of covariance (ANCOVA) are examined. Examples of each analysis are presented and explained. The total fitting of the model by estimation and statistically testing the significance of the estimators is also examined by both manual and computerized methods.

Finally, the chapter ends with a discussion of the widely used, less restrictive nonparametric ANOVA methods. The Kruskal–Wallis ANOVA is presented, with an example.

DISCUSSION QUESTIONS

1. Suppose the president of an advertising agency did an experiment on the perception of a set of ads for a grocery product. Each participant in the experiment was randomly drawn from the population and each was randomly assigned to one of three groups. Group 1 was a *control*

group that just watched a television show segment, then filled out a questionnaire about their perception of the product. Group 2 saw the television show segment, including a very inexpensive-to-produce commercial, then filled out the same questionnaire. Group 3 saw the television show segment, including a very expensive commercial, then filled out the same questionnaire as the other groups. We can symbolize the experiment as follows:

$$\boxed{R} \; X_1 \; 0$$
$$\boxed{R} \; X_2 \; 0$$
$$\boxed{R} \; X_3 \; 0$$

a. Assuming interval data, how would you instruct the researcher to analyze the data for the agency president?

b. Suppose you were not sure of the scale of the data. What would you do?

c. Suppose the main response measure was nominally scaled as 1 = I like the commercial; 2 = Not sure; 3 = Can't stand the commercial. What would you suggest?

2. Imagine that each treatment (or control) group in Question 1 has an equal number of males and females, and the agency president wants to know the differences across treatment groups according to gender. What could one do in this situation? What are we really trying to do with the error variance? What might happen to the between-(treatment) groups' sum of squares? Do you think this is representative of a large number of other situations?

   Set up a dummy table (that is, no numbers) for an *F*-test. Assume ratio data in the response measure.

3. A retail grocery chain owner had an experiment done to test the response (change in sales) to a set of price changes. He also felt that there were two different types of stores represented in his chain of stores—(1) highly advertised, large stores, and (2) small neighborhood stores. He used three levels of price (40% raised price, no change, 40% price cut). His niece suggested he should use analysis of covariance to analyze the results of his experiment because the variable, *overall store sales*, should be used as a covariate to reduce the variance due to outside effects on store traffic and to account for the different sizes of stores. He argued that he could understand that he had a two-way analysis of variance (that is, price change and store type), but could not see how it was analysis of covariance.

a. Explain to the store owner how analysis of covariance differs from analysis of variance.

b. Which model would you prefer using? Why?

c. Do you think your choice in part b would lead to a better decision on pricing policy if you were the chain owner?

d. If you divided the dependent variable by the covariate to give a standardized change score of sales, could you use a two-way ANOVA?

4. Discuss the types of situations where covariance analysis, analysis of variance, and dummy variable regression would be substitutable and contrast this with situations where they are not. In the case of non-parametric data, would your same reasoning apply?

5. Suppose a study was conducted to test the effectiveness of three types of incentive systems on employee productivity. Management felt that the differences in productivity could be different between male and female employees. The following table of results was obtained.

	Incentive Systems		
Gender	1	2	3
Male	26	51	52
	34	50	64
	46	33	39
	48	28	54
	42	47	58
Female	49	50	53
	74	48	77
	61	60	56
	51	71	63
	53	42	59

a. Assume that the dependent variable is intervally scaled. What type of design is best suited for this problem? How would you statistically analyze the results?

b. Using your proposed analysis, are there statistical differences across incentive systems? Any difference between males and females? What about interactions?

c. Using any statistical package you desire, recheck your calculations by running the data.

6. A sample survey of boat owners in southeast Florida has yielded the following data on boat costs and buyer income.

Respondent	Boat Cost, Y ($000's)	Buyer Income ($000's)
1	2.5	20.2
2	10.3	30.5
3	40.0	100.0
4	12.3	38.1
5	3.0	16.4
6	5.6	20.2

Respondent	Boat Cost, Y ($000's)	Buyer Income ($000's)
7	1.0	14.6
8	9.4	40.2
9	7.8	23.4
10	3.5	17.6
11	5.2	19.3
12	50.1	130.4

a. Compute a linear regression of Y on X. How do you interpret the results?

b. Compute the R^2 value. Interpret this measure.

c. Suggest possible uses of this regression equation if you were a boat seller in Florida.

7. Assume next that the study of boat owners also produced information on the age (Z) of the owner. The data on age is as follows:

Respondent	1	2	3	4	5	6	7	8	9	10	11	12
Age	25	33	45	35	28	26	39	21	31	29	25	55

a. Using the data in Question 6, manually compute a multiple linear regression of Y on X and Z. Interpret the results.

b. If you were told that a boat owner at age 35 had an income of $40,000, what boat cost would you predict?

c. Using the computer program of your choice, run a multiple regression. Interpret the R^2, standardized beta weights, and statistical significance.

NOTES

1. See Fred N. Kerlinger and Howard B. Lee, *Foundations of Behavioral Research,* 4th ed. (Belmont, CA: Wadsworth Publishers, 2000), Chapter 6.

2. Actually, error variance is made up of a variety of random and nuisance components, but these are omitted from the discussion to facilitate comprehension of the basic tenets of the decomposition equation, $T = B + W$ (Total variance = between variance + within variance).

3. The general linear model is discussed briefly here. The notation is generalized for all forms and uses of the model. Don't be confused by this. Try to understand the ideas behind the model. This basic exposure is adapted from Jeremy D. Finn, *A General Model for Multivariate Analysis* (New York: Holt, Rinehart, and Winston, 1974), pp. 1–18. The more advanced student should read Chapter 5 of F. A. Graybill, *An Introduction to Linear Statistical Models*, vol. 1 (New York: McGraw-Hill, 1961).

4. These assumptions and the use of the model are covered in a variety of books on design and analysis. See, for example, Geoffrey Keppel, *Design and Analysis*, 3rd ed. (Englewood Cliffs, NJ: Prentice-Hall, 1991); and Elazar J. Pedhazur, *Multiple Regression in Behavioral Research*, 3rd ed. (Orlando, FL: Harcourt Brace College Publishers, 1997).

5. See W. L. Hays, *Statistics*, 5th ed. (Fort Worth, TX: Holt, Rinehart, and Winston, 1997); and G. V. Glass and K. D. Hopkins, *Statistical Methods in Education and Psychology* (Englewood Cliffs, NJ: Prentice-Hall, 1984), pp. 350–353.

6. See Keppel, op. cit. note 4, Chapter 11.

7. Emanuel J. Mason and William J. Bramble, *Understanding and Conducting Research*, 2nd ed. (New York: McGraw-Hill, 1989), pp. 233–234. Also see Robert Rosenthal and Ralph L. Rosnow, *Essentials of Behavioral Research: Methods and Data Analysis* (New York: McGraw-Hill, 1991), pp. 365–391, for an entire chapter dealing with this critical issue.

8. For a discussion of interactions in factorials, see Paul D. Berger and Robert E. Maurer, *Experimental Design* (Belmont, CA: Wadsworth, 2002), pp. 271–273.

9. Ibid., pp. 245–249; Keppel, op. cit. note 4, pp. 173–181.

10. The topic is covered in all advanced research design books and statistical analysis books. See, for example, Hays, op. cit. note 5; or B. J. Winer, *Statistical Principles in Experimental Design* (New York: McGraw-Hill, 1971).

11. See Mark L. Berenson, David M. Levine, and Matthew Goldstein, *Intermediate Statistical Methods and Application* (Englewood Cliffs, NJ: Prentice-Hall, 1983), pp. 147–149.

12. Such topics as unequal cell size two-way ANOVA, random effects and mixed effects ANOVA, and repeated measures ANOVA are beyond the scope of this text. Interested readers should read Winer, op. cit. note 10.

13. See, for example, Norman Draper and Harry Smith, *Applied Regression Analysis*, 3rd ed. (New York: Wiley, 1998), specifically Chapters 1–3; and John Neter, William Wasserman, and Michael Kutner, *Applied Linear Statistical Models*, 4th ed. (Homewood, IL: Irwin, 1996).

14. See Draper and Smith, op. cit. note 13, Chapter 3.

15. For a good treatment of the assumptions for MLR, see F. A. Graybill and H. K. Iyer, *Regression Analysis* (Belmont, CA: Duxbury Press, 1994), pp. 232–234.

16. See Draper and Smith, op. cit. note 13, pp. 157–162.

17. See Draper and Smith, op. cit. note 13, Chapter 13.

18. See Berenson et al., op. cit. note 11, pp. 332–341.

19. See Pedhazur, op. cit. note 4, Chapter 5.

20. J. F. Hair, Jr., R. E. Anderson, R. L. Tatham, and W. C. Black, *Multivariate Data Analysis*, 5th ed. (Upper Saddle River, NJ: Prentice Hall, 1998).

21. Michael P. Allen and Sharon K. Panian, "Power Performance and Succession in the Large Corporation," *Administrative Science Quarterly* (1982), pp. 538–547; and S. Siegel and N. J. Castellan, Jr., *Nonparametric Statistics for the Behavioral Sciences*, 2nd ed. (New York: McGraw-Hill, 1988).

SUGGESTED READING

The Nature of Variance Decomposition

Kerlinger, Fred N., and Howard B. Lee, *Foundations of Behavioral Research*, 4rd ed. (Belmont, CA: Wadsworth, 2002).

This updated classic retains the look and feel of the original. An excellent discussion on variance and its control.

Linear Models

Draper, N. R., and H. Smith, *Applied Regression Analysis*, 3rd ed. (New York: Wiley, 1998).

The bible on all forms of regression analysis. Because its approach is sophisticated, a knowledge of matrix algebra is suggested. However, don't let this scare you away. It is very readable even with a limited math background.

Montgomery, Douglas C., Elizabeth A. Peck, and G. Geoffrey Vining, *Introduction to Linear Regression Analysis*, 3rd ed. (New York: John Wiley, 2001).

An up-to-date look at regression as a contemporary analysis method. A good introductory look at regression analysis and its components.

Pedhazur, Elazar J., *Multiple Regression in Behavioral Research*, 3rd ed. (Orlando, FL: Harcourt Brace Publishers, 1997).

This book is consistent with using the linear model in analysis. A well-written book that contains important research considerations in analysis and interpretation of data. Also examines the ANOVA/regression similarities using a general linear model.

Nonparametric ANOVA

Conover, W. J., *Practical Nonparametric Statistics* (New York: Wiley, 1980).

A very easy book to use and understand, covering the basics of nonparametric analyses. It has an applied focus and many easy-to-follow examples.

Sigel, S., and N. J. Castellan, Jr., *Nonparametric Statistics for the Behavioral Sciences*, 2nd ed. (New York: McGraw-Hill, 1988).

Provides an excellent overview of many nonparametric techniques. A true classic.

THE USE OF DUMMY AND EFFECT CODING TO EXAMINE GROUP DIFFERENCES USING MULTIPLE LINEAR REGRESSION

Most often one thinks of analysis of variance, linear regression, and analysis of covariance as independent methods with different capabilities and mathematics. Actually, mentioned in this chapter, the techniques are interwoven in the context of the general linear model. In fact, the method that subsumes all of the capabilities for model building and hypothesis testing for the parametric analysis in Chapter 13 is multiple linear regression. For example, the one-way ANOVA that was explained in this chapter (see Exhibit 14.1) can be conducted using a multiple regression analysis, where the categorical grouping variable X (size of company) can be broken down into $(p-1)$ (p = groups) independent variables. The dependent variable will still remain the same as in ANOVA. A categorical variable, as you remember, is a variable on which some units of analysis differ in type or kind. The values that the independent variables take on are called *dummy, effect,* or *orthogonal* codes. The specific coding of the categorical variable(s) represents some portion of the design of the study. The simplest way to code a categorical variable is to use dummy coding to reflect only differences in groups. Effect coding, on the other hand, reflects the effects of the treatments in the analysis (for example, post hoc comparisons). Orthogonal and nonorthogonal coding are used when comparisons are planned.

Exhibit 14A.1 shows the data from Exhibit 14.1 laid out for multiple regression using both dummy variables and effect coding.[1] Simply speaking, the X_1 variable is coded with all 1s to reflect scores in the group A_1. The X_2 variable is coded with all 1s to reflect scores in the group A_2. The X_3 variable is coded with all 1s to reflect the scores in group A_3. Usually, the final coding of the last group when using multiple regression to perform a one-way ANOVA is for reference purposes. Depending on the purpose of the analysis, it is coded with all 0s for dummy coding, or all –1s for effect coding. The number of independent categories codes is always $(p-1)$. No matter how many groups are used, a 1 is used to code the variable representing that group's scores, and the final group is coded as the reference group. It is important to note that one should always use $(p-1)$ categories to avoid singular matrices in computer routines. The results will be unintelligible if you do not use $(p-1)$ categories.

When multiple regression is run using both sets of codes, the summary table produced is the same as the ANOVA summary table, except it is reported in regression partitioning terms. In the dummy coded model, the regression equation is

$$Y' = a + b_1X_1 + b_2X_2 + \ldots + b_pX_p$$

where the constant term a is equal to the mean of group A_1 (largest company size). This group was coded with 0s throughout. Each b represents the

Group	Individual	Y	Dummy Coding			Effect Coding		
			X_1	X_2	X_3	X_1	X_2	X_3
Company size:								
A_1 (smallest)	1	37	1	0	0	1	0	0
	2	22	1	0	0	1	0	0
	3	22	1	0	0	1	0	0
	4	25	1	0	0	1	0	0
A_2	5	36	0	1	0	0	1	0
	6	45	0	1	0	0	1	0
	7	47	0	1	0	0	1	0
	8	23	0	1	0	0	1	0
A_3	9	43	0	0	1	0	0	1
	10	75	0	0	1	0	0	1
	11	66	0	0	1	0	0	1
	12	46	0	0	1	0	0	1
A_4 (largest)	13	76	0	0	0	−1	−1	−1
	14	66	0	0	0	−1	−1	−1
	15	43	0	0	0	−1	−1	−1
	16	62	0	0	0	−1	−1	−1

DUMMY OUTPUT

B	Source SS	MS	F
X_1 −35.25	Reg 3314.25	1104.75	7.34
X_2 −24.00	Res 1805.5	150.458	
X_3 −4.25			
C 61.75			$R_{xy}^2 = 0.647$

EFFECT OUTPUT

B	Source SS	MS	F
X_2 −19.37	Reg 3314.25	1104.75	7.34
X_2 −8.125	Res 1805.5	150.458	
X_3 11.62			
C 45.875			$R_{xy}^2 = 0.647$

where

$\quad$ Mean of groups $= \overline{Y}_{ij}$ or $\overline{A}_1 = 26.5$

$$\overline{A}_2 = 37.75$$

$$\overline{A}_3 = 57.50$$

$$\overline{A}_4 = 61.75$$

deviation of the mean of the groups assigned 1 in the analysis and the 0-coded group. For example, the mean of group $A_1 = 61.75$. Therefore, the b of group A_1 is equal to the mean of group A_1 minus the constant term. Thus,

$$\overline{A}_1 - C = b_1$$

$$26.50 - 61.75 = -35.25 \qquad \text{for all group means and } b \text{ values}$$

Also, the mean of any group is equal to the constant (a) plus the b associated with any group, where the X value was coded 1. For example, the mean of group A_1 is equal to

$$\overline{A} = C + b_1(X_1)$$

$$26.50 = 61.75 - 35.25(1)$$

$$= 26.50$$

The test of overall differences in the groups can be made by the F-test, where the calculated F for (p) and ($n - p$) degrees of freedom is compared to the critical F for (p) and ($n - p$) degrees of freedom. This is identical to the interpretation of the F-test in one-way ANOVA. The b values can be tested by t-tests. These t-tests reflect comparison of group means (the coded-1 values versus the reference group coded with 0s). The r_{xy}^2 is analogous to ω^2.

In the effect-coded model, the results reflect the fixed effects linear model for one-way ANOVA. The intercept (a) is equal to the grand mean, and the b values reflect a treatment effect (the β_j in the one-way linear model). Keep in mind that an individual score is a component of the grand mean, error, and the treatment effect of the group to which the subject belongs. The effect coding can also produce multiple comparisons in the form mentioned in the chapter.

NOTE

1. Adapted from Elazar J. Pedhazur, *Multiple Regression in Behavioral Research* (New York: Holt, Rinehart, and Winston, 1982), pp. 271–329. Also see Joseph F. Hair, Jr., Rolph E. Anderson, Ronald L. Tatham, and William C. Black, *Multivariate Data Analysis*, 5th ed. (Upper Saddle River, NJ: Prentice Hall, 1998).

15

Advanced Multivariate Analysis

Overview

Introduction to Multivariate Analysis

Selection of a Multivariate Technique

Analysis of Dependency

Multivariate Analysis of Variance

Multiple Discriminant Analysis

Conjoint Analysis

Covariance Structure Analysis

Analysis of Interdependency

Factor Analysis

Cluster Analysis

Multidimensional Scaling

Comment on Multivariate Techniques

Summary

Discussion Questions

Notes

Suggested Reading

OVERVIEW

Chapter 14 dealt with the analysis of relationships with one dependent variable and one or more independent variables. In such cases, the total variance in a specific model is partitioned so that an estimate can be made of the amount of variance in the dependent variable attributable to changes in the independent variable(s).

In the analytical world of the business researcher, relationships are often more complex. One dependent variable in an analysis may be supplanted by multiple dependent variables. Additionally, some analyses may require no specification of an independent and a dependent variable, yet have a multitude of variables with some underlying relationship. In any event, the researcher must be aware of the methods of analysis for any multivariable study situation.

Although multivariate methods are complex, their presentation in this chapter is kept simple. The aim is understanding potential uses and misuses of the procedures, so most technical properties are omitted from the discussion.

In addition, for students who choose to use a statistical package like SPSS, many of these statistical procedures are explained, with excellent accompanying supplementary materials.

We begin the chapter by examining the nature of multivariate relationships and their analysis. The importance of multivariate analysis is established, and various guidelines for choosing a specific technique are explored. Next, we discuss the multivariate techniques used when a dependent-independent relationship is examined: multivariate analysis of variance, multiple discriminant analysis, conjoint analysis, and covariance structure analysis. Finally, we examine the three multivariate procedures that can help summarize the information found in a large set of variables. These procedures (which analyze interdependencies among the variables) are factor analysis, cluster analysis, and multidimensional scaling.

INTRODUCTION TO MULTIVARIATE ANALYSIS

Obviously, business researchers have long known that a multitude of variables affect any decision outcome. Not until the advent of efficient computers and effective analytical software packages, however, could complex design structures be used in research.

Multivariate analysis can be considered in different ways. Two conceptualizations seem to be prevalent in modern-day research. First, it is a means of analyzing phenomena based on the realization that anything worth studying (whether for discovery or hypothesis testing) has more than a single facet. Second, it is the fitting of algebraic models to situations with multiple random variables (usually dependent variables) that are measures of the same sample units. In this latter role, multivariate analysis allows the properties of multiple correlations among the dependent variables to be considered in one analysis.

Both conceptualizations are valid. All multivariate techniques have one thing in common. They allow the researcher to investigate the interactions, dependencies, and commonalities among variables or sample units in a specific multivariate design. The techniques vary from very precise analytical models, with strict a priori hypothesis-testing features, to very loose, almost heuristic rules of thumb.

The prudent problem solver should not be intimidated by the complex mathematical formulations of multivariate procedures. Because computers will handle most of the mathematics and statistical testing in the models, the problem solver needs to know only three important aspects of multivariate analytic models: (1) how to select the right technique, (2) how to use the technique correctly, and (3) how to evaluate and use the results of the analysis.

For example, suppose one would like to examine differences in cohesiveness between random samples of two work groups, where the independent variable is size of group (large and small). In this instance, in a sampling situation, one would use a simple t-test for differences in means. The decisions to be made are obvious. First, make sure the assumptions of the test are considered. Next, make sure the dependent variable is at least intervally scaled. Third, perform a statistical test of hypothesis (a t-test) to see if there are differences.

Next, suppose the situation has changed with regard to the problem specification. Now you would like to see if there are differences between groups on a set of dependent variables whose effect you would like to consider together. What decisions do you now make concerning the analysis? Intimidated? Believe it or not, many univariate procedures have multivariate analogs. In this instance, a multivariate analog to the simple t-statistic is the Hotelling T^2 statistic. Given certain assumptions, at least intervally scaled dependent variables, and a multivariate statistical hypothesis, one could test the differences between groups on the entire set of dependent variables using this statistic.

This chapter describes several multivariate analytical techniques used in business research. We first discuss a general scheme for selecting a particular technique. Subsumed in each discussion of a specific technique is how to use the technique correctly and also how to interpret and evaluate the results.

SELECTION OF A MULTIVARIATE TECHNIQUE

Naturally, the choice of a multivariate analysis technique depends on the specification of the research problem and the subsequent choice of a study design. The choice of technique is analogous to the selection of a univariate technique except there are more variables to consider. Our selection framework considers three questions concerning the variables. How these questions are answered will determine the choice of the appropriate multivariate technique. These questions pertain to the specification of the relationship in the variables, the number of variables, and their level of measurement. The framework, presented in Figure 15.1, is not meant to cover all the possible techniques that could be used in any situation. Also note that many of the techniques that specify a level of measurement in the dependent variable can combine different levels of measurement in the independent set of variables (for example, multiple discriminant analysis, covariance analysis). Finally, some of the techniques mentioned can be considered extensions of a more basic technique and are related philosophically and mathematically (for example, ANOVA-MANOVA, multiple regression–multiple discriminant–conjoint analyses).

FIGURE 15.1

Selecting a Multivariate Analytical Technique

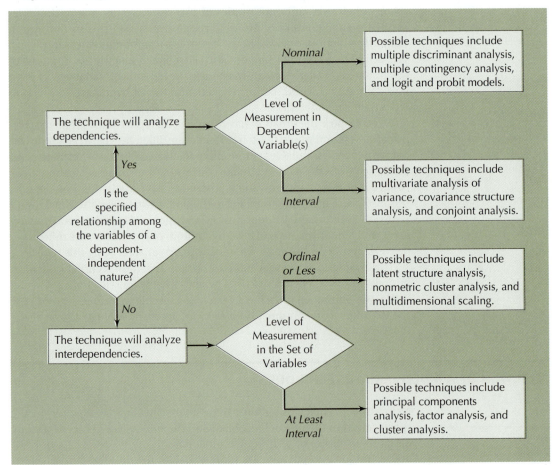

When a multivariate relationship is studied by proposing a dependency structure, $[Y = f(X)]$, data must include measures of a set of independent and dependent variables. Several techniques are available to analyze multivariate dependency structures. The most widely used of these techniques are discussed in this section: (1) multivariate analysis of variance, (2) multiple discriminant analysis, (3) conjoint analysis, and (4) covariance structure analysis.

Multivariate Analysis of Variance

Multivariate analysis of variance (MANOVA) is a technique that examines the effect of a treatment (or grouping variable) on two or more dependent variables. The technique is analogous to univariate ANOVA, but with two important differences. First, MANOVA provides more total information than can be gotten from performing single ANOVAs on each dependent variable in an analysis. Second, the use of MANOVA takes into consideration the fact that the dependent variables are correlated in some way. This in itself may hinder the analysis if separate ANOVAS are used.

For example, to examine the differences among three randomly selected groups of workers on the variables satisfaction (Y_1) and income (Y_2), two ANOVAs could be run independently. This would produce two independent F-tests of which, by chance alone, one could be statistically significant.[1] Also, the researcher could neither assess nor report the combined differences across groups on the two variables. Finally, the correlations among dependent variables (as facets of behavior) are very important in research. Multivariate techniques, especially MANOVA, take these correlations into account.

Multivariate analysis of variance uses the general linear model and follows a similar set of assumptions as in ANOVA (i.e., assumptions about additivity). However, a more complex multivariate normal distribution is involved, so the user of the technique should be aware of the assumptions and how to test for their adequacy if they seem to be violated.

Because the analysis is complex, all of the computations are completed by computers. Output information from most software packages is very similar, but interpretation of the output is crucial to understanding what is being accomplished by MANOVA. With this in mind, let us reconsider the example concerning the three groups of workers and the two variables, satisfaction (Y_1) and income (Y_2). Now we use a multivariate analysis of variance to examine differences across the groups. As you remember from Exhibit 14.1, the univariate null hypothesis was stated as follows: H_0: $\mu_1 = \mu_2 = \mu_3$, where sampling data is used to perform the statistical test among means on one dependent variable. With more than one dependent variable, the rationale of the null hypothesis remains the same. However, the single population parameter in the hypothesis now becomes a combined set of parameters representing the

1. Null hypothesis

   $H_0: \mu_1 = \mu_2 = \mu_3$

   $H_1: \mu_1 \neq \mu_2 \neq \mu_3$

   where μ represents the combined means of a group.

2. Data

   a. Three groups consisting of five subjects to a group

   b. Two dependent variables measured for each subject and each group, where k = groups, p = dependent variables, and n = group size

3. Dependent variable means for each group

$$\begin{bmatrix} \bar{Y}_{11} \\ \bar{Y}_{21} \end{bmatrix} \quad \begin{bmatrix} \bar{Y}_{12} \\ \bar{Y}_{22} \end{bmatrix} \quad \begin{bmatrix} \bar{Y}_{13} \\ \bar{Y}_{23} \end{bmatrix}$$

$$\begin{bmatrix} 3.60 \\ 1200 \end{bmatrix} \quad \begin{bmatrix} 5.40 \\ 920 \end{bmatrix} \quad \begin{bmatrix} 7.60 \\ 1380 \end{bmatrix}$$

   where $\bar{Y}_{(\text{variable, group})}$ is the mean of the dependent variable for the group (for example, $\bar{Y}_{11}$ = mean of satisfaction scores for group one).

4. Multivariate F-test of equality of mean vectors

   $F = 7.06$, d.f. = 4 and 22, p less than 0.001

Variable		Univariate F	p Less Than	Step-Down F	p Less Than
a. 1.	Satisfaction	6.96	0.01	6.96	0.01
2.	Income	4.67	0.05	3.99	0.05
b. 1.	Satisfaction	4.50	0.05	—	—
2.	Income	1.96	N.S.	—	—
c. 1.	Satisfaction	2.69	N.S.	—	—
2.	Income	3.67	N.S.	—	—

dependent variables in the analysis. The hypothesis, shown in Exhibit 15.1, is that there is no difference in the vector of means across the groups.

As usual, sampling data is needed to perform the test. In this example, there are five subjects per group, three groups, and two dependent measures for each subject in each group. The dependent variable sample means are shown in the example. The test statistic needed to reject the null hypothesis is a likelihood ratio, Λ (or a derivation thereof), that is transformed into a more appealing multivariate F statistic. The calculation and subsequent interpretation of this multivariate F statistic enables us to statistically test the null hypothesis. The tabled F statistic (from Appendix B, Table B.3), for $p(k-1)$ and $p(nk-k-1)$ degrees of freedom and an alpha level of ($\alpha = 0.01$), is $F_{0.99,4,22} = 4.31$. The calculated multivariate F statistic is $F = 7.06$. Therefore, if we use the rule for rejecting the null hypothesis that F calculated must be greater than F tabled, we reject the null hypothesis at $\alpha = 0.01$. Because $F = 7.06$ is greater than $F_{0.99,4,22} = 4.31$, we conclude, from the sample data, that the groups differ with respect to the *combined* dependent variable set.

When a null hypothesis is rejected in MANOVA (when the test is significant), the researcher must perform additional analyses in order to fully understand the data set. This is analogous to the further analysis in the univariate ANOVA, which you recall as post hoc multiple comparisons. There are four methods of post-analysis in MANOVA: (1) the univariate F procedure, (2) the step-down F procedure, (3) the multiple-comparison approach, and (4) the multiple discriminant analysis approach. Generally, all of these approaches help the researcher determine which dependent variable(s) contributed to the significant differences among the groups (the effects of the independent variable(s)).

The univariate F procedure involves examining the individual F statistics for each dependent variable in the analysis and the appropriate means for the significant univariate effects. In Exhibit 15.1, this analysis is presented in part 4. Here, the three possible outcomes could be where (1) both dependent variables are significant, (2) only one of the two is significant, or (3) neither is significant. In all instances, the multivariate F is significant.

In the first outcome, as shown in part 4a, both univariate F values are significant. In this case, one can interpret the findings by going back to the group means and inspecting the step-down F values, which will be discussed shortly.

In the second outcome (part 4b), only one variable—in this case, satisfaction—is statistically significant. Here, the variable "satisfaction" is found to be the major contributor to the group differences because of its significant univariate F.

In the last case (part 4c), the multivariate F is statistically significant, but none of the univariate F values is significant. This apparent contradiction can be reconciled if one examines the correlation between the dependent variables. If the correlation is high, the multivariate F might be significant, but the univariate F values will not be. Because individual F-tests ignore correlations among the dependent variables, they use less than the total information available for assessing group differences. On the other hand, MANOVA accounts for this by testing the linear combination of dependent variables that provides the strongest evidence of overall group differences.[2]

The step-down F procedure uses a logical or theoretical basis of ordering the dependent variables by importance in the analysis. Then each variable is tested sequentially. In Exhibit 15.1, the dependent variables could have been ordered by importance: Satisfaction is considered more important than income in the analysis. In this instance (see part 4a, Exhibit 15.1), one would first examine the multivariate F. If this is significant, a univariate F-test is performed on the first dependent variable. Then a univariate F-test is performed on the second variable to establish the additional effect on this variable due to the treatment (or grouping variable). This sequence is applied to all of the dependent variables. The step-down procedure allows the researcher to start with the dependent variable deemed most important in the set of dependent variables in the sequence. The test sequencing enables the analyst to study *additional* effects over and above those explained by prior dependent variables in the sequence.

The third procedure for estimating the effects of the treatment on the dependent variables is the application of multiple comparisons. In this instance, the researcher is interested in testing all $k(k-1)/2$ pairwise differences among k groups on each of the p dependent variables. The tests are performed on $p(k)(k-1)/2$ pairs, while strictly controlling for the probability of making a Type I error (alpha).

The final procedure involves the relationship between MANOVA and another multivariate technique called multiple discriminant analysis. The relationship deals with the general linear model. Both are methods for looking at multivariate differences among groups. The use of multiple discriminant analysis following MANOVA ensures that the researcher can find information about both the group separation and the underlying dimensionality of variables.

Multiple Discriminant Analysis

Multiple discriminant analysis (MDA) is a statistical analytical technique that examines the relationship between a nominally scaled dependent variable and a set of explanatory or independent variables. The set of independent variables must have interval properties or at least be transformable to an interval level (i.e., through dummy variables). It is analogous to multiple regression in its algebraic modeling format. Among the various reasons for using MDA to analyze data, one was alluded to in the last section (as a follow-up to MANOVA). Generally, the objectives of MDA are

1. To determine whether a set of independent variables can significantly differentiate among two or more groups of study units (i.e., whether the groups are different on the set of variables as a whole)

2. If so, then to determine which variables are contributing more to the discrimination (i.e., which variables are more important in the analysis)

3. To determine the specific set of variables that provides the most separation among the groups

4. To use this information to correctly reassign each study unit into its proper group

The mathematics of the model involves fitting a discriminant relationship, which is technically expressed as a discriminant function. The smaller of $(k-1)$ or p discriminant functions is always produced by the procedure. Each can be evaluated statistically and then evaluated for its substantive qualities and interpretability. Each function contains information about the discriminant coefficients that can be used to evaluate whether the objectives of MDA have been met. The amount of variance explained and the predictive ability of the functions are two evaluative criteria in MDA.

EXHIBIT **15.2**

Multiple Discriminant
Analysis: Two
Groups—Three
Independent
Variables

Groups

A_1 Good credit risk, $n_1 = 366$
A_2 Bad credit risk, $n_2 = 366$

Variables

	A_1 (GCR)	A_2 (BCR)
X_1: Age	33.2	25.6
X_2: Monthly income	1793	1282
X_3: Monthly bills	646	483

Standardized Discriminant Coefficients

X_1: 4.281
X_2: −2.493
X_3: 0.8585

Confusion Matrix

Actual Group	Predicted Group Membership		
	A_1	A_2	Total
A_1	281	85	366
A_2	31	335	366

Confusion Matrix Converted to Percent

Actual Group	Predicted Group Membership		
	A_1	A_2	Total
A_1	77%	23%	100%
A_2	9%	91%	100%

Classification Percent

A_1 $\quad ^{281}/_{366} \cong 77\%$
A_2 $\quad ^{335}/_{366} \cong 91\%$ $\qquad$ Average $\cong 84\%$

Multivariate F: 260.77 with 3 and 728 d.f.

To illustrate MDA in research, we will use an example of determining what might affect the good or bad credit rating of an individual. In a metropolitan area, 732 people were selected for study. Of those, 366 were previously determined to be bad credit risks, and the others were good credit risks. Data was also collected on the age, monthly income, and amount of monthly bills for the two groups of individuals. The purpose of the study is to assess key variables in the set that predict whether a person is a good or bad credit risk. The means of the variables for each group are presented in Exhibit 15.2, as is the subsequent analysis. The analysis produces $(k - 1)$ or only one discriminant function. After the discriminant coefficients are standardized, the discriminant function can be expressed by the linear compound:

$$Y' = \text{constant} + 4.281(X_1) - 2.493(X_2) + 0.8585(X_3)$$

The standardized coefficients can be interpreted as the measure of a variable's contribution to the discrimination between the groups. Generally, we can interpret the coefficients in the example as follows: We can conclude that the older the individual and the more monthly bills, the more likely the person will be a good credit risk. The negative sign for the coefficient (X_2) implies that people with lower monthly income tend to be bad credit risks. Another use of the discriminant function is to reclassify the original sample into groups based on the variables. This part of the analysis produces what is called the *confusion matrix*. In our example, the discriminant rule for classification produced by the analysis correctly classifies 84.15% of the individuals into their respective groups (good or bad credit risks). However, this procedure results in an upward bias in prediction accuracy of the discriminant function because the same individuals were used for deriving both the function and the subsequent classification procedure.

Better techniques are available for ensuring unbiased classification schemes. One is the *jackknife criterion,* which is available in some statistical software packages. Another procedure is the *split-sample* or *cross-validation* technique, where the total sample is randomly split into two groups. The first sample is then used to derive the discriminant function, and the second sample is for classification purposes.[3]

Finally, using a multivariate *F*-test, we can test for statistically significant group differences in the entire variable set. The multivariate *F*, with 2 and 728 d.f., is greater than the tabled $F_{.95,2,728} \cong 3.00$. Therefore, one can conclude that the populations are sufficiently separated. Additionally, the researcher can effectively use the discriminant model to classify individuals into a credit category.

Most often, MDA is a model-building effort that requires the determination of what variables or subsets of variables are best used in the classification of study units. In this instance, the stepwise MDA procedures are best suited to the task.[4]

Conjoint Analysis

As the name suggests, *conjoint analysis* is a statistical technique that is concerned with the joint effect of a combination of rank-ordered independent variables on a dependent variable.[5] The technique is similar to analysis of variance (main effects only) in the form of a multiple linear regression model using dummy coding to represent the independent variables. It is referred to as *monotone* regression because the coefficients must add to scores that preserve the same rank order as the input data. A computer program called MONANOVA can be used to estimate the coefficients or importance weights associated with each independent variable in the analysis.

Conjoint analysis has been used extensively in marketing research to determine which combinations of relevant product or service characteristics are valued by consumers in a target market(s). When consumers evaluate alternative offerings, the decisions they make are based on a complete analysis of all the product attributes and benefits taken together. The consumer's preferred choice is the combination that provides the greatest relative satisfaction or utility. Conjoint analysis can be thought of as statistically reproducing the consumer process of trading off attributes. This can assist researchers in decisions regarding new product design or the repositioning of existing products. Thus, a researcher may be interested in finding the utility values (called *partworths*) that consumers impute to various characteristics of products.

Suppose we were interested in using conjoint analysis to approximate the relative values that consumers place on salient attributes in purchasing a microcomputer. In studying the market, a data reduction technique like factor analysis or focus group interviews can be used to identify key attributes that significantly influence a purchase decision. For ease of interpretation, we list in Exhibit 15.3 only three key attributes or factors that might be important; obviously, others could be included. It is determined that three levels will be associated with each factor. In our analysis, three sizes of internal memory (128, 256, and 512 megabytes), three processing speeds (400, 533, and 800 megahertz), and three prices ($700, $1400, and $2400) are available for the microcomputer.

In symmetrical designs, all combinations of factor levels are presented for evaluation. The total number of combinations is calculated by multiplying the number of levels of each factor ($3 \times 3 \times 3 = 27$). However, it would be very difficult and tiresome for respondents to rank all of the possible combinations, even in our limited analysis, which contains a total of 27. Instead, tradeoff analysis is carried out using asymmetrical designs that present subsets of factor levels to reduce the number of rankings required. The tradeoff matrix in our example is divided into three subsets that compare two attributes at a time. Each paired combination of two attributes is ranked from 1 to 9 in order of preference. From Exhibit 15.3, a respondent has ranked a microcomputer with 512 MB of internal memory at a price of $700 as the most preferred, and one with 128 MB at a price of $2400 as the least preferred combination of those two attributes.

Given the conjoint analysis results for our hypothetical respondent, we can determine the combinations of attributes that provide different utility scores by examining the relative magnitude of the partworth utilities for each factor level. A microcomputer with 512 MB, 800 MHz, at $1400 would be the most preferred choice, producing the highest utility score:

$$69.9 = 47.2 + 10.7 + 12.0$$

Similarly, the least preferred would be a computer with 128 MB, 400 MHz, and $2400, with a utility score of

$$-91.1 = (-52.9) + (-14.4) + -23.8$$

Product Characteristics or Attributes with Assigned Level Numbers

Internal Memory	Processing Speed	Price
1 = 128 MB	1 = 400 MHz	1 = $700
2 = 256 MB	2 = 533 MHz	2 = $1400
3 = 512 MB	3 = 800 MHz	3 = $2400

Conjoint Tradeoff Analysis Matrix

	Memory				Speed				Price		
Price	128 MB	256 MB	512 MB	Memory	400 MHz	533 MHz	800 MHz	Speed	$700	$1400	$2400
$700	5	2	1	69 MB	9	6	3	400 MHz	7	8	9
$1400	6	4	3	128 MB	8	5	2	500 MHz	4	5	6
$2400	9	8	7	256 MB	7	4	1	600 MHz	1	2	3

Conjoint Analysis Results

Factor	Factor Importance Weights	Levels		
		1	2	3
1. Internal Memory	0.552	−52.9	5.7	47.2
2. Process Speed	0.237	−14.4	3.7	10.7
3. Price	0.211	11.8	12.0	−23.8
Total	1.000			

EXHIBIT **15.3**

Conjoint Analysis: Three Variables— Three Levels

Other combinations of these attributes can also be examined for total value by adding their respective partworth utilities. The assumption that attributes do not interact (that is, they are not additive) is critical when evaluating conjoint models. The importance weight for each factor (because they sum to 1.00) can be interpreted as a percentage of the total impact of the three attributes taken as a whole. By far the most important attribute is internal memory (55.2%), followed by processing speed (23.7%) and price (21.1%). Conjoint measurement lends itself to further segmentation analysis by clustering like respondents in terms of partworth similarities.

Covariance Structure Analysis

A class of analyses that has been gaining popularity in business research circles is causal modeling. Essentially, *causal modeling* attempts to test directional relationships according to some prespecified theoretical structure.

The mathematics of this set of techniques can be complex, but its use is nevertheless increasing.

One frequently used causal modeling technique is *covariance structure analysis*. In covariance structure analysis, the researcher can specify the relationships between theoretical constructs (called *latent* variables) and observables (called *manifest* variables). Essentially, the technique compares an estimated set of parameters [designated as $\Sigma(\Theta)$] from a sample to a proposed set of population parameters (designated Σ) hypothesized by theory.

Two general computer programs used for this purpose are EQS and LISREL® (acronyms that stand for structural equations and linear structural relations, respectively).[6] These programs are based on the statistical method of maximum likelihood. (Other analysis methods are also available—for example, weighted least squares, two-stage least squares, and so on.) Maximum likelihood estimates are those that are most likely to reproduce the hypothesized theoretical structure (that is, does $\Sigma(\Theta) = \Sigma$?). This method iteratively estimates a set of parameters from a sample that will maximize the joint probability of reproducing a population covariance matrix. In essence, empirical models that efficiently fit their proposed theoretical structures produce estimates with a high degree of fit.

For example, Figure 15.2 is a structural diagram in which latent variables are enclosed in circles and manifest variables in rectangles. In the diagram, relationships between variables in the causal model are depicted by arrows, and small Greek symbols are used to represent the individual parameters to be estimated. The construct "performance" is measured by three observable variables: sales volume (in dollars), number of production units, and rate of inventory turnover, as estimated by λ_4 through λ_6, respectively. Similarly, the "ability" construct is measured by its three indicators: achieved level of education (λ_1), management training completed (λ_2), and previous managerial experience (λ_3). In our recursive model, ability is shown by the solid one-way arrow (as estimated by γ_1) to directly influence managerial performance.

FIGURE **15.2**

Structural Diagram of the Relations Between Managerial Ability and Performance

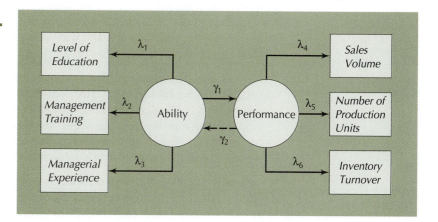

However, in covariance structure analysis, reciprocal causation may also be examined using nonrecursive models. For instance, suppose we hypothesize that managerial performance also directly influences ability. In that case, we could add the dotted arrow (γ_2) from "performance" back to "ability" to the model, and obtain simultaneous estimates of the two-way influence between constructs. However, students are encouraged not to complicate their initial covariance structure models with reciprocal feedback loops until they are quite familiar with their data and the analysis technique itself.

Needless to say, an assessment of the overall fit of a model and the interpretation of appropriate test statistics is quite complicated. Although structural models may seem easy to construct, many underlying conceptual, statistical, and operational assumptions must be met. Often the results of practical applications of the technique contain some ambiguity for either accepting or rejecting a model. To further complicate matters, many aspects of model assessment can be considered more art than science, requiring judgment on the part of the researcher. To date, some tentative guidelines for model assessment are available, and more formal rules and evaluative criteria are being developed.[7]

ANALYSIS OF INTERDEPENDENCY

When the focus of the analysis is on a whole set of interdependent relationships, and the interest of the researcher is to summarize the information in the set by using some smaller set of linear composite variates, the analytical procedure is one of several interdependency methods. Three widely used techniques are discussed in this section: factor analysis, cluster analysis, and multidimensional scaling.

Factor Analysis

In recent years, the multivariate statistical technique of factor analysis has found increased use in all realms of business research. Factor analysis uses a linear approach to the reduction and summarization of data.[8] It comprises a large set of techniques with a common purpose, briefly described as follows:

Identify a set of dimensions that are not easily observed in a large set of variables. These dimensions, called factors, summarize a majority of the information in the data, whereby the identification of these factors can help develop new business measures, identify appropriate variables for inclusion in subsequent analytical procedures, or partially or totally replace the original variables in a subsequent analysis.

Generally, the procedure systematically follows a series of steps to arrive at a solution. The first step involves generating a correlation matrix from the original set of data (variables). As can be expected, the initial factor structure is derived from the correlation matrix. Why? Factors are simply variables that share common variance, very similar to what correlation is.[9] Because the input variables may contain the same information, redundancy may exist in the data set. A set of these variables may be measuring, in part, the same underlying concept. This characteristic is analogous to a *factor*. The factor is inferred from other variables. The factor can also be considered a grouping of input variables that measure or are relative indicators of the factor.

At subsequent steps after the calculation of the correlation matrix, decisions must be made that will dramatically affect the final use and interpretation of the results of the analysis. These decisions fall into four categories:

1. The estimates of communality
2. The number of factors to keep
3. The indeterminacy problem (i.e., rotation to simple structure)
4. Theoretical considerations

Factor analysis of any kind should only be performed with very large samples. This is a statistical fact, because the correlation matrix must be developed from the data for the factor analysis to be performed. Communality is the variance that all the variables have in common. A measure of communality must be put on the diagonal of the correlation matrix (in place of the 1s, on the principal diagonal) to derive the initial factor structure. For example, one may decide to use the largest (absolute) off-diagonal element in each row and column as measures of communality for an analysis. Many other estimates of communality may be used.[10]

Another problem in factor analysis is deciding what factors to keep. Although there is an abundance of rules in this matter, the most commonly used is that if a factor has an eigenvalue greater than 1, it is a candidate for interpretation.

The indeterminacy problem involves the interpretation of the factors. There are numerous ways to reproduce the underlying dimensions in the original data set with factors. Therefore, the researcher must seek a simple structure that is easy to interpret. This is accomplished by some form of rotation of the initial factor structure. The most popular ways are the VARIMAX and oblique solutions. In any event, the main characteristics of simple structure are the following:[11]

1. Any column of the factor loading matrix should have mostly small values, as close to zero as possible.
2. Any given row of the matrix of factor loadings should have nonzero entries in only a few columns.
3. Any two columns of the matrix of factor loadings should exhibit a different pattern of high and low loadings.

A good factor solution should show invariance in structure when the factor loadings are derived from various solution techniques. Similarly, solutions should be cross-validated by using split samples to estimate and rotate the initial factor structure. Good solutions require that various techniques should produce similar patterns on cross-validated samples.

Although considered last, the final decision is very important when using factor analysis. Generally, factor analysis is a mathematical procedure, not a statistical one. It is often misused under this guise. Granted, the factor loadings are produced by sampling information, but they cannot be easily tested for statistical significance. The user of the procedure should have some idea of underlying patterns in the data before the analysis begins. It should not be used as a fishing expedition. However, this does not preclude using exploratory factor analysis to describe and gain insights into your data. In fact, exploratory factor analysis is extremely useful in the initial stages of research if done with a purpose. Subsequently, factor solution techniques should be used to confirm a hypothesized data structure.

For example, suppose we test 200 middle managers on six psychological variables, with the intent to summarize this information into a few factors. These factors will be used in a future analysis of the managers' performance. In part a of Exhibit 15.4, the correlation matrix for this data is reproduced. Because there are only six variables (usually factor analytical studies involve many more variables), we can intuitively combine the variables into factors by visually examining the correlation matrix. This intuitive approach is often impossible because of the size and complexity of the correlation matrix, but it is used here to highlight the factoring procedure. Two intuitive factors can be constructed from the correlation matrix. In Exhibit 15.4, part b, these factors are illustrated. The "intuitive" factor matrix contains an X where substantial relationships occur with a factor, and a dash where the variable is basically unrelated to the factor. The principle here is that when several variables are consistently found to be associated, they are grouped together. They can be thought of as representing a single facet.

Because a researcher usually works with much larger variable sets than we have in our example, computerized factor analysis is often implemented. The purpose of the analysis is to reduce the variable set to a set of factors so that most of the *information* found in the relationships of the variables is reproduced in the factor set. Also, to be useful, the number of factors must always be much less than the number of variables.

A principal components analysis with 1s on the diagonal of the correlation matrix (as a measure of communality) was used in the example. The factor extraction criterion was that the eigenvalues should be 1 or greater. (Don't let this statement throw you; it's a matrix algebra concept involving solution techniques.) The solution is shown in Exhibit 15.4, part c. Most often this form is hard to interpret, and we must rotate the factors to obtain a clear interpretation of the loadings. This is the quest for simple structure. In the example, VARIMAX rotation is used to create simple structure.

Simple structure is developed in Exhibit 15.4, part d. Variables that load on each factor are determined by examining the value of each factor loading,

EXHIBIT **15.4**

a. Correlations Among Six Psychological Variables Constructed from a Testing of 200 Middle Managers

Variables	1	2	3	4	5	6
1. Information	1.00					
2. Verbal ability	0.67	1.00				
3. Verbal analogies	0.43	0.47	1.00			
4. Ego strength	0.11	0.12	0.03	1.00		
5. Guilt proneness	−0.07	−0.5	−0.14	−0.41	1.00	
6. Tension	−0.17	−0.14	−0.10	−0.48	0.40	1.00

b. Intuitive Factor Matrix

	Factors	
Variables	A	B
1. Information	X	—
2. Verbal ability	X	—
3. Verbal analogies	X	—
4. Ego strength	—	X
5. Guilt proneness	—	X
6. Tension	—	X

c. Unrotated Principal Axis Loadings

	Factors	
Variables	A	B
1. Information	0.70	0.31
2. Verbal ability	0.73	0.36
3. Verbal analogies	0.53	0.24
4. Ego strength	0.39	−0.55
5. Guilt proneness	−0.34	0.48
6. Tension	−0.44	0.50

d. Simple Structure Factor Loadings (VARIMAX)

	Factors	
Variables	A	B
1. Information	(0.76)	−0.09
2. Verbal ability	(0.81)	−0.07
3. Verbal analogies	(0.58)	−0.07
4. Ego strength	0.06	(−0.67)
5. Guilt proneness	−0.05	(0.59)
6. Tension	−0.12	(0.66)

and then by assigning the variable to a factor. Usually, some rule of thumb (for example, loadings should exceed 0.40 or 0.50, positive or negative) is used for assignment purposes. Factor *A* contains variables 1, 2, and 3. Factor *B* contains variables 4, 5, and 6. It should also be noted that the sign of the loading is important for interpretation purposes. More specifically, the sign of the loading shows how the variable and factor covary. In addition, the other rules of simple structure stated earlier must also be taken into consideration. In our example, all these rules are satisfied. Obviously, this is not always the case.

Researchers then intuitively name the factors according to the variable loadings. In this case, the naming seems very simple. Factor *A* can be thought of as verbal comprehension and factor *B* as anxiety. (Maybe you can come up with better names.) In any case, these factors are interpreted and then most often used in a subsequent analysis.

Examples of factor analysis abound in the business research literature. One study sought to establish the underlying dimensions of the construct cosmopolitanism so that it could be used in a subsequent analysis. Two samples were chosen (for cross-validation), and eight variables were used in the factor analysis. The results (see Table 15.1) indicated four factors that would describe a cosmopolitan in each group (professionals and administrators of hospitals): (1) journal publications, (2) attendance at professional meetings, (3) offices held in professional associations, and (4) journal readership. These factors explain 68% of the variance in the linear model for both the administrator and professional respondent groups. Both sets of factors are relatively consistent (as assessed by a split-sample factor analysis for each group). The factor loadings reported are the values of the largest loadings in each column, following the general rules of simple structure.[12]

Cluster Analysis

Cluster analysis is a dimension-free classification procedure that attempts to subdivide or partition a set of different objects, variables, or both into relatively similar groups. Therefore, there are similarities within groups and differences between groups. Most often, cluster analysis is used when a preliminary study is undertaken to reveal natural groups of study units in a data matrix. Cluster analysis is mostly a tool of discovery. Generally, it is used to determine structure in a data set or to construct taxonomies for future classification of study objects.[13] Literally hundreds of clustering algorithms can provide structure or taxonometric solutions from a data matrix. Four basic decisions must be made in cluster analysis:

1. The number and types of variables on which to cluster
2. Whether the variables should be standardized
3. What measure of interobject similarity or dissimilarity to use
4. How many clusters are expected to exist

TABLE 15.1 Cosmopolitanism Factor Structure and Loadings

Items	Factor 1	Factor 2	Factor 3	Factor 4
Professionals				
Number of articles published last year	.86			
Number of articles ever published	.79			
Local professional association meetings		.79		
National professional association meetings		.68		
Offices held in national associations			.87	
Offices held in local associations			.80	
How extensively read journals				.77
Number of journal subscriptions				.71
Cumulative explained variance	(.27)	(.44)	(.57)	(.68)
Administrators				
Local professional association meetings	.83			
National professional association meetings	.80			
Number of articles published last year		.87		
Number of articles ever published		.87		
Offices held in national associations			.83	
Offices held in local associations			.78	
Number of journal subscriptions				.80
How extensively read journals				.42
Cumulative explained variance	(.25)	(.41)	(.56)	(.68)

Source: Thomas S. Robertson and Yoram Wind, "Organizational Cosmopolitanism and Innovativeness," *Academy of Management Journal* (1983), *26:2*, p. 335. Reproduced by permission from the Academy of Management.

All of these decisions relate to the problem specification and the purpose of the analysis. General clustering algorithms are part and parcel of the major statistical software packages and can be easily set up, run, and interpreted.

For example, a study to segment Florida residents' attitudes toward tourism illustrates one application of cluster analysis.[14] The problem addressed was to understand what local residents felt about tourists coming to Florida. Considerable public and private resources were being spent to attract visitors to the state, and negative resident attitudes might counteract the impact of promotional monies. The problem was to identify and segment Florida residents' attitudes toward tourism and make recommendations in light of the findings of the study.

No previous comprehensive studies of residents' attitudes toward tourism had been conducted in Florida, so little was known about their viewpoints. The study was designed to collect information on residents' attitudes toward tourism, their general knowledge of the tourism industry in Florida, and a number of classificatory variables for informational purposes. This data was then used to classify residents according to their attitudes and opinions toward tourism in Florida.

As a first step, 31 attitudinal variables were used to group respondents into five mutually exclusive clusters. The obtained clusters were validated using a split-sample approach and were parsimonious in both structure and content. The five clusters were named (based on residents' attitudes and opinions toward tourism) as haters (16%), cautious romantics (21%), in-betweeners (18%), love 'em for a reason (26%), and lovers (20%). The resultant clusters and their attribute and characteristic profiles are presented in Table 15.2. Table 15.3 shows the summary characteristics and potential policy appeals.

Multidimensional Scaling

Multidimensional scaling (MDS) involves the spatial representation of relations among the perceptions and preferences of individuals. Multidimensional space is broken down into components of objective and perceptual space. These dimensions of geometric space represent the characteristics along which objects may be compared. Objective space is defined by the actual characteristics of an object, whereas perceptual space refers to an individual's subjective perceptions. Ideal points, where one combination of perceived characteristics is preferred over all others, may also be identified.

For example, suppose Ford Motor Company conducts a study to identify major characteristics of a new product for a specific target market. The organization subsequently develops a new sport vehicle with a set of pre-ferred characteristics (ideal points). The new vehicle will not roll over, gets great gas mileage, and will be the next monster truck. However, if the potential consumer perceives that this new product is top-heavy, a gas guz-zler, and basically an inadequate truck, there would clearly be a large dis-crepancy between perceived and actual characteristics. How could we scale this multidimensional relationship?

Scaling multidimensional relations involves converting the characteris-tics represented as points in space to metric coordinates using a distance formula.[15] By comparing the relative distances between pairs of points in reference to an ideal point, preferences for certain combinations of charac-teristics can be determined. Similarly, we can assess the relative similarities of characteristics using interpoint distances; closer distances can be said to be more similar. These relationships can then be arranged by the proximity of their coordinates on a perceptual map.

For the purposes of scaling, computer programs using both metric and nonmetric algorithms are available. In many applied situations, metric and nonmetric methods yield similar results. Application of nonmetric MDS for discovering patterns in preference data has potential in many areas of business research. Marketing is a fertile area for applying MDS—in new product development, marketing segmentation, competitive posi-tioning, and advertising evaluation. Marketing objectives can be evaluated by seeing how a certain increase in the brand's image would change the

TABLE 15.2 Attribute Profile of Tourism Public Attitude Clusters

Attribute	Rank	Haters	Lovers	Cluster Cautious Romantics	In-Betweeners	Love 'Em for a Reason
Benefits	1	Employment (41%)	Employment (44%)	Employment (44%)	Employment (41%)	Employment (41%)
	2	Entertainment (29%)	Tax revenues (36%)	Tax revenues (36%)	Lower taxes (26%)	Lower taxes (27%)
	3	Property values (21%)	Lower taxes (34%)	Lower taxes (30%)	Entertainment (29%)	Entertainment (20%)
	4	Lower taxes (20%)	Entertainment (24%)	Property values (26%)	Tax revenues (19%)	Tax revenues (18%)
	5	Tax revenues (19%)	Property values (22%)	Entertainment (23%)	Property values (12%)	Property values (10%)
Disadvantages	1	Traffic (39%)	Traffic (42%)	Traffic (43%)	Increases crime (33%)	Traffic (33%)
	2	Overcrowding (36%)	Overcrowding (35%)	Increases crime (32%)	Traffic (29%)	Increases crime (27%)
	3	Increases crime (37%)	Increases crime (32%)	Increases prices (28%)	Increases prices (23%)	Increases prices (20%)
	4	Increases prices (26%)	Increases prices (23%)	Overcrowding (23%)	Higher taxes (21%)	Overcrowding (19%)
	5	Higher taxes (25%)	Higher taxes (15%)	Higher taxes (9%)	Overcrowding (19%)	Higher taxes (11%)
Born in state of Florida		40%	15%	21%	22%	23%
Scored high on knowledge of tourism's impact on Florida		12%	34%	20%	30%	31%

Source: Duane Davis, Jeff Allen, and Robert Cosenza, "Segmenting Local Residents by their Attitudes, Interests and Opinions Toward Tourism," *Journal of Travel Research* (Fall 1988), pp. 6–7.

TABLE 15.3 Summary Characteristics and Policy Appeals for the Five Segments

Segment	Characteristics and Appeals			
	Degree of Negativity Toward Tourism	Percent Scoring High on Knowledge	Percent Native Born	Public Policy Appeal
Haters (16%)	High	12%	40%	Educate to the positive impact of tourism on their lives, community, and state
Cautious romantics (21%)		20	21	Reaffirm tourism's benefits—assert positive impact of tourism growth
In-betweeners (18%)		30	22	Reaffirm tourism's benefits—demonstrate how tourism supports their community and has helped reduce taxes
Love 'em for a reason (26%)		31	23	Emphasize jobs created and the facilities it creates for residents
Lovers (20%)	Low	34	16	Reassure and remind

Source: Duane Davis, Jeff Allen, and Robert Cosenza, "Segmenting Local Residents by their Attitudes, Interests and Opinions Toward Tourism," *Journal of Travel Research* (Fall 1988), pp. 6–7.

brand's competitive positioning. MDS can also be used in problems ranging from employee's preferences for organizational policymaking efforts to customers' views of strategic plans compared to those of other competitors.

Many leading business researchers are now using an associated technique called *correspondence analysis* to summarize interdependencies at any level of data. This analytical technique describes relationships among row and column categories in crosstabulations by their proximity on a perceptual map. On a correspondence map, nearby categories are positively correlated, and distant categories are negatively correlated. For example, if the row variable is brands and the column variable is attributes, the proximity of brands and attributes describes brand images.

Correspondence analysis can describe relationships in crosstabulations between any type of object and characteristic and virtually any number of crosstabulated variables. The data may be expressed as frequencies, means, medians, percentages, probabilities of purchase, or top box scores. A *top box score* is an aggregation of the responses with high scores. For example, if we were to use a five-point importance scale (most, very, somewhat, little, and no importance at all), we would merely sum the responses in the "most" and "very" categories to obtain a top box score.

In Exhibit 15.5, a perceptual mapping program called MAPWISE is used to explore a typical problem of comparing brand images.[16] The respondents describe each of five brands (A–E) on five images. Our input data consists of the top box scores of each brand on each image. Correspondence analysis best summarizes relationships among all categories on the perceptual map.

EXHIBIT **15.5**

Correspondence
Analysis Using the
MAPWISE Program

Source: Used with permission
of Dr. Betsy Goodnow, Market
ACTION Research Software, 16
West 501 58th Street, Suite
21-A, Clarendon Hills, IL
60512-1740.

Input Data Table of Top Box Scores

Brand	Attribute				
	Style	Speed	Price	Size	Repair
A	55	57	34	47	12
B	37	32	45	67	24
C	78	36	47	51	46
D	71	16	47	38	24
E	13	54	61	37	78

Correspondence Analysis of Brand Images

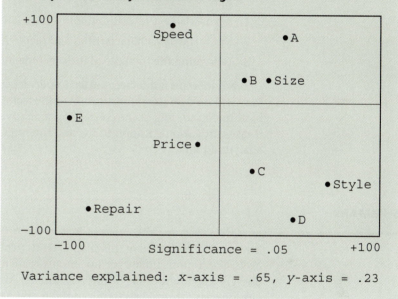

Variance explained: x-axis = .65, y-axis = .23

Because the solution is significant at the 0.05 level, it best distinguishes these relationships. The horizontal axis explains 65% of the relationships (or variance) among the brands and images, and contrasts style with ease of repair. The vertical axis explains 23% of the variance by contrasting the speed of the brands.

MAPWISE generates a point-to-point map, so axes and vectors are not needed to interpret the solution. The proximity of all categories on the map describes their mutual relationships. Relatively speaking, brand A is speedy, brand B is large, brand C is low-priced, brand D is stylish, and brand E is easy to repair. The close proximity of brands A and B suggests that they are direct competitors. The respondents in our hypothetical example reveal that price is the least distinguishing image, because it is located closest to the center of the map.

COMMENT ON MULTIVARIATE TECHNIQUES

Although the journey through the multivariate jungle of the business researcher's analytical world has been a cursory one, we hope you have gained some understanding of the usefulness and importance of the techniques. Most business decisions are difficult to make. The choice of an appropriate multivariate technique is difficult and challenging. We urge the proper technique selection by adherence to a few simple rules:

1. Acknowledge the assumptions of the technique.
2. Know the data characteristics and requirements.
3. Determine the final use of the outcome information.

It is important not to become technique-dependent or technique-scared. Very seldom does the sophistication of the techniques get in the way of the basic understanding of the concepts of a multivariate analysis. Most often, it takes little sophistication to use and interpret a very powerful statistical analytical tool.

SUMMARY

The material presented in this chapter tends to be overlooked by most managers. This is unfortunate because multivariate techniques for data analysis are very important to the business researcher. We have approached the discussions with three ideas in mind: the proper selection, use, and interpretation of multivariate analyses. The ideas are simple, but the knowledge is paramount.

The first section of the chapter is devoted to the analysis of dependencies. When a relationship is expressed in the form $Y = f(X)$, and when there is more than one dependent variable, MANOVA is the proper analytic technique. For the joint effect of a combination of independent variables on a dependent variable, conjoint analysis (MONANOVA) is suggested. A useful technique in the development and testing of theoretical relationships between variables is covariance structure analysis.

In the second section of the chapter, three methods of data reduction or summarization are examined. Factor analysis is a technique most often used as an input to a subsequent analysis. Cluster analysis is a procedure for grouping study objects for subsequent analysis. Finally, for representing the relations in perceptions and preference data, multidimensional scaling may be the proper technique to apply. Because there are no implied relationships (form), we call this set of procedures the analysis of interdependencies.

DISCUSSION QUESTIONS

1. Describe what is meant by multivariate techniques to a hypothetical employee who is confused by the term.

2. How do the inputs to factor analysis and cluster analysis differ?

3. A bank manager wanted to use discriminant analysis to find differences in the profiles of prompt loan payers (repayers) versus those who do not repay on time (deadbeats). How would you propose to do such a study? Which independent variables would you expect to use? How would you implement such a model if it were useful? How would you judge if the model was actually useful from a managerial perspective? *Hint:* Cost of bad loan versus good loan.

4. Suppose you are devising an experiment to evaluate various sales-force incentive plans. Your boss feels that three different plans should be tested by dividing the country into four geographical areas and using four dependent measures (change in sales, change in motivation, change in attitude toward the job, change in cold sales calls). This looks like four separate ANOVAs to him, but a colleague of yours argues that it is really MANOVA. Discuss this issue both in terms of design *and* extra variables that you feel can aid in the goals of this experiment.

5. Discuss the issues of data reliability and validity and how they can impact on the managerial usefulness of any multivariate technique you choose.

6. Suppose Bertram Boat Company conducted a research study to identify what product characteristics consumers would prefer in a sport-fishing yacht. The conjoint analysis results regarding consumer preferences for three levels of three salient product characteristics of speed (14 knots, 18 knots, 22 knots), comfort (spartan, functional, luxurious), and beam width (12 feet, 10 feet, and 8 feet), are as shown in the table.

Factor	Factor Importance Weights	Levels		
		1	2	3
1. Speed	0.267	−17.5	2.8	14.7
2. Comfort	0.168	12.3	47.8	−60.1
3. Beam width	0.565	22.7	35.1	−57.8

What combination of characteristics is

a. The most preferred choice for the product?

b. The least preferred choice for the product?

c. What is the most important product characteristic?

d. The least important?

7. In studying a problem, a business researcher is guided by a theoretical framework that specifies a relationship between the two latent constructs of performance and ability.

   a. How would a researcher diagram the causal model to be tested?

   b. What possible manifest variables could be used to measure each construct?

8. What if a study was conducted to compare three boat companies on the dimensions of reliability, innovativeness, quality, and stylishness. The results are presented in a perceptual map:

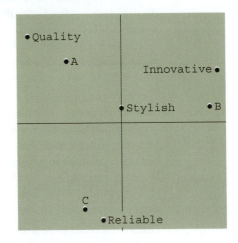

   a. Which boat companies are best associated with which dimensions?

   b. Which boat companies are most similar and why?

   c. Suppose the largest market segment of potential boat owners perceives that quality is the most important dimension (in fact, 66% of the market). How would companies B and C reposition their products to compete with company A?

9. Go to the Web sites of SPSS (http://www.spss.com) or SAS (http://www.sas.com) and identify what types of multivariate techniques are available. Briefly note the technique and provide a brief overview of how and when it could be used in business research.

NOTES

1. This notion is similar to performing a series of *t*-tests on a large group of variables. By chance alone, some *t*-statistics will be significant.

2. See Joseph F. Hair, Jr., Rolph E. Anderson, Ronald L. Tatham, and William C. Black, *Multivariate Data*

Analysis, 5th ed. (Upper Saddle River, NJ: Prentice-Hall, 1998).

3. See Hair et al., ibid.

4. See James M. Lattin, J. Douglas Carroll, and Paul E. Green, *Analyzing Multivariate Data* (Pacific Grove, CA: Brooks/Cole, 2003), Chapter 12.

5. See P. E. Green and V. R. Rao, "Conjoint Measurement for Quantifying Judgmental Data," *Journal of Marketing Research* (August 1971), *8*, pp. 355–363.

6. See P. M. Bentler, *Theory and Implementation of EQS: A Structural Equations Program* (Los Angeles: BMDP Statistical Software, 1988); and K. G. Joreskog and D. Sorbom, *LISREL: Analysis of Linear Structural Relationships by the Method of Maximum Likelihood* (Mooresville, IN: Scientific Software, 1984).

7. For discussions of practical issues and evaluating structural equation models, see P. M. Bentler and C. Chou, "Practical Issues in Structural Modeling," *Sociological Methods & Research* (August 1987), *1*, pp. 78–117; R. P. Bagozzi and Y. Yi, "On the Evaluation of Structural Equations Models," *Journal of the Academy of Marketing Science* (Spring 1988), *16*:1, pp. 74–94; and Kenneth A. Bollen, *Structural Equations with Latent Variables* (New York: John Wiley & Sons, 1989).

8. Two companion books that explain factor analytical techniques at a very nonsophisticated level are H. H. Harman, *Modern Factor Analysis* (Chicago: University of Chicago Press, 1976); and R. S. Gorsuch, *Factor Analysis* (Philadelphia: W. B. Saunders, 1983).

9. Fred N. Kerlinger and Howard B. Lee, *Foundations of Behavioral Research,* 4th ed. (Belmont, CA: Wadsworth, 2002).

10. These are R^2 values on the diagonal and an estimate of unique variance by a maximum likelihood procedure.

11. Originally the work of L. L. Thurstone, *Multiple Factor Analysis* (Chicago: University of Chicago Press, 1947), but can be found in both Harman and in Gorsuch, op. cit. note 8.

12. See, for example, Curtis Hardyck and Lewis F. Petinovich, *Understanding Research in the Social Sciences* (Philadelphia: Saunders, 1975), Chapter 10, for a treatise on interpretation of factor structure.

13. Ibid., Chapter 10; and Mark L. Berenson, David Levine, and Matthew Goldstein, *Intermediate Statistical Methods and Applications* (Englewood Cliffs, NJ: Prentice-Hall, 1983), Chapter 16.

14. This example is from Duane Davis, Jeff Alen, and Robert Cosenza, "Segmenting Local Residents by Their Attitudes, Interests and Opinions Toward Tourism," *Journal of Travel Research* (Fall 1988), pp. 2–8.

15. A discussion of distance functions is presented in P. E. Green, F. J. Carmone, and S. M. Smith, *Multidimensional Scaling: Concepts and Applications* (Boston: Allyn and Bacon, 1989), pp. 25–46.

16. The author would like to thank Dr. Betsy Goodnow and Market ACTION Research Software for this example. For more information on MAPWISE, call Dr. Goodnow at (708) 986–0830, or write to 16 West 501 58th Street, Suite 21-A, Clarendon, Hills, IL 60514–1740.

SUGGESTED READING

Introduction to Multivariate Analysis

Grimm, Laurence G., and Paul R. Yarnold, Eds., *Reading & Understanding Multivariate Statistics* (American Psychological Association, 1996).

The editors focus on providing conceptual understanding of the meaning of the statistics in the context of research questions and results.

Johnson, Dallas A., *Applied Multivariate Methods for Data Analysis* (Belmont, CA: Wadsworth, 1998).

Uses real-life examples to explain the when, why, how, and what of numerous multivariate methods, stressing the importance and practical applications of each method. SPSS and SAS output is provided.

Selection of a Multivariate Technique

Hair, Joseph F., et al., *Multivariate Data Analysis with Readings,* 5th ed. (Upper Saddle River, NJ: Prentice-Hall, 1998).

An interesting approach to presenting multivariate data analysis. It blends a how-to focus with an understanding of the theory and implications of using a particular multivariate technique.

Lattin, James M., J. Douglas Carroll, and Paul E. Green, *Analyzing Multivariate Data* (Pacific Grove, CA: Brooks/Cole, 2003).

A revision of a classic text that introduces multivariate methods. It maintains a practical orientation and presents complex concepts with a relatively informal style.

Analysis of Dependency

Pedhazur, Elazar J., *Multiple Regression in the Behavior Sciences,* 3rd ed. (Orlando, FL: Harcourt Brace, 1997).

This book is a must if you want clear, concise explanations of multivariate techniques. Also, provides an introduction to causal modeling, followed by a discussion of structural equations methodology using LISREL.

Analysis of Interdependency

Johnson, Richard A., and Dean A. Wichern, *Applied Multivariate Statistical Analysis* (Upper Saddle River, NJ: Prentice-Hall, 1998).

Provides a good overview of multivariate statistics. Good discussions on the analysis of interdependencies in a multivariate context.

Schumacker, Randall E., and Richard G. Lomax, *Beginner's Guide to Structural Equation Modeling* (New York: L. Earlbaum, 1996).

This book is designed to give students and researchers in all of those disciplines a good working knowledge of structural equation modeling. EQS5 and LISREL8-SIMPLIS are used in data examples throughout the book.

16

Ethical Considerations in Business Research

Overview

Ethical Issues in Business Research
Societal Rights
Subjects' Rights
Clients'/Managers' Rights
Researchers' Rights

Ethical Issues in a High-Technology Environment
Protection of Subjects' Rights
Quality of Research
Research Versus Direct Marketing

Codes of Ethics

Managerial Considerations

Summary

Discussion Questions

Notes

Suggested Reading

Appendix: Excerpts from the ICC/ESOMAR International Code of Marketing and Social Research Practice

OVERVIEW

No business research text would be complete without a discussion of ethics. Ethics in this context means the proper conduct of the research process in business inquiry—rights and responsibilities of the various parties involved in research. Ethics is important to all parties because it affects the rights of individuals and ultimately the quality of the data obtained from inquiry.

The chapter begins by discussing ethical issues and possible consequences in business research. A framework outlining the major ethical relationships is presented. Four parties have both rights and responsibilities in the conduct and use of research results: society, subjects, clients/managers, and researchers. Each has rights with regard to information and conduct that are the responsibilities of other members in the research endeavor. The implications of these rights are discussed from a managerial standpoint.

The next section takes a look at the ethical issues being created by the technological capabilities that have been discussed throughout this text. These issues center on the invasion of privacy and the fast-developing capabilities of the Internet. Here, it is noted that we must, as researchers, separate E-commerce from online research if we are to maintain our objectivity and credibility in the marketplace.

The final section deals with codes of ethics currently in use among researchers. The content and purpose of these codes are explained. The chapter ends with the major managerial considerations.

ETHICAL ISSUES IN BUSINESS RESEARCH

The law usually provides the background and sets the limits on most ethical conduct. In recent years, various legislative actions have been designed to ensure ethical conduct in research.[1] Issues dealing with telephone privacy, do-not-call listings, electronic monitoring, data privacy, Internet privacy, antispam legislation, and *mugging* (marketing under the guise of research) have all become topics that are or soon will be on various legislative dockets throughout the world. Further legislative action seems inevitable even though most serious researchers would prefer to set their own high standards of ethical conduct.

The conduct of business research requires the participation of the general public. How business inquiry is conducted can affect the willingness of the public to participate, so it is important to conduct research in an ethical manner. Figure 16.1 graphically presents the relationship among business inquiry, public acceptance, and research participation by the public. The ethical level of business inquiry is shown on the horizontal axis. Overall public acceptance and subsequent participation are plotted on the

FIGURE **16.1**

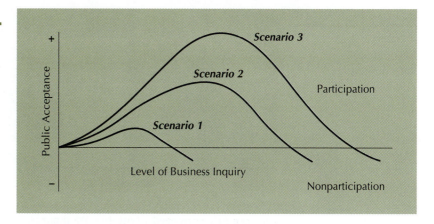

vertical axis. In the model, public acceptance may be positive, resulting in research participation, or negative, leading to nonparticipation.

The model suggests that the degree of public acceptance and participation can be attributed to ethical or unethical research behaviors. Three scenarios are presented in Figure 16.1. In scenario 1, general public acceptance is low because of exposure to unethical research methods. This results in the worst-case scenario, where participation is limited to unethical research experiences and public acceptance of research quickly becomes negative. In scenario 2, public acceptance is more positive, with participation at a higher level. Scenario 3 is the best-case situation, with the greatest level of public acceptance and participation. In this case, the overall acceptance of research is substantially higher because of the responsible and ethical conduct of business research.

The implications of the model are important for the conduct of business research. Although public acceptance of research appears to be positive but leveling off, unethical research abuse is increasing at a rapid rate.[2] To ensure the public's acceptance of business research, researchers must operate ethically, or increased regulation becomes a distinct possibility. This prospect has potentially negative effects in all areas of the business research process.

The present chapter deals normatively with the ethical issues that should be addressed in the conduct of research. A series of basic rights of the various parties in the process are asserted. To organize the issues, a framework is outlined in Figure 16.2. As this framework shows, the four participants in any research endeavor are society, the subjects of inquiry, the clients/managers requesting the research, and the researchers. All of these participants have responsibilities and corresponding rights in the conduct of ethical business research.

Of course, this is merely a normative framework. The complexity of the issues makes it impossible for any one individual or organization to ensure that research will be conducted ethically. We must all do our part to ensure that ethical considerations are incorporated into the research design and

FIGURE **16.2**

A Framework of
Major Ethical
Relationships in
Business Research

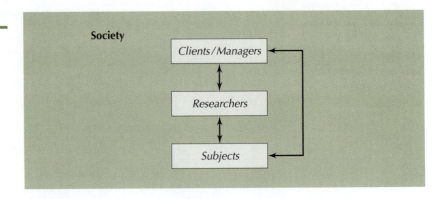

subsequently conveyed to other participants in the research process. With this introduction, let us begin our discussion of the rights of each party.[3] We begin with the largest: society.

Societal Rights

Business is a social phenomenon. Therefore, business research has certain responsibilities to society, whose rights are presented in Exhibit 16.1. There are three major rights that society can expect of business research: (1) the right to be informed of critical research results, (2) the right to expect objective research results, and (3) the right to privacy.

The Right to Be Informed. If business research uncovers factors that may affect the health and well-being (either negatively or positively) of the general public, society deserves to be informed of this finding. For example, if an automobile manufacturer found out through research that a certain make of car had a high probability of exploding if it was struck in a certain fashion, that organization would have the responsibility to inform the general public of this danger. Society would have the right to know of this potential threat.

The Right to Expect Objective Research Results. The second right is closely related to the first one. If results are publicly reported by an organization, the general public has the right to expect they will be objective,

EXHIBIT **16.1**

Societal Rights

The right to be informed of critical research results.
Awareness of research results that may affect the health and well-being of the general public.

The right to expect objective research results.

a. The right to expect complete reporting of research.

b. The right to expect research that is unbiased and scientifically sound.

The right to privacy.
The right of individuals in society to be let alone.

complete, unbiased, and scientifically sound. Sometimes, so-called research results have been used to confirm or deny a particular claim. However, these are often not truly research findings but rather statements designed to persuade some people to act in a particular fashion. In other words, the research study is not conducted with commonly accepted practices. This behavior is unethical; it violates a basic right of society.

The Right to Privacy. The third right, the right to be let alone, poses a dilemma for business research. Although the continued application of research can have a positive effect on the efficiency of business in meeting society's needs, it can also have a negative effect. For example, technology has made it possible to compile large household lists that include names, addresses, and a great deal of individual specific data. Ford Motor Company has a database estimated to include more than 50 million names; General Foods' database includes 25 million names; and Citicorp has in excess of 30 million names.[4] This database technology has the potential to threaten privacy in a free society. Society has the right to expect that personal information gained through research endeavors will not be disclosed. Organizations such as the Red Cross, AT&T, and Reader's Digest have recognized this right, and they refuse to sell personal information to other firms. However, as research endeavors and technological capabilities grow, so does the potential for the infringement on human privacy.

Ironically, individual privacy issues in the European Union have had a profound impact in the United States. In 1995, the EU Directive on Data Protection was passed, which stated members must incorporate research privacy safeguards to protect their citizens and provide a common law for the EU. This legislative action has forced the United States (because it does not have strict data privacy laws that incorporate all the principles in the EU Directive) to adopt *safe harbors* to provide "adequate privacy protection" to enable transmission of information by participating EU members of information to the United States. (For more information, go to http://www.esomar.org)

Subjects' Rights

Subjects' rights are often overlooked in the conduct of business research.[5] This is unfortunate and must be addressed, for three major reasons.

First, the individual subject is the cornerstone of social research. Without his or her full cooperation, the results obtained may be distorted. Cooperation and trust on the part of the subject are a necessary condition for the collection of relevant and accurate research data.

Second, business researchers are increasingly being faced with the problem of subjects refusing to respond. The guise of research is often used to gain entrance to homes for solicitation purposes.[6] The increasing

EXHIBIT **16.2**

Subjects' Rights

Source: This framework and exhibit are adapted from Alice M. Tybout and Gerald Zaltman, "Ethics in Marketing Research: Their Practical Relevance," *Journal of Marketing Research* (November 1974), *11*, p. 358. Used by permission.

The right to choose whether or not to participate in a study.

a. Awareness of the right not to participate.

b. Awareness of adequate information to make an informed choice.

c. Awareness of opportunity to end participation, if so desired.

The right to safety in terms of the avoidance of any physical or mental harm.

a. Guarantee of the protection of anonymity.

b. Freedom from stress.

c. Freedom from deception.

The right to be informed about all aspects of the research.

a. The right to debriefing.

b. The right to be informed as to the research results.

nonresponse rates in social research appear to result from unethical abuses of subjects by researchers.

Third, and most important, the subject has the right to be informed of his or her participation in an experiment. Research is a means of obtaining knowledge, but it is not philosophically above the rights of the people who participate. Exhibit 16.2 summarizes the major areas of privilege applicable to subjects' rights in research.

The Right to Choose. All subjects have the basic right of free choice. Subjects should be informed of the study they are participating in and any consequences of it. This right is succinctly stated in the Ethical Standards for Research with Human Subjects, adopted by the American Psychological Association.

> This psychologist's ethical obligation to use people as research participants only if they give their informed concern rests on well-established traditions of research ethics and on strong rational grounds. The individual's human right for free choice requires that the decision to participate be made in the light of adequate and accurate information.[7]

Three major areas are involved in this right to choose. First, all subjects must be aware of the right to choose. They must know that they can refuse further participation if they so desire. If the subject is obligated or required to continue, this right is grossly violated.

Second, all individuals must be given adequate information beforehand to decide whether or not participation in the study is desirable for them. In cases where full disclosure of the study procedures cannot be made, this should be so indicated and a basis found for anticipated informed consent. Peer group evaluation of the specifics of the investigation may also aid in ethically legitimizing the research design.

The third component of the right to choose involves having some say about the conditions a respondent will be subjected to. If the conditions are objectionable at any time during the research, the subject has the right to avoid that particular aspect of the study.

The Right to Safety. First and foremost here is the subject's right to avoid any physical and mental harm. Any procedure that may harm an individual in any way should be critically examined. If the possible benefits from the study are not greater than its consequences, and the individual is not fully informed of the procedure, the inquiry should be scrapped. Every person has the right to safety.

Safety also includes the guarantees of protection of anonymity, freedom from stress, and freedom from deception. All subjects should be guaranteed anonymity before they participate in any business research. Serious abuses of this right have occurred in the past.

All subjects have the right to be free from stress in any setting to which they are exposed. Any procedure that is designed to induce stress and cause the subject discomfort should be carefully evaluated for its necessity and usefulness. In most cases, the procedure should be abandoned, or an alternative method found for obtaining similar information.

Also included in safety is the right to be told the truth about the purpose of the investigation. Deception to obtain consent is totally unethical and should not be employed by any researcher. Deceiving a subject may place him or her in a stressful situation that might have been avoided if the individual had been aware of the true nature of the research project.

The Right to be Informed. The right to be informed is a carryover of the first two rights. Individuals have the right to be debriefed concerning the project they have participated in. Debriefing consists of informing subjects of all relevant information concerning the procedures in which they were involved. Debriefing is an important method for reducing experimental stress and assessing the effectiveness of the study's procedures.

The completed research data should be made available to individuals who participated in the experiment. Failure to inform the subjects can have negative consequences:

> Subjects may feel that they gain nothing from and are exploited by participating in research and consequently may distort their response and decline to participate in future research.[8]

If researchers do not inform respondents of their part in the overall research project, the next time research is performed, there will be distorted information or refusal to respond.

Clients'/Managers' Rights

Just as subjects have rights, so do clients: confidentiality and the expectation of high-quality research. These rights are presented in Exhibit 16.3; a description of them follows.

EXHIBIT **16.3**

Clients'/Managers'
Rights

The right of confidentiality of the working relationship.

a. Confidentiality of proprietary data.

b. Confidentiality or anonymity of client.

The right to expect high-quality research.

a. Protection against unnecessary research.

b. Protection against unqualified researchers.

c. Protection against misleading presentation of data.

The Right of Confidentiality. The nature of the research process requires open lines of communication between the client and vendor if the inquiry is to be effective. When a client enters into a working relationship with a research vendor, certain types of proprietary data will be discussed. The vendor has an ethical responsibility to hold the nature of their relationship confidential. In addition, the client has the right to expect complete anonymity, as long as it does not in some way harm or deceive respondents or others involved in the research endeavor.

Let us examine a hypothetical research project on management turnover in a medium-size manufacturing firm, Dolphin Industries. Suppose a research consultant is hired to examine ways to reduce turnover in Dolphin. Both the nature of the relationship and the identity of the client must be kept confidential if so requested by Dolphin's management. Confidentiality may be extremely important to Dolphin because competitors might be able to exploit this problem in the marketplace.

The Right to Expect High-Quality Research. Although this right may seem rather obvious, some important concerns must be addressed. Researchers use their own specialized language, which can be bewildering to the manager. This sometimes puts the client at the mercy of researchers. Therefore, researchers have certain ethical responsibilities to produce high-quality research.

First, clients have the right to expect protection from unnecessary research. Because researchers are experts, they should have some knowledge as to when and where research is necessary and will be valuable. If a researcher doubts that a particular research endeavor is feasible or productive, the researcher has an ethical responsibility to inform the client that the research is of questionable value.

The second dimension of this right involves protection from unqualified researchers. Again, the responsibility falls on the research community. If researchers find that they are unqualified to conduct a particular research endeavor, they have the moral responsibility to inform the client of this fact.

Finally, the client or manager has the right to be protected against misleading presentation of data. Researchers have the ethical responsibility to truthfully and objectively present the results of their investigations. It may be in the researchers' interest to portray the data in a slightly distorted fashion so as to project more positive research findings, but this is

EXHIBIT **16.4**

Researchers' Rights

The right to expect ethical client behavior.

a. The right to the honest solicitation of proposals.

b. The right to expect the accurate representation of findings.

c. The right to expect confidentiality of proprietary information.

The right to expect ethical subject behavior.

a. The right to expect accurate reporting of data.

b. The right to expect confidentiality of data.

completely inexcusable. Clients have the right to demand the truthful reporting of research findings.

Researchers' Rights

Thus far, we have discussed everyone's rights except those of researchers. Although it is true that researchers initiate inquiry and often intrude in others' lives at inconvenient times and places, their work can be of substantial benefit to those same individuals and to society as a whole. On the assumption that researchers act ethically, they too have certain rights in the conduct of their work. These rights, presented in Exhibit 16.4, center on the researcher–client and researcher–subject relationships.

The Right to Expect Ethical Client Behavior. Clients have three major responsibilities to researchers and their related organizations. First of all, they have the responsibility to honestly solicit proposals for research problems. Some unscrupulous clients may issue an RFP (request for proposals) with no intention of contracting externally for conduct of the study. Instead, the client uses the vendor-submitted proposals to design its research, using the vendor's design suggestions. This is clearly unethical.

The next responsibility relates to the accurate representation of findings by the client. For example, suppose we perform a study for an organization and conclude that the acquisition of a related business entity is a good idea if the economy stays healthy. The client then develops a loan proposal to acquire this business, stating that our research organization finds that it is a good idea to acquire this business, omitting the contingency clause attached to the recommendation. This behavior expressly violates our rights as researchers to have the results, as stated, accurately represented.

The final responsibility of the client relates to the confidentiality issue. Just as the client can expect confidentiality of certain types of data, so can the researcher. Researchers may reveal analytical techniques or solution strategies that are competitively sensitive. In these instances, the client has the responsibility and the researchers have the right to demand confidentiality of proprietary information.

The Right to Expect Ethical Subject Behavior. Subjects have certain ethical responsibilities in research also. If subjects demand and obtain their rights in the research process, they should reciprocate with equally ethical behavior. What researchers can expect here involves the accurate reporting and the guaranteed confidentiality of the data.

The subjects have a responsibility to report data truthfully and without bias, *as long as this does not conflict with some other ethical value or principle of the individual*. In other words, if the subject agrees to participate in an investigation, it would be highly unethical for him or her to deliberately distort or conceal data so as to mess up the study's results. Similarly, if the subject has the right to obtain the research results, he or she has the responsibility to respect the confidentiality of the data if the organization supplying the information so requests.

While the basic ethical issues discussed above are the foundation for any ethical dilemma in business research, new issues are being created by technological changes in research. These issues, which are discussed throughout this text, center on (1) the protection of subjects' rights, particularly in the area of privacy, (2) the quality of research, and (3) the E-commerce/research dilemma. These technological challenges are discussed in the next section.

ETHICAL ISSUES IN A HIGH-TECHNOLOGY ENVIRONMENT

As noted throughout this text, technology is thoroughly changing how researchers go about their business. These technological changes have created capabilities for researchers that have never been possible before. Although it is exciting that these changes are happening in our lifetime, it also poses a whole new set of ethical concerns for the business research community. Here we have, or will have, the ability to communicate instantaneously with people, which we have never had the opportunity to do before. At the same time, the potential for abuse of this technology is frightening from the ethical researcher's viewpoint.

Consider the possibility of having one-on-one, instantaneous, two-way contact with respondents throughout the world, with the ability to monitor and identify respondents from one research platform. Sound ideal? Think of the negative consequences of the issues brought up earlier in this chapter. We have the potential for an unprecedented invasion of privacy, unethical direct marketing in the guise of legitimate research endeavors, and a total breakdown in trust, and thus response, from our ultimate study subject, the individual. If this occurs, we are faced with Scenario I illustrated in Figure 16.1, where public trust is nonexistent, and participation rates fall precipitously lower than we are currently facing today.

As a result, all major research associations are attempting to be proactive in regard to Internet-based research and E-commerce. Although all of these issues ultimately fall under the major ethical issues in business research identified earlier, technology has brought new aspects to already existing ethical dilemmas in the research field. These issues seem to gravitate around three major concerns: the protection of subjects' rights, quality of research issues, and the research versus direct marketing quagmire.

Protection of Subjects' Rights

Protection of the rights of subjects is the key issue in ethics relating to the advancements of technology in business research. Here, the "cookies" of the Internet and the ability to track respondents electronically must not take precedence over subjects' right to be informed or their right to privacy. If we violate the basic trust here, the research field is doomed for any future endeavors.

Here, the issue is *informed consent,* which implies that subjects realize and give consent to the fact that they are being studied and that *anonymity,* if so requested, is respected at all times. Given the tracking capabilities on Internet usage, how should one be informed of international tracking devices? Should everyone that accesses a particular person's "data bank" be required to report to that individual the information he or she requested? If not, why not? Our standards in this regard are still being developed.

A related concern is that of requesting information from children without their parent's knowledge or consent. Given that the Internet is expected to be a major educational delivery device in the future, researchers must grapple with this problem. The questions are what is research and what is educational? Are we selling, or are we educating? If we question or observe, what is the notification procedure to parents? An organization can hit some pretty rough waters if it assumes that what it is doing is for the good of humanity.

In response to these and other subjects' rights, research associations such as CASRO (Council of American Survey Organizations, http://www.casro.org), ESOMAR (World Association of Opinion and Marketing Research Professionals, http://www.esomar.org), and IMRO (Interactive Marketing Research Organization, http://www.imro.org/code.htm) have been very proactive in creating guidelines and codes for research companies to show them how to meet the privacy requirements for personal data collection, storage, and dissemination. For example, CASRO has created CASRO 3P, a single-source membership service to show its members how to meet the privacy directives of such measures as U.S. Safe Harbor, the European Directive on Data Protection, Health Insurance Portability and Accountability Act (HIPAA), and the Children's Online Privacy Protection Act (COPPA). Other associations have developed similar privacy protection guidelines.

Quality of Research

The quality of research issue has been brought up in other chapters in this text. The ease of requesting samples on the Internet, coupled with the ability to almost instantaneously distribute these results has created an environment with immense potential for being abused. Firms, politicians, individuals, or others can create online biased questions that confirm what they want, or would like, to hear.

The early solution to this problem is to request the same scientific rigor from research on the Internet as that required from land-based probes. Although this solution appears intuitively satisfying, there are numerous "loose ends." Now anyone on the Internet can ask the question, interpret its meaning, and apply any name to it. The good, bad, and ugly of the Internet. The value of information sources varies wildly on the Internet. One must be extremely careful in accepting and using Internet data. This will continue to be an issue as long as the Internet is viewed, at the same time, as a gold mine and a junkyard.

Research Versus Direct Marketing

Although the protection of subjects' rights is viewed as the main issue with technology, the idea of trying to maintain the distinction between research and direct marketing is also important, and difficult at best.[9] The essence of this issue goes back to the reasons why the two needed to be separated in the first place. First, the validity of research depends on the willingness of subjects to participate in a nonthreatening and anonymous environment, without fear of being solicited or identified in any manner. If we lose the trust of respondents, research results will be worthless. Second, the legalistic and moral concerns are different for these two activities. If research professionals fail to separate themselves from "spammers," direct marketers, and other forms of E-commerce, the credibility is gone for the researcher and the organization sponsoring the research.

The ICC/ESOMAR International code (http://www.esomar.org) recommends a number of specific activities to avoid the research–direct marketing conflict. The major recommendations deal with distinguishing research activities from direct marketing efforts to the study subject. To start, whenever one is acting in a research capacity, one must not be involved in direct marketing activities. If a person presents himself or herself as part of a research company or affiliate, that person must not conduct direct marketing activities. Finally, if a client organization carries out both research and direct marketing activities as part of its business, the two must be separated organizationally so they do not become compromised.

Clearly, the issues in the ethical conduct of business research are complex. Several major associations concerned with business research have adopted codes of ethics to help researchers in determining ethical behavior in research, as discussed in the next section.

CODES OF ETHICS

The public is angry about abuses of research, so government will continue to legislate ethical conduct for researchers if the researchers do not adequately regulate themselves. Formal codes of ethics are now common among professional societies and other organizations involved in business research.

The American Psychological Association was a leader in the adoption of a code of ethics for research activities. Other professional organizations, such as the American Marketing Association, the American Association for Public Opinion Research, and the World Association for Public Opinion Research, have subsequently adopted codes of ethics for their research organizations.

The codes of ethics that are being adopted by various associations all involve aspects of the ethical relationships and obligations framework illustrated in Figure 16.2. The documents all purport to guide researchers in choosing ethical means to do research, while at the same time protecting those involved in the process itself. As an example, the appendix to this chapter presents the basics of the code of the International Chamber of Commerce/European Society for Opinion and Marketing Research (ICC/ESOMAR). An examination of this code should enable you to grasp the primary ethical concerns that researchers must face. The reader is encouraged to visit the Web sites of research associations to view their position on codes of ethics.

MANAGERIAL CONSIDERATIONS

The managerial considerations in the field of business research ethics are extremely important. Managers are responsible for setting the tone of ethical behavior in their organizations. Specifically, managers have five key concerns to help promote ethical research in their businesses:

Communicate the benefits of research to subjects and society, and adopt ethical research practices. Management should take a proactive stance in informing subjects and ensuring that ethical research practices are followed. These activities will help ensure public acceptance and participation in future research endeavors.

Develop or adopt a code of ethics that will serve as the corporate policy in the conduct of business research. Many associations and businesses have adopted codes of ethics as policy. Management should take the initiative to develop or adopt some code that will provide direction in the design, purchase, and implementation of research endeavors.

Communicate ethical behavior down through the organization hierarchy through the use of formal sanctions. Management sets the ethical tone within the organization. If subordinates do not perceive ethical concerns to be of interest to management, these concerns will be subordinated to other goals. Thus, management must promote ethical behavior through the use of rewards and sanctions, if necessary. Codes of ethics often become only pieces of paper in the personnel handbook if sanctions are not applied for failing to follow the standards of behavior.

Institute a formal procedure to check that all research performed or bought by the organization follows the highest ethical standards, as dictated by the code of ethics. A number of options are available. A review board can be assembled to oversee that all proposed research

EXHIBIT **16.5**

1. Is management taking a proactive stance in communicating the benefits of research to subjects and society by demanding the adoption of ethical research practices? If not, why not?

2. Does the organization currently follow a code of ethics in its conduct of business research? If not, why?

3. What efforts are being made to support ethical research behavior in the organization? Are management's ethical concerns being effectively communicated to those conducting or buying research?

4. Are there formal procedures to check that research being conducted or bought follows the highest ethical standards as dictated in the code of ethics?

5. Is there a formal procedure for conducting periodic ethical audits of the business research function?

6. What sanctions are you willing to impose if unethical activities are found in the research function? How will the company rectify this situation in the future?

conforms to the ethical standards set by the firm. There should be a list of specific questions to be answered before any investigation is begun. The objective of these procedures is to force researchers and vendors to deal with the ethical issues of conducting research.

Conduct periodic ethical audits to ensure that the research function is in fact being performed to the highest standards. An ethical audit entails a review of recent research endeavors to ensure that ethical standards of the firm are being met. The audit would consist of a review of procedures and a polling of the parties involved in the research function. If ethical problems are identified, corrective action should be undertaken.

These considerations should help guide the firm in becoming ethically oriented in its research endeavors. These basic concerns are restated in a series of key managerial questions in Exhibit 16.5.

SUMMARY

Ethics is an extremely important topic in the study of business research because it affects the rights of individuals and, ultimately, the quality of data obtained from the research process.

Business research ethics is the proper conduct of the business research process. Law usually sets the background and limits on ethical conduct. However, legislative action is not the only means to curb ethical abuses. Ethical and unethical research has significant effects on public acceptance and participation in business research. Many researchers prefer to set their own high standards of conduct so that government action is not needed. This premise forms the basis for the discussion of major issues in the chapter.

A framework of ethical relationships for business research is presented. Four participants have both rights and responsibilities in the conduct and use of research: society, subjects, clients/managers, and the researchers themselves.

Technology has drastically changed the research environment. New ethical challenges are surfacing. Three areas where this is obvious are in the protection of subjects' rights, the quality of research in general, and the research–direct marketing conflict.

Next, codes of ethics are briefly examined. A number of associations and organizations have recently adopted such guidelines for ethical behavior. An example of such a code is presented in an appendix to this chapter.

The chapter ends with a discussion of the managerial considerations in business research ethics. Managers should develop or adopt formal codes of ethics for their organizations, then support these codes through effective communication and the initiation of formal procedures to ensure compliance. Summary questions to guide the manager in this task end the chapter.

DISCUSSION QUESTIONS

1. What is business ethics? What relevance does this topic have to the field of business research?

2. What are the effects of ethical and unethical business research practice on the acceptance and participation of the general public? What about implications for the future of business research as a profession?

3. What are the major ethical relationships inherent in business research? When answering this question, be sure to identify the major participants in any research endeavor, and outline each participant's major rights.

4. What are codes of ethics? Discuss the pros and cons of adopting a code of business research ethics for a business firm. Be specific and provide examples to support your points.

5. Examine the basic requirements of the ICC/ESOMAR International Code of Marketing and Social Research Practice, presented in the chapter appendix. Do you feel this code covers the major ethical issues that one would face in the conduct of business research? What issues can you think of that are not specifically addressed in this code?

6. What can managers do to promote ethical behavior in their research organizations? Give specific activities managers can do to ensure that ethical research is being conducted in their organizations.

7. Suppose you are a research manager and you discover that subjects' rights are being violated in a specific research project. What would you do?

8. Using the Internet, conduct a search to identify the major ethical issues facing managers and researchers today. A good starting place for your search would be the Web sites of such research organizations as AAPOR (http://www.aapor.org) and ESOMAR (http://www.esomar.org) and other such entities. Be prepared to discuss your findings in class.

9. The distinction between direct marketing and marketing research can become blurred. Given the importance of maintaining the distinction between the two, research the differences and make recommendations to an organization of your choice about how to maintain this distinction. Use the Web to develop your search strategy.

10. Go online and find a code of ethics for an organization of your choice. Compare this code with the ESOMAR code in the appendix of this chapter. What are the differences and similarities?

NOTES

1. See, for example, the Privacy Act of 1974, which requires that all respondents in a federal government survey be explicitly informed, both in writing and verbally, of their rights of participation and refusal. Also see Howard Schlossberg, "Privacy Bill Alarms Researchers," *Marketing News* (3 January 1994), *28*:1, pp. 1ff; Diane Bowers, "Privacy and the Research Industry in the US," *ESOMAR Research World* (July/August 2001), no. 7, pp. 8–9; and "European Data Protection Legislation and How It Affects Market Research: A Plain Guide," *Research World* (March 2000), no. 7, downloaded from http://www.esomar.org/cotent/pdf/guidelines/data_protection.pdf on December 15, 2003.

2. See, for example Eleanor Singer and Felice J. Levine, "Research Synthesis: Protection of Human Subjects of Research: Recent Developments and Future Prospects for the Social Sciences," *Public Opinion Quarterly* (Spring 2003), *67*:11, pp. 148–165, and Thomas A. Wright and Vincent P. Wright, "Organizational Researcher Values, Ethical Responsibility, and the Committed-to-Participant Research Perspective," *Journal of Management Inquiry* (June 2002), *11*:2, pp. 173–188.

3. Much of the following section is drawn from Donald S. Tull and Del I. Hawkins, *Marketing Research*, 6th ed. (New York: Macmillan, 1993),

Chapter 23; and Alice M. Tybout and Gerald Zaltman, "Ethics in Marketing Research: Their Practical Relevance," *Journal of Marketing Research* (November 1974), *11*, pp. 357–368. These two sources provide excellent treatments of the major ethical issues in the conduct of business research.

4. Linda Morris and Steven Pharr, "Invasion of Privacy: A Dilemma for Marketing Research and Database Technology," *Journal of Systems Management* (October 1992), *43*:10, pp. 10ff.

5. The discussion of subjects' rights draws heavily from Tybout and Zaltman, op. cit. note 3, pp. 357–368.

6. See Lesley Young, "Researchers Fight Against Mugging," *Marketing Magazine* (January 15, 2001), *106*:2, p. 3.

7. Ellen Berscheid, Robert Baron, Marshall Deumer, and Mark Libman, "Anticipating Informed Consent: An Empirical Approach," *American Psychologist* (October 1973), *28*, p. 913.

8. Tybout and Zaltman, op. cit. note 3, p. 359.

9. See, for example, "ICC/ESOMAR Guideline on Maintaining the Distinctions Between Marketing Research and Direct Marketing" (http://www.esomar.org), 20 September 1998; and David Schwartz, "Sharing Responsibility for E-Commerce and the Privacy Issue," *Direct Marketing* (June 1998), pp. 48–98.

SUGGESTED READING

Ethical Issues in Business Research

Birch, Maxine, Melanie Mauthner, Julie Jessop, and Tina Miller, Eds., *Ethics in Qualitative Research* (Thousand Oaks, CA: Sage Publishing, 2002).

Explores the ethical research issues in qualitative research from a variety of perspectives. Provides a good overview of ethical considerations in this important area of research.

Buchanan, Elizabeth A., *Readings in Virtual Research Ethics: Issues and Controversies* (Hershey, PA: AAA Publishing, 2003).

This book looks at the field of online research and the corresponding ethical dilemmas associated with it. Issues such as subject selection and recruitment, informed consent, privacy, ownership of data, and research with minors are explored in the virtual realm.

Messick, David M., and Ann E. Tenbrunsel, Eds., *Codes of Conduct: Behavioral Research into Business Ethics* (New York: Russell Sage Foundation, 1997).

This compendium of articles explores the effects of social influence and group persuasion on individual and organizational behavior.

Ethical Issues in a High-Technology Environment

The very essence of technology is that it is evolving constantly. As a result, the best readings will be online. Check out the Web sites of major research organizations. Here are a few to start with.

American Association for Public Opinion Research (AAPOR), http://www.aapor.org

American Marketing Association (AMA), http://www.ama.org

European Society for Opinion and Marketing Research (ESOMAR), http://www.esomar.org

Interactive Marketing Research Organization (IMRO), http://www.imro.org

Codes of Ethics

As noted, check out the Web sites of the major research organizations. Their codes of ethics are generally stated prominently on their sites. Complete copies of the ICC/ESOMAR International Code of Marketing and Social Research Practice may be obtained from one of the following two sources:

International Chamber	*E.S.O.M.A.R*
of Commerce	*J. J. Viottastraat 29*
38 Cours Albert 1er	*1071 JP Amsterdam*
75008 Paris	*The Netherlands*
France	*Tel: 31-20-664.21.41*
Tel: 33-1-49.53.28.28	*Fax: 31-20-664.29.22*
Fax: 33-1-49.53.28.62	*http://www.esomar.org*

EXCERPTS FROM THE ICC/ESOMAR INTERNATIONAL CODE OF MARKETING AND SOCIAL RESEARCH PRACTICE

Introduction

1. Public opinion research—the study of people's attitudes and beliefs about political, social and other issues—forms a part of the total marketing and social research field. It is subject to exactly the same professional and ethical requirements as other forms of survey research. These requirements are set out in the ICC/ESOMAR International Code of Marketing and Social Research Practice.

2. However, public opinion research tends to be a specially 'sensitive' area. It deals with issues which arouse greater public interest and emotion than do most commercial market research projects. In addition, its findings are much more widely published and debated, and may sometimes be presented in a provocative or even tendentious way. ESOMAR has therefore set out specific recommendations about the publication of such research.

3. Opinion polls have a valuable role to play in present-day society. It is desirable that the general public, politicians, the media and other interested groups should through research have access to accurate and unbiased measures of public attitudes and intentions. We recognise that there are some sincerely-held worries about possible (but mostly unproven) effects which some polls could in theory have upon voting or other behaviour. However, the alternative is that the public is exposed only to unscientific and probably inaccurate assertions about the situation, in many cases presented by individuals or organisations who have an insufficient understanding of the nature of the information they are using or who take an extremely partisan approach to presenting the facts. The objective of this Code is to reduce the risk of the public being misled by research which is inadequate or badly presented.

4. The Parliamentary Assembly of the Council of Europe has examined this ESOMAR Code for the Publication of Opinion Polls and has given the Code its blessing. The Council of Europe has recommended the widespread application of this Code to govern the publication of polls.

5. The validity and value of public opinion polls depend on three main considerations:

(i) the nature of the research techniques used and the efficiency with which they are applied,

(ii) the honesty and objectivity of the research organisation carrying out the study,

(iii) the way in which the findings are presented and the uses to which they are put.

This Code concentrates primarily on the second and third of these issues. Guidelines on techniques and the conduct of opinion polls, and pre-election in particular, are given in the next section.

6. Major problems can arise when opinion poll findings are published and debated. It would clearly be unrealistic, and unreasonable, to expect the media to quote the full technical background of a survey when presenting its findings: they have limitations of space and must also hold the interest of their audience. However, there is certain basic information which must be provided if that audience is to have the opportunity of judging for itself the evidence presented and deciding whether or not it agrees with any conclusions drawn from the research. This Code is primarily concerned with trying to ensure that the public has reasonable access to this key information about the survey, and that published reports of the findings are not misleading. The Code tries to strike a realistic balance between what would be theoretically desirable and what is practicable.

7. All reputable research organisations apply the appropriate scientific methods and operate with professional objectivity. In doing so they conform to the ICC/ESOMAR International Code of Marketing and Social Research Practice. There is also general agreement among them on the principles which should underlie the publication of research results. However, normal professional practice varies between countries in some respects and in certain countries additional information to that specified in this Code will also customarily be provided as part of the standard key material.

8. Research organisations have a particular responsibility in the field of public opinion polls, to make sure that both the client and the public have a reasonable understanding of the special problems and limitations involved in measuring attitudes and beliefs as distinct from behaviour. Such research frequently deals with complex and sensitive issues about which respondents have varying degrees of knowledge and interest, and whether their views may often be half-formed, confused and inconsistent. High professional integrity and skill is essential if the research itself is to be unbiased and meaningful, and if the findings are to be presented and interpreted clearly and accurately. It is important also that the research budget available is sufficient to carry out a valid study. ESOMAR fully recognises that such considerations are vital if public opinion polls are to merit public confidence and support

9. Finally, if as a result of past experience, a research organisation has reason to believe that a particular client will not fairly present opinion poll results in his published version of the findings, the research organisation has a responsibility to stop carrying out polls for publication by that client.

Code

Basic Requirements of the ICC/ESOMAR International Code of Marketing and Social Research Practice

1. All research organisations which conduct public opinion polls **must conform** to the ICC/ESOMAR International Code of Marketing and Social Research Practice. Particular attention is drawn to the requirements of Article 13 (concerned with the misrepresentation or misuse of research techniques); Article 29 (concerned with the Publishing of Results); Articles 33 and 34 (concerned with Reporting Standards); and Article 35 (concerned with Implementation of the Code). These Articles are reproduced as Appendix 1 to the present document.

2. It is important to distinguish between the requirements which apply to the reporting of public opinion poll results by a research organisation to its original client, and those which apply to the subsequent publishing of any poll findings by that client to a wider audience. The first of these situations is largely covered by the existing International Code which specifies reporting requirements in detail. This supplementary Code is intended to clarify certain additional requirements which arise in connection with the wider publication of the findings, and therefore applies especially to the second situation.

Additional Requirements

3. When any public opinion poll findings are published in **print media** these should always be accompanied by a clear statement of:

(a) the **name of the research organisation** carrying out the survey;

(b) the **universe** effectively represented (i.e. who was interviewed);

(c) the **achieved sample size** and its **geographical coverage**;

(d) the **dates of fieldwork**;

(e) the **sampling method** used (and in the case of random samples, the success rate achieved);

(f) the **method by which the information was collected** (personal or telephone interview, etc.);

(g) the relevant **questions asked.** In order to avoid possible ambiguity the actual wording of the question should be given unless this is a standard question already familiar to the audience or it is given in a previous published report to which reference is made.

4. In the case of **broadcast media** it may not be possible always to give information on all these points. As a minimum, points (a)–(d) above should normally be covered in any broadcast reference to the findings of a public opinion poll, preferably in visual (written) form where practical.

5. The percentages of respondents who give 'don't know' answers (and in the case of voting-intention studies, **of those who say they will**

not vote) must always be given where they are likely to significantly affect the interpretation of the findings. When comparing the findings from different surveys, any changes (other than minor ones) in these percentages must also be indicated.

6. In the case of voting-intention surveys, it must always be made clear if **voting-intention percentages quoted include any of these respondents who answered 'don't know' or 'may not/will not vote'** in reply to the voting questions asked.

7. Whatever information may be given in the published report of the survey, the publisher and/or the research organisation involved must be prepared on request to supply the other information about the survey methods described in Article 34 of the International Code. Where the questions reported on have formed part of a more extensive or 'omnibus' survey, this must be made clear to any enquirer.

Arrangements Between the Research Organisation and Its Client

8. In order to ensure that these Code requirements are followed and to avoid possible misunderstandings, the research organisation must make clear in advance to its client:

(i) that the research organisation itself is bound by the requirements of the general International Code.

(ii) that subsequent wider publication of the research findings should be in accordance with this supplementary Code.

It is therefore the responsibility of the research organisation to draw its client's attention to the present Code on Publication of Results and to use its best endeavours to persuade the client to follow the Code's requirements.

9. The research organisation and the client each have a responsibility in the public interest to ensure that the published report on a public opinion poll does not misrepresent or distort the survey data. For example, misleading comments based on non-significant differences must be avoided. Special care must be taken to ensure that any graphs or charts used do not convey a misleading impression of the current survey's results or of trends over time. It is also important that the reader or listener should be able clearly to distinguish between the survey findings as such and any editorial or other comments based upon them. Particularly in the case of print reports, the research organisation must wherever feasible approve in advance the exact form and content of publication as required in Article 29(a) of the original International Code.

10. The research organisation cannot normally be held responsible for any subsequent use made of public opinion poll results by people other than the original client. It should however be ready to issue immediately such comments or information as may be necessary to correct any cases of misreporting or misuse of results when these are brought to its attention.

11. In the event that a client releases data from a survey which was not originally intended for publication, this Code of Conduct will apply to it as if it had originally been commissioned for publication.

APPENDIX 1

While all Articles of the International Code apply to Public Opinion Polls, the following Articles have special significance in this connection;

Relations with the General Public and the Business Community

Article 13. No activity shall be deliberately or inadvertently misrepresented as Marketing Research. Specifically, the following activities shall in no way be associated, directly or by implication, with Marketing Research interviewing or activites:

  a) enquiries whose objectives are to obtain personal information about private individuals *per se,* whether for legal, political, supervisory (e.g. job performance), private or other purposes.

  b) the compilation of lists, registers or data banks for any purposes which are not Marketing Research;

  c) industrial, commercial or any other form of espionage;

  d) the acquisition of information for use by credit-rating or similar services;

  e) sales or promotional approaches to the Informant;

  f) the collection of debts;

  g) direct or indirect attempts, including the framing of questions, to influence an Informant's opinions or attitudes on any issue.

Publishing of Results

Article 29. Reports and other records relevant to a Marketing Research project and provided by the Researcher shall normally be for use solely by the Client and his consultants or advisers. The contract between Researcher and Client should normally specify the copyright of the research findings and any arrangements with respect to the subsequent more general publication of these. In the absence of such a specific agreement, if the Client intends any wider circulation of the results of a study either in whole or in part:

  a) the Client shall agree in advance with the Researcher the exact form and contents of publication or circulation: if agreement on this cannot be reached between Client and Researcher, the latter is entitled to refuse permission for his name to be quoted in connection with the study;

  b) where the results of a Marketing Research project are given any such wider circulation the Client must at the same time make available the information listed under Article 34 about the published parts of the study. In default of this, the Researcher himself is entitled to supply this information to anyone receiving the above mentioned results;

c) the Client shall do his utmost to avoid the possibility of misinterpretation or the quotation of the results out of their proper context.

Reporting Standards

Article 33. Normally every report of a Marketing Research project shall contain an explanation of the points listed under Article 34, or a reference to a readily available separate document containing this explanation. The only exception to this Article is in the case where it is agreed in advance between the Client and the Researcher that it is unnecessary to include all the listed information in the formal report or other document. Any such agreement shall in no way remove the entitlement of the Client to receive any and all of the information freely upon request. Also this exception shall not apply in the case where any or all of the research report or findings are to be published or made available to recipients in addition to the original Client.

Article 34. The following information shall be included in the report on a research project:

Background

a) for whom and by whom the study was conducted;

b) the purpose of the study;

c) names of sub-contractors and consultants performing any substantial part of the work;

Sample

d) a description of the intended and actual universe covered

e) the size, nature and geographical distribution of the sample, both planned and achieved; and, where relevant, the extent to which any of the data collected were obtained from only a part of the sample;

f) details of the sampling method and of any weighting methods used;

g) where technically relevant, a statement of response rate and a discussion of possible bias due to non-response;

Data Collection

h) a description of the method by which the information was collected (that is, whether by personal interview, postal or telephone interview, group discussion, mechanical recording device, observation or some other method);

i) adequate description of field staff, briefing and field quality control methods used;

j) the method of recruitment used for Informants and the *general nature* of any incentives offered to them to secure their co-operation;

k) the time at which the fieldwork was done;

l) in the case of "Desk Research", a clear statement of the sources and their reliability;

Presentation of Results

m) the relevant factual findings obtained;

n) bases of percentages, clearly indicating both weighted and un-weighted bases;

o) general indications of the probable statistical margins of error to be attached to the main findings, and of the levels of statistical significance of differences between key figures;

p) questionnaires and other relevant documents used (or, in the case of a shared project, that portion relating to the matter reported upon).

Implementation of the Code

Article 35. Any person or organisation involved in, or associated with, a Marketing Research project and/or proposal is responsible for actively applying the rules of this Code in the spirit as well as the letter.

Public opinion research is subject to exactly the same professional and ethical requirements as other forms of survey research.

Source: ESOMAR Guide to Opinion Polls (Amsterdam, The Netherlands: ESOMAR, 1993), pp. 5–12. Reprinted with permission.

17

Research Reporting and Evaluation

Overview

Written Research Reports
The Outline
Guidelines

Oral Presentations

Research Evaluation

Communication Challenges in the Future

Managerial Considerations

Summary

Discussion Questions

Notes

Suggested Reading

As you should remember from Chapter 1, there are two focal points of communication between management (the users of research) and researchers (the doers): (1) the research proposal and (2) the final research report. This chapter deals with research reporting, both written and oral.

It would be hard to exaggerate the importance of effectively communicating the results of a research endeavor to management. Unless the results of a study are communicated to those who can use the findings for decision-making purposes, resources are wasted. The following pages suggest guidelines for the development, delivery, and evaluation of research reports, both written and oral.

The chapter begins with the written research report. The topics discussed include a general outline for the content of the report and some guidelines for ensuring that the written report effectively communicates the research results.

Next, five guidelines for oral presentations are suggested as a basis for developing effective oral research reports to supplement the formal written report. The importance of oral presentations is emphasized given the goal of effective communication to management.

Once the nature of reports is discussed, the meaningful process of research report evaluation is presented. All reports should be formally reviewed for quality before the results are used for decision-making purposes. A framework for evaluation is proposed.

The communication challenges that both the researcher and manager must face in the future are then viewed in light of multiculturalism and the globalization of business. These concerns coupled with other managerial considerations in research reporting and evaluation end both this chapter and the text.

WRITTEN RESEARCH REPORTS

The written research report is the end product of the research process. It is a formal statement of the background of the problem being investigated, the nature of the study itself, and the relevant findings and conclusions drawn from the research process. It is critically important to generate an effective report when the research is completed. The discussion that follows focuses on two aspects of generating good research reports: the outline of the report and the criteria for a good report.

EXHIBIT 17.1

Organization of the
Written Research
Report

1. Title page
2. Executive summary
3. Contents
4. Introduction
5. Body
   a. Methodology
   b. Results
   c. Limitations
   d. Conclusions and recommendations
6. Appendix
   a. Data collection instrument
   b. Tables
   c. Other relevant supporting materials

The Outline

Planning the written report is essential to realizing its full communication potential. Before writing begins, the main topics to be included in the document must be outlined. Initially, a preliminary outline should be prepared to help organize thoughts about the components needed in the report. Next, you should organize, and reorganize if necessary, the topics to ensure that all material is covered and presented in a logical and coherent fashion. There is no one universally accepted format, but the outline presented as Exhibit 17.1 includes the salient parts of a well-developed business research report.[1] The specific makeup of a report depends on the individuals to whom it is directed.

Title Page. The title page identifies the subject of the report, to whom it is directed, who prepared it, the date, and other relevant contact information (addresses, fax and telephone numbers, e-mail addresses, and so on). The title page is designed to convey critical information in a concise format.

The title of the report should be short, yet convey the essence of the material that follows. The title page should also show precisely to whom it is directed and provide information for contact and verification purposes. Without this information, the title page lacks completeness and clarity. An example of the first few pages of a report is presented in Exhibit 17.2,[2] which includes a sample title page, an executive summary, a table of contents, and an introduction.

Executive Summary. The executive summary is an important part of longer research reports. It highlights the nature of the project and succinctly summarizes the conclusions and recommendations of the study. The summary is important because many managers read only this section

A SURVEY OF COMMUNITY ATTITUDES
AND PERCEPTIONS REGARDING
ST. TIMOTHY'S MEDICAL CENTER

Prepared for

Mr. Donald K. Webster
Vice President
St. Timothy's Medical Center
1620 Ryan Street
Smithstown, CT 06001

Prepared by

THE DATAFAX COMPANY, INC.
861 West Morse Blvd., Suite 50
Winter Park, FL 32789

January 20, 2005

EXECUTIVE SUMMARY

This study consisted of a telephone survey of 406 randomly selected residents of the Smithstown, Connecticut area. The questionnaire was disguised in such a manner that the respondents were not aware of the identity of the client. Over 92% of the respondents indicated that they were very interested or somewhat interested in the general subject of health and medicine. When the respondents were asked if either they or a member of their immediate family had been admitted to a Smithstown area hospital during the last two years, 48% answered that they had. Smithstown Hospital accounted for 47.2% of the patients, and had the largest market share in the area. St. Timothy's Medical Center was second with a market share of 39.5%, while Daly City Hospital was a distant third at 11.8%.

EXHIBIT 17.2

(Continued)

The most frequently mentioned reason that patients gave for choosing the hospital they were admitted to was that their physician had recommended it; over 60% of the patients responded in this manner. Other factors included the proximity of the hospital to the patient's home, a preference for a particular hospital on the part of the patient, and admittance due to an emergency.

Patients that were admitted to St. Timothy's Medical Center tended to be younger than patients that were admitted to Smithstown Hospital. The market shares of these two hospitals were virtually identical among single respondents and divorced respondents, while Smithstown Hospital's market share among married and widowed respondents was higher than St. Timothy's.

All respondents who fell into the nonpatient category were asked to name the Smithstown area hospital that they knew the most about. Smithstown Hospital was mentioned by 47.4% of the respondents, and St. Timothy's Medical Center by 40.3% of the respondents. Both of those percentages are a very accurate reflection of each hospital's market share. Younger respondents were more likely to name St. Timothy's Medical Center as the hospital they knew the most about, while older respondents were more aware of Smithstown Hospital.

Patients were asked what they had seen or heard in the last six months about the hospital they had been admitted to, while nonpatients were asked the same question with respect to the hospital that they knew the most about. The most frequent comment received regarding St. Timothy's Medical Center was that former patients of the hospital recommend it. Personal experience and friends were the most frequently mentioned sources of information, with the newspaper coming in third. It was determined that St. Timothy's Medical Center was clearly behind Smithstown Hospital in the newspaper category. Respondents were more likely to hear something about St. Timothy's Medical Center from a former patient than read about it in the newspaper, while the opposite was true of Smithstown Hospital.

When respondents were asked the question, "Which hospital in the Smithstown area do you feel best serves the needs of the community?", 41.4% of the respondents answered, "Smithstown Hospital," St. Timothy's Medical Center was mentioned by 26.5% of the respondents, while 23.7% did not answer the question because they didn't know or couldn't decide. Reasons for choosing Smithstown Hospital included best equipment, serves everyone, burn unit, and clean, modern building. The most popular reasons for naming St. Timothy's Medical Center as the best area hospital were its good location and its good reputation. It was determined that married, widowed, and divorced respondents were more likely to name Smithstown Hospital than St. Timothy's Medical Center as the hospital that best serves the needs of the community, while St. Timothy's ranked ahead of Smithstown Hospital among single respondents. Respondents in the 21 to 30 age category were also more likely to name St. Timothy's Medical Center, while older respondents were more likely to name Smithstown Hospital. Smithstown Hospital was also more popular among female respondents than among male respondents.

Hospitals in the Smithstown area were evaluated by respondents on a series of scaling questions. St. Timothy's Medical Center generally received higher scores among patients than Smithstown Hospital, while nonpatients tended to rate the two hospitals very closely. Both patients and nonpatients rated St. Timothy's Medical Center higher than Daly City Hospital in most categories. It was determined, however, that both Smithstown Hospital and Daly City Hospital received higher scores than St. Timothy's Medical Center in categories dealing with the number of doctors and nurses available at each hospital. St. Timothy's Medical

(Continued)

EXHIBIT **17.2**

(Continued)

Center received similar scores from patients and nonpatients with the exception of categories dealing with the emergency room. Nonpatients provided much lower scores than patients did in this area. Patients and nonpatients were generally in agreement when evaluating Smithstown Hospital, while several differences were noted between the evaluations of patients and nonpatients discussing Daly City Hospital. Nonpatients discussing Smithstown Hospital were more likely to offer an opinion on the categories being discussed than nonpatients discussing St. Timothy's Medical Center, an indication that nonpatients are more aware of Smithstown Hospital than they are of St. Timothy's.

All respondents were asked about their willingness to participate in a series of special programs and seminars. The most popular, long-term programs were dealing with back injury, stress management, and weight control. The most popular seminars were assessing risk of cancer, assessing risk of heart disease, breast examination, and household first aid. When the willingness to participate in these special programs was cross-classified by demographic information, it was determined that the degree of interest for certain programs was varied among specific demographic groups.

CONTENTS

I. Introduction

II. Methodology

III. Results

   A. Market Shares of Smithstown Area Hospitals

   B. Perceptions of Smithstown Area Hospitals

   C. Likert Scale Ratings of Smithstown Area Hospitals

   D. Interest in Special Programs and Seminars

IV. Limitations

V. Conclusions and Recommendations

Appendix

   A. Questionnaire

   B. Sampling Plan

   C. Test of Statistical Precision

INTRODUCTION

This report summarizes the findings of a market research study conducted on behalf of St. Timothy's Medical Center. The overall goal of the research was to explore the perceptions held by the Smithstown community toward St. Timothy's Medical Center and its competitors. The specific objectives of the research were

1. to explore specific areas of strength and weakness of St. Timothy's Medical Center and its competitors as seen by the community;

2. to explore the community's attitudes toward and their interest in a series of special programs and seminars;

3. to determine the attitudes of respondents who have either been hospitalized or who have had a member of their immediate family hospitalized during the last twelve months toward the Smithstown area hospital they were admitted to;

EXHIBIT 17.2

(Continued)

4. to determine the reasons why patients chose the hospital they were admitted to;
5. to examine the perceptions of respondents who have not been hospitalized during the last twelve months toward the Smithstown area hospital they know the most about;
6. to determine the demographic characteristics of hospital patients that were admitted to St. Timothy's Medical Center and its competitors.

To accomplish these objectives, a telephone survey of 406 Smithstown area residents was conducted from October 6, 2004, to October 29, 2004. Details of the sampling plan utilized in this study are featured in Appendix A.

of the paper. Therefore, the summary must be able to stand by itself and convey the essentials of the report in one or two pages.

Although some managers prefer a one-page summary, it is not a hard-and-fast rule. The general rule is to make the summary as concise as possible while including as many details as necessary for the reader to grasp the highlights of the study. Of course, this rule does not apply if the client specifies a one-page summary.

Contents. The contents lists the main headings and secondary headings in the report, derived from the outline developed earlier. The contents helps the reader to follow the flow of the report and serves as a mechanism to locate critical information easily; it should always be included in longer, more complex reports.

In the contents, the main headings and subheadings should appear in order and be accompanied by page numbers for reference purposes. Always think of making it easier for the reader to assimilate and use the report for decision-making purposes.

Introduction. The introduction states the background information needed for the reader to appreciate the body of the report. Usually, the objectives of the research are stated here, along with any additional information that led to the conduct of the project. Often, much of this information can be obtained from the research proposal that was written before the project began.

After reading the introduction, the reader should be familiar with the purpose of the investigation, how the study came about, and any other pertinent information that helps him or her understand the body of the text.

Body. The body includes all significant details of the research. It contains specific information on how the research was conducted, actual study results, a statement of research limitations, and the conclusions and recommendations that are derived from the results. This is the heart of the report.

The *methodology* section of the report should outline the methods and procedures used to collect and analyze the data. Again, much of this information can be obtained from the research proposal, adjusted for any modifications due to unforeseen circumstances in the conduct of the investigation. The methodology section of the report should be detailed enough for the reader to evaluate the substantive issues of the project. However, if extremely detailed explanations are needed, they should be put in an appendix at the back of the report. This will improve readability and help the reader focus on the results and implications of the study itself.

The *results* section of the paper should clearly outline the findings of the study. This section is often the largest in the report. A complete and clear presentation of the data and data analysis must be given, organized in a logical fashion and presented in an easy-to-read format that will facilitate the reader's understanding. It often helps to use graphic displays that serve to summarize the findings. The choice of presentation format should be made with the characteristics of the reader in mind.

The body should also include a statement on the limitations of the study. The weaknesses of the study should be addressed so that the reader is aware of them. Similarly, the degree to which the results can be generalized should be stated explicitly. For example, a nonrepresentative sample should be noted in the body of the report, and the degree to which this affects the study's validity should be expressly stated.

From the decision maker's standpoint, the *conclusions and recommendations* section of the report is the valuable part of the research endeavor. The results provide the basis for this section. Here, the conclusions, and the rationale for them, should be unambiguously stated. The logical steps should be detailed enough that the reader can see how the conclusions were arrived at and why the recommendations were made. The writer should assess and reassess this section to ensure perfect clarity of meaning for the intended reader.

Appendix. The appendix should include any other information and materials necessary for a complete presentation of the research project. If a data collection instrument is used in the study, it should be included in the appendix. Similarly, large tables of results or materials that might otherwise disrupt the flow of the actual report are best placed in the appendix so as to increase the overall readability of the research report.

Guidelines

We consider the following five guidelines critical in the development of first-class reports.

> **Know your audience.** This guideline is crucial. A research report is a communication device. If you fail to consider the audience to whom you are writing, the information contained in the report will not be used to best advantage. For example, highly statistical treatments of results simply confuse many managers. The report should be presented with the reader in mind.

Organize the report logically. A poorly organized report can render the information within it useless. A well-organized report helps the reader follow the logic of the investigation and the conclusions. The flow of the report should be logically outlined before the writing begins, to ensure that the information is presented in an easy-to-follow fashion. Different organizational styles are preferred by various research organizations. Make sure the one you choose to follow presents the results logically.

Watch your writing style. This point is often overlooked. A research report that is written poorly and has gross grammatical errors, misspelled words, and typographical errors is a disgrace. It not only reflects a lack of care on the part of the researcher, but it also makes the reader suspect the quality of the research results. Do not embarrass yourself or the firm by producing a sloppy report.

In this regard, software writing aids can be of tremendous help in improving one's writing style. Note, however, that it is still up to the writer to produce the best document possible. Some suggestions of the software may not be appropriate for the audience or your purposes. The researcher must make intelligent decisions about writing style that are consistent with the goals and purpose of the study.

Note the limitations of the study. A good research report notes the study's limitations. It just makes good sense for limitations to be expressly stated. It is good for the researchers because it protects against the improper use or application of the study's findings. It is also good for managers because it allows them to see the imperfections or inadequacies in the results. In essence, it serves to protect both parties in the use of research for decision-making purposes.

Be succinct and visual. The final guidelines relate to the way in which the report is presented. A succinct report is always desirable over a long and wordy presentation. It is important here to stress that presentations should be easy to understand and are accompanied by conclusions that make sense. As we noted at the start of the text, managers are bombarded with all types of data. The shorter and more powerful the written presentation, the better. Similarly, the saying, "A picture is worth a thousand words," also holds true in formal research reports.[3] Use visuals and graphics, if appropriate, to convey important information to the reader.[4]

Graphics and visuals can come in a wide variety of forms: tables, diagrams, graphs, photographs. The primary purpose of visuals is to develop the reader's understanding of the information being presented. Beware of misleading and dishonest graphics that give the wrong impression of a situation. For example, Figure 17.1 shows two situations using the same data, but different

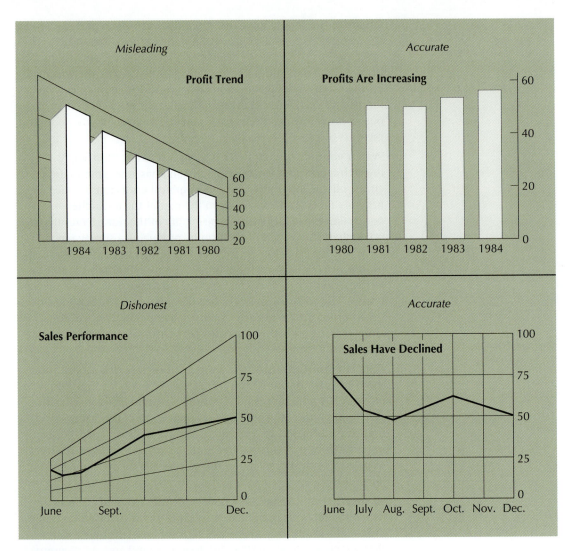

FIGURE 17.1

Examples of Misleading
and Accurate Graphical
Presentations

Source: Excerpted from Gene
Zelazny, *Say It with Charts*
(Homewood, IL: Business One
Irwin, 1991), pp. 76–77. Used
with permission.

graphics, to illustrate profit trend and sales performance for
a company. The graphics drastically change the impression
the reader gets from the data. It is crucial to be careful in constructing and evaluating graphics that accompany research
reports.[5]

Graphics software has improved dramatically in recent years.
Most of the popular word processing and desktop publishing programs have the ability to import and/or create graphics. Similarly,
most sophisticated data analysis packages have integrated graphics capabilities. The reader is encouraged to become familiar with

integrated graphic, statistical, and word processing packages that can enhance professional business research communications.[6]

ORAL PRESENTATIONS

Often, both oral and written research presentations are requested. The oral presentation is used to clarify and extend the research results and conclusions contained in the written research report. Oral presentations are desirable from the researcher's perspective because they allow the researcher to effectively communicate the essence of the study and to deal with any misconceptions or questions that may arise from reading the written report. An effective presentation can do much to ensure that the results of a study are interpreted accurately and adopted for decision-making purposes.

Oral presentations aim to convince the audience that the research conducted and the results obtained are useful and reliable. The art of making successful presentations requires skill, knowledge, a lot of practice, and a generous amount of experience. A full treatment of making successful presentations is beyond the scope of this chapter; the interested reader is encouraged to read one of the many books on the subject.[7]

An effective oral presentation should be molded around the following five guidelines.

Know your audience. This point cannot be overemphasized. An oral presentation should be tailored to the audience to which it is directed. If you do not know who you're speaking to, how can you communicate effectively? Entry-level managers have different concerns from CEOs; technical people have different reference points from "soft side" managers. Identify the needs and concerns of the audience; then develop your presentation to effectively communicate with them.

In addition, expect the unexpected. More often than not, a presentation will not go exactly as planned. For example, how would you handle the situation when the president of the company walks into your presentation? What happens when you have a hostile audience or an interrupter? Flexibility needs to be built into the presentation, and you should be mentally prepared to handle unexpected changes in audience composition.

Be organized. There is nothing more confusing than a poorly organized oral report. The oral presentation should be thoroughly organized so that the maximum amount of communication occurs during the set presentation period.

Frequently, the organization of the oral report is different from that of the written report because there is generally a limited amount of time to cover a vast amount of material. Outline the key objectives and issues you need to cover in the presentation; then design it to accomplish those objectives. As noted, however, expect the unexpected in your organization plan.

Use visual aids. Do not underestimate the importance of visual aids in oral presentations. They are effective from a communication perspective, and they are also useful to the presenter for organizing and pacing the flow of the presentation. In addition, visuals allow audiences to get more involved in the presentation and thus serve to focus their attention on the information being communicated, rather than on the presenters themselves. This can alleviate some of the pressure felt by speakers in difficult situations.

There are numerous visual aids to help the presenter of research information. Computerized slide shows, the Internet, overheads, flipcharts, posterboards, models, and movies are among the visual aids that can help direct the flow of the talk and keep the audience on track.[8] Creative use and integration of these tools will go a long way toward increasing the effectiveness of your oral presentation. Always think of how visuals can be used to help make your presentation better.

Do not read the report. This is a common mistake of novice presenters. The oral report is supposed to clarify and extend the written report, not duplicate it. Reading only serves to bore the audience and inhibit the spontaneous discussion that lends insight to the research results. The judicious use of note cards is acceptable, but it is much better to put critical information on visual aids, and use these stimulators of thought and organizing mechanisms to help you during the presentation.

In addition, overreliance on written text conveys to the audience that you do not feel comfortable enough with the material to talk about it without a structured text. This can also lead to a lifeless and uninspiring presentation, lacking in conviction or persuasive ability. Thus, the presenter should always feel comfortable enough with the material that reading is not necessary.

Practice, practice, practice. Adequate practice is essential to giving effective presentations. During practice, the presenter can test the flow, content, and communication effectiveness of the planned presentation. The presenter uses practice sessions to identify and remove distracting mannerisms from the talk, to develop an awareness of the time constraint, and generally to become comfortable with the content of the research investigation.

Start with a mental draft of your presentation. Think about how you want the presentation to proceed by integrating the topics, the visuals, and your personal plan for presenting the material. Then rehearse by speaking aloud. Finally, rehearse in front of colleagues who can give feedback. Videotapes and live audience practice are very useful in perfecting your delivery. Only through practice will you truly feel comfortable with the delivery of the information to a critical audience.

RESEARCH EVALUATION

As emphasized throughout this text, the primary purpose of business research is to provide information for decision-making purposes. Therefore, the research must be of sufficient quality to enable the manager to confidently use the data in making decisions. The primary means by which managers can ensure this is by familiarizing themselves with the components of the research process so that they can intelligently specify and evaluate specific types of research endeavors.

The specification of reserach was dealt with in Chapter 5 of this text. There, the research proposal was characterized as the formalized offer or bid for research, specifying the problem to be studied and the plan of action to solve the stated problem. The components of the research process were dealt with in the succeeding chapters.

Evaluation is closely related to specification because each procedure requires a basic knowledge of the research process. The questions posed throughout the book were designed to acquaint you with the terminology and actions essential to the design and conduct of good business research. Keep these questions in mind as you evaluate any research undertaking.

In essence, the evaluation procedure requires one to examine each aspect of the research process to ensure that errors are minimized and that the conclusions and implications of the study are supported by the data. Generally, some type of formal checklist or framework, such as that proposed in Exhibit 17.3, is recommended. Although other frameworks are possible (see the Suggested Reading section), this one outlines the major components usually found in any evaluation procedure. The primary advantage of a formalized evaluation procedure is that it helps ensure (1) that no major errors are present in the research methodology and (2) that the manager gains confidence in the study and its results.

Let us look at how this framework could be used to get an overall evaluative score for a study. First, the evaluator would go through each of the five dimensions (Problem Development, Research Design, Data Collection, Analytical Procedures, and Reporting) and would rate from weak to excellent each aspect of the research undertaking. Once the number of responses in each category is identified, the evaluator would multiply these by the scale values assigned to the evaluation dimensions and then divide by 20 (the number of items in the evaluation form). This would give a sort of grade-point average for the research report.

EXHIBIT **17.3**

A Sample Evaluation Framework for a Written Business Research Report

Institution	Principal Investigator					Project Number
Project Title						
Evaluative Dimension	**Check (✓) where applicable**					**General comments and reasons for the rating given**
	Weak	Poor	Fair	Good	Excellent	
1. Problem Development						
a. The background of the problem is fully developed.						
b. The research problem is clearly stated.						
c. Research objectives are explicitly stated.						
d. Important assumptions are identified.						
2. Research Design						
a. The research design is succinctly described.						
b. The research design is appropriate for the problem.						
c. The measurement instruments are appropriate.						
d. The sampling design is appropriate for the stated problem.						
3. Data Collection						
a. The collection procedures are described.						
b. The collection procedures are appropriate to the stated problem.						
c. The data collection procedures were implemented according to the research design.						
4. Analytical Procedures						
a. The procedures used to analyze the data are appropriate.						
b. The procedures were used correctly to analyze the data.						
c. The results of the analysis are stated clearly.						
5. Reporting						
a. Conclusions are unambiguously stated.						
b. The data support the conclusions.						
c. The report is written clearly.						
d. The report is logically organized.						
e. The limitations of the study are clearly stated.						
f. The report contains all the necessary data for evaluation purposes.						
6. Overall Evaluative Score						
Please provide any additional comments or remarks applicable to the research report.						
Reviewer's Name	Reviewer's Signature					Date

For example, if we assigned the scale values of 0 = weak, 1 = poor, 2 = fair, 3 = good, and 4 = excellent, our evaluative score would be as follows:

	Scale Value	×	Number of Responses	
Weak	0	×	0	= 0
Poor	1	×	2	= 2
Fair	2	×	4	= 8
Good	3	×	8	= 24
Excellent	4	×	6	= <u>24</u>
				58/20 = <u>2.9</u>

Referring to our original scale values, this would be considered a research report of good quality (3 = good).

This simple example illustrates the calculation of an unweighted evaluation score. Other evaluation schemes might reflect the manager's preferences on the importance of specific aspects of the business report. For example, if the evaluator felt that research design items 2a through 2d in Exhibit 17.3 were twice as important as other aspects in the evaluation, the items could be weighted by a factor of 2. Thus, the evaluation scheme presented can accommodate individual manager preferences.

Again, other types of evaluative schemes are possible. The important point is that an honest attempt has been made to assess the general quality of the research findings. You should no more use low-quality data for decision-making purposes than put low-quality fuel in your Porsche. *An evaluation of any research results should be made before the data is used to make decisions.* This has been the thrust of this text; it should be considered a basic principle in the use of research in the managerial setting.

COMMUNICATION CHALLENGES IN THE FUTURE

In the year 2000, multinational corporations controlled approximately half of the world's assets. In addition, the work force has become more culturally diverse in a complex economic environment. Increased awareness of differences in communication styles and cultures is necessary to ensure successful communication. Otherwise, valuable information will be lost because of an ethnocentric view of the world.

It is cultural differences that distinguish groups of people. The perception of time, the nature of trust, the display of emotion, and the meaning of silence are but a few of the cultural variables that can hinder effective communication. We all must become more aware of the meaning of words, actions, and symbols in cultures other than our own.

EXHIBIT 17.4

Examples of
International Business
Blunders

Source: Vern Terpstra and
Kenneth David, *The Cultural
Environment of International
Business* (Cincinnati, OH:
South-Western Publishing,
1991), p. 21; and Minerva
Neiditz, *Business Writing at Its
Best* (Homewood, IL: Richard
D. Irwin, 1994), pp. 209–222.

A major U.S. pharmaceutical company launched a new painkiller colored green
in a foreign country. The market associated the color green with disease. The
product did not sell.

Tomato paste became tomato *glue* when mistranslated into Arabic.

Ford had problems with Mexican perceptions of three of its models. Its Fiera
truck was perceived an "ugly old woman," its Caliente car was translated as
"streetwalker" in Mexican slang, and the Pinto as "a small, male appendage."

Most Americans consider a firm handshake desirable, whereas Japanese can view
this as aggressive and impolite.

The American "OK" sign (index finger and thumb together forming a circle) can
mean a bribe is asked for in Japan; it is an insulting gesture in Brazil.

It is strongly recommended that managers and researchers alike become
more familiar with cultural differences, particularly when conducting, interpreting, and implementing international business research. Otherwise, tremendous international blunders can occur, as exemplified in Exhibit 17.4.[9]

MANAGERIAL CONSIDERATIONS

The managerial considerations in the area of research reporting are particularly important. It is at this point that managers actually receive information that is potentially useful for decision-making purposes. Also, it is
the responsibility of managers to clearly specify how they want the research
reported. Four concerns should be addressed, relating to the quality and types
of information that are to be obtained from the research report:

Define what it is you expect in a research report. If you have a particular format or at least an outline of the types of information you want in
the research report, specify it. Business research exists to provide information for decision making. The manager should clearly define what
output is wanted from the research process. Do you want oral and written reports highlighting only the salient results, or would you prefer
comprehensive, in-depth reports that include detailed statements of the
methodological issues in the project?

Require that the limitations of the research be clearly stated. All
research has its limitations. Managers should require researchers to
explicitly state the limitations of their study because, if managers fail to
take these parameters into consideration when making decisions, they
could make a serious error. Thus, a statement of limitations should be
required in every research undertaking.

Demand excellence in research reporting. Only managers can ensure
that research reports are of the finest quality. Managers should make
known exactly what type of information they want from the research
function. Well-written, grammatically correct reports should be the

EXHIBIT **17.5**

Key Managerial
Questions Pertaining to
Research Reporting

1. Have you previously specified what it is you want in the research report?
2. Are the limitations of the study clearly noted so that effective decision making can ensue?
3. Are both oral and written research reports required to ensure that the maximum amount of information is extracted from the research endeavor?
4. Are the research results of sufficiently high quality to support decisions?
5. Have you considered if any cultural considerations will affect the interpretation and use of the study's results?
6. Has the research been formally evaluated to assess its usefulness for the designated decision-making situation?

norm rather than the exception. In other words, demand excellence in all aspects of research reporting so as to make the assimilation of the information as easy as possible.

Require that research be formally evaluated. Without formal evaluation, research is often accepted and acted upon without serious review of its merits. All research should be evaluated before being acted upon. This is particularly true in light of a whole new domain of problems associated with multiculturalism and globalism in the research arena. Many blunders have occurred because of the unchallenged acceptance of research results from very respected business researchers.

These basic concerns are relatively simple, but recognizing their importance is vital. You, as a manager, should become involved in the research reporting process to ensure that you obtain the best possible information from which to make decisions. These considerations, coupled with the questions posed in Exhibit 17.5, should enable you to obtain the most from the research requested for your organization.

SUMMARY

Research reporting is a primary communication link between managers and researchers. The research report, whether oral or written, communicates the details of the research project to managers who will ultimately use the information for decision-making purposes.

The written research report is the end product of the research process. It is a formal statement of the background of the problem being investigated, the nature of the study, and the relevant findings and conclusions drawn from the research. A basic outline for the business research report is presented.

Five major guidelines help in the development of the written report: (1) know your audience, (2) organize the report logically, (3) watch your writing style, (4) note the limitations of the study, and (5) be succinct and visual.

The oral research report is a communication device that is used to clarify and extend the research results and conclusions contained in the written report. Five guidelines are suggested in the development of effective oral reports: (1) know your audience, (2) be organized, (3) use visual aids, (4) do not read the report, and (5) practice.

The discussion then turned to the evaluation of research. All research should be formally evaluated for its value in the decision-making process. A framework was proposed and four managerial considerations were explored with regard to the research report and its evaluation.

In conclusion, previous chapters were designed to familiarize you with the process of business research. The various elements in the process were discussed and analyzed in terms of the basic concerns of managers. The text took a normative approach to the subject of research. We have raised significant managerial considerations with regard to the quality of data for decision-making purposes.

We hope that you have gained (1) an appreciation for the value of research itself, (2) an appreciation of the complexity and difficulty in conducting good research, and (3) the means to specify and evaluate research in a business setting. If these goals are accomplished, our initial objectives in writing this text will have been met.

DISCUSSION QUESTIONS

1. What are the major component parts of a well-thought-out and complete research report? Present an outline and a brief description of the components you believe are necessary in a business research report.

2. What is an executive summary? What information should be presented in an executive summary?

3. What guidelines could you propose that would help a researcher write a high-quality business research report? Be specific and enumerate all that you can think of to help the researcher.

4. What are the essentials of a good oral research report? Again, present some guidelines that you think are essential to the development and delivery of sound oral research reports.

5. Suppose you are a manager who is to receive a number of research reports on projects you requested. What can you do to ensure that the projects deliver the information that is required for decision-making purposes? Be specific.

6. The quality of the research report is often directly tied to the quality of the research proposal. Discuss the validity of this statement. Provide specific examples to support your contentions.

7. Use the evaluation framework shown in Exhibit 17.3 to evaluate any research report you choose, such as a state government report of a

consulting nature that is publicly available. Be sure to add any criteria you consider necessary for the type of report you are reviewing.

8. Communication challenges abound in a shrinking world. Identify four examples of how differences in culture can, or have, influenced the interpretation of a specific action or word.

9. With the explosive growth of the Internet, the possibility for delivering and producing "interactive" reports to clients is becoming a reality. Use the Internet to research this topic and produce a summary report outlining the possibilities and use of this new form of research reporting. Be sure to reference your work in this summary.

NOTES

1. For other suggested formats and in-depth discussions of the content of such reports, see Carol M. Lehman and Deborah Daniel Dufrene, *Business Communication: Anniversary Edition*, 13th ed. (Cincinnati, OH: South-Western, 2002); and Raymond V. Lesikar and Marie E. Flatley, *Basic Business Communication*, 9th ed. (New York: McGraw-Hill, 2002).

2. The names, addresses, and figures in this report have been disguised for confidentiality reasons. The authors wish to thank DATAFAX for permission to reproduce the front part of this research report.

3. See, for example, Kevin P. Lonnie, "Researchers Must Show Their Findings to Clients," *Marketing News* (11 May 1998), p. 4.

4. For sources on the use of graphics and visuals, see Edward R. Tufte, *The Visual Display of Quantitative Information*, 2nd ed. (Cheshire, CT: Graphics Press, 2001); and Robert L. Harris, *Information Graphics: A Comprehensive Illustrated Reference: Visual Tools for Analyzing, Managing, and Communicating* (New York: Oxford University Press, 1999).

5. For further discussion on choosing and using charts and graphics, see Gene Zelazny, *Say It with Charts*, 4th ed. (New York: McGraw-Hill Trade, 2001).

6. To check on the availability and compatibility of various software programs, see *The Software Encyclopedia 2000* (New Providence, NJ: R. R. Bowker, 2000). Updated regularly.

7. See Jackie Jankovich and Elaine LeMay, *Presentation Success: A Step-by-Step Approach*, (Cincinnati, OH: South-Western, 2001); and Kitty O. Locker and Stephen Kyo Kaczmarek, *Business Communication: Building Critical Skills*, 2nd ed. (New York: McGraw-Hill, 2004).

8. See Diane Diresta, *Knockout Presentations: How to Deliver Your Message with Power, Punch, and Pizzazz* (Worcester, MA: Chandler House Press, 1998).

9. See, for example, Michael R. Czinkota, Ilkka A. Ronkainen, and Michael H. Moffett, *International Business Update 2003*, 6th ed. (Cincinnati, OH: South-Western, 2003); and Charles Hampden-Turner and Fons Trompenaars, *Riding The Waves of Culture: Understanding Diversity in Global Business*, (New York: McGraw-Hill, 1998).

SUGGESTED READING

Written Research Reports

Lehman, Carol M., and Deborah Daniel Dufrene, *Business Communications: Anniversary Edition*, 13th ed. (Cincinnati, OH: South-Western, 2002).

This text focuses on the development of effective written and oral communication skills in a business setting. A continuation of a classic text.

Lesikar, Raymond V., and Marie E. Flatley, *Basic Business Communication*, 9th ed. (New York: McGraw-Hill, 2002).

This text presents an overview of contemporary research reporting integrating technology whereever applicable. Good reading and reference material for the active presenter.

Oral Presentations

Lehman, Carol M., *Creating Dynamic Multimedia Presentations*, 2nd ed. (Cincinnati, OH: South-Western, 2003).

    This text demonstrates how to use Microsoft's PowerPoint to create dynamic and effective oral presentations. A useful book when trying to maximize the impact of one's oral presentation.

Locker, Kitty O., *Business and Administrative Communication*, 6th ed. (New York: McGraw-Hill, 2003).

    The text presents an overview of both oral and written business communications. An engaging reading style with examples makes this a highly useful reference.

Research Evaluation

Katzer, Jeffrey, Kenneth H. Cook, and Wayne W. Crouch, *Evaluating Information: A Guide for Users of Social Science Research*, 4th ed. (Reading, MA: Addison-Wesley, 1997).

    This paperback is written from the research user's standpoint. It provides a very good supplement to this chapter for those who are interested in understanding and evaluating research reports.

Communication Challenges in the Future

Lewis, Richard D., *When Cultures Collide*, rev. ed. (London; Naperville, IL: Nicholas Brealey, 2000).

    This book provides a global and practical guide to working and communicating across cultures, with penetrating insights into how different business cultures accord status, structure their organizations, and view the role of leaders.

Morrison, Terri, *International Traveller's Guide to Doing Business in Europe* (London: Macmillan, 1997).

    The guide has information arranged in an easy-to-use who/what/when/where/why format. It provides vital data on each country in the European Union.

Morrison, Terri, Wayne A. Conaway, and Joseph J. Douress, *Dun & Bradstreet's Guide to Doing Business Around the World,* rev. ed. (Upper Saddle River, NJ: Prentice-Hall, 2000).

    An interesting guide that provides essential information about possible communication challenges when conducting business around the globe.

Research Cases

Amtech, Inc
Training Needs Analysis

Checker's Pizza
Employee Retention

Fastest Courier in the West
Selecting a Service

Glenco Bonus Program
Employee Absenteeism

Ryder Appraisal District
Property Tax Assessment

The Keels Agency
Advertising Media Selection

**River End Golf and
Entertainment Center**
Market Assessment Study

AMTECH, INC.

Training Needs Analysis

Amtech, Inc. is in the business of selling sophisticated computerized point-of-sale cash registers, cash register printers, computer-printer software and interfaces, printer supplies, and hands-on training to commercial accounts. Amtech's owners chose their market niche to be smaller companies that appear ready for a large scale expansion of their business. Smaller companies need, and are willing to pay for, the personal attention and customization of the hardware, software and training that Amtech offers. Once committed to a particular configuration of hardware and software, these clients seldom reconsider their alternatives when expansion occurs. This is a high margin business and no client is considered too small to deserve attention. Small clients often blossom into larger businesses, and their growth typically signals the expansion of Amtech's sales as well.

The typical marketing plan is centered on identifying customer needs and then developing a plan to satisfy those needs. Amtech's first contact with potential clients demonstrates their cash registers' capabilities and the extent to which software can be customized to meet a particular client's needs. Amtech account executives emphasize the importance of customizing both their software and training to meet clients' needs as the features that distinguish Amtech from its competitors. They then request an opportunity to distribute a questionnaire to key personnel in the client organization in order to develop a customization plan that "fits" the client's needs and to make a presentation to the client based on their findings.

The questionnaire asks participants about important business issues needed to customize the software, and the backgrounds of those persons who will use the high tech equipment in order to customize the training program. Part I of the questionnaire deals with technical business issues that reflect the nature of the client's business and business plan. For example, information is collected regarding their accounting system, the inventory monitoring, control and reordering functions, the extent to which the client wishes to create a data base of client purchases as a marketing tool, and the selection and printing of in-store coupons on the back of the sales receipt designed to elicit future purchases that complement current purchases. These business functions, among others, are discussed with the management of the organization and go into designing the software to meet client's business needs. All senior level management and headquarters' professional staff (*e.g.*, accountants) complete this part of the questionnaire.

The second part of the questionnaire is aimed at understanding the backgrounds of those persons who will use the equipment. This is important information, since the training program will have to be designed to meet the needs of those persons who will complete the questionnaire. If the typical employee has limited background in the use of computers, then a longer, more costly training program will need to be designed. On the other hand, if their background with computers is substantial, then a mere orientation to the use of this equipment might suffice. This is an important consideration from more than the training costs involved, since a great deal of evidence that exists suggests that poorly trained employees either become frustrated and quit, or if they stay, never help the company reap the full benefit of the technology it has purchased.

Wendy Cambridge began her career at Amtech less than two years ago and has risen quickly from marketing assistant to assistant account manager for some of Amtech's larger clients. She was recently given her first opportunity to co-lead a sales team for a new potential client, Grass-Roots, Inc.

Grass-Roots fits the profile of an ideal customer. They are setting up a small chain of plant and garden stores in a large Southern city and have plans to expand their company to ten additional cities over the next five years. Each store location will be surveyed to insure high levels of customer service. Operations staff will work individually with customers, from when they first walk in the store through check-out.

The Amtech sales team has made their initial presentation to Grass-Roots, who have agreed to allow Amtech to do a follow-up survey to see exactly how Amtech can help them. Amtech knows they have their "foot in the door" and can make a potentially lucrative sale depending on how good a job the sales team does from this point on.

Wendy Cambridge was put in charge of administering the questionnaire and preparing a presentation on Grass-Roots' training needs. This is her first time to head any part of a sales effort and she wants to do it well. While she has never interpreted survey results, written a proposal or made a sales presentation on her own, she has worked directly for Mary Gordon, Amtech's Corporate Sales Manager, on these assignments and feels confident in her ability to take on these responsibilities.

Wendy administered the Amtech questionnaire to all current management and operations employees in their headquarters and three retail store locations. In total, she met with and distributed the questionnaire to all six corporate management and professional staff and, in the stores, to a total of nine management and 33 operations staff. Each store employs a manager and two assistant managers, plus an operations staff of 11 persons. She asked only headquarters' managers and professional staff to complete Part I of the questionnaire; everybody was asked to complete Part II. All employees completed the questionnaire on company time, in small group settings. Wendy prepared a computerized file containing the survey data and then analyzed the data regarding employee background using a common PC statistical software package.

Wendy then examined the results from Part II of this questionnaire administration, and based on her analyses, developed a training proposal. In preparation for her presentation to Grass-Roots, she also prepared charts (shown below) that summarized her findings.

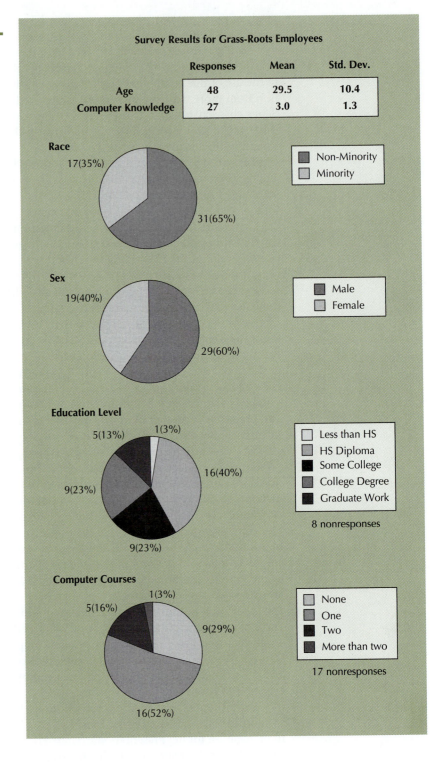

Survey Results for Grass-Roots Employees

	Responses	Mean	Std. Dev.
Age	48	29.5	10.4
Computer Knowledge	27	3.0	1.3

Race

17(35%)

31(65%)

- Non-Minority
- Minority

Sex

19(40%)

29(60%)

- Male
- Female

Education Level

1(3%)
5(13%)
16(40%)
9(23%)
9(23%)

- Less than HS
- HS Diploma
- Some College
- College Degree
- Graduate Work

8 nonresponses

Computer Courses

1(3%)
5(16%)
9(29%)
16(52%)

- None
- One
- Two
- More than two

17 nonresponses

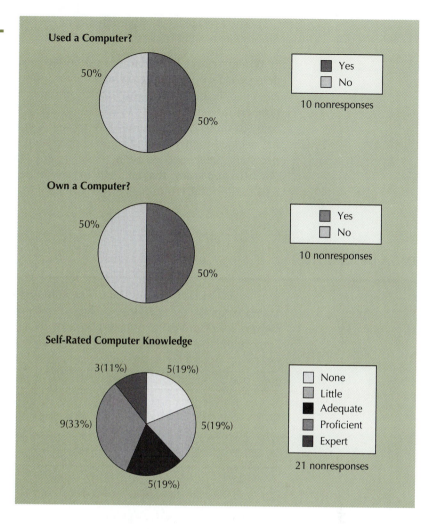

Used a Computer?

50%

50%

Yes

No

10 nonresponses

Own a Computer?

50%

50%

Yes

No

10 nonresponses

Self-Rated Computer Knowledge

3(11%) 5(19%)

9(33%) 5(19%)

5(19%)

None

Little

Adequate

Proficient

Expert

21 nonresponses

In brief form, Wendy suggested that employees of Grass-Roots were reasonably computer literate and, as a result, would need only minimal training—possibly little more than an orientation program, a brief tutorial on the use of the cash registers, and a question and answer session. She based her conclusion on survey results that she interpreted as follows:

(1) Employees have a relatively high level of education.

(2) More than half of the employees have taken at least one computer course.

(3) More than half of the employees use or have used a computer on their present or prior jobs.

(4) Half of the employees own a PC.

(5) The average level of self-rated computer sophistication is at the adequate level of understanding.

While Wendy's background is not in business, she has a track record of doing good work and of growing as her responsibilities are increased. Like most people, however, Wendy makes mistakes in her early efforts on new projects. When given feedback, she quickly "catches on" and can be counted on to learn from her lessons well. For this reason, Mary Gordon has decided to give Wendy some backup help to insure that she has the necessary "safety net" on her new assignment. Mary thinks a lot of Wendy's potential and knows that a little help early in her career will help make her a successful contributor for years to come.

DATA DESCRIPTION

File AMTECH.DAT on the Data Disk contains the coded data from the survey of Grass-Roots employees. Data are recorded in the manner depicted below. A "*" indicates non-response or missing information. (See the Important Note below.)

ID	JOB	RACE	AGE	SEX	ED	COURSES	USED	OWN	KNOW
1	3	2	24	1	3	1	2	2	2
2	1	1	42	2	5	1	2	2	4
3	3	2	19	1	*	*	*	*	*
4	3	1	22	2	2	*	1	1	*

The variables are defined as follows:

ID:	Survey identification number.
JOB:	1 = Headquarters Management and Professional Staff, 2 = Store Management, 3 = Store Operations Staff.
RACE:	1 = non-minority, 2 = minority.
AGE:	Age on last birthday.
SEX:	1 = female, 2 = male.
ED:	1 = less than a high school diploma, 2 = high school diploma, 3 = some college work,

4 = college degree,

5 = graduate work or degree.

COURSES: Number of computer courses taken.

USED: Has the employee used a computer?
0 = no,
1 = yes.

OWN: Does the employee own a computer?
0 = no,
1 = yes.

KNOW: Self-evaluation of computer knowledge using a rating scale, where
1 = no knowledge,
2 = little knowledge,
3 = adequate knowledge,
4 = better than adequate knowledge,
5 = expert level of knowledge.

IMPORTANT NOTE: The symbol "*" (asterisk) is used in the data file AMTECH.DAT to denote non-response or missing data. Consult your statistical software documentation for a list of valid missing data codes. If your statistical software package does not accept "*" as a valid missing code, you have two options. Your statistical software package may allow you to specify another missing code. Otherwise, you will need to change each occurrence of "*" in the data file to a missing symbol that is acceptable to your particular statistics package. This can be done by editing the file AMTECH.DAT within a word processing package. See your instructor if you need assistance.

ASSIGNMENTS

1. Conduct a secondary data search to more fully understand the nature of the research problem faced by Wendy? What should she be looking at and what should her search strategy be?

2. Using the framework outline in Exhibit 5.2, develop a complete research problem definition for obtaining the information needed from Grass-Roots for Amtech.

3. Develop a Type I research proposal (See Chapter 5) for the problem definition in assignment (2) above.

4. Is the study as conducted capable of answering Wendy's questions? If not, why not? How could the study be improved?

5. You have been assigned to be Wendy's "safety net." Look over the survey results and judge whether or not Wendy has correctly analyzed and interpreted the results from this survey. You'll find the data in

AMTECH.DAT on the Data Disk. A description of this data is provided. Prepare a brief report for Mary Gordon that reflects your interpretation of the survey's results and then make a recommendation regarding your support for Wendy's training proposal. Finally, prepare a chat that reflects the computer literacy of relevant Grass-Roots personnel. The questions below will assist you in your analysis of the data. Use important details from your analysis to support you recommendations.

a. Was Wendy's interpretation of the survey results flawed? Explain.

b. Describe the "computer literacy" of those employees who will use the new point-of-sale cash registers. Be sure to describe (a) the sample of employees on which you based your interpretation and (b) the statistical results of your analyses.

c. Based on your analyses, what is your recommendation regarded needed training for Grass-Roots employees?

d. Explain the business implications of your recommendations versus that of Wendy Cambridge.

e. Prepare a chart, in graphic or tabular form, that summarizes the results from your analyses of the survey data and that can be used to make key points in the sales presentation. Attach your chart to this assignment.

Employee Retention

Checker's Pizza is a chain of 40 pizza restaurants in the New England area. Checker's started in 1975 and specialized in making pizzas "from scratch." Each pizza was covered with a special sauce, made from a "secret" family recipe, that kept customers coming back for more. But Checker's, like other pizza restaurants in the 80's, had to respond to the wave of fast food pizza restaurants and delivery services that seemed to corner the market.

Checker's responded to this threat by changing the way they did business. Not only did they expand their product line and start delivering pizzas like their competitors, they also started making pizza using the same "fast food" procedures as their competitors. Old customers were acutely aware that the pizzas were different. But, while the pizzas are now made using a conveyer oven from dough that is made in a central kitchen, frozen and then delivered to each restaurant, the sauce had not changed at all. And that was the key to their growth. While old customers knew that the pizzas were not as good, new customers only knew that they were better pizzas than Checker's competitors made.

Even still, the pizza business had fallen on hard times by the late 90's. Competition for customers in the fast food business had soared and the weak economy at that time had slowed business for everybody. It was in this business climate that Terri Chester was hired as General Manager of the Checker's Pizza chain. She joined the company almost two years ago. Her job was to restore the market share and profit of this business.

Terri attempted to address her mission in a systematic fashion. She worked with her executive team to identify and prioritize key issues, and then set up a series of task forces to deal with those problems that required immediate attention. Among the key problems identified was one associated with employee turnover. The turnover rate for store personnel was considered extremely high. Terri knew that such a high turnover rate not only impacted expenses due to costs of replacing persons who quit, but also impacted customer service as well. It's hard to serve customers well with a continual flow of recently hired staff who have yet to fully learn their jobs.

Checker's Pizza, like all fast food restaurants, employs teenagers in most positions. They work as counter persons, wait staff, cooks, and delivery persons. As is typical with this employee population, turnover tends to be high. In the fast food industry, restaurant turnover for younger employees typically ranges from 200 to 600 percent per year! Given such high turnover rates in this industry, replacement costs can add up to a considerable sum. These costs reflect dollars actually spent on recruitment, interviewing, hiring and training, as well as non-tangible costs associated with having staff who haven't climbed the learning curve to fully pull their own weight on the job. The employee turnover task force estimated that it costs Checker's approximately $250 to replace someone who quits.

Over the past three years, turnover at Checker's Pizza has occurred at an average annual rate of 350 percent across all of its restaurants. This translates into newly hired store personnel who stay, on average, approximately 90 days before terminating. Thus, every time someone leaves, the position is refilled by a new employee who, on average, will stay only 90 days. This, in turn, means that on average each store replaces 48 employees over the course of a year since each store has twelve employees. Each time someone leaves, Checker's incurs the costs of recruiting, hiring, training, and developing his or her replacement. At a $250 replacement cost, this adds up to $12,000 in annual replacement costs per store and $480,000 per year across the restaurant chain!

Terri Chester hired a consultant, Dr. Shannon Train, from a local university to conduct a study to try to identify why people were leaving. The consultant contacted a sample of former employees and conducted a brief telephone interview with each. While a large list of specific reasons were identified, the one theme that recurred was the treatment employees received from store management. Leavers said they were often treated "like children" by their managers. Store managers reacted quickly and harshly whenever employees did anything that did not reflect store policy, even for newly hired persons who had yet to fully understand what store policy actually meant. Further, they seldom involved staff in any discussion about problems.

Dr. Train suggested that Terri focus her efforts on store management's treatment of staff, particularly newly hired staff. He said there are a number of approaches commonly used to deal with this problem, and recommended two in particular. First, he suggested that store managers receive a two-day human relations training program designed to not only impress upon them the importance of considering interpersonal relations, but to help managers develop important interpersonal skills that will facilitate good working relationships. The second alternative that Dr. Train recommended was to give store managers an incentive for keeping staff from quitting. He had heard that other companies in the fast food industry have used this approach with some success. He suggested that managers be given $100 (40% of the replacement costs that would have been spent) for keeping employees twice as long than they currently stay employed. Thus, store managers would receive a $100 bonus for each new hire who stayed at least six months (i.e., 180 days). The cost of providing the two-day interpersonal skills workshop would be $500 per manager, or $20,000 for all 40 store managers.

Dr. Train pointed out to Terri that she might want to consider doing both the training and bonus. He noted that a pilot study could help answer the questions without going into the same degree of expense as would be required in a full-scale implementation. As Dr. Train put it:

> If you really hope to put a dent in the retention problem, you might need to both conduct the training and provide the bonus. Think of it this way. To get managers to change their behavior, you need to not only give them a reason to change, but the skills to do so. As I tell my students, empowerment requires enablement. Now, managers may well know exactly how to treat their people to keep them from leaving in droves. If this is true, then you only need to give them some incentive to do so. On the other hand, while your managers may have good interpersonal skills in general, they may not have the needed interpersonal skills to work with teenagers. So, maybe all that's needed is the skills training. It's hard to say, without looking into it more carefully, whether either one or the other approach will be enough to do the trick or whether both might be necessary.

While Terri Chester liked both ideas, she wasn't ready to proceed simultaneously with both approaches. It would require a substantial commitment of money at a time when money wasn't easy to put into the budget. When she stated her conclusion to Dr. Train, he suggested that Terri run an experiment to see if either, or both, approaches he suggested made an impact on retention before she committed funds company-wide to either approach. That way, she'll know what might work before she commits the funds to both approaches. She liked the idea!

Dr. Train designed an experiment in which 40 store managers were randomly chosen to be in one of four "experimental" conditions that reflected all combinations of the training and bonus conditions: (1) no training-no bonus, (2) no training-bonus, (3) training-no bonus, and (4) training-bonus. Ten store managers were assigned to each condition.

Training and a discussion of the bonus program were done by the end of the month. The actual experiment began the following month. All restaurant staff hired during the month were included in the study. The tenure of each of these persons was followed for 12 months. For each employee in the study, tenure was recorded in days, from the date of hire until the date of termination or until the date the study ended. Thus, tenure could range from one day up to 365 days for those persons who were still employed when the study ended.

DATA DESCRIPTION

The CHECKERS.DAT file on the Data Disk contains the results from the experiment aimed at determining whether the training program and/or the bonus incentive has an effect on employee tenure. Data are recorded in the manner depicted below.

ID	Training	Bonus	Tenure
001	0	0	14
002	0	1	178
003	1	1	164
004	0	1	97
005	1	0	116

The variables are defined as follows:

ID: Sequential identification numbers assigned to employees included in the study.

Training: 1 = training given to employee's store manager, 0 = no training given.

Bonus: 1 = bonus given to employee's store manager, 0 = no bonus given.

Tenure: Number of days the person was employed, from date of hire to the date of termination or the date the study ended.

ASSIGNMENTS

1. Conduct a secondary data search to more fully understand the nature of the research problem faced by Terri? What should she be looking at and what should her search strategy be?

2. Using the framework outline in Exhibit 5.2, develop a complete research problem definition for obtaining the information needed from Checker's Pizza to help them make their decision.

3. Develop a Type I research proposal (See Chapter 5) for the problem definition in assignment (2) above.

4. Is the study as conducted capable of answering Terri's questions? If not, why not? How could the study be improved?

5. As of today, all data for this staff retention study have been collected and recorded in computer ready form. As a member of the retention task force, you have been asked by Terri Chester to examine these data. You will find the data in CHECKERS.DAT on the Data Disc. A description of this data set is provided. Your task is now to make a recommendation to Terri regarding whether or not to use either, or both, of the training and bonus interventions to stem the flow of turnover among restaurant staff. The questions below will assist you in your analysis of the data. Use important details from your analysis to support your recommendations.

a. Does either the training of the bonus program or both relate to employee retention? Explain your answer. Attach any supporting analysis.

b. Draw a graph depicting the effect of the training and bonus programs on employee tenure. Explain, in words, the nature of the relationship you observed.

c. Make specific recommendations to Terri regarding the use of the training and bonus programs.

FASTEST COURIER IN THE WEST

Selecting a Service

The law firm of Adams, Babcock, and Connors is located in the Dallas-Fort Worth metroplex. Randall Adams is the senior and founding partner in the firm. John Babcock has been a partner in the firm for the past eight years, and Blake Connors became a partner just last year. The firm employs two paralegal assistants and three secretaries. In addition, Bill Davis, the newly hired office manager, is in charge of day-to-day operations and manages the financial affairs of the law firm.

A major aspect of the law firm's business is the preparation of contracts and other legal documents for their clients. A courier service is employed by the firm to deliver legal documents to their many clients as they are scattered throughout the metroplex. The downtown centers of Dallas and Fort Worth are separated by a distance of approximately 30 miles. With the large sizes of these cities and their associated heavy traffic, a trip by car from the southwest side of Fort Worth to the northeast side of Dallas can easily take longer than an hour. Due to the importance of the legal documents involved, their timely delivery is a high priority item. At a recent partner's meeting, the topic of courier delivery came up.

Adams: "Recently, we have received a couple of complaints from some of our best clients about delayed contract deliveries. I spent the better part of an hour yesterday afternoon trying to calm down old man Dixon. He claims that if those contracts had arrived any later, his deal with the Taguchi Group would have fallen through."

Connors: "Well, it wasn't our fault. Anne had the contracts all typed and proofread before nine in the morning."

Adams: "No, no. Everything was handled fine on our end. The delay was the courier's fault. Something about a delay in the delivery . . ."

Babcock: "Metro Delivery has always done a good job for us in the past. I am sure that these are just a few unusual incidents."

Connors:	"On the other hand, it could be an indication that their service is slipping."
Adams:	"In any event, we cannot afford to offend our clients. No one is perfect, but it only takes one or two bad incidents to lose important clients. At least two new courier services have opened in the metroplex during the last two years. I hear good things about them from some of my friends. The question is: Should we keep using Metro or consider using one of these other services?"
Connors:	"How would you suggest that we make this decision?"
Babcock:	"Why not give each one a trial period and choose the best performer?"
Adams:	"Great idea! But how would you decide who's best?"
Babcock:	"Well, obviously the choice boils down to picking the fastest courier service. Given our recent problem, we also want to avoid the infrequent, but costly, delayed deliveries. Delivery cost is an important secondary criterion."
Connors:	"Why not let our new office manager run this little "contest" for a few weeks? Bill normally keeps fairly detailed information about contract deliveries anyway. As the need for deliveries arises, he can rotate through all three couriers."
Adams:	"Let's be sure not to let any of the couriers know about the contest; otherwise, we may not see their typical performance. We'll take up this topic again after Bill has collected and analyzed some data and is ready to make a presentation."

During the past month, Bill Davis has kept detailed records of the deliveries made by each of three courier services: DFW Express, Carborne Carrier, and Metro Delivery (the courier presently used by the law firm). Due to the importance of the documents delivered, a courier is required to phone the law office as soon as a delivery has been made at its destination. For each delivery, Bill's data set contains: the courier used, the pickup time, the delivery time, the mileage of the delivery, and the cost of the delivery. Each of the courier services charges a flat fee plus a mileage charge for each delivery. These charges vary from courier to courier.

DATA DESCRIPTION

For each delivery, Bill Davis' data set contains the courier used, the pickup time, the delivery time, the mileage of the delivery, and the cost of the delivery. The first few entries in the database are shown below. The entire database is in the file COURIER.DAT on the Data Disk. It contains information on 182 courier deliveries.

Courier	Pickup Time	Delivery Time	Mileage	Cost
1	13	14	7	16.55
3	20	51	20	26.50
2	22	33	12	19.00
3	11	47	19	25.60
3	17	18	8	15.70

The variables are defined as follows:

Courier:
1 = DFW Express,
2 = Carborne Carrier,
3 = Metro Delivery.

Pickup Time: Time in minutes from when the order is phoned in until a courier agent arrives.

Delivery Time: Time in minutes that it takes for the documents to be delivered to the destination from the firm.

Mileage: Distance in miles from the law firm to the destination.

Cost: Charge for the delivery. Each of the courier services charges a flat fee plus a mileage charge. These charges vary from courier to courier.

ASSIGNMENTS

1. Using the framework outline in Exhibit 5.2, develop a complete research problem definition for obtaining the information needed by Bill Davis to help him make his recommendation.

2. Develop a Type I research proposal (See Chapter 5) for the problem definition in assignment (1) above.

3. Is the study as conducted capable of answering Bill's question? If not, why not? How could the study be improved?

4. Develop a preanalytic plan for Bill. What statistics and variables should he use to make his decision?

5. As office manager, Bill Davis is responsible for making the decision as to which courier service will be given the exclusive contract. Using the data set stored in the file COURIER.DAT on your Data Disk, assist Bill in choosing among the three courier services and defending his choice to the partners of the firm by performing a statistical analysis of the data. Write a short report to the partners giving your final recommendation. The questions below will assist you in your analysis of the data. Use important details from your analysis of the data to support your recommendations.

a. Based on your analysis, which of the three courier services would you recommend to the law firm? Why?

b. Which variable (or variables) did you use as the main criterion for making your decision? Explain. Why?

c. Which statistical tools did you use to summarize the data? Attach any relevant output.

d. Which statistical inference method(s) is most useful for comparing courier services? Describe the results of this analysis.

e. Give a brief outline of the steps you took in your data analysis. Point to specific discoveries that you made during the analysis, which suggested what to do next and eventually, led to your conclusions and recommendation. Attach a copy of any relevant output.

Employee Absenteeism

Everybody at Glenco, a moderate-sized manufacturing company, is talking about the new incentive bonus program. Last August, John Marchant, CEO of Glenco, announced a new program designed to encourage the best efforts of persons on the production floor. While production employees could count on regular pay raises as usual, Glenco would now be offering a $500 bonus to the top 20 percent of all production workers. Thus, of the 155 employees, 31 (or so) persons could expect to get a nice surprise just in time for Christmas.

The reaction to Marchant's announcement was immediate, and mixed. Clearly, there were a number of employees who liked the new bonus plan. In fact, a few persons laid bets on being among the top 20 percent. Sam Miller, Pete Cravens, Mike Sanders and Tom Reeves decided to pool their bonuses on a new fishing boat that the four of them would share, and they actually made a down payment on it in early November. Even Harvey Fried, who seldom spoke out about anything, was overheard telling some of his friends about "... how nice it will be to finally get some recognition for all my good work around here."

Not all the talk, however, was positive. There were, of course, those persons who knew that they would never receive a bonus. They complained loudly and often about the proposed bonus plan and suggested any number of "better" uses for the money. For example, Frank Phillips thought that the money was better spent on underwriting dental insurance for everybody and Martha Renner thought the money would be better spent on day care benefits. Other employees had still other suggestions for use of the money.

Even management reaction was split. The most vocal negative reaction came from Mark Brown, one of the more senior foremen at Glenco. Mark said that he had seen incentive programs like this before on three previous jobs. He said that they never seemed to be fairly administered and, as a result, always caused more problems than they seemed to solve. Mark said, "Programs like this never seem to work out smoothly. When people who expect the bonus don't get it, they just get angry and make life unpleasant for the rest of us. I wish we had never opened this can of worms!"

In spite of the complaints, plans for implementing the new bonus plan were made. A new committee, named the Employee Bonus Committee, was formed to choose the bonus recipients. This committee was comprised of five manufacturing managers. They were to make their choices based on supervisory ratings of employee performance. This was the normal way of evaluating employees' performance.

Supervisors normally made their ratings in late November, using Glenco's standard performance appraisal form. Prior use of this form to assess performance has resulted in reasonably satisfactory results, although it appeared that some supervisors ignore the company's request to keep their average ratings near the scale midpoint on the 10-point rating scale (1 = poor performance, 10 = excellent performance). Even training did not seem to help equate supervisors on the standards they used to evaluate their subordinates. For the most part, however, this did not seem to cause any problems, since performance had not been used as the basis for making salary adjustments or for distributing other types of rewards to manufacturing employees in the past. In fact, evaluations served little more than to formally document poor performance for employees who were put on notice to improve their work or face a pink slip!

This year, the ratings were to be turned in directly to the Employee Bonus Committee. Their job would be to choose the employees who had earned the bonus and to explain how they were chosen (*i.e.*, be prepared to defend their choices to disgruntled employees who did not receive the bonus).

DATA DESCRIPTION

The data for the Glenco Manufacturing case are contained in the file GLENCO.DAT on the Data Disk. The file contains information on 155 employees. In addition to their ID number, the data file includes information on sex, race, and age for all employees as well as their department number and performance rating. A complete listing of the data on the disk is shown below. **Note, however, that the names of the employees are not included in the disk file GLENCO.DAT.** The alphabetical listing of Glenco employees given here can be used to find the names corresponding to the employees ID numbers in answering Case Question #1.

In the file GLENCO.DAT, the variables are defined as follows:

ID:	Employee identification number.
Sex:	1, if female, 2, if male.
Race:	1, if minority, 2, if non-minority.
Age:	Age in years.
Dept:	Department number (1–7).
Rating:	Employee performance rating (1–10) as given by the departmental supervisor, with 10 as the highest rating and 1 as the lowest rating.

ASSIGNMENTS

1. Using the framework outline in Exhibit 5.2, develop a complete research problem definition for obtaining the information needed by Tom Marlin to help him make his recommendations for bonuses.

2. Develop a Type I research proposal (See Chapter 5) for the problem definition in assignment (1) above.

3. Is the study as conducted capable of answering Tom's question? If not, why not? How could the study be improved?

4. Develop a preanalytic plan for Tom. What statistics and variables should he use to make his decision?

5. The chairman, Tom Martin, has asked each member of the committee to come to the first meeting with a list of "winners". You are a member of this committee. Use the data in the file GLENCO.DAT on the Data Disk. This file contains information on ID number, gender, race, age, department number and performance rating for each employee. Analyze these data and make your list of employees whom you believe are deserving of the $500 bonus (the list should include approximately 20% of the employees). Describe the statistical procedures you used to come up with this list and explain why your procedures will result in the fair treatment of all employees at Glenco (i.e., defend your reasoning for making those choices at the EBC meeting). (Hint: You might want to look at, and compare, the distributions of ratings given by the supervisors.)

   a. Make your list of employees that you would recommend for the $500 bonus. Your list should contain approximately 20% of the employees.

   b. Describe the statistical procedures you used and the results you obtained for developing your list of bonus recipients. Attach any relevant supporting output.

   c. Explain why your procedures will result in the fair treatment of all employees at Glenco (i.e., defend your reasoning for making those choices at the EBC meeting).

ALPHABETICAL LISTING OF GLENCO EMPLOYEES

Last Name	First Name	ID	Sex	Race	Age	Dept	Rating
Alexander	Charles	001	2	2	35	1	5
Ammann	Ray	002	2	2	60	1	5
Anderson	James	003	2	1	61	2	4
Babcock	Marilyn	004	1	1	36	2	5

(Continued)

Last Name	First Name	ID	Sex	Race	Age	Dept	Rating
Bagley	Jan	005	1	2	40	5	8
Baker	Lisa	006	1	2	59	3	6
Baldwin	Lois	007	1	1	35	4	5
Bartlett	Bill	008	2	1	26	4	4
Blair	Elizabeth	009	1	2	41	6	5
Bolden	Holly	010	1	2	50	6	4
Bowling	Frances	011	1	2	22	7	8
Brown	Charlotte	012	1	2	62	7	4
Bruton	Penny	013	1	2	23	7	4
Campbell	Delores	014	1	1	43	7	5
Cash	Bill	015	2	1	35	7	7
Chandler	Margo	016	1	2	42	4	7
Chang	Sally	017	1	2	25	4	6
Clark	John	018	2	2	29	5	7
Cook	Debbie	019	1	2	43	2	6
Cravens	Pete	020	2	1	55	5	8
Crowder	Thomas	021	2	1	20	2	5
Cummings	Jeanette	022	1	2	43	1	6
Cunningham	Sherry	023	1	2	48	3	8
Davis	Ruth	024	1	1	44	2	5
Dawson	Michael	025	2	2	24	2	5
Dickens	Jimmy	026	2	2	60	5	7
Dorsey	Patricia	027	1	2	44	3	8
Drake	Annette	028	1	2	20	7	5
Echols	Betty	029	1	2	41	7	5
Ellis	Betty	030	1	1	55	6	5
England	David	031	2	2	30	6	7
Ennis	Jerry	032	2	2	27	1	5
Ernst	Faye	033	1	2	43	3	6
Esterline	Laura	034	1	2	39	5	8
Ferguson	Sylvia	035	1	2	31	5	8
Fields	Anne	036	1	2	46	1	5
Flores	Bo	037	2	2	32	2	5
Fountain	Iris	038	1	2	61	2	5
Freed	Nancy	039	1	2	36	3	6
Freeman	Vic	040	1	1	45	4	6
French	Charles	041	2	2	30	4	6
Fried	Harvey	042	2	2	58	5	8
Garrett	Helen	043	1	2	32	3	6
Gebhardt	Virginia	044	1	2	57	6	9
Gerken	Denise	045	1	1	24	1	6
Gillespie	Robbert	046	2	2	42	7	6
Golden	Jeff	047	2	2	35	2	5
Goode	Brad	048	2	2	36	2	5

Last Name	First Name	ID	Sex	Race	Age	Dept	Rating
Green	Paula	049	1	2	45	5	7
Gregory	Debbie	050	1	1	23	5	10
Haley	George	051	2	2	47	1	5
Hanson	Greg	052	2	1	50	3	7
Harding	Teresa	053	1	2	50	3	7
Henderson	Beverly	054	1	1	36	2	5
Hicks	Catharine	055	1	1	36	2	6
Horn	Ernest	056	2	2	57	7	8
Horner	Wilbur	057	2	2	39	7	6
Howard	Brian	058	2	2	55	6	7
Hubbard	Cheryl	059	1	2	42	7	5
Ivie	Marsha	060	1	1	39	5	7
Jackson	Charlie	061	2	2	27	5	8
Johnson	Herb	062	2	2	30	1	5
Johnston	Edwin	063	2	2	60	2	4
Jones	Nancy	064	1	1	33	2	7
Jordan	Clifton	065	2	1	21	3	7
Kimeldorf	Emily	066	1	2	25	5	7
Leaf	Bernice	067	1	2	36	4	6
Lee	Liz	068	1	2	39	5	7
Leiter	Ricky	069	2	2	23	5	7
Leonard	Cheryl	070	1	2	42	7	6
Lewin	Ronnie	071	2	2	31	1	5
Little	Annie	072	1	1	26	6	4
Lockley	Pam	073	1	1	45	6	7
Long	Peggy	074	1	2	57	7	6
Lyle	Sunny	075	2	2	50	7	5
Malone	Barbara	076	1	2	55	7	5
Marshall	Frank	077	2	2	19	1	4
Matthews	Bob	078	2	1	44	2	7
McCormick	Barbara	079	1	2	57	3	5
McDonald	Vera	080	1	2	52	4	4
Miller	Diane	081	1	2	31	5	8
Miller	Sam	082	2	2	48	5	10
Milligan	Guy	083	2	1	47	5	7
Mitchell	Donna	084	1	1	62	6	7
Moore	Michele	085	1	2	48	7	4
Morton	Jack	086	2	2	54	1	5
Nelson	Cynthia	087	1	2	60	2	7
Neugent	Will	088	2	2	20	3	6
Nielsen	Dudley	089	2	1	38	4	6
Nunn	Phil	090	2	1	35	5	7
Oliver	Dick	091	2	2	28	6	5
Owen	Monica	092	1	2	44	1	4

(Continued)

Last Name	First Name	ID	Sex	Race	Age	Dept	Rating
Page	Dorothy	093	1	2	50	1	5
Pearson	Calvin	094	2	2	40	2	7
Perry	Claude	095	2	2	45	3	7
Pettit	Janice	096	1	2	26	2	5
Phillips	Frank	097	2	2	38	4	5
Powell	Angie	098	1	1	27	5	8
Powers	Jerri	099	1	2	47	1	5
Rahn	Dean	100	2	2	59	2	4
Reeves	Tom	101	2	2	36	5	9
Renner	Martha	102	1	2	58	2	4
Roberts	Margie	103	1	2	56	2	4
Robertson	Helen	104	1	2	40	3	6
Robinson	Harvey	105	2	2	31	4	9
Rodman	Wes	106	2	2	50	6	6
Rogers	Kent	107	1	2	38	1	5
Rose	Debbie	108	1	2	46	5	9
Ross	Fran	109	1	2	20	1	4
Sanders	Mike	110	2	2	32	5	9
Saunders	Edith	111	1	2	46	1	5
Schmidt	Carrie	112	1	1	24	3	9
Schuster	Jean	113	1	1	52	5	7
Scott	Jay	114	2	2	56	3	7
Seagraves	Abraham	115	2	2	47	3	7
Sharpe	Alice	116	2	2	23	3	8
Shein	Ralph	117	2	2	33	6	7
Sherman	Myra	118	1	2	40	3	5
Simmons	Alex	119	2	2	28	3	8
Sims	Denise	120	1	2	53	2	5
Smith	Margie	121	1	2	40	4	5
Stillman	Billie	122	2	2	28	1	4
Stowe	Paula	123	1	2	52	1	6
Stuart	Jean	124	1	2	54	7	6
Sullivan	Joan	125	1	2	32	3	7
Summers	Linda	126	1	2	26	4	5
Taylor	Carol	127	1	2	28	2	6
Thomas	Jeff	128	2	2	45	3	5
Thompson	Lori	129	1	1	51	3	8
Tinsley	Brian	130	2	1	37	1	4
Tobias	Ron	131	2	2	41	5	8
Townsend	Anne	132	1	2	52	6	6
Tucker	Kay	133	1	2	45	4	9
Tupper	Jean	134	1	1	34	4	8
Van Cleve	Edith	135	1	2	31	4	7
Van Zandt	Howard	136	2	2	49	1	6

Last Name	First Name	ID	Sex	Race	Age	Dept	Rating
Vandell	Debbie	137	1	2	45	3	8
Wadley	Carla	138	1	2	22	3	8
Walker	Earl	139	2	2	59	4	6
Ward	Ron	140	2	1	55	5	9
Waters	Alan	141	2	1	24	1	4
Watson	Karen	142	1	2	43	6	8
Webster	Sam	143	2	2	21	1	5
Welch	Jeffrey	144	2	2	36	3	8
White	Suzanne	145	1	2	38	3	7
Whitt	Nancy	146	1	2	29	6	6
Wiggins	Harvey	147	2	2	59	2	6
Williams	Claudia	148	1	2	24	2	8
Wilson	Kathy	149	1	2	58	6	4
Wilson	Lee	150	2	2	20	7	4
Wolf	Jess	151	2	2	42	6	4

Property Tax Assessment

Ryder Appraisal District (RAD) is the county agency responsible for property taxation in Ryder County, South Carolina. RAD is charged with the appraisal of property for tax purposes, the determination of the amount of tax owed by a property owner, and the collection of these property taxes. There are approximately 12,000 residential properties in Ryder County.

James Bradford was elected last year to the position of Ryder County Tax Commissioner. For many years, taxpayers have complained about the inequities in residential property tax assessments in Ryder County. Bradford promised county residents that, if elected, he would do what he could to make the property taxing system more equitable.

The current residential property taxing system in Ryder County has evolved over several decades. Unimproved property, *i.e.*, a residential lot without a house, is currently taxed at a flat amount of $150. Owners of improved residential property are charged a flat amount plus an additional amount based on the characteristics of the house built on the property, such as the total square footage of heated/cooled floor space, the number of bedrooms, the number of bathrooms, etc. If the owner actually lives in the house, he is given a homestead exemption which reduces his property tax bill by a fixed 10%. If that homeowner is also retired, he is entitled to a retirement exemption that reduces his property tax bill by another 10%. While it would seem that the current system of taxing is fair, several factors have led to inequities.

Each year, new property assessments are mailed out to property owners. These assessments tell the property owners the value of their property as determined by RAD for taxing purposes. After the new assessments are mailed, the property owners are given the opportunity over a period of one month to contest their assessments. A panel known as the County Review Board hears the appeals and makes any adjustments they deem necessary to property assessments. The County Review Board has been known in the past for being "too responsive" in changing assessed values. Those homeowners who complain to the Board are usually granted their reduction. Unfortunately, many of these decisions have led to taxing inconsistencies and discrepancies across the county.

RAD is supposed to reappraise each piece of residential property once every three years. In this way, the RAD data base is kept up to date and includes recent improvements to residential property, such as swimming pools or other additions. However, due to poor economic conditions, county governments have been receiving less and less funding from the state and federal governments. The appraisal staff at RAD has dwindled in the past few years to one full-time appraiser and there are no funds available for hiring professional appraisers on a part-time basis. Over time, the data base has become outdated and inaccurate. This, of course, has led to further inequities in the taxing system.

James Bradford has proposed a simplified property tax assessment plan that he hopes will avoid the problems and inequities of the past and cost less to implement than the current system. His plan is to tax unimproved property in the same way as before using the current rate based on lot size. For improved properties, the new tax will consist of a flat tax plus an amount based on the square footage of heated/cooled space in the house. Both the flat tax and the variable tax components are yet to be determined. Bradford has decided to maintain the current 10% reductions for homestead exemptions and retired homeowners. The square footage measurement would be made once for each house, with an appraisal update occurring whenever an addition is made to the house. This will reduce the workload of RAD appraisers and the amount of necessary recordkeeping. The simplified taxing system should also result in more consistency in taxation and fewer complaints from homeowners about incorrect assessments.

There have been two major concerns expressed about switching to the simplified tax assessment plan. First, it is important to try to keep individual taxes under the new plan as close as possible to what they are currently, otherwise there will be problems. Everyone understands that some property owners will have higher taxes and others will have lower taxes, but the size of the changes should be minimized as much as possible. Secondly, the new tax plan must be revenue-neutral, that is, the total amount of property taxes under the new structure should be roughly the same as before. James Bradford has said that attention to the first concern will largely mitigate this second concern.

The current database contains information on the square footage of each house in Ryder County. Bradford believes that these data may not be accurate. In most cases, the recorded square footage is reported by the builder and is never checked by RAD. Before launching a massive effort to determine the square footage of each house in Ryder County, Bradford has decided to see whether or not the current data are sufficiently accurate.

In fact, James Bradford believes that he can kill two birds with one stone. He has collected a random sample of 80 residential properties from around the county. Over the course of the past two months, the staff appraiser has examined the houses on these properties and their blueprints to come up with an accurate estimate of the actual square footage of heated/cooled space of each house. Other information from the RAD data base was also included in the sample information. From this sample, James Bradford wants to do two things. First, he wants to know whether or not the square footage estimates in the current data base differ significantly from the actual square footage values. Secondly, he wants to use the sample data to estimate the flat tax and the tax rate per square foot that must be charged in order to minimize the changes from the current tax system for improved property owners.

DATA DESCRIPTION

The data for the Ryder Appraisal District case is contained in the file RYDER.DAT on the Data Disk. The file contains information on a sample of 80 residential properties in Ryder County. A partial listing of the data is shown below.

SUB	BLK	LOT	IMP	HSX	RET	REC SQFT	ACT SQFT	TAX
7	3	34	1	1	1	1595	1598	922.48
2	21	47	1	1	1	1855	1876	1047.90
11	8	52	1	1	0	1538	1528	967.03
9	21	8	1	1	0	1687	1713	985.22
2	21	51	1	1	0	1785	1764	1031.35
5	19	66	1	1	1	2084	2107	1093.42

These data are coded as follows:

SUB: Subdivision in which the property is located.

BLOCK: Block in the subdivision in which the property is located.

LOT: Lot number of the property. (SUB, BLOCK, and LOT together describe the location of the property.)

IMP: 1, if improved (*i.e.*, a house is built on the property), 0, if unimproved.

HSX: 1, if homestead exemption (*i.e.*, owner lives in house), 0, otherwise.

RET: 1, if owner retired and HSX = 1, 0, otherwise.

REC SQFT: Square footage recorded in the RAD data base.

ACT SQFT: Actual square footage as measured by RAD appraiser.

TAX: Current county tax on the property under the existing taxing system.

ASSIGNMENTS

1. Using the framework outline in Exhibit 5.2, develop a complete research problem definition for obtaining the information needed by James Bradford to help him accomplish his goals and make his decision.

2. Develop a Type I research proposal (see Chapter 5) for the problem definition in assignment (1) above.

3. Is the study as conducted capable of answering James Bradford's questions? If not, why not? How could the study be improved?

4. Develop a preanalytic plan for James. What statistics and variables should he use to make his decision?

5. The data from the sample of 80 residential properties are contained in the file RYDER.DAT on the Data Disk. A description of this data is provided. Using this data set and other information given in the case, first help James Bradford decide whether or not the square footage information in the database is significantly different from the actual values. After that, determine the flat tax amount and the tax charged per square foot that will accomplish his goals. The questions below will assist you in your analysis of the data. Use important details from your analysis to support your recommendation.

    a. Examine the data for the 80 residential properties using appropriate statistical summaries. Are the data collected on the 80 residential properties suitable for analysis? If not, describe any adjustments that need to be made.

    b. Based on the data collected, are the square footage figures in the RAD database reasonably accurate? Explain your answer and support your conclusions from the data.

    c. Perform an appropriate statistical test to determine if the mean recorded square footage in the entire RAD database is significantly different from the mean actual square footage for all houses represented in the database. What recommendations would you make to James Bradford based on your answers to this question and the previous question? Explain.

    d. Use an appropriate statistical technique with the data to determine the fixed and variable (with respect to square footage) tax components for improved properties in Ryder County. Explain how your approach to estimating these components will meet the concerns of James Bradford about the new taxing system being "revenue neutral" (i.e., keeping property taxes about the same as under the current system, both on an individual property basis and on a total basis).

    e. Were there any unusual data that had a large effect on you estimates of the fixed and variable tax components? How did you deal with these influential data in your analysis? Provide the supporting logic behind your approach to dealing with these influential data.

THE KEELS AGENCY

Advertising Media Selection

The Keels Agency (TKA) is a small advertising agency in Portland, Oregon that helps clients get the biggest return on their advertising dollars. TKA specializes in working with companies that are looking to advertise their products and services for the first time. Such companies are typically newer businesses that have begun to grow and now have the revenues to take the next step by investing in advertising. TKA has a good track record of helping these companies feel comfortable with their expenditure of advertising dollars. As pointed out by Beth Keels, founder and CEO of this agency, the costs incurred with advertising can be considerable and are always perceived as a relatively high percentage of clients' revenues. For first-time clients, the thought of investing in advertising, no matter how much sense it might make, always leads to questions about whether the expense will be worth the investment.

Companies like TKA typically try to identify the particular market segments that are most likely to buy their clients' goods and services and then locate an advertising outlet that will reach this particular market group. Client groups require considerable explanation about how this "matching" occurs. Beth Keels typically explains it like this:

> We collect a lot of information on clients' actual sales over a two to three month period and on the people who make those purchases. We get this information from a variety of sources, including surveys, interviews, credit records, mailing lists, contests, and so forth. Our goal is to learn as much as we can about our clients' customers to see whether there might be a distinct "profile" of the typical customer for a particular product or service. If a distinct profile emerges from our research, then we try to match that profile to advertising outlets, such as TV, radio, newspapers, and magazines known to be watched, listened to or read by people with this particular profile. In this way, we target advertising directly to high potential customers. This procedure goes a long way in helping our clients feel more comfortable that at least the money spent on advertising is putting their products and services in front of the right audience. We've been doing it this way for years and have a long track record of being successful.

TKA recently signed a new client, Cycle World, in nearby Seattle. Cycle World markets, under its own name, three lines of racing and mountain bikes, made by several bicycle manufacturers. Cycle World currently sells its bikes in their six retail outlets in major cities throughout the Northwest. Cycle World is now ready to launch a direct sales campaign of their products by advertising bicycles in nationally distributed magazines.

This direct sales effort will rely on reaching potential customers by placing half-page, two-color ads in popular magazines that have large, national subscription bases. The marketing campaign would attempt to (1) create name recognition for Cycle World's products based on placing five ads in each issue of chosen magazines, and (2) offer customers savings that result from eliminating the "middle-man." Thus, it is clear that choosing target magazines for each product is crucial in order to insure that Cycle World's new venture will be successful. They have set aside $240,000 to advertise their products in this manner. In addition to the costs of placing the ads, this budget must also cover TKA's separate charges to Cycle World for the creation and production of the advertising copy as well as their fee and overhead charges. Choosing the wrong magazine not only means that this total budget is being spent on multiple ads to reach the wrong audience, but that the real potential customers would still go unreached.

Cycle World sells three lines of bicycles. The lower line includes "basic" racing and mountain bikes. These bicycles, made by the largest bicycle manufacturer in the U.S., tend to be heavy as far as bikes go, have relatively few features and offer few customer options. Their middle line, made by a popular West Coast manufacturer, includes bicycles that are made of light-weight metals, have many features that serious bikers want and that provide a modest number of options to help buyers customize their bikes. The upper line is made by one of Europe's leading bicycle manufacturers, and includes bicycles that are made of ultra-light alloy metals, that have all the "bells and whistles" that can be put on a bike. Customers are allowed to choose among a number of options to customize their purchases from the upper line of bicycles.

Beth Keels put together a market research team to identify the profile of the typical customer for each product line. To collect this information, the market research team collected information from persons who purchased bicycles at Cycle World's six retail stores. A random sample of customers during a two-month period was asked to complete a short survey that contained descriptive questions about themselves. To encourage customers to complete the survey, each was offered as a gift for their participation, a biker's helmet, a mileage meter, or a bicycle tire pump. Over 90% of the sampled customers completed the survey. Questions were chosen to get an understanding of the demographic background (*i.e.*, age, gender, marital status, education) and the interest level in biking (*i.e.*, extent of use, fitness level, and self-rated interest) of customers.

Based on these data, a profile of the "typical" customer for each product line of merchandise needed to be created and compared to the "typical" subscriber profile for a list of magazines. The list of potential magazines was chosen to reflect three issues: (a) the subscriber base needed to be a national one, (b) the subscriber list needed to fall in the moderate-size category for nationally distributed magazines, and (c) the magazine needed to focus on a particular topic or theme.

Cycle World very specifically wanted to reach a national market in their first attempt to enter the direct sales arena. They reasoned that this was the best way to guard against the problems created by unpredictable, cyclical, regional economic downturns. The choice of looking at magazines in the moderate-sized national subscription base would mean that ads would be similar in costs and within Cycle World's advertising budget. Finally, the typical subscriber was expected to represent a larger proportion of the subscriber base for magazines with a particular focus or theme. The list of potential magazine outlets, and the profile of the typical customer for each, are provided in Table 1. The magazines in Table 1 were chosen as possible advertising outlets because they have a moderately large national subscription base and focus on particular themes.

The data in Table 1 reflect the average age, salary, activity level of subscribers (1 = none to 5 = very active), the percent of subscribers who are

TABLE 1 Potential Advertising Outlets for Cycle World

Magazines	Age	% Male	Education	Salary	Level
ALIVE!	26	45	BA	26,000	5
Business World	30	70	BA	50,000	4
Chinese Cooking	38	30	HS/BA	34,000	3
Computer Technician	34	92	Tech/BA	37,000	2
Country Cookin'	32	20	HS	20,000	2
Crafters	32	30	HS/BA	34,000	3
Creative Projects	28	20	HS/BA	32,000	4
Cycle Time	29	65	BA	60,000	5
Electronics Today	42	90	Tech/BA	42,000	2
Entrepreneur's Day	26	90	HS	27,000	3
Family Living	30	55	HS/BA	31,000	3
Fashion Flair	20	10	HS	14,000	4
Fisherman's Line	50	90	HS	34,000	3
Gourmet's Kitchen	46	60	BA	56,000	3
Outdoor Fun	27	55	HS/BA	30,000	3
Naturalists	38	60	BA	45,000	3
Parent's Digest	28	50	HS/BA	29,000	2
Runners' World	43	70	BA	38,000	5
RV Country	57	69	HS/BA	28,000	2
Software Review	28	60	BA	48,000	4
Sporting World	28	52	HS/BA	31,000	4
Sports Line	35	76	HS	28,000	4
Today's Cyclist	25	10	HS	22,000	2
Today's Home Video	32	40	BA	36,000	2
Traveler's Digest	46	60	HS/BA	44,000	4
Who's Hot In The Movies	29	45	HS/BA	29,000	2
Who's Hot In Music	22	30	HS	18,000	3
Who's Hot In Sports	25	80	HS	22,000	3
Woman's World Today	28	10	BA	34,000	3
Wood Crafters	42	85	HS/BA	42,000	3

male, and the modal educational background of subscribers (HS = high school diploma, BA = bachelor's degree, and Tech = technical certificate from a trade school). The average cost of a half-page ad in these magazines is approximately $2,000 per issue.

DATA DESCRIPTION

File KEELS.DAT on the Data Disk contains the coded data from the survey of Cycle World customers. Data are recorded in the manner depicted below.

Product Line	Age	Sex	Education	Marital Status	Income	Times/ Week	Miles/ Week	Fitness
3	22	1	4	1	26000	3	120	5
2	50	1	4	2	41000	2	60	3
1	50	2	4	2	47000	3	70	3
2	19	1	3	1	18000	3	60	2
3	48	1	4	1	26000	3	120	5

These data are coded as follows:

Product Line:	1 = low product line, 2 = middle product line, 3 = high product line.
Age:	Age on last birthday.
Sex:	1 = male, 2 = female.
Education:	1 = no high school diploma, 2 = high school diploma, 3 = some college-level work, 4 = college degree, 5 = graduate work or degree.
Marital Status:	1 = single, 2 = married.
Income:	Annual family income, rounded off to the nearest $1000.
Times/Week:	Average number of times the person uses or plans to use bicycle each week.
Miles/Week:	Average number of miles completed or planned each week.
Fitness:	Self-rated fitness level, based on scale ranging from 1 = poor shape to 5 = excellent shape.

1. Using the framework outline in Exhibit 5.2, develop a complete research problem definition for obtaining the information needed by TKA to help them make their media recommendations.

2. Develop a Type I research proposal (See Chapter 5) for the problem definition in assignment (1) above.

3. Is the study as conducted capable of answering TKA's dilemma? If not, why not? How could the study be improved?

4. Was the sampling procedure appropriate? If not, why not?

5. Develop a preanalytic plan for TKA. What statistics and variables should they use to make their decision?

6. As a member of the market research team, your job is to examine the data that contains descriptive information about Cycle World's customers. You will find the data in KEELS.DAT on the Data Disk. A description of this data set is provided.

   You will need to analyze these data in order to create a profile of the "typical customer" for each line of merchandise. Once done, compare your understanding of the typical customer for each product line to the "typical customers" subscribers given in Table 1. Based on this comparison, make recommendations about which magazines should be targeted as advertising outlets for each product line. Finally, estimate the total cost for advertising all products. Assume that each ad will take a half-page and run five times in each issue and in four issues a year. The questions below will assist you in your analysis of the data. Use important details from your analysis to support you recommendation.

   a. Describe the profile of the "typical" customer for each product line. Be sure you use statistical indices that are appropriate for the data being analyzed.

   b. Compare each of these product-line profiles to the profiles of typical subscribers of the magazines listed in Table 1. Recommend the two most appropriate magazine outlets for advertising each separate product line.

   c. Based on matching product-line and subscriber profiles, how many different magazine outlets are needed to insure that each product line is advertised twice? List the final set of magazine outlets that you recommend.

   d. Estimate the total cost of advertising bicycles in the chosen magazines. Assume that each half-page ad will run five times in each issue and in four issues a year.

   e. What percent of the total advertising budget does your recommendation represent?

Market Assessment Study

Bill Cate is an enterprising entrepreneur who is interested in developing a new business concept called River End Golf and Entertainment Center. This center is a proposed new sports related concept that would include a variety of golf and entertainment related activities. The center would have a distinctive golf focus and would include entertainment and learning activities for both the family and the golfer. It is anticipated the center would include a miniature golf course, a driving range, a putting course, a golf instruction center, a nine hole pitch and putt, a nine hole executive golf course, a nationally recognized golf and recreational retailer and other entertainment related activities. The proposed location of this concept would be adjacent to the Mall at River End.

Bill is very optimistic about the prospects of this venture, but prudently, he needs more information before he and his financial partners make the leap and commit their investment money to the project. More specifically, Bill wants to know who is the market(s) for such a venture and what would be their perceptions of the center as proposed. In addition, the financial feasibility of this project relates not only to the acceptance of the components by the marketplace, but their willingness to pay certain prices for the various activities that are proposed.

As a result, Bill sought out Go Business, a local consulting firm specializing in survey market research, to undertake a market assessment of the proposed concept. Go Business met with Bill and discussed the plans and concerns of his business partners to design a study that would address their information needs. After a brief orientation period, Go Business produced a basic outline of the proposed objectives of the study and how it would be undertaken. This outline is shown in Table 1.

After considerable discussion, it was agreed the study should proceed. This resulted in the construction of the attached survey. Data collection was subsequently undertaken and the primary task that remained was to make sense of the data. Go Business is now charged with reevaluating the merits of the study and analyzing the results to make recommendations to Bill and his investors.

The data is contained in a file called *Golf Data-SPSS* on the Data Disk. The codebook for this data can be found in the Data View tab of the SPSS Data Editor. This data presents one with a number of assignments.

TABLE 1 Discussion Outline for the River End Golf and Entertainment Center Study

> ### River End Golf and Entertainment Center
> **Market Assessment Study**
> **January 2005**

Primary Objectives

1. Identify the primary market segments and their perceptions of the *River End Golf and Entertainment Center* concept.

2. Assess the perceived value, trial and usage of various components of the Center concept among these identified market segments.

3. Identify the importance of planned and potential features of the *River End Golf and Entertainment Center* concept.

Study Methods

1. The attached survey needs to be personally administered to ensure that the sequential sampling design objectives are met. In essence, this sampling procedure will assure adequate representation of the potential market segments for the golf center concept and its components. The procedure is iterative, and it requires that assessments be made of specific characteristics of the populations of interest so that one can have confidence in the findings and reasonable projections can be made from the data. The procedure would work as follows:

   a. Four locations would be chosen for the first sampling iteration. I would suggest (unless you feel the local market is better represented by some other locations) the Mall, locations in Byron and Biloxi, and local location(s) that would insure adequate capture of the various populations for the center are represented. Each location would be responsible for generating 75 random adult respondents matching the characteristics of the area. The data would then be tabulated in terms of various classifications of interest to the study. More specifically, the major classifications would be as follows: residents (further broken down by drive times from work and home), non-residents; golfers, non-golfers; interested and not-interested to ensure adequate representation of the various cells. This first sample will be used to break out the participation and interest rates for the general sample for projection purposes. In addition, the cell numbers will be checked in these categories for their adequacy. For example, if golfers (as should be expected the first round because of a 19% participation rate) are under-represented, then they will be specifically sampled to insure a cell size of at least 100 exists for increased confidence in the study results. Other cells will be filled as appropriate to meet the study objectives.

   b. Since quite a bit of information is being asked of these respondents, an incentive (small gift, $5, or some other small incentive of value) should be given to those that respond. In essence, the interviewers will have to approach potential respondents to ask them for their opinions about a new Alabama golf and entertainment center promising them a small incentive for their cooperation. Are you going to be able to develop some basic concept pictures to illustrate the Center? If not, we need to adjust the instrument.

TABLE **1** *(Continued)*

River End Golf and Entertainment Center
Market Assessment Study
January 2005

c. As noted on the first page, those individuals that definitely will not visit are exited from the survey very early on. However, it is important to get answers to the first five questions in the first iteration so that an estimate can be made of the markets interest levels in the concept. This is extremely important up front. Later iterations will lead to useful segmentation and design information for the smaller pieces of the market that will use the center.

River End Golf and Entertainment Center

The *Golf and Entertainment Center* at River End is a new sports related concept that will include a variety of golf and entertainment related activities. The center would have a distinctive golf focus and would include entertainment and learning activities for both the family and golfer. It is anticipated the Center would include a miniature golf course, a driving range, a putting course, a golf instruction center, a nine hole pitch and putt, a nine hole executive golf course, a nationally recognized golf and recreational retailer and other entertainment related activities. This Center would be located on property adjacent to the Mall at River End.

1. Do you golf? ❏ No ❏ Yes, if so, how many rounds per year? _____

2. Based on what you know now, how interested would you be in visiting this golf center if it was located adjacent to the Mall at River End? *(Check one box)*

   ❏ Extremely interested

   ❏ Very interested

   ❏ Somewhat interested

   ❏ Not very interested

   ❏ Not at all interested

3. How likely are you to visit this attraction based on the above description? *(Check one box)*

   ❏ Definitely will visit

   ❏ Probably will visit

   ❏ Might or might not visit

   ❏ Probably will not visit

   IF ❏ Definitely will not visit—***STOP,*** *Thank you for your time and consideration, otherwise, please continue.*

River End expects to have a number of basic options for this center. Please think about each option, then answer the questions that follow.

Option 1: Adventure Golf—A themed, 18 hole miniature golf course with synthetic and natural areas which includes decorative landscaping, waterfalls, rock formations and other amenities. Intended for speedy play and total family enjoyment. Priced at $6.00 per person.

4a. Based on the $6.00 price, how would you rate this option's value for the money? *(Check one box)*

- ❏ Excellent value
- ❏ Very good value
- ❏ Good value
- ❏ Fair value
- ❏ Poor value

4b. Based on the $6.00 price, how likely are you to try this option? *(Check one box)*

- ❏ Definitely will try
- ❏ Probably will try
- ❏ Might or might not try
- ❏ Probably will not try
- ❏ Definitely will not try

4c. Based on the $6.00 price, do you feel the price is fair?

- ❏ Yes ❏ No, *then what would you expect as a fair price?* _____ Price

Option 2. Driving Range—A double ended driving range with natural and synthetic hitting areas for all skill levels of golfers. Includes 30 all weather, lighted hitting and landing areas with natural target greens. Simulated rough and fairway areas with quality golf balls. Price would be $7.00 for 75 balls.

5a. Based on the $7.00 price, how would you rate this option's value for the money? *(Check one box)*

- ❏ Excellent value
- ❏ Very good value
- ❏ Good value
- ❏ Fair value
- ❏ Poor value

5b. Based on the $7.00 price, how likely are you to try this option? *(Check one box)*

- ❏ Definitely will try
- ❏ Probably will try
- ❏ Might or might not try
- ❏ Probably will not try
- ❏ Definitely will not try

5c. Based on the $7.00 price, do you feel the price is fair?

❑ Yes ❑ No, *then what would you expect as a fair price?*_____ Price

Option 3: Putting Course—A professionally designed 18 hole challenge putting course with natural grass playing characteristics. Surrounded by natural landscaping, the course is designed to put the player into realistic putting situations with length of holes varying from 30 to 100 feet. Price is $8.00.

6a. Based on the $8.00 price, how would you rate this option's value for the money? *(Check one box)*

❑ Excellent value

❑ Very good value

❑ Good value

❑ Fair value

❑ Poor value

6b. Based on the $8.00 price, how likely are you to try this option? *(Check one box)*

❑ Definitely will try

❑ Probably will try

❑ Might or might not try

❑ Probably will not try

❑ Definitely will not try

6c. Based on the $8.00 price, do you feel the price is fair?

❑ Yes ❑ No, *then what would you expect as a fair price?* _____ Price

6d. Would you prefer ❑ natural grass or ❑ artificial turf on this challenge course?

Option 4: Nine Hole Pitch and Putt—A nine hole challenge course with shots ranging from thirty (30) to ninety (90) yards. Multiple tee positions will be available to simulate realistic approach shots that a golfer could experience during a round of golf. Price: $9.00 per round.

7a. Based on the $9.00 price, how would you rate this option's value for the money? *(Check one box)*

❑ Excellent value

❑ Very good value

❑ Good value

❑ Fair value

❑ Poor value

7b. Based on the $9.00 price, how likely are you to try this option? *(Check one box)*

❏ Definitely will try

❏ Probably will try

❏ Might or might not try

❏ Probably will not try

❏ Definitely will not try

7c. Based on the $9.00 price, do you feel the price is fair?

❏ Yes ❏ No, *then what would you expect as a fair price?* _____ Price

Option 5: Nine Hole Executive Golf Course—A professionally designed and challenging 9 hole executive course with bunkers, water traps and all amenities. Designed expressly for enjoyment and to improve one's golfing game. Price: $14 for 18 holes.

8a. Based on the $14 price, how would you rate this option's value for the money? *(Check one box)*

❏ Excellent value

❏ Very good value

❏ Good value

❏ Fair value

❏ Poor value

8b. Based on the $14 price, how likely are you to try this option? *(Check one box)*

❏ Definitely will try

❏ Probably will try

❏ Might or might not try

❏ Probably will not try

❏ Definitely will not try

8c. Based on the $14 price, do you feel the price is fair?

❏ Yes ❏ No, *then what would you expect as a fair price?*_____ Price

Option 6: Learning Center—A golf academy and learning center will provide individual instruction with certified PGA and LPGA instructors. The center will have state-of-the-art computer assisted video instruction, workshops, club fitting, and other fundamentally driven instructional aids. Price: $45 for 45 minutes of instruction.

9a. Based on the $45 price, how would you rate this option's value for the money? *(Check one box)*

❏ Excellent value

❏ Very good value

❏ Good value

❏ Fair value

❏ Poor value

9b. Based on the $45 price, how likely are you to try this option? *(Check one box)*

- ❏ Definitely will try
- ❏ Probably will try
- ❏ Might or might not try
- ❏ Probably will not try
- ❏ Definitely will not try

9c. Based on the $45 price, do you feel the price is fair?

❏ Yes ❏ No, *then what would you expect as a fair price?*_____ Price

9d. How important would the following endorsements be in your determination of choosing to go to a golfing academy such as this?

	Extremely	Somewhat	Not at All
Golf Digest Pitch and Putt Center	❏	❏	❏
A Professional Teaching Pro	❏	❏	❏
A Touring Pro	❏	❏	❏
The National Golfing Institute	❏	❏	❏

10. Would you be willing to pay a daily fee of $45 to have access to all the golfing facilities (learning academy excluded) ❏ Yes If no, ❏ What would be a fair price? _____

11. Would you be willing to pay a membership fee of $250 for unlimited use of the golfing facilities for a year? (academy excluded) ❏ Yes If No, ❏ What would be a fair price? _____

Listed below are a number of features that could be included in the *Golf and Entertainment Center*. Please rate the importance of these features to you personally.

	Extremely	Somewhat	Not at All
12. a. mini golf	❏	❏	❏
b. driving range	❏	❏	❏
c. lighted driving range	❏	❏	❏
d. all weather, heated driving range	❏	❏	❏
e. double tiered driving range	❏	❏	❏
f. pitch and putt	❏	❏	❏
g. golf academy	❏	❏	❏
h. putting course	❏	❏	❏
i. 9 hole executive course	❏	❏	❏

> Finally, can you please answer a few demographic questions for classificatory purposes only? We thank you for your time and consideration.

13. How many minutes driving time from Mall at River End, do you live? _____ Work? _____

14. Please indicate your sex and age. ❑ Male ❑ Female Age? _____

ASSIGNMENTS

1. Using the framework outline in Exhibit 5.2, develop a complete research problem definition for obtaining the information needed by Bill Cate to help him in this endeavor.

2. Develop a Type II research proposal (See Chapter 5) for the problem definition in assignment (1) above.

3. Is the study as conducted capable of answering Bill's questions? If not, why not? How could the study be improved?

4. Was the sampling procedure appropriate? If not, why not?

5. Develop a preanalytic plan for River End. What statistics and variables should be used to solve the problem at hand?

6. Analyze the data, and create a presentation that will address the objectives stated in the research plan. You are responsible for analyzing, interpreting, and presenting the results to the investors in this project.

APPENDIX

A

A Practitioner's Guide to Secondary
Business Information Sources

Overview

**Research Guides
and Bibliographies**

**Encyclopedias, Directories,
Dictionaries, and Handbooks**

Indexes

Indexes of Periodicals

Indexes of Newspapers

Indexes of Documents

Statistical Sources

**References for Individual
Companies**

**Other Sources: International
Business**

Handbooks

Almanacs, Dictionaries,
and Encyclopedias

Directories

Yearbooks

**A Note on the Standard
Industrial Classification (SIC)
and the North American
Industry Classification System
(NAICS)**

OVERVIEW

A manual search for secondary data usually begins in a library. To make the search process a little easier, we have provided this index, which is a compendium of references. Although this guide is not an exhaustive listing of what is available in any library, it is a practical collection of what we think might be useful to the business practitioner.

As we noted in Chapter 4, the boundaries among online databases, Internet-based databases, and other, more traditional sources that require the thorough researcher to use multiple search strategies are beginning to blur. As a result, a sampling of Internet-based resources and more traditional print documents are interleaved throughout this Guide to help the practitioner get started in their search for information.

In addition to following existing information leads, researchers should always use the major indexes such as ABI/INFORM and *Applied Science and Technology Index*, available online on WILSONLINE and on CD/ROM as WILSONDISC, to help gather subject information. Numerous new sources are appearing as well (for example, see http://www.brint.com) and are available on the Web for the interested business researcher. As a result, the field is very dynamic and changing all the time. Use this guide as a start to your search.

The guide is divided into six categories (not independent by any means) that correspond to the discussions in Chapter 4 pertaining to the manual search for secondary data. In addition, we have included a short note on the SIC (Standard Industrial Classification) and NAICS (North American Industry Classification System), which has replaced the SIC code. This coding system is an invaluable aid for data searching.

RESEARCH GUIDES AND BIBLIOGRAPHIES

(AMA) American Marketing Association Bibliography Series.

Critical bibliographies of material available on specific marketing subjects. Each bibliography has its own author and title, which can be located in the card catalog under the series title. (http://ama.org)

Commodity Yearbook. New York: St. Martin's Press, Annual.

Covers the 110 basic commodities, giving trends of prices, supplies, and distribution. Also contains special fact-filled, authoritative, and timely research studies.

Daniels, Lorna M. *Business Information Sources.* 3rd ed. Berkeley: University of California Press, 1993.

A select, annotated list of business books and reference sources. Includes the basic kinds of business reference

resources such as bibliographies, indexes and abstracts, directories, statistical and financial sources, and data on current business and economic trends.

Directories of Information of specific industries—for example, *Executive Directory of the U.S. Pharmaceutical Industry.*

These may be located by checking the reference card catalog under the industry name. Do online searches for Web-based industry directories; see, for instance, http://www.asgenet.org.

Dissertation Abstracts International. Ann Arbor, MI: University Microfilms, 1938–. Frequency varies.

A listing by author, discipline, and key word of most of the dissertations done in the United States and some foreign universities. Includes pagination, date, and school.

Dow Jones-Irwin Business Almanac. Homewood, IL: Dow Jones-Irwin, annual.

    Brings together a great deal of useful and widely scattered information and data on many aspects of business, investments, finance, and economics.

Editor and Publisher Market Guide. New York: Editor and Publisher Co., annual.

    This guide contains market data collected from more than 1600 daily newspaper markets in the United States and Canada. It provides estimates of total and per household income and retail sales; population estimates; and market-by-market listings of food and drug chains, supermarkets, and discount and department stores.

Encyclopedia of Banking and Finance. 10th ed. Rolling Meadows, IL: Bankers Publishing Co., 1993.

    An authoritative, comprehensive encyclopedia covering banking and finance; indispensable for students of applied economics.

Encyclopedia of Business Information Sources. 19th ed. Detroit: Gale Research Company, 1998.

    A detailed listing of primary subjects of interest to managerial personnel, with a record of source books, periodicals, organizations, directories, handbooks, bibliographies, and other sources of information on each topic.

Fairchild's Financial Manual of Retail Stores. New York: Fairchild Publications, frequency of new editions varies.

    Provides listing of major U.S. retailers. Includes data on business activities, directors, officers, transfer agents, sales and earnings, number of stores, stocks, assets and liabilities, and profit and loss. Arranged alphabetically by name.

Harvard University. Graduate School of Business Administration, Baker Library. *Core Collection: An Author and Subject Guide.* Boston, annual.

    A guide to Baker Library's open-shelf collection of about 3500 recent English-language business books. A useful guide for identifying basic material.

Management Information Guides. Detroit: Gale Research Company.

    Provides a comprehensive and organized inventory of the resources available for fact finding in a particular field. Each volume is devoted to a specific topic and is designed to direct the user to key sources by arranging, describing, and indexing published sources as well as the programs and services of organizations, agencies, and facilities. These guides are as useful as the American Marketing Association bibliographies.

"New Market Data" issue of *Advertising Age,* annual (usually appearing in a late May issue).

    A descriptive list of more than 1300 items available from media, trade associations, and other sources. The material is arranged in eight sections: National Markets, Farm Markets, Regional and Local Markets, Canadian Markets, International Markets, Distribution Markets, Professional Markets, and Industrial Markets.

Official Congressional Directory. Washington, D.C.: U.S. Government Printing Office, annual.

    Biographical sketches of members of Congress, arranged by states; committee membership, commissions, joint committees and boards; executive, judiciary, diplomatic, and independent agency personnel listed. Individual index.

Progressive Grocer's Marketing Guidebook. New York: Progressive Grocer Company, annual.

    Provides a breakdown of the wholesale and retail grocery market into seven regions and 79 specific marketing areas throughout the country. This includes distribution center profiles, market area boundaries (with maps), and sales and market characteristics for each area.

Rand McNally Commercial Atlas and Marketing Guide. New York: Rand McNally, annual.

    Marketing data, with maps for each state, include population, basic census data, auto registrations, and so on, for counties and Standard Metropolitan Statistical Areas (SMSA); added statistics for cities. Tables at front include population rankings and basic trading area statistics.

Standard Rate and Data Service, 1979–. Skokie, IL, frequency varies.

    Published in separate sections with various titles, these are an excellent source of information on rates and advertising media. Included are Business Publication Rates and Data, Newspaper Rates and Data, *and* Spot Television Rates and Data. *These separate sections include Market Data Summary estimates and metropolitan area rankings for population, households, consumer spendable income, retail sales by store groups, and so on. State, county, and metropolitan area data also appear at the beginning of the section for each state.*

Thomas's Grocery Register. New York: Thomas Publishing Co., annual.

    Volume I provides a comprehensive listing of U.S. supermarket chains and wholesalers, brokers, exporters, and warehouses. Volume II includes a directory for manufacturers, importers, and other sources for food and nonfood products, equipment and machinery, and industry-related services. Volume III provides an alphabetical listing of companies in Volumes I and II and lists brand names and trademarks.

Thomas's Register of American Manufacturers. New York: Thomas Publishing Co., annual.

    Lists companies that manufacture products or supply services by products and services, with brand name index, company listing, and catalogs in separate volumes. (http://www.thomasregister.com)

Annual lists in journals: lists of largest U.S. companies, ranked by sales or asset size, appear annually in some business or financial journals.

A summer issue of Advertising Age, *for example, provides a listing of top national advertisers. Other examples are "Agencies Ranked by Billings," in* Advertising Age *(usually the third issue in February); "Electronics Top 100," in* Electronic News *(usually Section 2 of the third issue in July); "Annual Report of Housing Giants," in* Professional Builder *(July each year, available in most libraries); "The Major Independent Finance Companies," in* Bankers Monthly *(April 15 issue).*

Dun and Bradstreet. *Dun's Reference Book of Corporate Management.* New York, annual.

Lists director and selected officers of approximately 2400 companies with annual sales of $20 million or more and/or 1000 or more employees. Data on directors includes date of birth, education, business position presently and previously held.

Dun and Bradstreet. *Middle Market Directory.* New York, annual.

The directory lists more than 30,000 U.S. companies with an indicated worth of $500,000 to $999,999. Gives officers, products (if manufacturer), SIC, approximate sales, and numbers of employees. The yellow pages list companies geographically, and the blue pages list companies by SIC industries.

Dun and Bradstreet. *Million Dollar Directory.* New York, annual (with semiannual cumulated supplements).

Lists approximately 42,000 U.S. companies with an indicated worth of $1 million or more. Coverage and information given is similar to the Middle Market Directory. *An alphabetical list of officers and directors is also provided.*

Fortune directories, annuals.

Part 1 (May) ranks the 500 largest U.S. industrial corporations; the 50 largest banks, merchandising, transportation, life insurance, and utilities companies (Fortune 500). Part 2 (June) ranks the second 500 largest U.S. industrial corporations. Part 3 (August) ranks the 200 largest foreign industrial corporations and 50 largest foreign banks. Tables in each part include rank by sales, assets, net income, invested capital, and number of employees. (http://www.fortune.com and http://www.pathfinder.com)

Jane's Major Companies of Europe. New York: Franklin Watts, 1970–, annual.

Helpful for locating vital information about European firms. Many directories are also compiled for specific industries. Some include information on officers, equipment, capacity, and so on. Consult subject card catalog under headings such as "Aerospace Industries—United States—Directories."

Standard Directory of Advertisers. Skokie, IL: National Register Publishing Company, annual (with monthly cumulated supplements).

Directory of 17,000 companies that advertise nationally, arranged by industry grouping, with alphabetical index. Gives officers, products, agency, advertising appropriations (in some cases), media used, and so on. Includes a trademark index. Geographic index is in a separate volume.

Standard and Poor's Register of Corporations, Directors and Executives. New York: Standard & Poor's, annual (with three cumulated supplements).

Alphabetical list of approximately 37,000 U.S. and Canadian corporations, giving officers, products (if manufacturer), standard industrial classification, sales range, and number of employees. The second volume gives brief information on about 75,000 executives and directors. Index of companies by SIC number is in Volume 3, as is its geographical index.

Thomas's Register of American Manufacturers. New York: Thomas Publishing Company, annual.

Volumes 1–6 list manufacturers by specific product; Volume 7 is an alphabetical list of companies, giving address, branch offices, subsidiaries, products, estimated capitalization, and occasionally principal officers. Volume 8 is an index to product classifications and includes a list of leading trade names (pink pages), Boards of Trade, Chambers of Commerce. Volumes 9–10 are catalogs of companies.

Trade Names Dictionary. Detroit: Gale Research Co., published about every three years, with a periodic supplement.

This is a two-volume guide to more than 160,000 "consumer-oriented" trade names, with addresses of their manufacturers, importers, marketers, or distributors.

Wasserman, Paul. *Encyclopedia of Business Information Sources.* Detroit: Gale Research Company, updated regularly.

A detailed listing of primary subjects of interest to managerial personnel, with a record of source books, periodicals, organizations, directories, handbooks.

Indexes of Periodicals

Accountant's Index, 1921–. New York: American Institute of Certified Public Accountants, quarterly (with cumulations).

> *A subject index to periodicals, books, pamphlets, and government documents in the fields of accounting, auditing, financial statements, finance and investments, management, taxation, and specific businesses, industries, and trades.*

Applied Science and Technology Index (formerly *Industrial Arts Index*), 1913–. New York: H. W. Wilson Company, monthly except July (with periodic cumulations).

> *A subject index to more than 200 selected journals in the fields of aeronautics, automation, chemistry, construction, electricity and electronics, engineering, geology and metallurgy, industrial and mechanical arts, machinery, physics, telecommunications, transportation, and related subjects. From 1913 to 1957, entitled* Industrial Arts Index. *In 1958, this was divided into* Applied Science and Technology Index *and* Business Periodicals Index.

Business Periodicals Index, 1958–. New York: H. W. Wilson Company, monthly except August (with periodic cumulations).

> *A subject index to selected periodicals in the fields of accounting, advertising and public relations, automation, banking, communications, economics, finance and investments, insurance, labor, management, marketing, taxation, and specific businesses, industries, and trades.*

Business Publications Index and Abstracts (BPIA). Prepared by Management Contents. Detroit: Gale Research Co., monthly, cumulating annually.

> *A two-part monthly subject/author index to all articles in more than 700 English-language business/management periodicals.*

Economic Abstracts: Semi-Monthly Review of Abstracts on Economics, Finance, Trade, Industry, Foreign Aid, Management, Marketing, Labour. The Hague, Netherlands: Library of the Economic Information Service, semimonthly.

Funk & Scott Index of Corporations and Industries, 1960–. Cleveland, OH: Predicasts, weekly (with monthly and annual cumulations).

> *An index to articles on companies and industries that have appeared in more than 250 foreign and U.S. trade journals. It is arranged in three parts: by SIC number, by region and county, and by company.* F&S Index International *follows the same format.*

Index of Economic Articles in Journals and Collective Volumes, 1886/1924–. Homewood, IL: Richard D. Irwin, Inc., irregular.

> *Classified index to journal articles and collections in the fields of economic theory, contemporary economic conditions, economic fluctuations, consumer economics, business organization, managerial economics, natural resources and land economics, and numerous related fields.*

International Bibliography of Economics/Bibliographie internationale de science economique, 1952–1979. Paris: UNESCO, 1981.

> *An extensive, classified list of books, pamphlets, periodical articles, and official government publications in various languages. Indexes by author and subject.*

"Marketing Abstracts," in *Journal of Marketing*, 1936 to date.

> *Each issue has an annotated bibliography covering selected articles of interest to marketers. It is arranged under 22 broad headings such as Consumer Behavior, Industrial Marketing, International Marketing, and so on.*

Psychological Abstracts. Washington, D.C.: American Psychological Association, 1947–.

> *Issued monthly, this source provides nonevaluative summaries of the world's literature in psychology and related disciplines. Abstracts are listed under 16 major classification categories, with some categories having subsections as shown in the table of contents. A subject and author index is published every six months.*

Public Affairs Information Service, 1915–. New York: Public Affairs Information Service, Inc., weekly (with periodic cumulations).

> *A selective subject list of books, pamphlets, government publications, reports of public and private agencies, and periodical articles, relating to economic and social conditions, public administration, and international relations, published in English throughout the world.*

Note: Many individual periodicals, such as *Management Science*, have annual and/or cumulative indexes. The annual indexes usually appear in the last issue for the year.

Indexes of Newspapers

New York Times Index, 1851–. New York: New York Times Co., semimonthly (cumulated annually and quarterly).

Wall Street Journal (Indexes). 1958–. New York: Dow Jones, monthly (with annual cumulations).

Each issue is divided into (1) Corporate News and (2) General News. (http://www.wsj.com)

Indexes of Documents

American Statistics Index (ASI): Annual and Retrospective Edition. Washington, D.C.: Congressional Information Service.

Abstracts: *Complete bibliographic information and concise descriptions of a full range of federal statistical publications, including social, socioeconomic, business, economic, and natural resource statistics.*

Almost all statistical publications issued by federal agencies currently in print are included in ASI, as well as the most significant publications issued since the early 1960s.

Index: *A comprehensive index to the Abstracts arranged by subjects and names, by categories, by titles, by agency report numbers, and by selected standard classification. This scheme enables the user to locate statistical information contained in government publications by several methods.*

Supplement Numbers (monthly, with semiannual cumulations): Monthly supplements to the Abstracts and Index provide up-to-date information on changes in publications included in ASI: Annual and Retrospective Edition, as well as descriptions. Also cites older publications that were not available for inclusion in the annual edition. Index supplements are arranged by subject and name.

Congressional Index Service. Chicago: Commerce Clearing House, 1970–.

An index by agency, subject, and name to all the hearings in both houses of Congress. The hearings are abstracted and analyzed for content.

GPO Sales Publication Reference File. Washington, D.C.: U.S. Government Printing Office, bimonthly (with monthly supplement).

Catalog, usually on microfiche, of all government publications in print and for sale. Arranged alphabetically by agency, author, title, subject, key words, series, and SIC number.

U.S. Superintendent of Documents. *Monthly Catalog of United States Government Publications*, 1895–. Washington, D.C.: monthly (with annual cumulations).

A current bibliography of publications issued by all branches of the government, including both the congressional and the department and bureau publications.

Note: Federal, state, and city documents, in general, are not listed in the public catalog. Special indexes for federal documents are usually found in the Federal Documents Room.

STATISTICAL SOURCES

Guide to Consumer Markets. New York, annual.

Currently issued by the Conference Board, this annual guide presents statistical data on population, employment, income, expenditures, production and distribution, and prices.

Robert Morris Associates. *Annual Statement Studies.* Philadelphia, annual.

Contains composite financial data on over 300 lines of business, manufacturers, wholesalers, retailers, services, and contractors. Many widely used financial ratios are computed for every industry appearing in the studies.

Sales and Marketing Management. Annual statistical issues.

Estimates for population; effective buying income (EBI); retail sales for U.S. states, counties, cities; per capita and per household income, percentage of U.S. Also, for metropolitan areas the estimates include population by age groups; income; retail sales for nine retail store groups. Similar data for Canada.

"Annual Survey of Television and Newspaper Markets" (September issue).

For each television market area, tables give Nielsen's estimated number of TV homes, percentage households with color TV, and TV buying power index (BPI); also estimates of population, effective buying income, and sales for the five major retail store groups. For each newspaper market area, tables include percentage of households reached; also population, effective buying income, and sales for the five major retail store groups.

"Marketing on the Move" (November issue).

Includes tables for "Metropolitan Area Projections," estimating populations, effective buying income, and retail sales for U.S. counties and metropolitan areas. Similar tables for Canada.

"Survey of Buying Power, Part 1" (July issue).

Part I (July issue). Gives population, income, and retail sales estimates for states, counties, and metropolitan areas.

"Survey of Buying Power, Part 2" (October issue).

Part II (October issue). Gives data for television and newspaper markets in the United States and Canada. The data for television markets outlines the areas of dominant influence, whereas the newspaper markets information identifies both dominant and effective coverage areas. The issue also includes top radio market demographics. Annual statistical issue.

Standard and Poor's Corporation. *Industry Surveys*. New York, quarterly.

Current and basic analysis in two volumes of 39 industries, with financial comparisons of leading companies within each industry.

Standard and Poor's Corporation. *Standard and Poor's Trade and Securities Statistics*. New York, irregular.

Gives current and basic statistics in the following areas: banking and finance; production and labor; price indexes; income and trade; building and building materials; transportation and communications; electric power and fuels; metal; autos, rubber, and tires; textiles, chemicals, paper; agricultural products; security price index record. (Loose-leaf binder.)

Troy, Leo. *Almanac of Business and Industrial Financial Ratios*. Englewood Cliffs, NJ: Prentice-Hall, annual.

By industry, supplies the most complete, most recent corporate performance facts and figures currently available.

For the Bureau of the Census, start with http://www.census.gov.

U.S. Bureau of the Census. *Annual Survey of Manufacturers*. Revised annually for years not covered by *Census of Manufacturers*.

Covers basically the same data provided by the Census of Manufacturers, but with considerably less depth.

U.S. Bureau of the Census. *Census of Agriculture*. Five-year intervals.

Nationwide census of agriculture featuring data for all farms in the United States. Findings of the census are organized in six volumes: area (state) reports, general subject reports, agriculture services, irrigation, special reports, drainage of agriculture lands.

U.S. Bureau of the Census. *County Business Patterns*. Annual.

Statistics provide information on reporting units, payroll, and employment by industry classification and county location. Useful for analyzing market potential, measuring the effectiveness of sales and advertising programs, setting sales quotas and budgets, analyzing the industrial structure of regions, making basic economic studies and small area studies serving other business uses.

U.S. Bureau of the Census. *Current Business Reports*. Frequency varies.

Provides the most current statistics available on retail and wholesale trade sales, inventories, and accounts receivable.

U.S. Bureau of the Census. *Current Industrial Reports*. Frequency varies for each industry group.

Provides the most current statistics available for more than 50 industrial groupings.

U.S. Bureau of the Census. *Economic Censuses*. Revised every five years.

The Economic Censuses *consist of the following major sections, each providing statistical analysis of the business or industrial area named:* Census of Retail Trades, Census of Wholesale Trade, Census of Selected Service Industries *(the preceding reports were previously issued under the single title* Census of Business*),* Census of Construction Industries, Census of Manufacturers, Census of Mineral Industries, Census of Transportation, Enterprise Statistics.

U.S. Bureau of the Census. *Guide to Foreign Trade Statistics*. Revised irregularly.

This guide contains an explanation of the coverage of foreign trade statistics and sample illustrations of the content and general arrangement of the data presented in the individual trade reports.

U.S. Bureau of the Census. *Miniguide of the 1990 Economic Censuses*. Revised irregularly for each census year.

This miniguide is designed to provide information on the 1990 economic censuses program for users and potential users of Census Bureau data. Also included is a classification system used by the Census Bureau for arrangement of its data.

U.S. Bureau of the Census. *Statistical Abstract of the United States*. Annual Supplements include *Historical Statistics, City and County/State*, and *Metropolitan Data Books*.

Published annually since 1878, the abstract is the standard summary of statistics on the social, political, and economic organization of the United States. It is designed to serve as a convenient volume for statistical reference and as a guide to other statistical sources. The latter function is served by an introductory text to each section and source notes appearing below each table. (http://www.census.gov/stat_abstract/)

U.S. Bureau of the Census. *U.S. Commodity Exports and Imports as Related to Output*. Annual.

This report is prepared to meet the need for a systematic body of statistical information on the relationship between U.S. domestic output on the one hand and U.S. commodity exports and imports on the other.

U.S. Bureau of Economic Analysis. *Business Conditions Digest*. Monthly.

Almost 500 economic indicators are presented in a form convenient for analysts with different approaches to the study of current business conditions and prospects (for example, the national income model, the leading indicators, and anticipations and intentions), as well as for analysts who use combinations of these approaches. Various other types of data and analytical measures, such as the balance of payments and the diffusion indexes, are provided to facilitate complete analysis.

U.S. Bureau of Economic Analysis. *Long-Term Economic Growth, 1869–1970*. Revised irregularly.

This publication provides a statistical and narrative overview of all phases of the American economy for over 100 years.

U.S. Bureau of Economic Analysis. *Population and Economic Activity in the U.S. Historical and Projected: A Supplement to Survey of Current Business*.

This volume presents projections of personal income, employment, and population for 1950–2020 for 426 geographic areas.

U.S. Bureau of Economic Analysis. *Survey of Current Business*. Monthly.

The Survey *is the Commerce Department's regular statistical and narrative report on the nation's economy. Each monthly issue features a section on "Current Business Statistics." In addition, a number of supplements are issued, both regularly and irregularly.*

Business Statistics: The biennial supplement is issued in odd-numbered years; statistics cover a 25-year period for each edition.

Special issues: some issues of the Survey *provide in-depth statistical analysis of various economic categories. July usually covers "National Income and Product Accounts." The May issue contains "Local Area Personal Income."*

Supplemental compilations of statistics, such as "The National Income and Product Accounts of the United States," appear in monthly issues.

U.S. Bureau of Labor Statistics. *Monthly Labor Review*. Monthly.

Articles on employment, labor force, wages, prices, productivity, unit labor costs, collective bargaining, worker satisfaction, social indicators, and labor development abroad. Regular features include a review of developments in industrial relations, significant court decisions in labor cases, book reviews, and current labor statistics. (http://stats.bls.gov)

U.S. Congress, Joint Economic Committee. *Economic Indicators*. Monthly.

Prepared by the Council of Economic Advisors. Presents in chart and tabular form statistics collected by various government agencies on total output; income and spending, employment, unemployment, and wages; production and business activity; prices; currency, credit, and security markets.

U.S. Domestic and International Business Administration. *Utilizing Government Data for Marketing Research*.

Discusses how to use governmental statistical sources for marketing resources. (Slide set.)

U.S. Federal Reserve Board. *Federal Reserve Bulletin*. Monthly.

Presents articles of general and special interest, policy and other official statements issued by the Board, and statistical information relating to domestic, financial, and business developments and to international financial development. (http://www.ny.frb.org)

U.S. Office of the President. *Economic Reports of the President Together with Annual Report of the Council of Economic Advisors*. Revised annually.

In addition to the narrative report of the president and his advisors, the Economic Report *usually contains some statistical background on the condition and direction of the economy.*

Value Line Investment Survey. New York: A. Bernhard, weekly.

A weekly service that analyzes trends in the stock market and recommends specific stocks for purchase. Contains charts and graphs. Other services in personnel practice, environment, taxation, and so on are available in the reference collection of the library.

For government publications, a library usually has the following:

Guide to U.S. Government Publications. *McLean, Va.: Documents Index, 1973–.*

U.S. Bureau of the Census. Statistical Abstract of the United States. *Washington, D.C.: U.S. Government Printing Office, 1879–.*

U.S. Office of Business Economics. Survey of Current Business. *Washington, D.C.: U.S. Government Printing Office, 1879–.*

Note: Many of the preceding titles are indexed in the *American Statistics Index*.

REFERENCES FOR INDIVIDUAL COMPANIES

Annual Reports.

Issued annually by individual corporations, these reports generally contain detailed information on company operations and affiliates, assets, income and expenditures, management, and marketing decisions. They are arranged alphabetically by company name. They usually can be found in the reference room of a library. Not all companies are usually represented.

Directory of Corporate Affiliations. Skokie, IL: National Register Publishing Co. Annual (with bimonthly supplements).

Section 2 lists more than 4000 major American parent companies, giving lines of business, approximate sales, number of employees, ticker symbol, top officers, subsidiaries, and divisions or affiliates. Section 1 is a cross-reference index for all subsidiaries, divisions, and so on.

Moody's Investors Service. *Moody's International Manual (Moody's Municipal and Government Manual, Moody's Bank and Finance Manual, Moody's Industrial Manual, Moody's Public Utility Manual, Moody's Transportation Manual, Moody's OTC Industrial Manual).* New York, six parts issued annually (with semiweekly or weekly supplements).

These manuals are designed as an aid to the bond and securities investor. The Municipal and Government Manual *includes information on the U.S. debt, government bonds, internal and external economic and financial data. Following this there is a state-by-state analysis of individual municipalities' wealth, finances, public debts, and bonds. Other volumes include a brief history of each company, its operations, subsidiaries, plants, products, officers and directors, comparative income statements, balance sheets, and selected financial ratios.*

Predicasts F&S Index United States (F&S). Cleveland, OH: Predicasts, Inc. Weekly (with monthly and annual cumulations). Online counterpart is PTS indexes.

This is the best periodical index to use when searching for current U.S. company, product, and industry information in a wide selection of more than 750 business, industrial, and financial periodicals.

Standard & Poor's. *Standard Corporation Records.* Updated annually.

Descriptions of corporations in which there is wide public interest, including all companies listed on the New York and American Stock Exchanges. The reports, arranged alphabetically by company name, include annual or interim reports, affiliated information, number of employees, stock data, earnings, finances, company officers, and more. Seven volumes. (Loose-leaf binders.)

Value Line Investment Survey. New York: A. Bernhard & Co., weekly additions. Online counterpart is Value Line Data Base II. (Loose-leaf binder.)

This popular investment service continuously analyzes and reports on 1700 companies in about 95 industries.

10-K reports.

Annual reports filed by all stock companies with the Securities and Exchange Commission. These reports are primarily company balance sheets. They include detailed information on sales, assets, income, expenditures, and stock transactions.

Note: See also Directories, including "Listings Ranked by Size."

OTHER SOURCES: INTERNATIONAL BUSINESS

Handbooks

Maffry, Ann D. *Foreign Commerce Handbook.* 17th ed. Washington, D.C.: International Division, Chamber of Commerce of the United States, 1981.

Part 3 of this book consists of a bibliography of reference works, books, pamphlets, and periodicals of interest to foreign traders.

Reference Book for World Traders. 2nd ed. Queens Village, NY: Croner Publications, 1986. Kept up to date with amendments.

A loose-leaf handbook kept up to date with monthly supplements. It includes a listing of information sources in the United States and all foreign countries, consular requirements, shipping and air services, marine insurance and postal services, industry and professional associations. Three volumes.

Almanacs, Dictionaries, and Encyclopedias

Dun & Bradstreet International. *Principal International Businesses*. New York, annual.

> *Section 1 lists approximately 50,000 leading companies alphabetically by 133 countries, giving principal officers, major products/activities, and financial information.*

European Financial Almanac. New York: R. R. Bowker, updated periodically.

> *Covers 13 European countries. Part 1: Finance markets review. Part 2: Directory of major organizations. Part 3: Who's who in European finance.*

Exporter's Encyclopedia. New York: Dun & Bradstreet International, published periodically.

> *Dun & Bradstreet's world marketing guide. Entries for each country include market information, import and exchange regulations, shipping services, and so on. Also includes general information on export law and insurance, reference data on communications, weights and measures, and information about U.S. ports.*

Major Companies of Europe: Companies in the European Economic Community. London: Graham & Trotman, annual.

> *Volume 1 contains data on the top 4500 companies in the EC: principal officers, subsidiaries, trade names, principal activities, telex and cable address, financial figures, employees, and principal bankers. Two volumes.*

Note: These almanacs, dictionaries, and encyclopedias are updated regularly.

Directories

Directory of American Firms Operating in Foreign Countries. New York: Simon & Schuster, 1987. Updated periodically.

> *Volume 1 presents in alphabetical order, the 4300 U.S. firms operating overseas. For each entry, it gives the firm's address and telephone number, chief executive officer, principal products or services, and foreign country of operation. Volumes 2 and 3 list each firm by the country in which it has a subsidiary and give similar information to that in Volume 1.*

Who Owns Whom: Continental Europe. London: Dun & Bradstreet, Ltd., 1988. Updated periodically.

> *Volume 1 contains an alphabetical listing of appropriate companies, by country. There is also a listing of foreign parent companies. Volume 2 contains an alphabetical index to subsidiary and associate companies, showing their parent companies.*

Yearbooks

Europa Year Book. London: Europa Publications.

> *For each country, data includes recent history, basic economic statistics, media, financial institutions, chambers of commerce, trade and industrial organizations, and important international organizations. Two volumes published annually. (See http://europa.cu.int.)*

Jane's Major Companies of Europe. London: S. Low, Marston & Co.

> *Contains information on the important companies in 16 European countries. For each company, includes information on activities, chief executive officers' names, capital structure, principal subsidiaries, revenue, dividends and earnings, investments and assets.*

New African Yearbook. London: I. C. Magazines, Ltd.

> *Provides essential statistics on Africa and an overview of its economy. Includes for each of the 53 countries, statistics on production, revenue, balance of trade, and imports and exports.*

UNESCO Statistical Yearbook. Louvain, Belgium: UNESCO.

> *Detailed statistics on a wide range of variables for most countries.*

Yearbook of International Organizations. Vol. I: Organization Descriptions and Index. Munich and New York: K. G. Saur, annual. Can be purchased in U.S. from Gale Research Co., Detroit, MI.

> *An excellent, comprehensive directory of more than 19,000 international organizations of all types, usually giving for each: year founded, aim, structure, officers, consultative status, activities, publications, countries in which there are members.*

Yearbook of International Trade Statistics. New York: United Nations Publishing Service.

> *A two-volume publication containing statistics on individual countries, regions, and commodities. For U.N. publications, see http://www.un.org/pubs/.*

Note: These yearbooks are updated regularly.

A NOTE ON THE STANDARD INDUSTRIAL CLASSIFICATION (SIC) AND THE NORTH AMERICAN INDUSTRY CLASSIFICATION SYSTEM (NAICS)

SIC

The Standard Industrial Classification system is used in the classification of companies, established by type of activity in which they are engaged. It facilitates the collection, tabulation, presentation, and analysis of data relating to establishments. The SIC promotes uniformity and comparability in the presentation of statistical data collected by various federal government agencies, state agencies, trade associations, and private research organizations.

The SIC divides the nation's economic activities into 11 broad industrial divisions (generally identified by the first digit of the codes), two-digit major groups, three-digit industry subgroups, and four-digit detailed industries. For example:

Division	2	Manufacturing
Major group	22	Textile mill products
Industry subgroup	225	Knitting mills
Detailed industry	2254	Knit underwear mills

In some instances, an even more detailed classification has been devised explicitly for census purposes so that additional industries, kinds of business, or specific products can be identified within the SIC categories. For example, the Census Bureau has developed a system of classifying manufactured products into approximately 1000 five-digit product classes and 10,000 seven-digit products, consistent with the SIC system.

From the Office of Management and Budget, *Standard Industrial Classification Manual,* 1987:

01 to 09	Agriculture, forestry, and fishing
10 to 14	Mining
15 to 17	Construction
20 to 39	Manufacturing
40 to 49	Transportation and public utilities
50 to 51	Wholesale trade
52 to 59	Retail trade
60 to 67	Finance, insurance, and real estate
70 to 89	Services
91 to 97	Public administration
99	Nonclassifiable establishments

NAICS

On April 9, 1997, the Office of Management and Budget (OMB) announced the adoption of a new industry classification, replacing the Standard Industrial Classification (SIC). Both systems are identified here because the transition time will take years for the new system to be fully adopted. The updating of industry classifications is nothing new. Since its origination in the 1930s, the SIC system has been revised or updated every 10 or 15 years to reflect new developments in the American economy and to address problems identified by data users and statistical agencies.

The objectives for the 1997 revision were much broader. Not only was the system to identify new industries, but the process also sought to reorganize the system according to a more consistent economic principle—according to types of production activities performed—rather than the mixture of production-based and market-based categories in the SIC. That reorganization would allow for the presentation of more detail for the rapidly expanding service sector, which accounts for most economic activity but for only 40% of SIC categories. Further, the system was redefined jointly with Canada and Mexico so that comparable statistics could be obtained for the three NAFTA trading partners.

NAICS groups the economy into 20 broad sectors, up from the 10 divisions of the SIC system. Many of the new sectors reflect recognizable parts of SIC divisions, such as the Utilities and Transportation sectors, broken out from the SIC division Transportation, Communications, and Utilities. Similarly, the SIC division for Service Industries has been subdivided to form several new sectors with longer names: Professional, Scientific and Technical Services; Management, Support, Waste Management, and Remediation Services; Education Services; Health and Social Assistance; Arts, Entertainment, and Recreation; and Other Services except Public Administration.

Other sectors represent combinations of pieces from more than one SIC division. The new Information sector includes major components from Transportation, Communications, and Utilities (broadcasting and telecommunications), Manufacturing (publishing), and Services Industries (software publishing, data processing, information services, motion picture and sound recording). The Accommodation and Foodservices sector puts together hotels and other lodging places from Service Industries and eating and drinking places from Retail Trade.

NAICS industries are identified by a 6-digit code, in contrast to the 4-digit SIC code. The longer code accommodates the larger number of sectors and allows more flexibility in designating subsectors. It also provides for additional detail not necessarily appropriate for all three NAICS countries. The international NAICS agreement fixes only the first five digits of the code. The sixth digit, where used, identifies subdivisions of NAICS industries that accommodate user needs in individual countries. Thus, 6-digit U.S. codes

Figure A.1 — Netscape browser window showing "New Code System in NAICS" at http://www.census.gov/epcd/www/naicscod.htm

Examples of NAICS Hierarchy

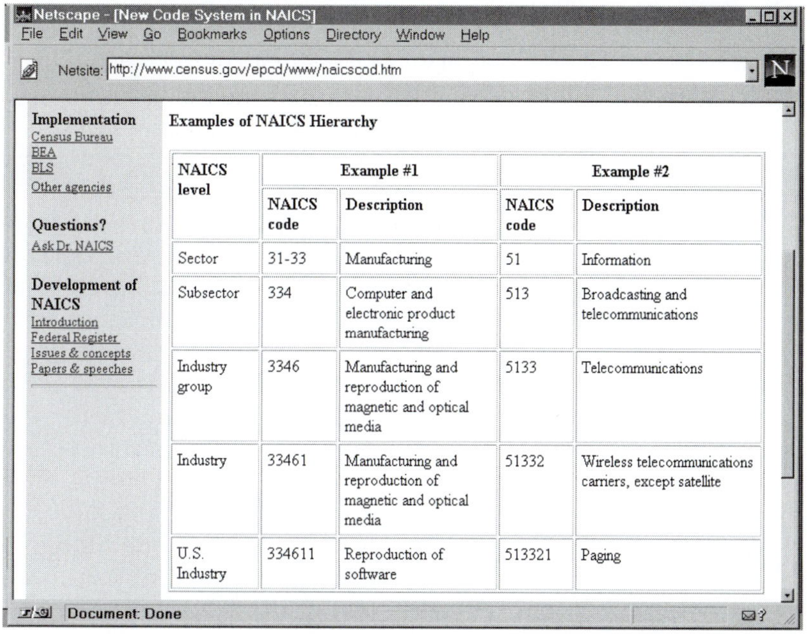

NAICS level	Example #1		Example #2	
	NAICS code	Description	NAICS code	Description
Sector	31-33	Manufacturing	51	Information
Subsector	334	Computer and electronic product manufacturing	513	Broadcasting and telecommunications
Industry group	3346	Manufacturing and reproduction of magnetic and optical media	5133	Telecommunications
Industry	33461	Manufacturing and reproduction of magnetic and optical media	51332	Wireless telecommunications carriers, except satellite
U.S. Industry	334611	Reproduction of software	513321	Paging

may differ from counterparts in Canada or Mexico, but at the 5-digit level they are standardized. Figure A.1 is a screen capture from the Census Bureau's Web pages, illustrating this numbering system. The reader is encouraged to explore this system by going to http://www.census.gov/epcd/www/naisusr.html, the source of this information.

APPENDIX

B

Selected Statistical Tables

B.1 Areas Under the Normal Curve

B.2 Distribution of t

B.3 F Distribution

B.4 Distribution of χ^2

TABLE B.1 Areas Under the Normal Curve

z	.00	.01	.02	.03	.04	.05	.06	.07	.08	.09
.0	.0000	.0040	.0080	.0120	.0160	.0199	.0239	.0279	.0319	.0359
.1	.0398	.0438	.0478	.0517	.0557	.0596	.0636	.0675	.0714	.0753
.2	.0793	.0832	.0871	.0910	.0948	.0987	.1026	.1064	.1103	.1141
.3	.1179	.1217	.1255	.1293	.1331	.1368	.1406	.1443	.1480	.1517
.4	.1554	.1591	.1628	.1664	.1700	.1736	.1772	.1808	.1844	.1879
.5	.1915	.1950	.1985	.2019	.2054	.2088	.2123	.2157	.2190	.2224
.6	.2257	.2291	.2324	.2357	.2389	.2422	.2454	.2486	.2518	.2549
.7	.2580	.2612	.2642	.2673	.2704	.2734	.2764	.2794	.2823	.2852
.8	.2881	.2910	.2939	.2967	.2995	.3023	.3051	.3078	.3106	.3133
.9	.3159	.3186	.3212	.3238	.3264	.3289	.3315	.3340	.3365	.3389
1.0	.3413	.3438	.3461	.3485	.3508	.3531	.3554	.3577	.3599	.3621
1.1	.3643	.3665	.3686	.3708	.3729	.3749	.3770	.3790	.3810	.3830
1.2	.3849	.3869	.3888	.3907	.3925	.3944	.3962	.3980	.3997	.4015
1.3	.4032	.4049	.4066	.4082	.4099	.4115	.4131	.4147	.4162	.4177
1.4	.4192	.4207	.4222	.4236	.4251	.4265	.4279	.4292	.4306	.4319
1.5	.4332	.4345	.4357	.4370	.4382	.4394	.4406	.4418	.4429	.4441
1.6	.4452	.4463	.4474	.4484	.4495	.4505	.4515	.4525	.4535	.4545
1.7	.4554	.4564	.4573	.4582	.4591	.4599	.4608	.4616	.4625	.4633
1.8	.4641	.4649	.4656	.4664	.4671	.4678	.4686	.4693	.4699	.4706
1.9	.4713	.4719	.4726	.4732	.4738	.4744	.4750	.4756	.4761	.4767

	.00	.01	.02	.03	.04	.05	.06	.07	.08	.09
2.0	.4772	.4778	.4783	.4788	.4793	.4798	.4803	.4808	.4812	.4817
2.1	.4821	.4826	.4830	.4834	.4838	.4842	.4846	.4850	.4854	.4857
2.2	.4861	.4864	.4868	.4871	.4875	.4878	.4881	.4884	.4887	.4890
2.3	.4893	.4896	.4898	.4901	.4904	.4906	.4909	.4911	.4913	.4916
2.4	.4918	.4920	.4922	.4925	.4927	.4929	.4931	.4932	.4934	.4936
2.5	.4938	.4940	.4941	.4943	.4945	.4946	.4948	.4949	.4951	.4952
2.6	.4953	.4955	.4956	.4957	.4959	.4960	.4961	.4962	.4963	.4964
2.7	.4965	.4966	.4967	.4968	.4969	.4970	.4971	.4972	.4973	.4974
2.8	.4974	.4975	.4976	.4977	.4977	.4978	.4979	.4979	.4980	.4981
2.9	.4981	.4982	.4982	.4983	.4984	.4984	.4985	.4985	.4986	.4986
3.0	.49865	.49869	.49874	.49878	.49882	.49886	.49889	.49893	.49897	.49900
3.1	.49903	.49906	.49910	.49913	.49916	.49918	.49921	.49924	.49926	.49929
3.2	.49931	.49934	.49936	.49938	.49940	.49942	.49944	.49946	.49948	.49950
3.3	.49952	.49953	.49955	.49957	.49958	.49960	.49961	.49962	.49964	.49965
3.4	.49966	.49968	.49969	.49970	.49971	.49972	.49973	.49974	.49975	.49976
3.5	.49977	.49978	.49978	.49979	.49980	.49981	.49981	.49982	.49983	.49983
3.6	.49984	.49985	.49985	.49986	.49986	.49987	.49987	.49988	.49988	.49989
3.7	.49989	.49990	.49990	.49990	.49991	.49991	.49992	.49992	.49992	.49992
3.8	.49993	.49993	.49993	.49994	.49994	.49994	.49994	.49995	.49995	.49995
3.9	.49995	.49995	.49996	.49996	.49996	.49996	.49996	.49996	.49997	.49997

Entry represents area under the standardized normal distribution from the mean to Z.

TABLE **B.2** Distribution of t

Degrees of Freedom	Probability													
n	.9	.8	.7	.6	.5	.4	.3	.2	.1	.05	.02	.01	.001
1	.158	.325	.510	.727	1.000	1.376	1.963	3.078	6.314	12.706	31.821	63.657	636.619
2	.142	.289	.445	.617	.816	1.061	1.386	1.886	2.920	4.303	6.965	9.925	31.598
3	.137	.277	.424	.584	.765	.978	1.250	1.638	2.353	3.182	4.541	5.841	12.924
4	.134	.271	.414	.569	.741	.941	1.190	1.533	2.132	2.776	3.747	4.604	8.610
5	.132	.267	.408	.559	.727	.920	1.156	1.476	2.015	2.571	3.365	4.032	6.869
6	.131	.265	.404	.553	.718	.906	1.134	1.440	1.943	2.447	3.143	3.707	5.959
7	.130	.263	.402	.549	.711	.896	1.119	1.415	1.895	2.365	2.998	3.499	5.408
8	.130	.262	.399	.546	.706	.889	1.108	1.397	1.860	2.306	2.896	3.355	5.041
9	.129	.261	.398	.543	.703	.883	1.100	1.383	1.833	2.262	2.821	3.250	4.781
10	.129	.260	.397	.542	.700	.879	1.093	1.372	1.812	2.228	2.764	3.169	4.587
11	.129	.260	.396	.540	.697	.876	1.088	1.363	1.796	2.201	2.718	3.106	4.437
12	.128	.259	.395	.539	.695	.873	1.083	1.356	1.782	2.179	2.681	3.055	4.318
13	.128	.259	.394	.538	.694	.870	1.079	1.350	1.771	2.160	2.650	3.012	4.221
14	.128	.258	.393	.537	.692	.868	1.076	1.345	1.761	2.145	2.624	2.977	4.140
15	.128	.258	.393	.536	.691	.866	1.074	1.341	1.753	2.131	2.602	2.947	4.073
16	.128	.258	.392	.535	.690	.865	1.071	1.337	1.746	2.120	2.583	2.921	4.015
17	.128	.257	.392	.534	.689	.863	1.069	1.333	1.740	2.110	2.567	2.898	3.965
18	.127	.257	.392	.534	.688	.862	1.067	1.330	1.734	2.101	2.552	2.878	3.922
19	.127	.257	.391	.533	.688	.861	1.066	1.328	1.729	2.093	2.539	2.861	3.883
20	.127	.257	.391	.533	.687	.860	1.064	1.325	1.725	2.086	2.528	2.845	3.850

df												
21	.127	.257	.391	.532	.686	.859	1.063	1.323	1.721	2.080	2.518	2.831	3.819
22	.127	.256	.390	.532	.686	.858	1.061	1.321	1.717	2.074	2.508	2.819	3.792
23	.127	.256	.390	.532	.685	.858	1.060	1.319	1.714	2.069	2.500	2.807	3.767
24	.127	.256	.390	.531	.685	.857	1.059	1.318	1.711	2.064	2.492	2.797	3.745
25	.127	.256	.390	.531	.684	.856	1.058	1.316	1.708	2.060	2.485	2.787	3.725
26	.127	.256	.390	.531	.684	.856	1.058	1.315	1.706	2.056	2.479	2.779	3.707
27	.127	.256	.389	.531	.684	.855	1.057	1.314	1.703	2.052	2.473	2.771	3.690
28	.127	.256	.389	.530	.683	.855	1.056	1.313	1.701	2.048	2.467	2.763	3.674
29	.127	.256	.389	.530	.683	.854	1.055	1.311	1.699	2.045	2.462	2.756	3.659
30	.127	.256	.389	.530	.683	.854	1.055	1.310	1.697	2.042	2.457	2.750	3.646
40	.126	.255	.388	.529	.681	.851	1.050	1.303	1.684	2.021	2.423	2.704	3.551
60	.126	.254	.387	.527	.679	.848	1.046	1.296	1.671	2.000	2.390	2.660	3.460
120	.126	.254	.386	.526	.677	.845	1.041	1.289	1.658	1.980	2.358	2.617	3.373
∞	.126	.253	.385	.524	.674	.842	1.036	1.282	1.645	1.960	2.326	2.576	3.291

Source: From R. A. Fisher and F. Yates, *Statistical Tables for Biological, Agricultural and Medical Research*, 6th ed. Published by Longman Group UK Ltd., London, 1974. Used by permission of the authors and publishers.

TABLE **B.3** *F* Distribution

Percentage Points of the *F* Distribution

	Upper 5% Points ($\alpha = 0.05$)						
v_2 \ v_1	1	2	3	4	5	6	7
1	161.4	199.5	215.7	224.6	230.2	234.0	236.8
2	18.51	19.00	19.16	19.25	19.30	19.33	19.35
3	10.13	9.55	9.28	9.12	9.01	8.94	8.89
4	7.71	6.94	6.59	6.39	6.26	6.16	6.09
5	6.61	5.79	5.41	5.19	5.05	4.95	4.88
6	5.99	5.14	4.76	4.53	4.39	4.28	4.21
7	5.59	4.74	4.35	4.12	3.97	3.87	3.79
8	5.32	4.46	4.07	3.84	3.69	3.58	3.50
9	5.12	4.26	3.86	3.63	3.48	3.37	3.29
10	4.96	4.10	3.71	3.48	3.33	3.22	3.14
11	4.84	3.98	3.59	3.36	3.20	3.09	3.01
12	4.75	3.89	3.49	3.26	3.11	3.00	2.91
13	4.67	3.81	3.41	3.18	3.03	2.92	2.83
14	4.60	3.74	3.34	3.11	2.96	2.85	2.76
15	4.54	3.68	3.29	3.06	2.90	2.79	2.71
16	4.49	3.63	3.24	3.01	2.85	2.74	2.66
17	4.45	3.59	3.20	2.96	2.81	2.70	2.61
18	4.41	3.55	3.16	2.93	2.77	2.66	2.58
19	4.38	3.52	3.13	2.90	2.74	2.63	2.54
20	4.35	3.49	3.10	2.87	2.71	2.60	2.51
21	4.32	3.47	3.07	2.84	2.68	2.57	2.49
22	4.30	3.44	3.05	2.82	2.66	2.55	2.46
23	4.28	3.42	3.03	2.80	2.64	2.53	2.44
24	4.26	3.40	3.01	2.78	2.62	2.51	2.42
25	4.24	3.39	2.99	2.76	2.60	2.49	2.40
26	4.23	3.37	2.98	2.74	2.59	2.47	2.39
27	4.21	3.35	2.96	2.73	2.57	2.46	2.37
28	4.20	3.34	2.95	2.71	2.56	2.45	2.36
29	4.18	3.33	2.93	2.70	2.55	2.43	2.35
30	4.17	3.32	2.92	2.69	2.53	2.42	2.33
40	4.08	3.23	2.84	2.61	2.45	2.34	2.25
60	4.00	3.15	2.76	2.53	2.37	2.25	2.17
120	3.92	3.07	2.68	2.45	2.29	2.17	2.09
∞	3.84	3.00	2.60	2.37	2.21	2.10	2.01

v_2 \ v_1	8	9	10	12	15	20
	Upper 5% Points ($\alpha = 0.05$)					
1	238.9	240.5	241.9	243.9	245.9	248.0
2	19.37	19.38	19.40	19.41	19.43	19.45
3	8.85	8.81	8.79	8.74	8.70	8.66
4	6.04	6.00	5.96	5.91	5.86	5.80
5	4.82	4.77	4.74	4.68	4.62	4.56
6	4.15	4.10	4.06	4.00	3.94	3.87
7	3.73	3.68	3.64	3.57	3.51	3.44
8	3.44	3.39	3.35	3.28	3.22	3.15
9	3.23	3.18	3.14	3.07	3.01	2.94
10	3.07	3.02	2.98	2.91	2.85	2.77
11	2.95	2.90	2.85	2.79	2.72	2.65
12	2.85	2.80	2.75	2.69	2.62	2.54
13	2.77	2.71	2.67	2.60	2.53	2.46
14	2.70	2.65	2.60	2.53	2.46	2.39
15	2.64	2.59	2.54	2.48	2.40	2.33
16	2.59	2.54	2.49	2.42	2.35	2.28
17	2.55	2.49	2.45	2.38	2.31	2.23
18	2.51	2.46	2.41	2.34	2.27	2.19
19	2.48	2.42	2.38	2.31	2.23	2.16
20	2.45	2.39	2.35	2.28	2.20	2.12
21	2.42	2.37	2.32	2.25	2.18	2.10
22	2.40	2.34	2.30	2.23	2.15	2.07
23	2.37	2.32	2.27	2.20	2.13	2.05
24	2.36	2.30	2.25	2.18	2.11	2.03
25	2.34	2.28	2.24	2.16	2.09	2.01
26	2.32	2.27	2.22	2.15	2.07	1.99
27	2.31	2.25	2.20	2.13	2.06	1.97
28	2.29	2.24	2.19	2.12	2.04	1.96
29	2.28	2.22	2.18	2.10	2.03	1.94
30	2.27	2.21	2.16	2.09	2.01	1.93
40	2.18	2.12	2.08	2.00	1.92	1.84
60	2.10	2.04	1.99	1.92	1.84	1.75
120	2.02	1.96	1.91	1.83	1.75	1.66
∞	1.94	1.88	1.83	1.75	1.67	1.57

(Continued)

TABLE **B.3** *(Continued)*

v_2 \ v_1	24	30	40	60	120	∞
			Upper 5% Points ($\alpha = 0.05$)			
1	249.1	250.1	251.1	252.2	253.3	254.3
2	19.45	19.46	19.47	19.48	19.49	19.50
3	8.64	8.62	8.59	8.57	8.55	8.53
4	5.77	5.75	5.72	5.69	5.66	5.63
5	4.53	4.50	4.46	4.43	4.40	4.36
6	3.84	3.81	3.77	3.74	3.70	3.67
7	3.41	3.38	3.34	3.30	3.27	3.23
8	3.12	3.08	3.04	3.01	2.97	2.93
9	2.90	2.86	2.83	2.79	2.75	2.71
10	2.74	2.70	2.66	2.62	2.58	2.54
11	2.61	2.57	2.53	2.49	2.45	2.40
12	2.51	2.47	2.43	2.38	2.34	2.30
13	2.42	2.38	2.34	2.30	2.25	2.21
14	2.35	2.31	2.27	2.22	2.18	2.13
15	2.29	2.25	2.20	2.16	2.11	2.07
16	2.24	2.19	2.15	2.11	2.06	2.01
17	2.19	2.15	2.10	2.06	2.01	1.96
18	2.15	2.11	2.06	2.02	1.97	1.92
19	2.11	2.07	2.03	1.98	1.93	1.88
20	2.08	2.04	1.99	1.95	1.90	1.84
21	2.05	2.01	1.96	1.92	1.87	1.81
22	2.03	1.98	1.94	1.89	1.84	1.78
23	2.01	1.96	1.91	1.86	1.81	1.76
24	1.98	1.94	1.89	1.84	1.79	1.73
25	1.96	1.92	1.87	1.82	1.77	1.71
26	1.95	1.90	1.85	1.80	1.75	1.69
27	1.93	1.88	1.84	1.79	1.73	1.67
28	1.91	1.87	1.82	1.77	1.71	1.65
29	1.90	1.85	1.81	1.75	1.70	1.64
30	1.89	1.84	1.79	1.74	1.68	1.62
40	1.79	1.74	1.69	1.64	1.58	1.51
60	1.70	1.65	1.59	1.53	1.47	1.39
120	1.61	1.55	1.50	1.43	1.35	1.25
∞	1.52	1.46	1.39	1.32	1.22	1.00

			Upper 2.5% Points ($\alpha = 0.025$)				
v_2\v_1	1	2	3	4	5	6	7
1	647.8	799.5	864.2	899.6	921.8	937.1	948.2
2	38.51	39.00	39.17	39.25	39.30	39.33	39.36
3	17.44	16.04	15.44	15.10	14.88	14.73	14.62
4	12.22	10.65	9.98	9.60	9.36	9.20	9.07
5	10.01	8.43	7.76	7.39	7.15	6.98	6.85
6	8.81	7.26	6.60	6.23	5.99	5.82	5.70
7	8.07	6.54	5.89	5.52	5.29	5.12	4.99
8	7.57	6.06	5.42	5.05	4.82	4.65	4.53
9	7.21	5.71	5.08	4.72	4.48	4.32	4.20
10	6.94	5.46	4.83	4.47	4.24	4.07	3.95
11	6.72	5.26	4.63	4.28	4.04	3.88	3.76
12	6.55	5.10	4.47	4.12	3.89	3.73	3.61
13	6.41	4.97	4.35	4.00	3.77	3.60	3.48
14	6.30	4.86	4.24	3.89	3.66	3.50	3.38
15	6.20	4.77	4.15	3.80	3.58	3.41	3.29
16	6.12	4.69	4.08	3.73	3.50	3.34	3.22
17	6.04	4.62	4.01	3.66	3.44	3.28	3.16
18	5.98	4.56	3.95	3.61	3.38	3.22	3.10
19	5.92	4.51	3.90	3.56	3.33	3.17	3.05
20	5.87	4.46	3.86	3.51	3.29	3.13	3.01
21	5.83	4.42	3.82	3.48	3.25	3.09	2.97
22	5.79	4.38	3.78	3.44	3.22	3.05	2.93
23	5.75	4.35	3.75	3.41	3.18	3.02	2.90
24	5.72	4.32	3.72	3.38	3.15	2.99	2.87
25	5.69	4.29	3.69	3.35	3.13	2.97	2.85
26	5.66	4.27	3.67	3.33	3.10	2.94	2.82
27	5.63	4.24	3.65	3.31	3.08	2.92	2.80
28	5.61	4.22	3.63	3.29	3.06	2.90	2.78
29	5.59	4.20	3.61	3.27	3.04	2.88	2.76
30	5.57	4.18	3.59	3.25	3.03	2.87	2.75
40	5.42	4.05	3.46	3.13	2.90	2.74	2.62
60	5.29	3.93	3.34	3.01	2.79	2.63	2.51
120	5.15	3.80	3.23	2.89	2.67	2.52	2.39
∞	5.02	3.69	3.12	2.79	2.57	2.41	2.29

(Continued)

	Upper 2.5% Points ($\alpha = 0.025$)					
v_2 \ v_1	8	9	10	12	15	20
1	956.7	963.3	968.6	976.7	984.9	993.1
2	39.37	39.39	39.40	39.41	39.43	39.45
3	14.54	14.47	14.42	14.34	14.25	14.17
4	8.98	8.90	8.84	8.75	8.66	8.56
5	6.76	6.68	6.62	6.52	6.43	6.33
6	5.60	5.52	5.46	5.37	5.27	5.17
7	4.90	4.82	4.76	4.67	4.57	4.47
8	4.43	4.36	4.30	4.20	4.10	4.00
9	4.10	4.03	3.96	3.87	3.77	3.67
10	3.85	3.78	3.72	3.62	3.52	3.42
11	3.66	3.59	3.53	3.43	3.33	3.23
12	3.51	3.44	3.37	3.28	3.18	3.07
13	3.39	3.31	3.25	3.15	3.05	2.95
14	3.29	3.21	3.15	3.05	2.95	2.84
15	3.20	3.12	3.06	2.96	2.86	2.76
16	3.12	3.05	2.99	2.89	2.79	2.68
17	3.06	2.98	2.92	2.82	2.72	2.62
18	3.01	2.93	2.87	2.77	2.67	2.56
19	2.96	2.88	2.82	2.72	2.62	2.51
20	2.91	2.84	2.77	2.68	2.57	2.46
21	2.87	2.80	2.73	2.64	2.53	2.42
22	2.84	2.76	2.70	2.60	2.50	2.39
23	2.81	2.73	2.67	2.57	2.47	2.36
24	2.78	2.70	2.64	2.54	2.44	2.33
25	2.75	2.68	2.61	2.51	2.41	2.30
26	2.73	2.65	2.59	2.49	2.39	2.28
27	2.71	2.63	2.57	2.47	2.36	2.25
28	2.69	2.61	2.55	2.45	2.34	2.23
29	2.67	2.59	2.53	2.43	2.32	2.21
30	2.65	2.57	2.51	2.41	2.31	2.20
40	2.53	2.45	2.39	2.29	2.18	2.07
60	2.41	2.33	2.27	2.17	2.06	1.94
120	2.30	2.22	2.16	2.05	2.94	1.82
∞	2.19	2.11	2.05	1.94	1.83	1.71

TABLE **B.3** (Continued)

$v_2 \backslash v_1$	Upper 2.5% Points ($\alpha = 0.025$)					
	24	30	40	60	120	∞
1	997.2	1001	1006	1010	1014	1018
2	39.46	39.46	39.47	39.48	39.49	39.50
3	14.12	14.08	14.04	13.99	13.95	13.90
4	8.51	8.46	8.41	8.36	8.31	8.26
5	6.28	6.23	6.18	6.12	6.07	6.02
6	5.12	5.07	5.01	4.96	4.90	4.85
7	4.42	4.36	4.31	4.25	4.20	4.14
8	3.95	3.89	3.84	3.78	3.73	3.67
9	3.61	3.56	3.51	3.45	3.39	3.33
10	3.37	3.31	3.26	3.20	3.14	3.08
11	3.17	3.12	3.06	3.00	2.94	2.88
12	3.02	2.96	2.91	2.85	2.79	2.72
13	2.89	2.84	2.78	2.72	2.66	2.60
14	2.79	2.73	2.67	2.61	2.55	2.49
15	2.70	2.64	2.59	2.52	2.46	2.40
16	2.63	2.57	2.51	2.45	2.38	2.32
17	2.56	2.50	2.44	2.38	2.32	2.25
18	2.50	2.44	2.38	2.32	2.26	2.19
19	2.45	2.39	2.33	2.27	2.20	2.13
20	2.41	2.35	2.29	2.22	2.16	2.09
21	2.37	2.31	2.25	2.18	2.11	2.04
22	2.33	2.27	2.21	2.14	2.08	2.00
23	2.30	2.24	2.18	2.11	2.04	1.97
24	2.27	2.21	2.15	2.08	2.01	1.94
25	2.24	2.18	2.12	2.05	1.98	1.91
26	2.22	2.16	2.09	2.03	1.95	1.88
27	2.19	2.13	2.07	2.00	1.93	1.85
28	2.17	2.11	2.05	1.98	1.91	1.83
29	2.15	2.09	2.03	1.96	1.89	1.81
30	2.14	2.07	2.01	1.94	1.87	1.79
40	2.01	1.94	1.88	1.80	1.72	1.64
60	1.88	1.82	1.74	1.67	1.58	1.48
120	1.76	1.69	1.61	1.53	1.43	1.31
∞	1.64	1.57	1.48	1.39	1.27	1.00

(Continued)

TABLE **B.3** *(Continued)*

v_2 \ v_1	1	2	3	4	5	6	7
Upper 1% Points ($\alpha = 0.01$)							
1	4052	4999.5	5403	5625	5764	5859	5928
2	98.50	99.00	99.17	99.25	99.30	99.33	99.36
3	34.12	30.82	29.46	28.71	28.24	27.91	27.67
4	21.20	18.00	16.69	15.98	15.52	15.21	14.98
5	16.26	13.27	12.06	11.39	10.97	10.67	10.46
6	13.75	10.92	9.78	9.15	8.75	8.47	8.26
7	12.25	9.55	8.45	7.85	7.46	7.19	6.99
8	11.26	8.65	7.59	7.01	6.63	6.37	6.18
9	10.56	8.02	6.99	6.42	6.06	5.80	5.61
10	10.04	7.56	6.55	5.99	5.64	5.39	5.20
11	9.65	7.21	6.22	5.67	5.32	5.07	4.89
12	9.33	6.93	5.95	5.41	5.06	4.82	4.64
13	9.07	6.70	5.74	5.21	4.86	4.62	4.44
14	8.86	6.51	5.56	5.04	4.69	4.46	4.28
15	8.68	6.36	5.42	4.89	4.56	4.32	4.14
16	8.53	6.23	5.29	4.77	4.44	4.20	4.03
17	8.40	6.11	5.18	4.67	4.34	4.10	3.93
18	8.29	6.01	5.09	4.58	4.25	4.01	3.84
19	8.18	5.93	5.01	4.50	4.17	3.94	3.77
20	8.10	5.85	4.94	4.43	4.10	3.87	3.70
21	8.02	5.78	4.87	4.37	4.04	3.81	3.64
22	7.95	5.72	4.82	4.31	3.99	3.76	3.59
23	7.88	5.66	4.76	4.26	3.94	3.71	3.54
24	7.82	5.61	4.72	4.22	3.90	3.67	3.50
25	7.77	5.57	4.68	4.18	3.85	3.63	3.46
26	7.72	5.53	4.64	4.14	3.82	3.59	3.42
27	7.68	5.49	4.60	4.11	3.78	3.56	3.39
28	7.64	5.45	4.57	4.07	3.75	3.53	3.36
29	7.60	5.42	4.54	4.04	3.73	3.50	3.33
30	7.56	5.39	4.51	4.02	3.70	3.47	3.30
40	7.31	5.18	4.31	3.83	3.51	3.29	3.12
60	7.08	4.98	4.13	3.65	3.34	3.12	2.95
120	6.85	4.79	3.95	3.48	3.17	2.96	2.79
∞	6.63	4.61	3.78	3.32	3.02	2.80	2.64

TABLE **B.3** *(Continued)*

			Upper 1% Points ($\alpha = 0.01$)			
v_2 \ v_1	8	9	10	12	15	20
1	5982	6022	6056	6106	6157	6209
2	99.37	99.39	99.40	99.42	99.43	99.45
3	27.49	27.35	27.23	27.05	26.87	26.69
4	14.80	14.66	14.55	14.37	14.20	14.02
5	10.29	10.16	10.05	9.89	9.72	9.55
6	8.10	7.98	7.87	7.72	7.56	7.40
7	6.84	6.72	6.62	6.47	6.31	6.16
8	6.03	5.91	5.81	5.67	5.52	5.36
9	5.47	5.35	5.26	5.11	4.96	4.81
10	5.06	4.94	4.85	4.71	4.56	4.41
11	4.74	4.63	4.54	4.40	4.25	4.10
12	4.50	4.39	4.30	4.16	4.01	3.86
13	4.30	4.19	4.10	3.96	3.82	3.66
14	4.14	4.03	3.94	3.80	3.66	3.51
15	4.00	3.89	3.80	3.67	3.52	3.37
16	3.89	3.78	3.69	3.55	3.41	3.26
17	3.79	3.68	3.59	3.46	3.31	3.16
18	3.71	3.60	3.51	3.37	3.23	3.08
19	3.63	3.52	3.43	3.30	3.15	3.00
20	3.56	3.46	3.37	3.23	3.09	2.94
21	3.51	3.40	3.31	3.17	3.03	2.88
22	3.45	3.35	3.26	3.12	2.98	2.83
23	3.41	3.30	3.21	3.07	2.93	2.78
24	3.36	3.26	3.17	3.03	2.89	2.74
25	3.32	3.22	3.13	2.99	2.85	2.70
26	3.29	3.18	3.09	2.96	2.81	2.66
27	3.26	3.15	3.06	2.93	2.78	2.63
28	3.23	3.12	3.03	2.90	2.75	2.60
29	3.20	3.09	3.00	2.87	2.73	2.57
30	3.17	3.07	2.98	2.84	2.70	2.55
40	2.99	2.89	2.80	2.66	2.52	2.37
60	2.82	2.72	2.63	2.50	2.35	2.20
120	2.66	2.56	2.47	2.34	2.19	2.03
∞	2.51	2.41	2.32	2.18	2.04	1.88

(Continued)

TABLE **B.3** *(Continued)*

v_2 \ v_1	24	30	40	60	120	∞
			Upper 1% Points ($\alpha = 0.01$)			
1	6235	6261	6287	6313	6339	6366
2	99.46	99.47	99.47	99.48	99.49	99.50
3	26.60	26.50	26.41	26.32	26.22	26.13
4	13.93	13.84	13.75	13.65	13.56	13.46
5	9.47	9.38	9.29	9.20	9.11	9.02
6	7.31	7.23	7.14	7.06	6.97	6.88
7	6.07	5.99	5.91	5.82	5.74	5.65
8	5.28	5.20	5.12	5.03	4.95	4.86
9	4.73	4.65	4.57	4.48	4.40	4.31
10	4.33	4.25	4.17	4.08	4.00	3.91
11	4.02	3.94	3.86	3.78	3.69	3.60
12	3.78	3.70	3.62	3.54	3.45	3.36
13	3.59	3.51	3.43	3.34	3.25	3.17
14	3.43	3.35	3.27	3.18	3.09	3.00
15	3.29	3.21	3.13	3.05	2.96	2.87
16	3.18	3.10	3.02	2.93	2.84	2.75
17	3.08	3.00	2.92	2.83	2.75	2.65
18	3.00	2.92	2.84	2.75	2.66	2.57
19	2.92	2.84	2.76	2.67	2.58	2.49
20	2.86	2.78	2.69	2.61	2.52	2.42
21	2.80	2.72	2.64	2.55	2.46	2.36
22	2.75	2.67	2.58	2.50	2.40	2.31
23	2.70	2.62	2.54	2.45	2.35	2.26
24	2.66	2.58	2.49	2.40	2.31	2.21
25	2.62	2.54	2.45	2.36	2.27	2.17
26	2.58	2.50	2.42	2.33	2.23	2.13
27	2.55	2.47	2.38	2.29	2.20	2.10
28	2.52	2.44	2.35	2.26	2.17	2.06
29	2.49	2.41	2.33	2.23	2.14	2.03
30	2.47	2.39	2.30	2.21	2.11	2.01
40	2.29	2.20	2.11	2.02	1.92	1.80
60	2.12	2.03	1.94	1.84	1.73	1.60
120	1.95	1.86	1.76	1.66	1.53	1.38
∞	1.79	1.70	1.59	1.47	1.32	1.00

Source: E. S. Pearson and H. O. Hartley, Eds., *Biometrika Tables for Statisticians,* vol. 1, 3rd ed. (London: Biometrika Trustees, 1966), pp. 655–659. Reproduced by the kind permission of E. S. Pearson and the trustees of Biometrika.

TABLE **B.4** Distribution of χ^2

| Degrees of Freedom | Probability | | | | | | | | | | | | | |
n	.99	.98	.95	.90	.80	.70	.50	.30	.20	.10	.05	.02	.01	.001
1	.03157	.03628	.00393	.0158	.0642	.148	.455	1.074	1.642	2.706	3.841	5.412	6.635	10.827
2	.0201	.0404	.103	.211	.446	.713	1.386	2.408	3.219	4.605	5.991	7.824	9.210	13.815
3	.115	.185	.352	.584	1.005	1.424	2.366	3.665	4.642	6.251	7.815	9.837	11.345	16.266
4	.297	.429	.711	1.064	1.649	2.195	3.357	4.878	5.989	7.779	9.488	11.668	13.277	18.467
5	.554	.752	1.145	1.610	2.343	3.000	4.351	6.064	7.289	9.236	11.070	13.388	15.086	20.515
6	.872	1.134	1.635	2.204	3.070	3.828	5.348	7.231	8.558	10.645	12.592	15.033	16.812	22.457
7	1.239	1.564	2.167	2.833	3.822	4.671	6.346	8.383	9.803	12.017	14.067	16.622	18.475	24.322
8	1.646	2.032	2.733	3.490	4.594	5.527	7.344	9.524	11.030	13.362	15.507	18.168	20.090	26.125
9	2.088	2.532	3.325	4.168	5.380	6.393	8.343	10.656	12.242	14.684	16.919	19.679	21.666	27.877
10	2.558	3.059	3.940	4.865	6.179	7.267	9.342	11.781	13.442	15.987	18.307	21.161	23.209	29.588
11	3.053	3.609	4.575	5.578	6.989	8.148	10.341	12.899	14.631	17.275	19.675	22.618	24.725	31.264
12	3.571	4.178	5.226	6.304	7.807	9.034	11.340	14.011	15.812	18.549	21.026	24.054	26.217	32.909
13	4.107	4.765	5.892	7.042	8.634	9.926	12.340	15.119	16.985	19.812	22.362	25.472	27.688	34.528
14	4.660	5.368	6.571	7.790	9.467	10.821	13.339	16.222	18.151	21.064	23.685	26.873	29.141	36.123
15	5.229	5.985	7.261	8.547	10.307	11.721	14.339	17.322	19.311	22.307	24.996	28.259	30.578	37.697
16	5.812	6.614	7.962	9.312	11.152	12.624	15.338	18.418	20.465	23.542	26.296	29.633	32.000	39.252
17	6.408	7.255	8.672	10.085	12.002	13.531	16.338	19.511	21.615	24.769	27.587	30.995	33.409	40.790
18	7.015	7.906	9.390	10.865	12.857	14.440	17.338	20.601	22.760	25.989	28.869	32.346	34.805	42.312
19	7.633	8.567	10.117	11.651	13.716	15.352	18.338	21.689	23.900	27.204	30.144	33.687	36.191	43.820
20	8.260	9.237	10.851	12.443	14.578	16.266	19.337	22.775	25.038	28.412	31.410	35.020	37.566	45.318
21	8.897	9.915	11.591	13.240	15.445	17.182	20.337	23.858	26.171	29.615	32.671	36.343	38.932	46.797
22	9.542	10.600	12.338	14.041	16.314	18.101	21.337	24.939	27.301	30.813	33.924	37.659	40.289	48.268
23	10.196	11.293	13.091	14.848	17.187	19.021	22.337	26.018	28.429	32.007	35.172	38.968	41.638	49.728
24	10.856	11.992	13.848	15.659	18.062	19.943	23.337	27.096	29.553	33.196	36.415	40.270	42.980	51.179
25	11.524	12.697	14.611	16.473	18.940	20.867	24.337	28.172	30.675	34.382	37.652	41.566	44.314	52.620
26	12.198	13.409	15.379	17.292	19.820	21.792	25.336	29.246	31.795	35.563	38.885	42.856	45.642	54.052
27	12.879	14.125	16.151	18.114	20.703	22.719	26.336	30.319	32.912	36.741	40.113	44.140	46.963	55.476
28	13.565	14.847	16.928	18.939	21.588	23.647	27.336	31.391	34.027	37.916	41.337	45.419	48.278	56.893

(Continued)

TABLE **B.4** (Continued)

| Degrees of Freedom | | | | | | | Probability | | | | | | | |
n	.99	.98	.95	.90	.80	.70	.50	.30	.20	.10	.05	.02	.01	.001
29	14.256	15.574	17.708	19.768	22.475	24.577	28.336	32.461	35.139	39.087	42.557	46.693	49.588	58.302
30	14.953	16.306	18.493	20.599	23.364	25.508	29.336	33.530	36.250	40.256	43.773	47.962	50.892	59.703
32	16.362	17.783	20.072	22.271	25.148	27.373	31.336	35.665	38.466	42.585	46.194	50.487	53.486	62.487
34	17.789	19.275	21.664	23.952	26.938	29.242	33.336	37.795	40.676	44.903	48.602	52.995	56.061	65.247
36	19.233	20.783	23.269	25.643	28.735	31.115	35.336	39.922	42.879	47.212	50.999	55.489	58.619	67.985
38	20.691	22.304	24.884	27.343	30.537	33.992	37.335	42.045	45.076	49.513	53.384	57.969	61.162	70.703
40	22.164	23.838	26.509	29.051	32.345	34.872	39.335	44.165	47.269	51.805	55.759	60.436	63.691	73.402
42	23.650	25.383	28.144	30.765	34.157	36.755	41.335	46.282	49.456	54.090	58.124	62.892	66.206	76.084
44	25.148	26.939	29.787	32.487	35.974	38.641	43.335	48.396	51.639	56.369	60.481	65.337	68.710	78.750
46	26.657	28.504	31.439	34.215	37.795	40.529	45.335	50.507	53.818	58.641	62.830	67.771	71.201	81.400
48	28.177	30.080	33.098	35.949	39.621	42.420	47.335	52.616	55.993	60.907	65.171	70.197	73.683	84.037
50	29.707	31.664	34.764	37.689	41.449	44.313	49.335	54.723	58.164	63.167	67.505	72.613	76.154	86.661
52	31.246	33.256	36.437	39.433	43.281	46.209	51.335	56.827	60.332	65.422	69.832	75.021	78.616	89.272
54	32.793	34.856	38.116	41.183	45.117	48.106	53.335	58.930	62.496	67.673	72.153	77.422	81.069	91.872
56	34.350	36.464	39.801	42.937	46.955	50.005	55.335	61.031	64.658	69.919	74.468	79.815	83.513	94.461
58	35.913	38.078	41.492	44.696	48.797	51.906	57.335	63.129	66.816	72.160	76.778	82.201	85.950	97.039
60	37.485	39.699	43.188	46.459	50.641	53.809	59.335	65.227	68.972	74.397	79.082	84.580	88.379	99.607
62	39.063	41.327	44.889	48.226	52.487	55.714	61.335	67.322	71.125	76.630	81.381	86.953	90.802	102.166
64	40.649	42.960	46.595	49.996	54.336	57.620	63.335	69.416	73.276	78.860	83.675	89.320	93.217	104.716
66	42.240	44.599	48.305	51.770	56.188	59.527	65.335	71.508	75.424	81.085	85.965	91.681	95.626	107.258
68	43.838	46.244	50.020	53.548	58.042	61.436	67.335	73.600	77.571	83.308	88.250	94.037	98.028	109.791
70	45.442	47.893	51.739	55.329	59.898	63.346	69.334	75.689	79.715	85.527	90.531	96.388	100.425	112.317

For odd values of n between 30 and 70, the mean of the tabular values for n − 1 and n + 1 may be taken. For larger values of n, the expression $\sqrt{2\chi^2} - \sqrt{2n-1}$ may be used as a normal deviate with unit variance, remembering that the probability for χ^2 corresponds with that of a single tail of the normal curve.

Source: From R. A. Fisher and F. Yates, Statistical Tables for Biological, Agricultural and Medical Research, 6th ed. Published by Longman Group UK Ltd., London, 1974. Used by permission of the authors and publishers.

INDEX

A. C. Nielsen, 297
A priori hypothesis testing, 49, 433
AAPOR (American Association for Public Opinion Research), 471
ABI/INFORM index, 544
Access, personal interview condition, 277–272
Accuracy, of samples, 232
Active (primary) data collection
    defined, 271, 273
    in-depth interviews, 310–314
    offline methods, 274–284
    online methods, 284–292
    qualitative research, 310–316
Administration methods, personal interview errors, 276
Advanced experimental designs, 165–170. *See also* Experimental designs
    analysis of covariance (ANCOVA), 170, 386, 417, 419–420
    completely randomized design, 166
    emphasis, 165–166
    factorial design, 169–170
    Latin square design, 168–169
    randomized block design, 166–168
Advanced multivariate analysis, 431–458
Affordability, online research design, 158
AI (artificial intelligence) research, 33
Algebraic models, 386. *See also* Linear models
Almanacs, international business secondary sources, 552
Alternative forms method, reliability, 188, 191
Alternatives
    evaluating, decision-making process, 5, 6
    formulation of, and secondary data, 71, 72–73
    response, 209–210
AMA (American Management Association), 233
AMA (American Marketing Association), 101, 471
American Association for Public Opinion Research (AAPOR), 471
American Management Association (AMA), 233

American Marketing Association (AMA), 101, 471
American Psychological Association (APA), 464, 471
Analysis, 325–358. *See also* Data analysis, planning for
    advanced multivariate, 431–458
    contingency table, 368
    correspondence, 452
    data analysis planning, 325–350
    errors, 137, 138, 140
    multiple linear regression, 413–419
    multiple-variable, 384
    in research process, 17, 18
    sample size, 232, 234–235, 245
    variance and regression techniques, 383–429
Analysis, basic methods
    about, 352, 374–375
    association assessment, 360–368
    classification by purpose, 352–359
    covariation, 381
    difference association assessment, 368–374
    discussion questions, 375–377
    exploratory data analysis (EDA), 360
    mean, calculation, 380
    normal distribution, 379–380
    standardized data array, 380–381
    suggested reading, 378
    variance, calculation, 380
Analysis of covariance (ANCOVA), 170, 386, 417, 419–420
Analysis of dependency, 435–444
    conjoint analysis, 440–442
    covariance structure analysis, 442–444
    dependency structure, 435
    multiple discriminant analysis (MDA), 438–440
    multivariate analysis of variance (MANOVA), 435–438
Analysis of interdependency, 444–454
    cluster analysis, 448–450, 451
    factor analysis, 444–448
    multidimensional scaling, 450, 452–454
Analysis of variance (ANOVA), 383–429. *See also* Multivariate analysis
    about, 384, 422

analysis of covariance (ANCOVA), 170, 386, 417, 419–420
    defined, 386, 392
    discussion questions, 422–425
    factorial, 395–396
    linear models, 387–420
    linear regression, 403–419
    multiple linear regression, 427–429
    nonparametric, 358, 420–422
    one-way, 387–395
    suggested reading, 426
    tables, 393–394, 396, 399, 400–401
    two-way, 395–403
    variance decomposition, nature of, 384–387
Analytic information, 25
Analytical errors, 137, 138
Analytical software, 328–333
ANCOVA (analysis of covariance), 170, 386, 417, 419–420
Anheuser-Busch, 8, 283
Anonymity, 469
ANOVA. *See* Analysis of variance (ANOVA)
APA (American Psychological Association), 464, 471
Appendix in written research reports, 490
Applied Science and Technology Index, 544
Arbitron, 287
Area under the normal curve, statistical tables, 558–559
ARPANET, 86
Artificial intelligence (AI) research, 33
ASEAN (Association of Southeast Asian Nations), 11
Assessment
    of association, 360–368
    of difference, 368–374
    of research value, 105–107, 126–131
Association, methods of assessing, 360–368
Association of Southeast Asian Nations (ASEAN), 11
AT&T, 10–12, 106, 312, 463
Attitude
    concept/construct definitions, 46–47
    scale for, 210, 211
Audience, research reports, 490, 494
Authority, obtaining information through, 9–10

Bar charts, 360
Bartlett-Box test, one-way ANOVA, 394
Basic research, 105–106
Behrens-Fisher test, 374
Bell-shaped curve, 237, 379–380, 558–559
BI (business intelligence), 38, 39
Bias, survey, 282, 283, 284
Bibliographies, as secondary sources, 544–545
Bic Corporation, 315
Binomial variable, 180
Body, written research report, 489–490
Box plots, 360
BR. *See* Business research (BR)
Bramble, William J., 396
Brand comprehension, concept/ construct definitions, 46–47
Budget, as proposal control mechanism, 117, 121–122
Buick Park Avenue automobile, 73
Business intelligence (BI), 38, 39
Business research (BR). *See also* Ethics, in business research; Knowledge management (KM); Research; *specific topics*
    basic analytical framework, 344–346
    data analysis planning framework, 344–346
    and decision making, 1–22
    defined, 8, 19
    discussion questions, 19–20
    domestic vs. international, 11–12
    evaluation, 495–497
    "fuzzy" interrelationships with KM and IS, 31, 32, 38, 39
    global marketplace, 11–12
    knowledge management (KM), 31, 32, 36–39
    manager-researcher relationship, 12–14
    nature of decision making, 2–7
    process of, 15–18
    relevance of scientific inquiry, 55–63
    research proposal typology, 108–110
    role in decision making, 7–12
    state of the art, scientific inquiry, 57–63
    suggested reading, 21–22
Buyer Behavior, Howard and Sheth Theory of, 50, 51, 55

Callbacks, 275
Canada, NAICS, 554, 555
Canon AE-1 camera, 146
CAPI (computer-assisted personal interviewing), 282, 284
Carrier Corporation, 283
Case studies, experimental design, 154–155

CASRO (Council of American Survey Organizations), 469
Caterpillar Tractor Company, 60, 61
CATI (computer-assisted telephone interviewing), 28, 284
CATS (completely automated telephone surveys), 282, 284
Causal modeling, 442–444
Causality, 141, 143
CD-ROMs, 78, 81
CD-ROMs in Print (Gale Research), 78, 81
Census, defined, 231
Census Bureau, U.S., 72
Central Limit Theorem, 237, 265–267, 379–380
Central tendency, 342, 344
Cheating, personal interviews, 276
Chevron Chemical, 72
Chi-square test, 368–370, 571–572
Children's Online Privacy Protection Act (COPPA), 469
China, multinational research, 295
Choice, right of free, 464–465
Chrysler, 14
CIS (customer information system), 72
Citicorp, 463
Classification by purpose, basic analysis, 352–359
Classification variables (factors), 387
Clients' rights, 465–467
Closed-ended item phrasing, 208
Cluster
    analysis of, 448–450, 451
    sampling of, 240, 250
Clustering technique, 297–298
Coca-Cola, 8, 202
Code of ethics, 470–472, 476–482. *See also* Ethics, in business research
Codebook development, 339
Coding
    data preanalysis, 337–340
    dummy codes, 427–429
    effect codes, 427–429
    errors, 339, 340, 341
    measurement, 175
    multiple linear regression, 427–429
Coefficients
    contingency correlation, 362–364
    Pearson product-moment (PM) cor- relation coefficient, 365–368
    phi, 363–364
    of reliability, 189
    Spearman rank correlation, 364–365
Collection errors, 137, 138, 139–140
Common database, 77
Communality, 445, 446
Communication challenges, 497–498
Company references as secondary sources, 551
Comparative rating scales, 211, 212

Comparison, mail nonresponse bias method, 284
Completely automated telephone sur- veys (CATS), 282, 284
Completely randomized experimental design, 166
Complex scales, 179
Computer-assisted personal interview- ing (CAPI), 282, 284
Computer-assisted telephone interview- ing (CATI), 282, 284
Computers. *See also* Knowledge manage- ment (KM); Online and Internet; Technology
    data analysis planning software, 327, 328–333
    interviews and surveys, 282–284
    MONANOVA computer program, 440
Concepts, defined, in scientific inquiry, 45, 46, 47
Concerns, research problem develop- ment, 102–103
Conclusion section, written research report, 490
Concurrent technique, mail response rate inducements, 283
Concurrent validity, 187
Condescriptive analysis, 342
Conference Board, 27
Confidence, and sample size, 234
Confidentiality
    right of, 466
    vendor choice, 295–296
Confusion matrix, 440
Conjoint analysis, 213, 440–442
Connecticut Mutual Life Insurance, 58
Constitutive definitions, 46, 47
Construct
    defined, in scientific inquiry, 45, 46, 47
    validity in, 186–187
Consumer Surveys, 313, 314
Contact rate, telephone interviews, 280–281
Content validity, 185–186
Contents, written research report, 489
Contingency correlation, 362–364
Contingency table analysis, 368
Control
    experimental, 165
    level of, scientific endeavor, 60, 61
    as management function, 117
    statistical, 165–166
Control group, 147
Convenience sampling designs, 241, 251
Convergent validity, 186–187
Cooperation rate
    sampling, 254–255
    telephone interviews, 280–281, 282, 283

COPPA (Children's Online Privacy Protection Act), 469
Corporate development life cycle (SCLC), 420
Correlation
    contingency, 362–364
    Durbin-Watson test for serial, 417
    Spearman rank, 364–365
    spurious, 143
Correlation matrix, 444–445
Correspondence analysis, 452
Cost
    affordability, online research design, 158
    assessment of research value, 105–107, 126–131
    budget as proposal control mechanism, 117, 121–122
    concept/construct definitions, 46–47
    Internet data searches, 81
    primary vs. secondary data collection, 71
    of research, 105–106
    sample design, 235, 238
    secondary data purchase, 91–93
    telephone interviewing, 278
Council of American Survey Organizations (CASRO), 469
Covariance
    analysis of (ANCOVA), 170, 386, 417, 419–420
    defined, 381
    structure analysis, 442–444
Criterion-related validity, 186, 187–188
CRM (customer relationship management), 34
Cronbach-Alpha method, reliability, 192
Cross-break, 361–362
Cross-sectional design, 152, 154
Cross-validation technique, 413, 440, 445
Crosstabulation, 361–362
Customer information system (CIS), 72
Customer relationship management (CRM), 34

DARPA (Defense Advanced Research Projects Agency), 86
Data. *See also* Information; Primary data collection (PDC); Secondary data collection
    collection of, 16–18, 271, 291–292, 310, 311
    defined, 24
    editing, preanalysis, 335–336
    external, 77–78
    fact collection, 45, 46
    information systems management (ISM), 31–33
    information technology, 11
    internal, 77

knowledge management (KM), 24–25
    quality and relevance, 81, 83, 93, 158
    structure generation, 339, 381
Data analysis, planning for, 325–350
    about, 326, 348
    analytical software, 328–333
    basic framework for BR, 344–346
    discussion questions, 348–349
    exploratory, 360
    online research supplies, 333, 334
    outline of process, 326–328
    preanalytical process, 333, 335–344
    software selection, 328–333
    suggested reading, 350
Data array, standardizing, 381
Data marts, 35
Data mining, 33, 36–38
Data-mining tools, 36, 38
Data silos, 31
Data transformation, pyramid of, 25
Data warehouses, 34, 35, 36–37
Database
    common vs. unique, 77
    distributors, 84
    internal secondary data, 77, 79
    online, 80–85
    online industry, 82–85
    producers, 82–83
    types of, 78
Database management (DBM), 35–36
DATASTAR, 87
DBM (database management), 35–36
Decision Analysts, Inc., 258, 315, 318
Decision making, overview, 1–22. *See also specific topics*
    about, 2, 19
    business research process, 15–18
    business research role in, 7–12
    discussion questions, 19–20
    levels of organizational decisions, 3–5
    as linear process, 5–7
    make-or-buy, 292
    manager-researcher relationship, 12–14
    nature of, 2–7
    process of, 5–7
    role of research in, 7–12
    secondary data compilation or purchase, 91–93
    suggested reading, 21–22
Decision support system (DSS), 32, 33–34
Decomposition of variance, 384–387
Deduction, defined, 54
Deductive theory, 53, 54–55
Defense Advanced Research Projects Agency (DARPA), 86
Definitions, constitutive or operational, 46, 47, 177, 178
Degree of disguise, PDC, 273, 310, 311

Degree of structure, PDC, 273, 310
Degrees of freedom, 370
Dependency, analysis of. *See* Analysis of dependency
Dependent variable, 140, 151, 165. *See also* Analysis of variance (ANOVA)
Depth interviews, 310–311, 312–313
Description level, scientific endeavor, 60
Descriptive research, 147
Design, research. *See* Research design
Destructive measurement, sampling, 232
Determinant cause, 141
DIALOG, 85, 87
Diary, defied, 296
Dichotomous variable, 180
Dictionaries, as secondary source, 546, 552
Difference
    assessment methods, 368–374
    association assessment, 368–374
    chi-square test, 368–370
    other tests of, 374
    t-test for differences in means, 371–374, 433
    z-test for differences in proportion, 370–371
Dilemma, researchers', 357
Direct marketing, research vs., 470
Directories, as secondary source, 546, 552
Discovery, data analysis, 346
Discriminant validity, 187
Discussion questions
    analysis, basic methods, 375–377
    analysis of variance (ANOVA), 422–425
    business research (BR), 19–20
    data analysis planning, 348–349
    decision making, overview, 19–20
    ethics, in business research, 473–474
    instrument design, 223
    knowledge management (KM), 39–40
    measurement, 195–196
    multivariate analysis, 455–456
    primary data collection (PDC), 301, 321–322
    problems and proposals, 123–124
    research proposals, 123–124
    research reporting and evaluation, 500–501
    sampling designs, 261–262
    scientific inquiry, 63–64
    secondary data collection, 95
Disguise, degree of, PDC, 273, 310, 311
Disney, 33
Disproportionate stratified sampling, 240, 247–250
Distribution, normal, 379–380
Distribution statistics, 358
Distributors of databases, 84
Document indexes, as secondary sources, 548

Dodge, 14
Domestic business research, 11–12
Dow Chemicals, 270
Dow Jones News/Retrieval, 85
Downloadable interactive surveys, 286
DSS (decision support system), 32, 33–34
Duke Power Company, 73
Dummy codes, 427–429
Dummy variables, 417
Duncan's multiple range test, 394
DuPont, 11
Durbin-Watson test for serial
    correlation, 417
DVD CD-ROMs, 78, 81

E-commerce ethics, 468–470
E-mail survey, 282, 286
Ease of use, online research design, 158
EDA (exploratory data analysis), 360
Editec, 84
Effect codes, multiple linear regression,
    427–429
Efficiency, 231
Element, sampling, 228, 229
Empirical events, measurement process,
    175, 176
Encyclopedias as secondary source, 546,
    552
Enterprise Miner, 38
Enterprise Resource Planning (ERP)
    software, 34
Entrepreneurship multinational study,
    209
Environmental monitoring, strategic,
    71–72
EQS (structural equation), 443
ERP (Enterprise Resource Planning) soft-
    ware, 34
Errors. See also Type I errors; Type II
    errors
    in analysis, 137, 138, 140
    in collection, 137, 138, 139–140
    data coding, preanalysis, 339, 340,
        341
    design reduction, 136–140
    managerial strategies for addressing,
        139–140
    nonresponse, with personal inter-
        viewing, 275
    personal interviews, 274–276
    in planning, 136–138, 139
    in reporting, 137, 139, 140
    response, with personal interviewing,
        275–276
    sampling, 230
    telephone interviews, 282–284
    variance analysis, 385–386, 411–412
ESOMAR (World Association of Opinion
    and Marketing Research
    Professionals), 469
ESS (expert support system), 33

Estimation
    prediction, 404–405
    simple linear regression (SLR),
        403–413
Estimation
    subjective, mail nonresponse bias
        method, 284
    unbiased estimators, 370
Ethical Standards for Research with
    Human Subjects, APA, 464
Ethics, in business research, 459–482
    about, 460, 472–473
    clients'/manager's rights, 465–467
    code of ethics, 470–472, 476–482
    defined, 8–9
    discussion questions, 473–474
    in high-technology environment,
        468–470
    ICC/ESOMAR International Code of
        Marketing and Social Research
        Practice, 470, 476–482
    managerial considerations, 471–472
    normative model for, 460–462
    research quality, 469–470
    research vs. direct marketing, 470
    researchers' rights, 467–468
    societal rights, 462–463
    subject's rights, 463–465, 467–468,
        469
    suggested reading, 475
EU (European Union), 11, 295, 463, 469
European Directive on Data Protection,
    463, 469
European Union (EU), 11, 295, 463, 469
Evaluating alternatives, decision-
    making process, 5, 6
Evaluation
    of measurement scales, 184–192
    of oral presentations, 493–495
    program evaluation and review
        techniques (PERT), 117, 120–121
    of research process, 14, 18, 104,
        495–501
    of research proposals, 118–122
    of research reports, 484–495
    of written reports, 484–493
Ex post facto research design, 140,
    144–147, 152, 153, 154
Excel spreadsheet software, 35, 122,
    328–333, 360
Executive report, written research
    report, 485, 489
Expectancy theory, 50, 52
Expected benefits, knowledge manage-
    ment (KM), 27–31
Experience, obtaining information
    through, 10
Experimental control, 165
Experimental designs. See also Research
    design
    advanced, 165–170

ANCOVA, 170, 386, 417
ANOVA, 387
    completely randomized design, 166
    defined, 144
    emphasis of advanced, 165–166
    factorial design, 169–170
    field experimental design, 145,
        147–148
    laboratory experiments, 145,
        148–149
    Latin square design, 168–169
    randomization in, 151–152, 166–168
    randomized block design, 166–168
    specific designs, 154–157
    validity, 153
Expert support system (ESS), 33
Expertise
    for personal interviews, 277–272
    vendor choice, 295–296
Explained variance and variance decom-
    position, 385–386
Explanation level, scientific endeavor,
    60–61
Exploratory data analysis (EDA), 360
Exploratory research, 146
External data location, 77
External validity, 149, 150–151, 153
Extrapolation, mail nonresponse bias
    method, 284
Exxon Corporation, 82

F-distributions, 370, 562–570
F-procedure, 437–438
F-ratio, 394–395
F-statistics, 416–417
F-test, 393–394, 412
Factor matrix, intuitive, 446, 447
Factorial design, 169–170
Factors, 387, 395, 444–445. See also
    Analysis of variance (ANOVA)
Facts, defined, 45, 46
Fax-fax/mail-back surveys, 282
FGI Research, 258
Field experimental design, 145, 147–148
Field studies, as ex post facto design,
    144–146
Files, data structure, 341
FIND/SVP databases, 84
Fixed effects model, 387–388
Florida Power & Light Co., 284
FMC Corporation, 308
Focus groups, 272, 312, 313–315
Follow-up technique, mail response rate
    inducements, 283
Forbes 500, 312
Ford Motor Company, 58, 74, 75, 82, 463
Forecast International/DMS, 71
Formulation of alternatives, and sec-
    ondary data, 71, 72–73
Fortune, 79
France, spaghetti study, 202–203

Free choice, right of, 464–465
Freedom, degrees of, 370
Frequency, 342, 343–344, 360
Frequency analysis, 342
Frequency tables, 360
Functional theory, 53, 55

Gale Group, 78, 81
Gantt charts, 117, 120, 121
General Foods, 463
General linear model, 384, 386
General Mills, 157, 216
General Motors (GM), 48, 49, 50, 312
Geodemographic segmentation
    system, 73
Germany, spaghetti study, 202–203
Global marketplace, business research,
    11–12. *See also* International
    issues
GM (General Motors), 48, 49, 50, 312
Goodwill, personal interviews, 277–272
Graphic rating scales, 210–212
Graphics, written research reports,
    491–493
Graphs, cell means for two-way ANOVA,
    402
Group analysis. *See* Analysis of variance
    (ANOVA)
GTE, 33
Guttman's Scale-Analyzing method, 212

Hahn, Harley, 85
Handbooks as secondary source, 546, 551
Harley-Davidson Motor Company,
    Internet research example, 87–91
Harris Interactive, 258, 315, 318
Harvard Business Review, 308
Health Canada, 28–31
Health Insurance Portability and
    Accountability Act (HIPAA), 469
Hidden issue questioning, 312
High-quality research, right to expect,
    466–467
High-technology environment, and
    ethics, 468–470
HIPAA (Health Insurance Portability and
    Accountability Act), 469
Histograms, 360
History, validity, 150, 153
Hoescht-Rousel, 34
Hotelling t statistic, 433
Howard and Sheth Theory of Buyer
    Behavior, 50, 51, 55
HTML surveys, 286
Hypothesis, 48–49, 53. *See also* Null
    hypothesis
Hypothesis testing
    as basic analytical method, 346,
        354–357
    one-way ANOVA, 389, 393–394
    a priori, 49, 433

statistical test of hypothesis (STOH),
    354–355
two-way ANOVA, 400–401

IBM, 11, 33, 36–37
ICC/ESOMAR International Code of
    Marketing and Social Research
    Practice, 470, 476–482
ICH (Information Clearing House), 84
IMRO (Interactive Marketing Research
    Organization), 469
In-depth interviews, 310–311, 312–313
Incidence, sampling, 254–255
Independent variable, 140, 151, 165. *See
    also* Analysis of variance (ANOVA)
Indeterminacy problem, 445
Indexes, as secondary sources, 547–548
Individual company references as sec-
    ondary sources, 551
Inductive theory, 53, 55
Inferential statistics, 236–237, 354
Information. *See also* Primary data col-
    lection (PDC); Secondary data
    collection
    customer information system (CIS),
        72
    database users, 85
    decision-making process, 5, 6
    defined, 25
    explosion of, 58–60
    knowledge management (KM), 25
    methods of obtaining, 9–11
    online and Internet information
        explosion, 58–60, 80
    quality of, and decision types, 4–5
    search for, 5, 6, 46–47
    technological changes, 11
Information Age, 70
The Information Catalog (FIND/SVP of
    ICH), 84
Information Clearing House (ICH), 84
Information Resources, 287
Information systems (IS), 31–33, 38, 39,
    72
Information systems management
    (ISM), 31–33
Information technology (IT), 36–39
Informed, right to be, 462, 465
Informed consent, 569
InsightExpress, 286
Instrument design, 199–225
    about, 200, 222–223
    defined, 200, 213, 222
    discussion questions, 223
    example, River End Golf and
        Entertainment Center, 203–207
    guidelines, 219–220
    managerial considerations, 220–222
    and measurement process, 201–202
    nature of, 200–202
    online design aids, 214–218

pretest and correction, 219–220
scale development, 202–213
scale sequencing and layout,
    213–214
suggested reading, 225
validity, 150, 153
Intelligent Miner, 38
Interaction effects, 395–396
Interaction validity, 151, 153
Interactive Marketing Research
    Organization (IMRO), 469
Interactive surveys, downloadable, 286
Interdependency, analysis of. *See*
    Analysis of interdependency
Internal consistency method, reliability,
    188, 191
Internal data location, 77–78
Internal validity, 149–150, 153
International Chamber of Commerce/
    European Society for Opinion
    and Marketing Research.
    (ICC/ESOMAR International
    code), 470, 476–482
International issues
    design considerations, 159
    ethics, ICC/ESOMAR, 470, 476–482
    global marketplace, 11–12
    primary data collection (PDC),
        298–299
    sampling process, 256–257
    secondary sources of data, 551–552
Internet. *See* Online and Internet
The Internet Complete Reference (Hahn), 85
Internet service provider (ISP), 87
Interval scales, 180, 181, 183–184
Intervening variable, 142, 143
Interviews
    computerized, 282–284
    in-depth, 310–311, 312–313
    mail surveys, 279, 282–284
    personal surveys, 274–278
    telephone surveys, 278–281
Introduction, written research report,
    489
Intuition, obtaining information
    through, 10
Intuitive factor matrix, 446, 447
IS (information systems), 31–33, 38, 39,
    72
ISM (information systems manage-
    ment), 31–33
ISP (Internet service provider), 87
IT (information technology), 36–39
Italy, spaghetti study, 202–203
Item phrasing in scale development,
    203–209
Itemized rating scales, 211, 212

J. Walter Thompson, 254
Jackknife criterion, 440
Japan, 146, 295

J.C. Penney Company, 72
Judgment sampling designs, 241, 251

Kaplan, Abraham, 49
KDD (knowledge discovery in
    databases), 33
Key questions for management
    data analysis planning, 347
    ethics, in business research, 472
    instrument design, 222
    measurement process, 194
    primary data collection (PDC), 300
    problem screening, 106
    qualitative research, 320
    research design, 160
    research evaluation, 104, 122,
        498–499
    research proposal, 104, 122
    research reporting, 498–499
    sampling, 260
    secondary data collection, 94
Kiechel, Walter, III, 79
KM. *See* Knowledge management (KM)
Knowledge, defined, 25
Knowledge discovery in databases
    (KDD), 33
Knowledge management (KM), 23–42
    about, 24, 39
    basics, 24–25
    benefits expectations, 27, 31
    business research (BR), 31, 32,
        36–39
    database management (DBM),
        35–36, 38
    decision support systems (DSS),
        33–34
    defined, 26–27
    discussion questions, 39–40
    expected benefits of, 27–31
    Health Canada example, 28–31
    information systems (IS), 31–33, 38,
        39, 72
    online analytical processing (OLAP),
        35–36, 38
    suggested reading, 41–42
KR-20 formula, reliability, 192
Kruskal-Wallis ANOVA, 420, 422
Kurtzman/Slavin/Linda, 315

Laboratory experimental design, 145,
    148–149
Laddering, 312
LAIA (Latin American Integration
    Association), 11
Latent variables, 443
Latin American Integration Association
    (LAIA), 11
Latin square design, 168–169
Law, defined, scientific inquiry, 49
Least-squares procedure, regression,
    410

Levels
    of measurement, 179–184
    of organizational decisions, 3–5
    of scientific research, 60–62
Levene's test, one-way ANOVA, 394
LEXIS-NEXIS, 85, 87
Likert scale, 211, 212
Limitations of study, in written research
    report, 490, 491
Linda, Gerry, 315
Linear models. *See also* Analysis of vari-
    ance (ANOVA)
    general, 384, 386
    multiple linear regression, 413–419
    one-way ANOVA, 387–395
    randomized designs, 166
    simple linear regression (SLR), 366,
        403–413
    two-way ANOVA, 395–403
Linear regression, 403–419
    multiple, 413–419
    simple, 336, 403–413
Linear structural relations (LISREL), 443
LISREL (linear structural relations), 443
Logic and theory construction, 53
Longitudinal design, 155
Lotus spreadsheet software, 329
Lucent Technologies, 82
Lutz, Robert A., 14

Mail interviews as PDC method, 279,
    282–284
Make-or-buy decision, 292
Management information system (MIS),
    72
Managerial considerations
    control function, 117
    data analysis planning, 346–347
    errors, strategies for addressing,
        139–140
    ethics, in business research, 471–472
    instrument design, 220–222
    knowledge management (KM)
        importance, 27
    managers' rights, 465–467
    measurement scales, 192–194
    nature of decision making, 2–7
    primary data collection (PDC),
        298–300, 318–320
    problems, defined, 47–48
    qualitative research, 318–320
    relationship with researchers, 12–14
    research design, 158–160
    research problem development,
        102–103
    research proposals, 117–122
    research reporting and evaluation,
        498–499
    sampling, 259–260
    scientific inquiry, 47–48
    secondary data collection, 91–94

Managers' rights, 465–467
Manifest variables, 443
Mann-Whitney U test, 374
MANOVA (multivariate analysis of
    variance), 435–438
Manual retrieval methods, secondary
    data, 75–76
Mapping rules, measurement process,
    175–176
MAPWISE, 452–453
Market Facts, Inc., 294, 296
Market Research Corporation of
    America (MRCA), 294, 296
MarketTool, 157, 215–218, 286, 287–289
MarketTool Zoomerang, 215–218
Marriott, 308
Marx, Melvin, 53
Mason, Emanuel J., 396
Matsushita, 146
Maturation, validity, 150, 153
McNemar test, 374
MDA (multiple discriminant analysis),
    437, 438–440
MDS (multidimensional scaling), 213,
    450, 452–454, 488
Mean
    calculation, 380
    as central tendency measurement,
        344
    normal distribution, 237
    one-way analysis of variance
        (ANOVA), 388, 389–392
    population, 237
    sample, 380
    standard error of, 237
    t-test for differences, 371–374, 433
Measurement, 173–197
    about, 174, 194–195
    of central tendency, 342, 344
    components of, 175–176
    defined, 174, 194
    destructive, in sampling, 232
    discussion questions, 195–196
    evaluation of measurement scales,
        184–192
    instrument design and measure-
        ment process, 201–202
    interval scales, 183–184
    levels of, 179–184
    managerial considerations, 192–194
    nature of, 174–179
    nominal, 180, 182
    ordinal, 182–183
    process of, 175–179, 201–202
    ratio, 184
    suggested reading, 197
Measurement scales
    conversion of, 182
    defined, 179
    evaluation of, 184–192
    managerial considerations, 192–194

reliability, 188–192
simple vs. complex, 179
validity, 185–188
Median, 342
Median test, 374
Medium, Internet as, 82
META Group, 35
Metacrawlers for Internet search, 78, 80
Method of summated ratings, 212
Methodology section, written research
report, 489–490
Mexico, NAICS, 554, 555
Mexico City, telephone penetration, 279
Microsoft, 33, 315
Microsoft Excel spreadsheet software,
35, 122, 328–333, 360
Microsoft Office, 122
MIS (management information
system), 72
Mixed effects design, 387
Models and modeling. See also Linear
models
causal, 442–444
defined, 50
ethics, normative model for, 460–462
fixed effects, 387–388
model-based theory, 54
normative model for ethics, 460–462
predictive, 404–405
scientific inquiry, 50, 54
simple linear regression (SLR), 366,
403–413
Moderating variable, 141–143
MONANOVA computer program, 440
Monitoring, 71–72, 277
Monotone regression, 440
Monsanto Corporation, 308
M.O.R.-PACE, Inc., 316–317
Mortality, validity, 150, 153
Motivation, personal interviews, 277–272
MRCA (Market Research Corporation of
America), 294, 296
Multicollinearity, 416–417
Multidimensional scaling (MDS), 213,
450, 452–454, 488
Multiple comparison F-procedure, 437,
438
Multiple discriminant analysis (MDA),
437, 438–440
Multiple linear regression, 413–419
Multiple sampling, 240, 250
Multiple-variable analysis, 384
Multistage random sampling, probability
designs, 239, 247, 248–249
Multivariate analysis, 431–458
about, 432, 454–455
analysis of dependency, 435–444
analysis of interdependency,
444–454
comment on, 454
conceptualization of, 432–433

discussion questions, 455–456
selecting technique, 433–434
suggested reading, 457–458
Multivariate analysis of variance
(MANOVA), 435–438

NAFTA (North American Free Trade
Agreement), 11
NAICS (North American Industry
Classification System), 534–555
Nash Finch Company, 73
National Family Opinion (NFO), 258,
294, 296, 298
National Information Infrastructure
(NII), 86
National Research Council, 200
National Science Foundation, 70
National Semiconductor, 6, 7
NCR, 33
New York Times Information Bank, 85
Newspaper indexes, as secondary
sources, 548
NFO (National Family Opinion), 258,
294, 296, 298
NII (National Information
Infrastructure), 86
Nominal measurement, 180, 181, 182
Nomograph, simple random sample, 246
Nonparametric ANOVA, 358, 420–422
Nonprobability designs, 236, 241,
250–251
Nonresponse, errors, 275
Nonresponse bias, mail surveys, 282,
284
Nonresponse errors, interviews, 275,
279
Nonscience vs. science, 56–57
Nonscientific business research (NSBR),
345–346, 352, 368
Normal curve, 237, 558–559
Normal distribution, 379–380
Normative model for ethics, 460–462
North American Free Trade Association
(NAFTA), 11
North American Industry Classification
System (NAICS), 534–555
NSBR (nonscientific business research),
345–346, 352, 368
Nuisance variables, 165
Null hypothesis
chi-square test, 370
multivariate analysis of variance
(MANOVA), 435, 436
nonparametric ANOVA, 422
one-way analysis of variance
(ANOVA), 392–394
researchers dilemma, 357
statistical test of hypothesis (STOH),
354–355
Numbers, in measurement process,
175, 176

Objectivity, vendor choice, 295–296
Oblique solutions, 445
Observation
defined, 45, 46
observable empirical events, 175, 176
passive PDC, 271, 308, 310–311
Office of Management and Budget
(OMB), 554
Offline PDC methods, 274–284
mail interviews, 279, 282–284
vs. online, 291–292
personal interviews, 274–278
telephone interviews, 278–281
OMB (Office of Management and
Budget), 554
Omnibus panel, defined, 296
One-group pretest-posttest design, 155
One-shot case study design, 154–155
One-way ANOVA linear models,
387–395
Online analytical processing (OLAP),
35–36, 38
Online and Internet
background, 86–87
caution on resources on, 86
computer-assisted surveys, 282–284
data analysis planning research supplies, 327, 333, 334
databases, 82–85
ethics, 468–470
focus groups, 313–315
Harley-Davidson research example,
87–91
information explosion, 58–60, 80
instrument design aids, 214–218
metacrawlers, 78, 80
online analytical processing (OLAP),
35–36, 38
outsourcing research suppliers, 333
PDC methods, 273, 284–292
research design, 157–158
sampling, 257–258
search engines, 78, 80
searching on, 79–81
for secondary data collection,
79–91
secondary data retrieval methods,
75, 76
worldwide usage, 58, 257–258
Open-ended item phrasing, 208
Open-ended questions, data coding,
337–338, 340
Operational definitions, 46, 47, 177–178
Optimum allocation stratified sampling,
240, 247–250
Oral research presentations, 493–495
Ordinal measurement, 180, 181,
182–183
Organization
levels of decisions, 3–5
oral research reports, 494–495

Organization (*continued*)
    orientation, research problem development, 102–103
    written research reports, 491
Orthogonal codes, 427
Outsourcing research suppliers, 333

Panel, defined, 296
Panel vendors of PDC, 296–298
Parameters, defined, 230
Parametric statistics, 357–358
Partialing method, 418
Partworth utilities, 441
Passive methods, primary data collection (PDC), 271, 306, 308–311
PDC. *See* Primary data collection (PDC)
Pearson product-moment (PM) correlation coefficient, 365–368
Pearson's r, 365–368
PeopleSoft, 216, 286, 287–289, 294
Periodical Indexes, as secondary sources, 547
Perseus Development, 286, 333, 334
Personal interviews as PDC method, 274–278, 336
Personnel, sample size, 235
PERT (program evaluation and review techniques), 117, 120–121
Pfizer Inc., 286
Phi coefficient, 363–364
Pitney Bowes, 286
PlaceWare, 315
Planning. *See also* Data analysis, planning for
    errors in, 136–138, 139
    for personal interviews, 276–278
    sampling, 231, 232, 235, 256
PM (Pearson product-moment) correlation coefficient, 365–368
"Point and shoot" software, 328–329
Population
    defined, in sampling, 228–229
    international sampling, 256–257
    sample, 230
    selecting, 232–233
    statistics, 236–237
Post hoc tests, for one-way ANOVA, 392
Posttest-only control group design, 156
Power, statistical, and sample size, 234
Practice, oral research reports, 495–496
Preanalytical process, data analysis planning, 327, 333, 335–344
Precision, and sample size, 234
Precoded questionnaires, 337, 338
Prediction, 60–61, 404–405
Predictive validity, 187
Prenotification technique, mail response rate inducements, 283
Pretest-posttest control group design, 156
Pricewaterhouse Coopers, 27

Primary data collection (PDC), 269–324
    about, 270, 300, 306, 320–321
    active methods, 267–292, 310–316
    decision framework, 272
    defined, 70, 270
    discussion questions, 301, 321–322
    focus groups, 272, 312, 313–315
    international considerations, 298–299
    managerial considerations, 298–300, 318–320
    method comparison, 291–292, 310, 311
    nature of, 270–274
    offline methods, 274–284, 291–292
    online methods, 284–292
    panel vendors, 296–298
    passive methods, 271, 306, 308–311
    qualitative research, 306–308
    in research process, 16–18
    vs. secondary data, 71, 92
    suggested reading, 304, 323–324
    survey instruments, 274–292
    vendors, 292–298
Privacy, right to, 463
PRIZM, 297–298
Probabilistic cause, 141
Probability designs, 237–250. *See also* Research design
    choice considerations, 237–241
    cluster sampling, 250
    defined, 236
    multistage random sampling, 239, 247, 248–249
    simple random sampling (SRS), 242–246
    stratified sampling, 239–240, 247–250
    systematic sampling, 246–247
    types of, 239–240
Probability distribution, normal, 379
Problems
    alternative formulation, 71–73
    analysis, 5, 6
    assessment of research value, 105–107, 126–131
    clarification, 71, 72
    coding, 339, 340, 341
    discussion questions, 123–124
    formulation, 71–73, 104–106
    identification, 101–103
    recognizing, 5, 6, 71–72
    secondary data use, 71–73
    selection, 104, 106
    solving, 71, 73
    suggested reading, 125
Process
    of business research (BR), 15–18
    data collection, 16–18
    of decision making, 5–7
    evaluation of research, 14, 18, 104, 495–501
    of measurement, 175–179, 201–202

preanalytical, 327, 333, 335–344
sampling, 232–235, 256–257
Procter & Gamble, 157, 216
Producers of databases, 82–83
Program evaluation and review techniques (PERT), 117, 120–121
Projective qualitative research techniques, 316
Proportion
    stratified sampling, 239–240, 247–250
    z-test for differences in proportions, 370–371
Proposals. *See* Research proposals
Protocol qualitative research techniques, 316
Psychological Motivations, Inc., 307, 309
Public opinion research, ICC/ESOMAR, 470, 476–482
Purchase
    concept/construct definitions, 46–47
    of secondary data, 91–92
Purpose and secondary data, 93

Quaker Oats, 82
Qualitative research
    active methods, 271, 272, 310, 312–316
    defined, 270, 306
    focus groups, 272, 312, 313–315
    in-depth interview, 313–314
    managerial considerations, 318–320
    nature and uses, 306–308
    passive methods, 271, 306
    projective techniques, 316
    protocol techniques, 316
    vs. quantitative, 306–307
    vendors, 316–318
    word association techniques, 316
Quality
    of data, 81, 83, 93, 158
    high-quality research, right to expect, 466–467
    of information and decision types, 4–5
    of research, and ethics, 469–470
Quantitative research, 270, 271, 272, 306–307
Questioning. *See also* Active (primary) data collection; Key questions for management
    errors in, 276
    hidden issue, 312
    open-ended, data coding, 337–338, 340
    symbolic, 313
Questionnaires, 214, 337–338
Quota samples designs, 241

R-square, 417–418
Random-digit dialing (RDD), 230, 252–253, 257, 279

Random effects design, 387
Random factors, 387
Random sampling, 239, 247–249
Randomization, experimental designs, 151–152, 166–168
Range, 342, 343–344
Rank-order scale, 212
Rating scales, 210–213
Ratio measurements, 180, 181, 184
RCA, 11
RDD (random-digit dialing), 230, 252–253, 257, 279
Reader's Digest, 463
Reading, additional. *See* Suggested reading
Reading the report, oral research reports, 495
Recommendation section, written research report, 490
Red Cross, 463
Regression, linear, 403–419
   multiple, 413–419
   simple, 403–413
Regression least-squares, monotone, 440
Relevance of data, 81, 83, 92, 93
Reliability, measurement scales, 188–192
Repetitive sampling, 240, 250
Reports and reporting, 483–502
   about, 484, 499–500
   communication challenges, 497–498
   data analysis planning, 327
   discussion questions, 500–501
   errors in, 137, 139, 140
   evaluation of, 484–495
   managerial considerations, 498–499
   oral presentations, 493–495
   in research process, 14, 18
   scientific endeavor levels, 60
   suggested reading, 501–502
   written reports, 484–493
Requests for proposal (RFP), 117, 294, 467
Research. *See also* Business research (BR); Qualitative research
   association web sites, 59
   basic, 105–106
   defined, 7–8
   descriptive, 147
   design of, 16
   vs. direct marketing, 470
   ethics, 8–9, 469–470
   evaluation, 18, 495–501
   exploratory, 146
   quality, 469–470
   quantitative, 270, 271, 272, 306–307
   scientific inquiry problems, 48
Research design, 133–171. *See also* Advanced experimental designs; Experimental designs; Probability designs; Sampling designs

causality, 141, 143
data collection process, 16–18
defined, 16, 134
error reduction through, 136–140
ex post facto, 140, 144–147, 152, 153, 154
international considerations, 159
major types, 140–149
managerial considerations, 158–160
nature of, 134–136
nonprobability, 236, 241, 250–251
secondary data for, 72
specific designs and notational system, 151–157
validity concerns, 149–151
Research guides, as secondary sources, 544–545
Research Library and Information Services (RLIS), Ford, 74
Research proposals
   about, 100, 123
   assessment of research value, 105–107, 126–131
   data collection, 14, 16, 17
   defined, 100, 107
   developing, 107–122
   discussion questions, 123–124
   elements of well written, 107–108
   evaluation, 118–122
   extended proprietary (Type IV), 110
   in-house proprietary (Type I), 109
   internal vs. external origins, 109
   managerial considerations, 117–122
   online research design issues, 157–158
   public domain (Type III), 109
   research, 14, 16, 17, 100
   research contractor (Type II), 109, 111–117
   sample, 111–117
   structure of, 110–111
   suggested reading, 125
Research reporting. *See* Reports and reporting
Researcher-management relationship, 12–14
Researchers' dilemma, 357
Researchers' rights, 467–468
Residual sum of squares, regression, 411
Resource constraints, sampling, 231–232
Resources, vendor choice, 295–296
Response
   errors, in personal interview, 275–276
   formats for, scale development, 209–210
   rates for, 254–255, 280–281, 282, 283
   screening, 255
Results section, written research report, 490
Reuter, Paul Julius, 33

RFP (requests for proposal), 117, 294, 467
Rigby, Paul, 56
Rights
   of clients/managers, 465–467
   researchers, 467–468
   societal, 462–463
   of subjects, 463–465, 467–468, 469
River End Golf and Entertainment Center instrument design example, 203–207
RLIS (Research Library and Information Services), Ford, 74
Roadnet, Inc., 73
Robustness, defined, 374
Russell, Bertrand, 141
Russia, multinational research, 295

Safe harbors, right to privacy, 463
Safety, right to, 465
Safeway plc, 36–38
Sample, defined, 230
Sample and sampling, 237–251. *See also* Probability designs; Research design; Sampling designs
   about, 228, 260–161
   discussion questions, 261–262
   efficiency of, 231
   elements of good, 236
   errors, 230
   frame for, 229–230, 232, 256
   incidence, 254–255
   international, 256–257
   managerial considerations, 259–260
   mean, 380
   nature of, 228–235
   online, 257–258
   philosophy of, 235–237
   planning for, 231, 232, 235, 256
   practical considerations, 244–257
   process for, 232–235, 256–257
   rationale for, 221–222
   representative, 236
   response rate, 254–255
   size of sample, 232, 234–235, 245, 256, 307
   stratified sampling, 239–240, 247–250
   suggested reading, 263–264
   systematic, 246–247
   terminology, 228–231
   units for, 229, 232, 234, 256
   variance in, 380
Sample-free statistics, 358
Sampling designs, 227–267. *See also* Research design
   about, 227, 260–261
   Central Limit Theorem, 237, 265–267, 379–380
   choice considerations, 237–241
   defined, 237
   discussion questions, 261–262

Sampling designs (*continued*)
  nonprobability designs, 236, 241, 250–251
  online issues, 257–260
  probability designs, 242–250
  selecting, 232, 233–234
  suggested reading, 263–264
  types of designs, 239–241
SAS, 38
Satisfaction, concept/construct definitions, 46–47
Sawtooth Technologies, Inc., 283, 285
SBR (scientific business research), 8, 345, 346, 352, 368
Scale design. *See* Instrument design
Scale development, 202–214. *See also*
    Measurement scales
  common considerations, 202–203
  item phrasing, 203–209
  rating scales, 210–213
  response formats, 209–210
  sequencing and layout, 213–214
Scanning, UPC, 287
Scatterplots, 360
Scheffe's test, one-way ANOVA, 394
Schick, 286
Science, 53, 56–57
Scientific business research (SBR), 8, 345, 346, 352, 368
Scientific inquiry, 43–67
  about, 44, 63
  business research, 55–57
  concepts, 45, 46
  constitutive definitions, 47
  constructs, 45–46
  deductive theory, 54–55
  definitions, 45–52
  discussion questions, 63–64
  facts, 45
  functional theory, 55
  hypotheses, 48–49
  inductive theory, 55
  laws, 49
  managerial considerations, 47–48
  model-based theory, 54
  models, 50
  observations, 45
  operational definitions, 47
  relevance in BR, 55–57
  research problems, 48
  scientific method, 57
  state of the art in BR, 57–63
  suggested reading, 66–67
  theory construction methods, 49–50, 53–55
  value of scientific research, 57–58, 105–107, 126–131
  variable, 47
Scientific method, 56, 57
Scientific research, 7–8, 60–62
Scientific theories, 49–50, 52–55

SCLC (corporate development life cycle), 420
Screens
  problem, 106
  research problem development, 102, 103, 104
  response, 255
  as sampling tool, 254
Search engines, Internet, 78, 80
Secondary data collection, 69–97
  about, 70, 94
  in business research, 70–74
  compilation or purchase decisions, 91–92
  defined, 70
  discussion questions, 95
  Internet, 79–91
  locations of data, 77–79
  managerial considerations, 91–94
  vs. primary data, 71, 92
  retrieval, 74–77
  search strategy development, 74–85
  suggested reading, 97
  uses of, 71–74
Secondary sources, practitioner's guide
    to, 543–555
  bibliographies, 544–545
  CD-ROMs, 78, 81
  dictionaries, 546
  directories, 546
  encyclopedias, 546
  handbooks, 546
  indexes, 547–548
  individual company references, 551
  international business, 551–552
  research guides, 544–545
  statistical, 548–550
Selection
  data analysis software, 328–333
  multivariate analysis technique, 433–434
  of population, 232–233
  problems in, 104, 106
  sample size, 232, 234–235, 245
  validity, 150, 151, 153
Semantic differential, rating scales, 211, 212
Sensus, 283, 285
Sequential sampling, 240, 250
Serial correlation, Durbin-Watson test
    for, 417
Setting, validity, 151, 153
Shared variance, 385
Shell Oil & Chemical, 308
SIC (Standard Industrial Classification), 553
Significance
  t-test for differences, 371–374, 433
  Tukey's honestly significant
    difference, 394, 403

SILOF (Solar Insistence League of Florida), 242–245
Simple linear regression (SLR), 366, 403–413
Simple random sampling (SRS), 239, 242–246
Simple scales, 179
Simulations, virtual reality, 59, 289
Skinner, B. F., 65
SLR (simple linear regression), 366, 403–413
Snowball sampling design, 241, 251
Societal rights, 462–463
Software, data analysis planning, 327, 328–333. *See also specific software*
Solar Insistence League of Florida (SILOF), 242–245
Solomon four-group design, 156–157
Sony, 146
Spaghetti consumption study, 202–203
Spearman rank correlation, 364–365
Speed, online research design, 158
Split-ballot technique, 210
Split-half technique, 192
Split-sample technique, 440
SPSS (Statistical Package for the Social Sciences) software, 35, 328–331, 355, 358, 389, 405
Spurious correlation, 143
SRS (simple random sampling), 239, 242–246
SSI (Survey Sampling, Inc.), 251–253, 257, 279
Standard deviation, 342, 344, 379
Standard error of the mean, 237
Standard Industrial Classification (SIC), 553
Standardized data array analysis, 380–381
Staple Scales, 212
STATA (Statistical Analysis Program), 360
Static-group comparison design, 155–156
Statistical Analysis Program (STATA), 360
Statistical Package for the Social Sciences (SPSS) software, 35, 328–331, 355, 358, 389, 405
Statistical power, and sample size, 234
Statistical tables, 557–572. *See also* Tables
  area under the normal curve, 558–559
  chi square distribution, 571–572
  F distribution, 562–570
  t distribution, 560–561
Statistical test of hypothesis (STOH), 354–355
Statistics. *See also* Mean
  central tendency, 342, 344
  control, 165–166

data analysis planning framework, 327, 344–346
defined, 230
inferential, 236–237, 354
median, 342
nonparametric, 358
normal curve, 237, 558–559
normal distribution, 379–380
parametric, 357–358
secondary sources, 548–550
statistical efficiency, 231
Statistics Coach, 358–359
Step-down F procedure, 437
Stepwise multiple regression, 417
Stevens, S. S., 180
STOH (statistical test of hypothesis), 354–355
Strata, defined, 247
Strategic decisions, 4, 5
Strategic environmental monitoring, 71–72
Stratified sampling, 239–240, 247–250
Structural equation (EQS), 443
Structure, degree of, PDC, 273, 310
Study sample, 229
Subjective estimates, mail nonresponse bias method, 284
Subjects' rights, 463–465, 467–468, 469
Succinctness, written research reports, 491
Suggested reading
    analysis, basic methods, 378
    analysis of variance (ANOVA), 426
    business research (BR), 21–22
    data analysis planning, 350
    decision making, overview, 21–22
    ethics, in business research, 475
    instrument design, 225
    knowledge management (KM), 41–42
    measurement, 197
    multivariate analysis, 457–458
    primary data collection (PDC), 304, 323–324
    problems, 125
    reports and reporting, 501–502
    research proposals, 125
    research reporting and evaluation, 501–502
    sampling designs, 263–264
    scientific inquiry, 66–67
    secondary data collection, 97
Summated ratings, method of, 212
Survey frequency, telephone interviews, 280–281
Survey Sampling, Inc. (SSI), 251–253, 257, 279
Surveys
    active PDC instruments, 274–292
    computer-assisted surveys, 282–284
    ex post facto designs, 145, 146–147

in-depth interviews, 310–311, 312–313
mail interviews, 279, 282–284
online and Internet, 284–292, 333
personal interviews, 274–278
telephone interviews, 278–281
Sylvester, Alice K., 254
Symbolic questioning, 313
Systematic probability sampling, 246–247

T distribution, statistical tables, 560–561
T-test for differences, mean, 371–374, 433
Tables. See also Statistical tables
    contingency, 368
    frequency tables, 360
    one-way ANOVA, 391, 393–394
    two-way ANOVA, 393–394, 396, 399, 400–401
Tabulation, data preanalysis, 341–344
Taco Bell Corporation, 34
Tactical decisions, 4
Technical decisions, 4
Technology. See also Computers; Knowledge management (KM); Online and Internet
    computerized interviews, 282–284
    data analysis planning software, 327, 328–333
    ethics and high-technology environment, 468–470
    information systems management (ISM), 31–33
    information technology (IT), 36–39
Test-retest method, reliability, 188, 189
Tests and testing. See also Hypothesis testing
    Bartlett-Box test, one-way ANOVA, 394
    Behrens-Fisher test, 374
    Chi-square test, 368–370, 571–572
    Duncan's multiple range test, 394
    Durbin-Watson test for serial correlation, 417
    Levene's test, one-way ANOVA, 394
    Mann-Whitney U test, 374
    McNemar test, 374
    Median test, 374
    One-group pretest-posttest design, 155
    pretest instrument design, 219–220
    reliability, 188, 189
    Scheffe's test, one-way ANOVA, 394
    t-test for differences, mean, 371–374, 433
    univariate tests for difference, 368–374
    validity, 150, 151, 153
    Wilcox on matched-pairs signed-ranks test, 374, 420
    Z-test for differences in proportions, 370–371

Texas Instruments, 101
Theory
    defined, 49–50
    expectancy, 50–52
    Howard and Sheth Theory of Buyer Behavior, 50, 51, 55
    modes of construction methods, 53–55
    scientific inquiry, 49–52
3Com, 31, 33
Thurstone's Equal-Appearing Interval Scale, 212
TIGER (Topologically Integrated Geographic Encoding and Referencing), 72–73, 81
Time
    Internet data searches, 81
    personal interviews, 275
    primary vs. secondary data collection, 71
    sample design, 235, 238
    secondary data, 92, 93
Time series design, 155
Title page, written research report, 485
Tools
    for data-mining, 36, 38
    Internet as, 82
    screens as sampling tools, 254
Topologically Integrated Geographic Encoding and Referencing (TIGER), 72–73, 81
Total variance and variance decomposition, 385–386
Toyota, 146
Training, personal interviews, 276–278
True panel, defined, 296
Trust, personal interviews, 277–272
Two-way ANOVA linear models, 395–403
Type I errors, 357
Type II errors, 357

Unbiased estimators, 370
Unique database, 77
Unit of analysis (UOA), 175, 220–221, 228–229
United Airlines, 59
United Kingdom, Safeway grocery chain, 36–38
United States
    research infrastructure, 295
    survey researcher's view, 280–281, 289, 290
United Way of America, 72
Units, sampling, 229, 232, 234, 256
Univariate F procedure, 437
Univariate tests for difference, 368–374
Universal Product Code (UPC), 287
Unlisted rates, telephone interviews, 280–281

UOA (unit of analysis), 175, 220–221, 228–229
UPC (Universal Product Code), 287
Urban Decision Systems, 73
U.S. Census Bureau, 72
U.S. Department of Defense, 71, 86

Validity
 concurrent, 187
 construct, 186–187
 content, 185–186
 convergent, 186–187
 criterion-related, 186, 187–188
 discriminant, 187
 external, 149, 150–151, 153
 internal, 149–150, 153
 of measurement scales, 185–188
 predictive, 187
Value of scientific research, 57–58, 105–107, 126–131
Variables
 binomial, 180
 classification, 387
 defined, 47
 dependent, 140, 151, 165
 developing, preanalysis, 336–337
 dichotomous, 180
 dummy, 417
 independent, 140, 151, 165
 intervening, 142, 143
 latent, 443
 manifest, 443
 moderating, 141–143
 nuisance, 165
 personal interview, 275–276

Variance. *See also* Analysis of variance (ANOVA)
 calculating, 380
 decomposition of, 384–387
 defined, 384
 population, 237
VARIMAX, 445, 446, 447
Vendors
 primary data collection, 284–298
 qualitative research, 316–318
 statistical analysis, 328–329
Virtual reality simulations, 59, 289
VISION, 73
Visuals, research reports, 491–493, 494
Voice recognition software, 328–329

Web sites
 analytic software, 331, 333
 data mining tools, 36
 ICC/ESOMAR International Code, 470
 instrument design aids, 215
 knowledge management (KM), 26
 MarketTools, 157
 NAICS, 555
 online sampling, 258
 research associations, 59
 research proposal aids, 111
 right to privacy, 463
 search engines, 78, 80
 subjects' rights, 469
West Germany, spaghetti study, 202–203
Wheel of science, 53

Wilcox on matched-pairs signed-ranks test, 374, 420
Winch, Mike, 36–37
Windows and menus, analytical software, 329–331
Wisdom, defined, 25
Word association qualitative research techniques, 316
Wording, in item phrasing, 208–209
World Association for Public Opinion Research, 471
World Association of Opinion and Marketing Research Professionals (ESOMAR), 469
World Economic Forum, 27
Writing style, and written research reports, 491
Written reports, 484–493
 guidelines, 490–493
 outline and organization, 485, 489–490
 research reporting and evaluation, 484–493
 sample, 486–489

Xerox, 312

Yearbooks, international business secondary sources, 552

Z-test for differences in proportions, 370–371
Zoomerang, 215–218, 286, 287–289, 294